# Contemporary Ergonomics 1984–2008

# Contemporary Ergonomics 1984–2008

## Selected Papers and an Overview of the Ergonomics Society Annual Conference

Coordinating Editor:

### Philip D. Bust

*Loughborough University*

CRC Press
Taylor & Francis Group
Boca Raton  London  New York  Leiden

CRC Press is an imprint of the
Taylor & Francis Group, an **informa** business

A BALKEMA BOOK

The Ergonomics
society

Typeset by Macmillan Publishing Solutions, Chennai, India
Printed and bound in Great Britain by Antony Rowe (A CPI-group Company), Chippenham, Wiltshire

ISBN 978-0-415-80434-9 (Pbk)

# Contents

# Selected Papers and an Overview of the Ergonomics Society Annual Conference

## The Early Days

The Ergonomics Society conference originally was in a single session format and run by a single individual. Selected presentations from the conference were converted into paper form and were published in the journal Ergonomics and other scientific journals, or not at all.

In the early 80's the format of the conference changed with the introduction of parallel sessions to enable the full diversity of ergonomics to be represented at the conference, and poster presentations for work that was either not ready for a full scientific paper or to report small discrete projects. The Society felt that these changes would make the conference more attractive to all ergonomists, whatever their interests, and to people outside the Society who had an interest in the subject.

With the growth in the annual conference, and to make the work of society members quickly available to members and non-members alike, a published proceedings was instigated by the Meetings Committee with the support of Council.

This first issue was put together at somewhat short notice between November 1982 and February 1983, and would not have succeeded without the support of Taylor & Francis. Karenna Coombes was the editor of the first Proceedings of the Ergonomics Society Annual conference. Copies of the proceedings were supplied to all conference delegates at the 1983 Annual Conference in Cambridge and available on the open market. Following the successful launch of the first proceedings, the Contemporary Ergonomics series was born in 1984 and has been a freestanding publication in its own right over the subsequent years.

Over the years the roles and titles of the roles of people associated with the production of the conference have changed. In 1983 the Chair of Meetings was Ted Megaw; the Honorary Meetings Treasurer was Heather Ward; the Honorary Meetings Secretary was Jan Hart; the Conference Secretary was Rachel Birnbaum and the Programme Secretary was Karenna Coombes.

# The Editors of Contemporary Ergonomics
## 1984–2008

## Ted Megaw    1984 and 1987 to 1989

Year 1984 was very much a period of consolidating the efforts made in the previous year by the Meetings Committee and particularly by Karrena Coombes to ensure the continuation of the publication of the Society Annual Conference Proceedings. Although computer technology was then beginning to make an impact in the workplace, the effects were not that obvious in the proceedings themselves; word processing and data transfer were in their infancies so that the publishers, Taylor and Francis, were dependent on the supply of camera-ready copy by authors, many of whom found it difficult to adhere to the instructions provided by Ted Megaw, the Programme Secretary. Since then, ergonomists have become much aware of the significance of errors and violations!

Coincidently, it was the significance of human error that tended to dominate the proceedings edited by Ted Megaw from 1987 to 1989. There is little doubt that three major disasters contributed to this – the Zeebrugge Ferry disaster in March 1987, the Kings Cross underground fire in November 1987, and the Piper Alpha offshore platform disaster in July 1988. In many ways, these three disasters had an impact on the development of ergonomics and ergonomists that paralleled the impact of the Three Mile Island incident in the USA some ten years earlier. Nowhere better can you appreciate this impact than in the keynote papers of James Reason in the 1987 proceedings and of David Canter in the 1989 proceedings. A considerable effort was made by the Meetings Committee in the years 1987 to 1989 to increase the society members' participation at the conferences by ensuring that up to three parallel sessions were offered for most of the conferences' duration, and by inviting keynote addresses of the highest quality and of relevance to ergonomics issues of the time.

## D.J. Oborne    1985 and 1986

Dave Oborne was the programme secretary for the conferences held at Nottingham University 1985 and the University of Durham 1986. He was a senior lecturer in Psychology at Swansea University at that time and stayed at Swansea until his retirement a few years ago.

The conference in 1985 had some competition with the 9th Congress of the International Ergonomics Association which was also held in the UK at Bournemouth in September of that year.

At Durham in 1986 the conference annual dinner had a Transport Minister in attendance to give the after dinner speech. Delegates at this conference were surprised to be participants in an olfactory acuity experiment when substances with strong odours were placed under their seats during one of the coffee breaks.

## Ted Lovesey   1990 to 1993

Ted Lovesey joined The Ergonomics Research Society in 1965 and soon became a member of the Meetings Committee. Having organised the first Military Ergonomics one day meeting in London in 1969 he then went on to be Programme Secretary for the Annual Conference in Edinburgh in 1976. Pressure of ergonomics aviation research work at the Royal Aircraft Establishment caused him to leave the Annual Conference committee until he rejoined as Programme Secretary in 1990 to organise the conference at Leeds University. The theme was "Ergonomics – setting standards for the '90s". An appropriate keynote speech was given by Christopher Wickens, University of Illinois which outlined the problems of displaying complex 3-dimensional information to pilots in an understandable way.

The 1991 Annual Conference was held at the University of Southampton with a theme of "Ergonomics - Design for Performance". The keynote speech was given by David Oborne on tipping the balance towards ergonomics.

The 1992 Annual Conference, with a theme "Ergonomics for Industry", appropriately was held at Aston University, Birmingham. Michael Griffin's keynote speech on "Causes of Motion Sickness" is still a much used paper.

The 1993 Annual Conference returned to Edinburgh, but this time to Heriot-Watt University. The theme "Ergonomics and Energy" was well covered by Geoff Simpson's keynote address on lessons from the mining industry.

## Sandy Robertson    1994 to 1997

Sandy Robertson was the programme secretary for the conferences held at Warwick University 1994, University of Kent 1995, University of Leicester 1996 and Stoke Rochford Hall in 1997. He was a Research Fellow at the Centre for Transport Studies at UCL, investigating various aspects of road safety and driver behaviour.

He had been long associated with the conference and had spent a number of years as a member of the secretariat and held the post of Honorary Meetings Finance Secretary on Council until changes to the structure of Council and the Meetings committee made that post obsolete. He continued to actively contribute to setting the conference budgets and organising/helping the secretariat. Curiously, the first Conference he attended was before he became an ergonomist when in 1984 where he attended and was interviewed for a place on the Loughborough M.Sc. course.

During the period 1994–1997, popular topics for papers included manual handling, musculoskeletal disorders, drivers and driving, HCI, and Health and Safety.

In 1995 there was a change to the design of the covers of Contemporary Ergonomics with the introduction of a more modern style. This change in style was coincident with change to the format of the programme of the conference. For some years there had been considerable debate amongst the conference team about what format of conference that would best meet the needs of the conference or delegates. The conference had, in the years leading up to 1994, run with a 4 day format with 2 or 3 parallel sessions. On one evening there was a social event with a different theme each year. In 1995 this was changed to a 3 day format with 4 parallel sessions and no social event except the Annual Dinner.

At this stage in the history of the conference events were booked 1–2 years in advance in order to get the venue we wanted on the dates we wanted. This meant that there was a distinct planning cycle to the conferences and that changes to the conference tended to operate on geological timescales.

The universities were going through some painful changes and looking at the commercial aspect of conferences. Unfortunately they could not always come up with the associated quality of venue. There was considerable debate about the advantages and disadvantages of university venues versus commercial ones including the philosophical argument that as a learned society we should be supporting the universities. However, given that the universities were becoming much more commercially oriented and less able to support learned societies it was decided to look elsewhere.

1997 marked another big change in that the conference was first held at a non-university venue; Stoke Rochford Hall. The difference in quality was a big surprise and the food was good, but the downside was that the cost was greater.

Until then there had been a policy to move the conference around the country so as that all members would have a conference near them every three or four years. The downside of that was that there was a new venue each year of unpredictable quality. The meetings committee decided that it would be sensible to use 3 venues of proven quality in different parts of the country. This was the pattern that followed.

## Margaret Hanson     1998 to 2001

Margaret Hanson was the Programme Secretary for the conferences at the Royal Agricultural College, Cirencester (1998 and 2001), the University of Leicester (1999), and Stoke Rochford Hall (2000). The conference continued with the three day format with plenary sessions, and 4 parallel presentations, plus workshops. The conference was opened with the Donald Broadbent Memorial Address, being given by Prof Tom Cox, University of Nottingham (1998), on work related stress; Peter White, the BBC Disabilities Affairs Correspondent (1999), on designing for real needs; Prof James Reason (2000), on human variability at work; and Andrew Summers, Chief Executive of the Design Council (2001) on designing for our future selves.

The range of subjects covered by the papers and posters continued to reflect the breadth of ergonomics study and application. Subjects of long standing interest, such as driving, manual handling, and musculoskeletal disorders continued to be well covered, with increased interest in topics such as alarm design, training and psychosocial risks. The emphasis on workshops was developed, with 8–9 being run throughout the conference. Improvements in IT allowed less stressful presentations (no more OHPs!), and the professional appearance of Contemporary Ergonomics to be developed. Approximately 180–260 delegates attended the conferences during 1998–2001.

**Delegates at the Ergonomics Society 50ᵗʰ Anniversary Conference**

The Ergonomics Society launched a new logo in 1998, which incorporated a subtle ES within a circle; the circle represented the soft human forms with the bars representing the three human sciences of anatomy, physiology and psychology. In 1999 the Ergonomics Society celebrated its Fiftieth Anniversary, with events around the country. The main focus was an exhibition at the Science Museum in London in 1999, and at the Manchester Museum of Science and Industry in 2000.

During the time of her role of Programme Secretary, Margaret worked as an ergonomics researcher and consultant for the Institute of Occupational Medicine, Edinburgh. She gratefully acknowledges the assistance of Richard Graveling (IOM), Ted Lovesey, Paul McCabe and Sandy Robertson in her role as Programme Secretary in 1999/2000.

## Paul McCabe    2002 to 2005

Paul McCabe was Programme Secretary for the conferences held at Homerton College Cambridge (2002), Heriot-Watt University Edinburgh (2003), Swansea University (2004) and Hatfield University (2005). He began his association with the Conference as a member of the Secretariat in 1997 at Stoke Rochford Hall and continued to assist in this capacity until he became Programme Secretary.

During his time as Programme Secretary he worked for Atkins Ltd, providing consultancy services to the high hazard industries. During this period there was some experimentation with the format of the conference. The conference in Cambridge saw five parallel, one of which included a symposium on Hospital Ergonomics. The inclusion of symposia gave groups with an interest in a particular subject the opportunity to have a meeting of their own, under the aegis of the Annual Conference and this was to become a feature of conferences during this period.

In Edinburgh, Manual Handling and Rail were featured topics. The HSE launched their Manual Handling Assessment Charts to a wider audience and the Rail session drew together Network Rail's and Rail Safety and Standards Board's sponsored research. Though only two papers were presented on Slips, Trips and Falls at Heriot-Watt, at Swansea and Hatfield the number of papers had grown to enable the topic to be a featured symposium. This was due to the efforts of Wen-Ruey Chang of Liberty Mutual and Roger Haslam from Loughborough University.

The Hatfield event benefited from the support of BAE Systems. Laird Evens of BAE's Advanced Technology Centre not only arranging sponsorship but also organising numerous papers and speakers. Credit for the 2005 programme and conference at Hatfield should also go to Phil Bust who played a major role in the organisation. This experience stood him in good stead for the following year when he took the reins as Programme secretary.

**The Conference Secretariat 2007**

## Phil Bust    2006 to 2008

Phil Bust was the Programme Secretary for the conferences held at Robinson College Cambridge (2006) and the University of Nottingham (2007–2008). During this time he worked as a researcher at Loughborough University, in the civil and building engineering department, investigating aspects of health and safety in the construction industry. In Cambridge the conference's emphasis was on inclusive design and the Secretariat received special training before the event in order that they could assist any delegates with special needs. The inclusive design papers covered the three specific areas of the built environment, society and transport. The conference also hosted an Oil, Gas and Chemical Industries Symposium, an HCI Symposium and the Design Engage Project.

For the first time in its history the conference was held for two consecutive years at the same venue, Nottingham University. In 2007 there were no special collections of papers but there were a great number of papers on patient safety and medical ergonomics, a growing area of research in ergonomics. The Slips, Trips and Falls Symposium returned to the conference in 2008. There was also the first Health and Well-Being of Construction Workers Symposium and a sizeable number of papers on rail ergonomics.

During this period, in addition to the conference, the Society organised a very successful event looking at human and organisational factors in the Oil, Gas and Chemical Industries. The Ergonomics Society gained and lost a chief executive and having established roots in the offices at Elms Court, the Society made preparations in its move towards gaining chartered status and – another historic landmark – its first name change since the Ergonomics Research Society dropped the Research and became the Ergonomics Society.

# 1984

| Exeter University | 2nd–5th April |
|---|---|
| **Conference Secretary** | |
| **Honorary Meetings Secretary** | Jan Hart |
| **Honorary Meetings Treasurer** | Heather Ward |
| **Chair of Meetings** | Ted Megaw |
| **Programme Secretary** | Ted Megaw |

| Chapter | Title | Author |
|---|---|---|
| **Application of Ergonomics** | Jingle bells | M.A. Sinclair, P.G. Stroud, M.N.Thomas and P.A. Parsons |
| **Poster Papers** | "PADAS" – An ambulatory electronic system to monitor and evaluate factors relating to back pain at work | E.O. Otun, I. Heinrich, J.A.D. Anderson and J. Crooks |
| **Human Variability** | Ergonomics is kid's stuff: The ability of primary school children to design their own furniture | M.J. Callan and I.A.R. Galer |
| **Transport Control** | Ergonomics in the lighthouse | J. Spencer |

# JINGLE BELLS

## M. A. SINCLAIR*, P. G. STROUD**,
## M. N. THOMAS*** and P. A. PARSONS***

## * Loughborough University of Technology, Loughborough
## ** Institute for Consumer Ergonomics, Loughborough
## *** Loughborough Consultants Ltd, Loughborough

Over a decade there has been 10 man years of investigation into the causation of false alarms from burglar alarms. These studies have concentrated on integrated communication systems (i.e. no 'bells-only' systems). It was known that over 95% of alarms were false; the aim was to discover why, and how this might be remidied by ergonomics action, to the benefit of police resources. Two major studies took place, with some advisory work from time to time. The findings indicate: (1) 97% of alarms are flase, (2) about 75% of these are human-induced, (3) the alarm industry, not the police, are the agents of change, (4) the industry lacks the technology for human-based problems, (5) and so does ergonomics, in certain respects.

## 1.  INTRODUCTION

The Home Office Technical Working Party on Intruder Alarms commissioned the work discussed below; reports are available (Anon, 1974; Bozeat et al., 1976; Anon, 1983).

Several lines of enquiry had been followed into the false alarm problem; in 1973 it was decided that the 'human error' aspects should be investigated. At that time the problem was stated as follows:

1. Over 95% of alarm calls were false (defined as 'an alarm activation which does not require police response').

2. About 1,000,000 police man-hours were spent attending flase calls.

3. Some officers with 5 years experience had never attended a genuine call.

4. Only 2% of premises in England and Wales had alarm systems, and the market was expanding.

For these studies intruder alarm systems are defined as electronic and electromechanical systems to detect automatically the presence of a human within premises when it should not be there, and to signal this to a monitoring station. This excludes 'bells-only' systems, since it is extremely difficult to obtain data about these (even whether they exist), but includes the majority of systems protecting commercial property. A system comprises 3-100 detectors grouped into circuits linked to a control box. This allows the operator at the end of the day to test each circuit prior to departure. If the test is satisfactory the system is activated, save for an exit route which is inactive for a short period to allow the operator to leave. The procedure is reversed at opening time. Alarm systems are installed by private companies and monitored for compliance by the National Security Council for Intruder Alarms.

## 2.   THE 1973 STUDY

The main aim was to establish those ergonomic aspects which give rise to false alarms. This study lasted for 6 months, in one division of what was the Birmingham City Police. Over a period of 6 weeks 100 installations giving a false alarm were investigated. Within a day a team visited the installation with a questionnaire. Data was obtained from operators, police, keyholders, and alarm company personnel and records. In addition, a statistical survey of records in the Birmingham Police computer for the previous 31 months produced data for over 3000 installations giving false alarms. There were many ergonomic problems in this; the team was initially naive, much of the data was condensed and incomplete, shift systems produced inevitable delays between the alarm and the interviews, and many of the causes of a false alarm are ephemeral (e.g. a rattling door, or birds in a warehouse). However three major characteristics could be identified.

1) Time of day effects. There were very large incidences of false alarms at opening and closing times. If these peaks could be clipped, 30% of false alarms would be eliminated.

2) Repetitivity of false alarms. There is evidence of a heavy skew towards short intervals between repeat false alarms. There is a time-of-day effect as well; discounting this there is still evidence of non-random sequences of false alarms at a given installation.

3) False alarm susceptibility (Table 1). The membership of the 3 groups was stable, indicating that there would be

a benefit in directing attention to 'rogue' installations.

| Group | False alarms/ Installation/ 31 months | % of installations | % of all false alarms |
|---|---|---|---|
| III: 'Rogues' | 18 or more | 8 | 32 |
| II: 'Middling' | 9 to 17 | 22 | 34 |
| I: 'Best' | Up to 9 | 70 | 34 |

TABLE 1. *Installations Pareto analysis, 1973 study*

It was possible to group the causes of false alarms into 4 classes as follows:

1) Human error/negligence (e.g. delay in leaving the building after setting the alarm) 25% of causes.

2) Poor alarm system design (e.g. positioning a sensor where it could be accidentally damaged) 23% of causes.

3) Equipment-environment mismatches (e.g. movement detectors sensitive to movement outside the premises) 28% of causes.

4) Equipment failure. 24% of causes.

Clearly, about 75% of causes are amenable to ergonomics. A number of recommendations were made, under 3 main headings.

1) Changes in operations, including liaison committees between the Police and other parties, more formal channels of communication, changes in service call policies, and better training for all alarm company personnel and system operators.

2) Changes to alarm systems hardware and relevant British Standards.

3) Inspection of alarm systems prior to commissioning by the police to cover hardware, procedures, and the training of operators.

## 3. SEQUELAE TO THE 1973 STUDY

In 1974 the report was accepted; since most of the recommendations were in accord with thinking in the

industry, many were adopted.  BS4737:1978 appeared, including most of the technical changes but excluding most
training aspects; changes in 'approved' alarm company
management structures were agreed, and the NSCIA assumed a
greater role in inspecting systems.  The West Midlands
Police were the first to introduce a formal policy for
intruder alarms in 1978.  This provided for the inspection
of new systems; BS4737:1978 specification by approved
installers; a probationary operation period; and thresholds
for false alarm rates.  Should a system exceed any threshold
action would be taken culminating in the withdrawal of
Police response.  This usually nullifies insurance cover,
and is an effective spur to action.


4.   THE 1981 STUDY

   If the 1973 study was valid, there should have been a
relative decrease in false alarms in the West Midlands
Police region.  A study was commissioned in 1981 to
investigate this, among other things.  A Division of the
West Midlands Police was chosen, with a comparable
Division from the Greater Manchester Police Region.  The
latter had a force policy resembling that of the West
Midlands prior to 1978.  During 1981 all systems in these
two Divisions were monitored for 9 months, with a selected
sample investigated more thoroughly.  In addition, 7 other
Police regions were sampled to provide a statistical base
for other comparisons.  The major findings were as follows.
   1) The West Midlands Force Policy appears to have
achieved some of its aims, but not all.  Table 2 gives
various indices indicating an improvement.  Row (5) is the
important index.  There appears to be a cumulative saving
of 12 man years per annum in the West Midlands.  However,
the improvement is not as great as might be hoped; the
opening and closing peaks were halved, but still significant, and the short-cycle repetitivity was still present.
   2) The revised BS4737 also appears to be effective, but
not as much as was hoped.  Specific amendments were made
to eliminate time of day and repetitivity effects, but they
are still clearly present.  An interesting aspect is that
many old 'rogue' systems have been upgraded to BS4737:1978
specification because of the Policy.  Net, there has been
a reduction in false alarms; however 1978BS systems as a
group now appear to be worse than 1972BS and non-BS systems.
One inference is that the amended BS is incomplete, or misdirected.  An analysis of the selected samples indirectly
supports this; the probability of modification to a system
after installation is 0,4.  The probability of subsequent

modifications is also 0,4. System components are known to be reliable; the inference is that it is system design and installation that are at fault, BS4737:1978 does not address these aspects sufficiently.

| Variable | W. Midlands | G. Manchester |
|---|---|---|
| (1) No. of installations | 1226 | 1120 |
| (2) No. of installations giving false alarms | 758 | 748 |
| (3) % of installations giving false alarms | 61.8 | 66.8 |
| (4) No. of false alarms | 2501 | 2903 |
| (5) Mean flase alarm rate (4) : (1) | 2.04 | 2.59 |
| (6) Mean false alarm rate per installation giving false alarms (4) ÷ (2) | 3.30 | 3.88 |

TABLE 2. *Division comparison indices, 1981 study*

3) The existence of the Force Policy and its vigorous implementation has had beneficial effects on the alarm companies in the West Midlands. However, until the alarm companies maintain full, integrated data bases on their systems little more improvement can be expected.

4) 'Intelligent' control boxes would enable system designers to change from a philosophy of 'tripwire' systems with indiscriminate sensing to a more sophisticated approach involving overlapping sensor fields and error detection. How this should be implemented awaits suitable databases, and some ergonomics input.

5) Training of all personnel into the interactions between people alarm systems, and their environment is clearly necessary. What form this should take is unspecified.

## 5. CRITIQUE

The final reprot has been accepted.  Informal comment
indicates that its main points are being considered
actively; fortunately many of these again coincide with
industry thinking.  In summary the ergonomics contribution
has assisted in producing some amelioration of the
problems, but to a certain extent it has been misdirected.
The ergonomics recommendations are treating symptoms, and
are only indirectly affecting causes.  The latter will
require changes in the technology and in the organisation
of alarm companies.  This was known before the 1981 study,
but evidently was not communicated satisfactorily.
Secondly the areas of ergonomics knowledge required do not
seem to be well documented.  These cover the behaviour
patterns of innocuous people in buildings, the attributes
of buildings that cause these, the movements and changes
in the environment of intruders, and the reliability of
naive, unmotivated operators of control equipment.

REFERENCES

Anon, 1974, Investigation of the ergonomic aspects of
    false alarm calls from intruder alarms.  Rpt. DEC/HO/1,
    P.R.S.D.B., Woodcock Hill, St. Albans.

Anon, 1983, Study of the West Midlands and Greater
    Manchester Force Policies on intruder alarms.
    Rpt. HS/HO/3, P.R.S.D.B., Woodcock Hill, St. Albans.

Bozeat, N., Johnson, M.A., Penn, R.F. & Sinclair, M.A.,
    1976, Some ergonomic and other aspects of the causes of
    false alarms from intruder alarm systems.  Proc. IV
    Cong. I.E.A., Univ. Maryland, 11.16 June.

# "PADAS" - AN AMBULATORY ELECTRONIC SYSTEM TO MONITOR AND EVALUATE FACTORS RELATING TO BACK PAIN AT WORK

## E. O. OTUN, I. HEINRICH, J. A. D. ANDERSON and J. CROOKS

**A.R.C. Occupational Health Research Unit
Department of Community Medicine
Guy's Hospital Medical School
London Bridge, London SE1 9RT**

ABSTRACT

Absence from work due to back pain is a major source of industrial absenteeism. A physiological ambulatory data acquisition system (PADAS) has been developed to record back muscle electromyogram (EMG), intra-abdominal pressure (IAP) and spinal flexion as an objective non-invasive method of monitoring jobs at work, in particular the component tasks which may be responsible for the development or aggravation of back pain.

The system is small, robust and unobtrusive causing a minimum of interference to normal work activities. It is capable of simultaneously recording 8 separate channels of physiological data over a period of 10 hours, requiring a minimum of supervision during its use. The equipment is versatile and can be applied to many situations where ambulatory continuous monitoring is required.

## 1. INTRODUCTION

Back pain is a major cause of industrial absenteeism and imposes a serious economic burden on industry in terms of lost production, and a handicap on the sufferer with many social implications. Kelsey and White (1980) concluded that up to 80% of people some time during their lives suffer from this debilitating syndrome. Attempts have therefore been made to show a cause and effect relationship between a man's work and back pain.

Such studies require the recording of a number of physiological response variables related to back stress and work load to gain objective and reproduceable measures of the various tasks within a job at work.

This paper describes a physiological ambulatory data

acquisition system (PADAS) which has been developed to
record left and right surface electromyogram (EMG) of the
back at levels L4 and L5, intra-abdominal pressure (IAP)
and spinal flexion at T12 in the antero-posterior and
lateral planes as an objective non-invasive method of
monitoring the physical stress to which muscles and the
spine are subjected in different postures and movements at
work.

The methodologies for these techniques are based on
previous research by Ortengren et al (1981), Anderson et
al (1977), Anderson (1980), Sweetman et al (1976), Davis
(1981), Stubbs (1981) and others.


## 2. INSTRUMENTATION

The system is small, robust and light causing minimum
interference with normal work activities.  It is capable
of simultaneously recording 8 separate channels of
physiological data over a period of 10 hours, requiring a
minimum of supervision during its use.

Figure 1 shows the main components of the system, two
miniature 4 channel analog cassette tape recorders with
integral amplifiers, a miniature radio telemetry pressure
sensitive pill, receiver and aerial to measure intra-
abdominal pressure, 3 miniature inclinometers to measure
spinal flexion, 2 pairs of Ag/AgCl electrodes to record
the back muscle electromyogram and a timer to synchronise
both tape recordings.

The complete system is contained in three pouches worn
round the waist of the volunteer (Figure 2).

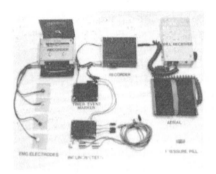

Figure 1.   PADAS Instrument          Figure 2.   A Volunteer at
            Layout                                work with PADAS

## 3. VALIDATION STUDY

As part of the validation experiment a series of 6 full
stoop lifts, lift height being 80 cm, were performed by
volunteers with weights ranging from 10 kg to the
volunteers individual maximum. Each series was preceded
by a rest period and each sequence of rest and lift series
was replicated 9 times.

All data recorded by PADAS was digitized and analysed
by a motorola 6800 based microcomputer.

## 4. RESULTS

As an example of the PADAS system output the results of
one volunteer performing the validation experiment are
summarized in figures 3-6 showing the means and standard
errors over 9 replicates for EMG spike counts/second
(SCPS), Intra-abdominal pressure, Anteroposterior and
Lateral flexion of the spine against the weights lifted.

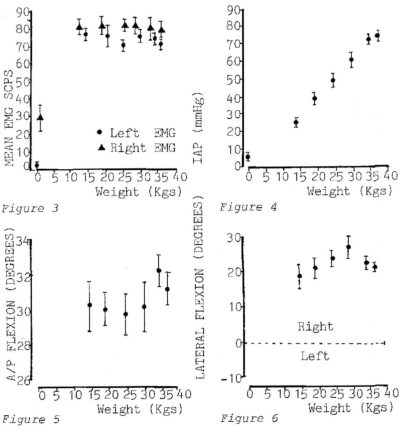

Figure 3

Figure 4

Figure 5

Figure 6

## 5. CONCLUSION

This project sets out to measure in a simple manner some physiological response variables which have been shown by previous research to be related to stress on the spine. In proposed industrial surveys we hope to discriminate between various response patterns within a job and between jobs considering age-sex differences and body build with the ultimate aim of seeking associations between occupational factors and back pain.

## 6. ACKNOWLEDGEMENTS

We would like to thank Dr. B. Kaye and Mr. H. Denyer for their invaluable assistance in this project.

The study is financed by a grant from the Arthritis and Rheumatism Council.

REFERENCES

Anderson, J. A. D., 1980, Occupational aspects of Low Back Pain. In Clinics in Rheumatic diseases Vol. 6, No. 1, edited by R. Grahame (W. B. Saunders Company Ltd.), p. 17-35

Andersson, G. B. J., Ortengren, R. & Nachemson, A., 1977, Intra-discal pressure, intra-abdominal pressure and myoelectric back muscle activity related to posture and loading. Clinical Orthopaedics, 129, 156-164

Davis, P. R., 1981, The Use of Intra-abdominal pressure in evaluating stresses on the lumbar spine. Spine, 6(1) 90-92

Kelsey, J. L. & White, A. A., 1980, Epidemiology and impact of Low back pain. Spine, 5(2), 133-142

Ortengren, R., Andersson, G. B. J. & Nachemson, A., 1981, Studies of relationships between lumbar disc pressure, myoelectric back muscle activity and intra-abdominal pressure. Spine 6(1), 98-103

Stubbs, D. A., 1981, Trunk stresses in construction and other industrial workers. Spine 6(1) 83-89

Sweetman, B. J., Jayasinghe, W. J., Moore, C. S. & Anderson, J. A.D., 1976, Monitoring work factors relating to back pain. Postgraduate medical journal, 52, 151-155

# ERGONOMICS IS KID'S STUFF
## THE ABILITY OF PRIMARY SCHOOL CHILDREN TO DESIGN THEIR OWN FURNITURE

## M. J. CALLAN and I. A. R. GALER

## Department of Human Sciences, University of Technology Loughborough, Leicestershire LE11 3TU

The ability was examined of primary school children to design and evaluate their own school furniture. Two groups of children were guided through a six-phase design and evaluation programme. The results suggested that they are able to make a realistic contribution, could offer rational criticism of their own and others' designs, and could reach sensible compromises of design features. Future equipment development programmes should, where appropriate, not ignore the contribution that this group of users can make.

## 1. INTRODUCTION

A fundamental principle of ergonomics is that the user of a product or facility should be involved at least in its evaluation, and preferably also in its design. Considerable efforts are made to select the participants in a design or evaluation exercise so as to represent the user population.

One group which seems to be used only infrequently in this way is children. It is true that children have contributed important data to the evaluation of a product (as with pharmaceutical containers) but the number of such cases is small in comparison with the number of products that children use. It is even more unusual for children to participate in the product design process itself, rather than in evaluation of a prototype at the end of this process.

It may be that the administrative problems of recruiting and using children are not seen as justifiable; or, perhaps more likely, that children are assumed to be incapable of furnishing useful answers to the questions put by an ergonomist or designer.

A recent project by one of the authors addressed the design, provision and maintenance of furniture for use by

children in primary schools. It was found that several of
the children were capable of making reasoned criticisms of
the equipment they were using. They also generated a
number of potentially useful suggestions for the design of
new furniture. The present study investigated further the
ability of children to participate in a design and evaluation process, and to make best use of their ideas for
furniture design for primary schools.

## 2.  STUDY DESIGN

The study was divided into six phases, as follows:
1.  Awareness phase;
2.  Innovation phase;
3.  Compromise phase;
4.  Mock-up phase;
5.  Evaluation phase;
6.  Report phase.

These phases are detailed below. They correspond to the
initial stages of a design process wherein the designer
reviews the field, generates ideas for a new design, takes
account of limitations to the design to reach a compromise,
builds a mock-up of this compromise solution, and then
evaluates it and reports the findings.

The study was carried out in two primary schools. At
the first school, fourteen children aged ten and eleven
took part, and at the second, twelve children aged nine to
eleven. There were about equal numbers of girls and boys
in each group. The work at the two schools was carried out
independently. Project work was carried out during school
hours.

## 3.  DESIGN PHASES

*Awareness phase*

This phase had two aims: to create a rapport with the
children; and to increase their awareness about the type
of product under consideration. Two methods were used:
first, an informal but guided discussion was held, where
children demonstrated the use of existing furniture for
various activities whilst the rest of the group commented
on this. Second, the class was split into groups and each
group wrote down at least ten different, preferably novel,

uses for a specific piece of furniture (cushion, chair, table, table and chair together, etc.) The children then chose the three ideas from each group that they would most like to use.

## Innovation phase

The aim of this phase was to allow the children to produce ideas for new furniture designs. Each child was asked to design ten pieces of school furniture. The children were encouraged not to feel constrained by convention; emphasis was placed on workplaces rather than on specific and familiar items of equipment.

This phase generated 259 ideas from the two schools. They could be categorised into four groups: adaptations of conventional chairs (83 ideas); variations of tables or desks, often with sloping tops or storage spaces (61 ideas); ideas involving cushions on the floor, often with an associated board to work at (75 ideas); and others, often having nothing to do with school equipment! (40 ideas).

The children produced a wealth of innovative ideas. Of interest was the number of designs which used a cushion to lie on and a sloping surface to work on. Using a cushion in this way had also been rated highly in the Awareness phase. Informal observation had shown that children in primary schools often sit or work on the floor, and the number of ideas using a cushion reflects this.

## Compromise phase

The aim of this phase was to reduce the large number of ideas from the previous phase to a small number that could be produced as mock-ups; and to do this in a manner that the children would see as fair.

At each school, the children were divided into four groups, and each group was assigned one of the four categories defined above. They worked through all the designs in this category and made lists of the specific features in each design. With the children together as one group, each category was taken (the 'other' category was discarded as it was not fruitful), and starting with a blank board a picture of a joint design was developed by reading through the list of features and asking the children whether they wanted to retain each one. Gradually a picture was developed with which everyone was happy.

The logic which went into each aspect of the design was

pleasantly surprising: all aspects were there for a good
reason. The children were not inhibited about expressing
their views: if an idea was disliked there was little
regard for the feelings of the child who had suggested it!
On the whole the designs were multi- or at least dual-
purpose, and were often adjustable: it seemed that the
children wanted their furniture to be used for many things
by many people. It was also clear that comfort was
important, to judge by the use of cushions and padding.
Figure 1 shows some of the designs.

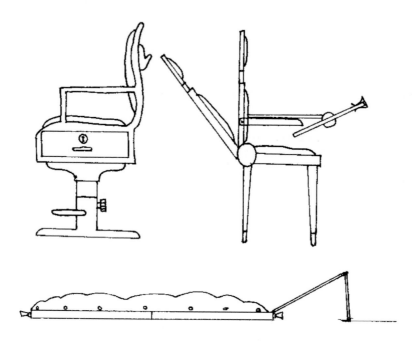

FIGURE 1.  *Examples of designs from the Compromise phase*

*Mock-up phase*

The aim of this phase was to produce mock-ups of the
five designs finally produced.
Each group of children produced one mock-up. Cardboard,
foam, tape, and other easily-available and cheap materials
were used. One child in each group made a suggestion box
for receiving comments about the design.
This phase required a great deal of effort on the part
of the children and the researcher. It was probably the
least successful of the six phases because the construction

of a realistic mock-up is not easy. Ideally, the job would
have been assigned to an engineering workshop, but this was
not possible because of cost and time constraints.

## Evaluation phase

The mock-ups were displayed in their respective schools,
and all pupils were invited to write down their comments
and put them in the suggestion box. In addition, two
questionnaires were developed by the researcher, one for
chair designs and one for cushion designs, which asked
simple questions about safety, efficiency, reliability,
comfort, and ease of maintenance (Kirk and Ridgway, 1970).
No questionnaire was developed for a table design, as a
table was produced at only one school and the quality of
the mock-up was very poor. These questionnaires were
distributed to the children involved in the design exercise,
who were told that this was an exam (thus minimising
disruption when the forms were being filled in).

The suggestion box at one school produced 195 comments,
whilst that at the other producedonly 17. The location of
the mock-ups, and knowledge of their existence, explains
this discrepancy. The comments were examined by the
children, and were classified as to whether they were
positive towards the design, neutral, or negative. The
questionnaires were analysed by the researcher.

The quality of the mock-ups clearly affected the
evaluations; nevertheless, this phase produced some useful
results. It emerged that safety, usefulness and comfort
were the three most important aspects of furniture design
to the children; this should be contrasted with criteria
of cost and stackability which are probably more important
to designers and purchasers of current school furniture.

## Report phase

The aim of this phase was to give the children the chance
to produce arguments in favour of their designs; and to
show them that any piece of work requires a proper report.

Each child first made a description of his or her
favourite workplace, based on the mock-ups and accompanied
by a diagram. Second, the child wrote a letter to a
manufacturer or to a Director of Education, making arguments
for the production or purchase of the furniture.

The letters showed that children were able to describe
coherently what they had done, and that they were often

able to give reasons for each aspect of their chosen design.

## 4. CONCLUSION

It was obvious that the children relished the opportunity to express their opinions about their working environment, and that they felt it worthwhile to contribute towards the workplaces in schools of the future. The study showed that children of this age, if guided carefully through a structured process, could develop innovative design ideas, and could evaluate these ideas critically. It is unlikely that all of the designs developed in this project would be viable from a manufacturing, economic, or even educational viewpoint. Three things are important, however: first, the children used criteria relevant to themselves, and it is unreasonable to expect such users, or indeed any users, to recognise all criteria. Second, even if a design as a whole is not viable, specific aspects of that design might be important and might be used in other prototypes and manufactured products. Third, quite apart from the main interest in this project, the process of exposing primary school children to ergonomics and design is felt to have been useful. Other ergonomists and designers might in future more readily take the opportunity to consult this important user group.

## ACKNOWLEDGEMENTS

The authors would like to thank the Leicester Education Department, and the staff and children at Sherrard School, Melton Mowbray and Woolden Hill School, Anstey, for their cooperation in this project.

## REFERENCES

Kirk, N.S. and Ridgway, S., 1970, Ergonomics testing of consumer products: 1. General considerations. *Applied Ergonomics*, *1,5*, 295-300.

# ERGONOMICS IN THE LIGHTHOUSE

## J. SPENCER

### Department of Applied Psychology
### UWIST, Penylan
### Cardiff CF3 7UX

ABSTRACT

Lighthouses offer an unimpeachable navigational reference for all types of vessels at sea in reasonable
visibility. The advent of reliable electronic position
fixing systems has not yet served to render lighthouses
obsolete but it has called into question the cost of
operating a manned lighthouse service. In the Scottish
lighthouse organisation lighthouse manpower accounts for
about 20% of the operating costs. This raises the question:
what do lightkeepers contribute to the organisation which
might be missed if the lights were made completely
automatic?
This study examines the work of the lightkeeper and
discusses the problem of how best to deploy human skills
in an organisation which is faced by rapidly increasing
navigational sophistication among its clients.

Lighthouse keeping is a profession which has developed
in the U.K. over the last three hundred years or so. Like
many other jobs however it is now faced with changes on a
major scale. Paraffin and acetylene light sources are
being replaced by mercury vapour, tungsten filament or
quartz - halogen sources. Remote control systems are
already in being and it is not too far-fetched to claim
that in the near future manned lighthouses might become
another monument to obsolescence. Sweden automated all its
lighthouses during the 70's. Should the U.K. adopt the
same policy? The study reported in this paper is part of
an attempt to find an answer to this question. The study
was conducted in Scotland where lighthouses are managed by
the Northern Lighthouse Board with headquarters in
Edinburgh.

Lighthouses vary over a wide range in their siting and architecture. For example, the altitude of the base of the light above M.S.L. among a sample of Scottish lighthouses varied from 5m to 190m. This affects the climate experienced by the lightkeepers, especially wind strength. Architecturally the height from the entrance door at the base of the tower to the light room varied from 2m to 33m which has a major effect on the energy expenditure during a watch when ascents of the tower have to be made. When the average number of ascents is taken into account the range in mechanical energy expenditure on ascents only is some 36 to 1 and if an estimate of overall physiological expenditure is attempted then the ratio becomes closer to 50 to 1 with an upper energy expenditure of some 200 K.cals per watch.

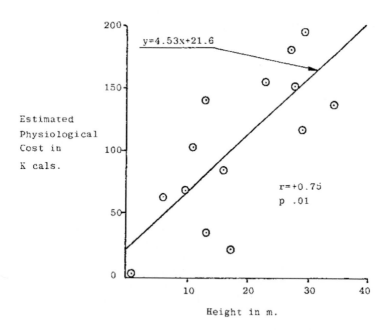

FIGURE 1. *Relation between Height of Lighthouse and Estimated Physiological Energy Cost of Ascents and Descents made during a Watch.*

Notes.

a) 150 Kcals represents about 5% of the total
   daily energy expenditure of a physically
   active adult male
b) The straight-line relation is a statistical
   'best-fit' and should pass through the
   intersection of the x and y axes of the graph
c) The graph assumes a lightkeeper whose
   clothed body weight is 77kg.

The equipment varies for both the lights and, where
installed, the fog signal. Lights may be acetylene gas,
paraffin vapour, tungsten filament or some variety of
halide lamp. Each involves different treatments and is
variously reliable. Fog signals may be compressed-air
driven which necessitates the operation of diesel driven
compressor engines or they may be electric driven which
usually necessitates diesel driven generators or
alternators. By the nature of their siting it is clear
that the equipment, the buildings and the precincts are
under almost constant atmospheric attack – salt spray
laden wind. So the lightkeeper is always at work on the
variety of tasks needed to keep everything 'ship-shape'.
Furthermore the daily duty cycle is highly variable with
the season of the year. For example at latitude 58°N
(about mid-ships on Lewis and Harris) the light is 'on' for
approximately 3h 25m in mid-Summer and for 17h 40m in mid-
Winter.

The lightkeeper normally spends about three years at
any one site before being moved to another. Consequently
all the variables as between stations are completely
'scrambled' if one attempts to measure separately their
possible effects on selected individuals. Nevertheless
even if one examines accidents, for example, on a
'scrambled' basis the results are extremely interesting.
Figure 2 illustrates the annual accident rate over a nine
year period. Although the examination is crude it is
quite clear that any sort of accident happens very, very
rarely at lighthouses. Is this perhaps because the
treatment room or the hospital is sixty miles away and
perhaps 48 hours in time? The risks are no less than in
many industrial settings from which rates an order of
magnitude higher would be found.

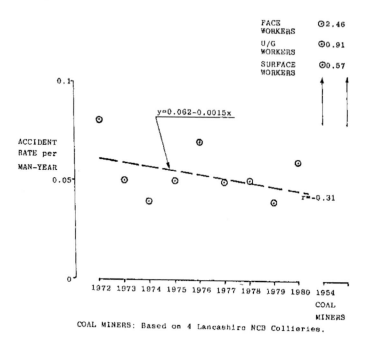

COAL MINERS: Based on 4 Lancashire NCB Collieries.

FIGURE 2. *Rates for All ACCIDENTS to LIGHTKEEPERS during the period 1972-1980 inclusive.*

The results of an examination of a sample of light-keepers using two well-known psychometric instruments show them to be much as other workers in the U.K. They differ very significantly in one respect. On the 16 P.F. Test they show a much higher mean score on Factor Q2 - self sufficiency.

Many other aspects of the lightkeeper's work were studied which space prevents a report here. However it became very clear that while the lightkeeper's job may be threatened by developments in mechanisation so also is the very existence of the organisation which manages those jobs. The development and 'mechanisation' of radio aids is perhaps the greatest threat to the lightkeepers and to their organisation unless some 'intelligent' synthesis between equipment and its human charges can be created. The suggestion is to create a mechanised lighthouse network

with augmented nav-aid facilities in which manned lighthouses are embedded.  These latter would, using VHF/UHF
links, carry out 'preventive' monitoring of the automatic
lights and their associated equipment.  Daily reports to
H.Q. in Edinburgh by either M/F or telephone links could
then determine the check and repair programme for
centrally based maintenance facilities.

# 1985

| Nottingham University | 27th–29th March |
|---|---|
| Conference Secretary | Rachel Birnbaum |
| Exhibitions Officer | |
| Honorary Meetings Treasurer | Heather Ward |
| Chair of Meetings | Ted Megaw |
| Programme Secretary | D.J. Oborne |

| Chapter | Title | Author |
|---|---|---|
| Information Technology | Touch screens: A summary report of an evaluation of improved screen layout designs | W.I. Hamilton |
| Cognitive Ergonomics | How intelligible is English spoken by non–native English speakers? | D. Irvine |
| Accident Behaviour | Slips and mistakes: Two distinct classes of human error? | J. Reason |
| Ergonomics and Police Detection Methods | Ergonomics and police detection methods | W.R. Harper |

# TOUCH-SCREENS: A SUMMARY REPORT OF AN EVALUATION OF IMPROVED SCREEN LAYOUT DESIGNS

**W. I. Hamilton**    Ergonomics Unit,
University College, London.

    Now at: Human Factors Department, FPC 267,
British Aerospace, Sowerby Research
Centre, Bristol. BS12 7QW.

The optimization of touch-screen switch displays, for use in function selection input tasks, was approached from the point of view of enhancing visual search performance when searching for a key. Three proposed screen layout designs had previously been developed for this purpose. They were intended to structure key search by the association of keys to performance cue features such as colour coding, semantically grouped key blocks and spatially separated/defined key blocks. These three principal cue feature designs were evaluated under an experimental simulation of a function selection task. Additionally subsidiary cue feature effects, such as those of key position information and key to code association strength, were examined.

The results indicated that among the principal cue features, only semantic grouping represented an improvement over a non-organised design acting as a control. Further, the combination of principal and subsidiary cue features within any one design layout was seen to have modified its effectiveness as a visual search performance aid. In particular, a knowledge of key position was seen to have been rapidly acquired and facilitated key search. Also, stronger associations between keys and cue features were seen to have enhanced their facilitative effect on performance. Recommendations are made concerning the use of visual search cues for the optimization of touch-screen displays.

## INTRODUCTION

Many modern computer systems provide a set of pre-programmed function options amongst which the user may select. These function alternatives have typically been

27

presented on a key pad type input device. More recently, however, the touch-screen device has been shown to be suitable for this kind of selection input task (Whitefield 1984). Further, Pfauth and Priest (1981) indicated that the touch-screen has enormous potential for application in many work station designs. They went on to argue that in order to realize the full benefits of this application, the design of the switch displays must be optimized. Essentially, selection type input involves the user having to search for the desired function switch in order to visually target the input action. What is needed, therefore is a set of screen layout designs which will facilitate this search task.

Fletcher (1982) proposed a set of "Improved layout designs" wich were intended to facilitate visual search performance when using a function keypad. These improvement designs essentially involve the coding of individual keys in order to reduce the total number of keys which it is necessary to search for a single input. The following "Improved layout designs" were evaluated in research conducted by the British Telecom Human Factors department (note 1).
1) A spatially separated design, in which the keys had been arranged in discrete physical sub-units.
2) A semantically organized design, in which key labels had been grouped into blocks according to their superordinate semantic category, for example all the keys with animal noun legends had been grouped into a block and so on.
3) A colour coded design, in which key labels had been assigned to coloured keys according to their semantic category.

The semanticaly organized and colour coded designs obtained significantly (P < 0.05) faster keying times when compared to a non-coded design acting as a control. The spatially separated design obtained slightly faster times than the control design but the difference was not significant. The increase in the overall size of the key area for the spatially separated design was believed to have negated any facilitative effect the layout may have had.

The results of this evaluation were taken as evidence for the general utility of the improvement designs for facilitating keying speed by untrained users in a selection type input task. It was further concluded that the correspondence between a key label and some cue feature acted to structure the search for the key. Without such organization, search was said to be more random and extended.

The purpose of the present study was to evaluate the

utility of these "Improved layout designs" for use with the touch-screen display. At the same time it was intended that the particular cue feature effect of each design layout should be examined in detail.


ADDITIONAL CUE FEATURE EFFECTS

The authors of the previous evaluative study (Note 1) had assumed that the only cue features which had affected performance had been the principal cue features of spatial separation, colour coding and semantic organisation. However, additional or subsidiary feature effects may also be relevant in explaining the overall effects of the "Improved layout designs", these will now be described.
1) Without principal cue features, the visual search for a key was assumed to have been essentially random and exhaustive. Search time had been, in fact, shown to be a function of the number of keys on a keyboard (Note 2). However, this is only likely to be true for initial exerience. After only a short time using a keyboard one will have learnt some of the key positions. This key position information need not be complete in order for it to facilitate performance and may be useful after only a few key presses. Key position information, therefore, represents a non-organized cue feature since, unlike the principal cue features, it is idiosyncratically acquired through experience and not imposed by design. Its effect on performance ought to be assessed.
3) With colour coding the association of a key to a colour provides a visual search cue. The nature of the link between the key and the colour code would obviously affect the strength of the association and in turn the utility of the colour code for visual search. A semantic key to colour assignment rule, such as that used in the British Telecom study, represents a very simple and efficient association. It is likely that the absence of such an effective asssignment rule, and its replacement with some arbitrary key to colour assignment, would result in a reduction of the performance effectiveness of a colour coded layout design. This effect ought to be examined. At the same time the colour coding cue feature could be evaluated independently of the effects of key position information and location cues.
The cue feature effects described above were examined for a set of "Improved layout designs" which had been adapted for the touch screen. For the sake of clarity the word key will continue to be used in reference to touch-screen switch positions.

METHOD

Forty subjects, recruited from British Telecom's Ipswich
subject panel, were allocated to 4 screen layout designs in
a balanced experimental design. Additionally each subject
was tested on 2 versions of the layout design to which they
had been allocated. All testing was conducted using the
same touch-screen display system. Four sets of eight
common English words were drawn from the superordinate
categories of animal, vehicle, number and human name.
These were employed as key legends in order to simulate a
function selection task with 32 function options.
The task. Subjects were required to locate and touch the
switch position labelled with a stimulus word which they
had been given. Response time and input error were taken
as dependent variables. Prior to testing subjects had to
learn the particular organizatin of the key layouts to a
criterion level of accuracy, assessed on a probe item test.

There were two versions of each layout design. Version
1 was the low feature content arrangement and version 2 was
the high feature content arrangement. The feature content
arrangements for each layout design tested will now be
described.

Non-organized design. Version 1, was a non-organized
design without principal cue features. In addition key
position information was inhibited by having a variable key
position manipulation (VKP) by which key positions were
changed between every trial. This arrangement is denoted
by VKP.
Version 2, as above except that key positions were fixed so
as to enable key position information to function as a
performance cue. The arrangement is denoted by FKP.

Spatially separated design. Version 1, the key matrix
was halved vertically and horizontally to create four
physically discrete quadrants/sub-units. An equal number
of key labels from each semantic category were contained
within each sub-unit. Also, within each sub-unit key
positions were changed between trials (L(VKP)).
Version 2, as above except that key positions within each
sub-unit were fixed for all trials (L(FKP)). In order to
avoid any spurious effects on performance arising from
changes in switch matrix area, the overall area of each of
the spatially separated layouts was made equal to that of
other layouts.

Semantically grouped design. Version 1, key labels were
grouped into blocks together with other labels from the
same semantic category. The blocks were not separated
physically. Key positions within each block were changed
between trials (S(VKP)).
Version 2, as above except that within each block key

positions were fixed for all trials (S(FKP)).

Colour coded design. Version 1, employed a variable key position manipulation between all trials. Key labels were allocated to colour codes in a balanced manner with an equal number of labels from each category represented within each colour. Keys of the same colour were not grouped together (CVKP).

Version 2, as above except that key labels were assigned to colour codes according to their semantic category. For example all animal nouns were presented on red keys, vehicle nouns on black keys, human names on white keys and numbers on blue keys (C(S)VKP).

## RESULTS AND DISCUSSION

Response time and error scores were submitted to a Single Factor ANOVA of screen layout design. Separate analyses were performed for each level of cue feature content.

No significant effects were obtained from the analysis of the error data and therefore it was not considered further. The analysis of response times showed significant effects of screen layout for both the high feature content and the low feature content arrangements ($F = 8.78$, df = 4,45, $P < 0.01$ and $f = 22.65$, df = 4,45, $P < 0.01$ respectively). See Figure 1.

Comparison between the low cue feature content arrangements using the Tukey method revealed that only the semantic arrangement (S(VKP)) represented a significant ($P < 0.05$) improvement over the non-organized VKP arrangement. The colour coded arrangement (CVKP) obtained significantly slower keying times than the VKP arrangement. Keying times obtained on the spatially separated (L(VKP)) arrangement were not significantly different from those obtained on the non-organized (VKP) arrangement.

The most likely explanation for the superiority of the semantic arrangement is that it provided the user with a location association for a key. That is, the user could associate a key with a semantically defined key block which had a known location on the screen. Therefore only this location needed to be searched to obtain the required key. The spatially separated design also provided location cues by the association of keys to sub-units, yet performance on the L(VKP) arrangement was not superior to the VKP arrangement. Since the subjects on both the L(VKP) and the S(VKP) arrangements had learnt the coding assignments to an equal level of accuracy, as tested, it is not immediately clear why the location association enhanced performance on only the semantic arrangement. A possible explanation is

that the effectiveness of the spatially separated layout in
providing location cues was limited by the reliability of
the subject's memory for the key to sub-unit associations.
The semantic layout did not suffer from this problem since
it also provided a fall back mnemonic in the form of the
semantic assignment rule.  All the subjects had to remember
was the category to location associations - a much less
demanding task since only four associations had to be
memorized instead of 32.

FIGURE 1.  Showing mean response times for each design
arrangement.  $\bar{X}$ = the mean values for the low cue feature
content versions; $\bar{X}$ = the mean values for the high cue
feature content versions.

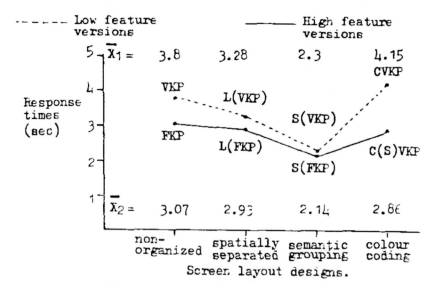

Screen layout designs.

Among the high cue feature content arrangements the
semantic layout had once again obtained significantly
faster RT's compared to all other layouts.  No other
differences had been significant for this group.


CUE FEATURE EFFECTS

Cue feature content effects within each layout type were
examined by t-test.
Non-organized layout. Keying times for the FKP
arrangement were significantly faster than those obtained
on the VKP arrangemet (t = 4.53, df = 9, P < 0.01).  The
superiority of performance on the fixed key position
arrangement supported the view that, even in the absence of

principal cue features, visual search can be organized by key position information.

Spatially separated layout. The addition of key position information to the spatially separated layout was seen to have significantly enhanced keying performance ($t = 3.26$, $df = 9$, $P < 0.01$). Coupled with the fact that spatial separation alone did not improve performance beyond the level obtained on the VKP arrangement, this suggests that spatial separation on its own is not sufficient to ensure improved performance.

Semantically organized layout. The addition of key position information to the semantically organized layout did not further enhance keying speed ($t = 1.47$, $df \approx 9$, $P > 0.01$). This suggests that for visual search, effective *location associations*, as provided by semantic organization, will be used in preference to key position information when both are available.

Colour coded layout. Comparison of the CVKP and the C(S)VKP arrangements indicated that the C(S)VKP arrangement had obtained significantly faster key press times ($t = 4.95$, $df = 9$, $P < 0.01$). In other words an arbitrary key to colour assignment rule was seen to have severely impaired the performance effectiveness of the low cue feature content arrangement. With a reliable (semantic) assignment rule, however, colour coding represented a substantial visual search cue.

SUMMARY AND RECOMMENDATIONS

Among the principal cue features examined only semantic grouping was shown to have *significantly improved* performance when tested independently of other feature effects. Additionally, the subsidiary cue features of key position information and coding association strength were shown to have influenced the overall effectiveness of "Improved Layout designs" in facilitating keying performance on a touch-screen. Certain design recommendations can be drawn from this work.

With a screen display of up to 32 items (as used here) with a fixed format, key position information was seen to have been rapidly acquired by the user and was sufficient to *improve performance over that for a variable format layout.* Therefore applications of the touch-screen incorporating small fixed format displays may obtain satisfactory performance levels without the extra expense of designing in performance cue features.

Where an application has a great many function alternatives, performance can be facilitated by the provision of location associations. In order to be effective, the associations must be reinforced by some reliable allocation rule.

Colour coding may also be used to facilitate search within a large switch matrix. Here too the reliability of the coding association must be ensured. The effectiveness of colour coding, when used in this way, even in the absence of key position information, suggests that it may be most successfully employed on displays which have a changing format.

Given that subsidiary cue features have been shown to modify the effects of principal cue features on keying times. Further work could assess the effect of combining principal cue features. It is likely that spatial separation, when combined with semantic grouping, may act to enhance the distinctiveness of screen locations. Also, arbitrary assignments of keys to colour may enhance key to location associations in the absence of suitably cohesive semantic categories.

ACKNOWLEDGEMENT

I would like to thank David Thomson, John Fletcher and Leo Borwick of the British Telecom Human Factors Dept, Mortlesham Heath, Ipswich; for their help in conducting this work.

NOTATIONS
1) Borwick, R.L.  An experimental investigation of the use of Colour Coding, Spatial Separation and Conceptual Grouping in the Labelling of Function Keys. British Telecom R19.2.1 Group Memo No. 83/19.
2) Borwick, R.L.  A summary of an investigation into the simulated use of function keys.  British Telecom R19.2.1 Group Memo No. 82/25.

REFERENCES
Fletcher, J.H.  Designing keyboards for the USER Proceedings of the I.E.E. Conference on man/machine systems, July 1982 (I.E.E. Conf. Publ. No.212).
Pfauth, M. and Priest, J. Person computer interface using touch screen devices.  Proceedings of the HUMAN FACTORS SOCIETY 25th Annual Meeting, October 1981.
Whitefield, A.  Human Factors aspects of Pointing as an input technique in Interactive Computer Systems. ERGONOMICS UNIT, University College London. (Unpublished).

# HOW INTELLIGIBLE IS ENGLISH SPOKEN BY NON-NATIVE ENGLISH SPEAKERS?

**D. Irvine**       School of Management Studies,
Polytechnic of Central London,
35 Marylebone Road, London. NW1 5LS

ABSTRACT

Recordings of sub-tests of the "Answer in sentence Test"
(AIS) were made by five non-native English speakers. They
were played back to a group of non-native English speakers,
and to a group of Native English speakers. The native
English speakers had significantly higher scores than the
non-native English speakers. There were also significant
differences in the ease with which the voices were
understood. There was also evidence that native English
speakers are better at understanding the speech of
non-native English speakers than the converse. From these
and earlier experiments a theory of speech intelligibility
was developed. It is based on the proposition that there
exist separate written and spoken "internal dictionaries".

INTRODUCTION

Although there are numerous papers on speech, and many
on speech intelligibility, relatively few have been
concerned with the intelligibility of spoken English to
non-native English speakers. This is perhaps surprising
when one considers that English is so widely used outside
English-speaking countries and is used as an official
language in most internationl organisations, and also is
used as a second official language in countries such as
India. Indeed one of the relatively few studies is that
of Bansal (1969), who studied Indian English. Another
study is that of Elanani (1968) who investigated the
intelligibility of Jondanian English for native English
speakers. Unfortunately both of these studies may be
criticised on the grounds that their test material is
clumsy and lacking in validation statistics, while the

experimental method is weak.  Lane (1963), on the other
hand, used acceptable methods but his research was
concentrated on the effects of masking and filtering on
English spoken by people whose native languages were
American, English, Serbian, Punjabi and Japanese.  His
listener subjects were six male and six female American
undergraduates.  The test material consisted of the Harvard
Word-Lists.  Research of a similar nature on languages
other than English is lacking.  Previous research by Irvine
(1977), Haug (1981), and Haug & Irvine (1983) concentrated
on the problems which non-native English (nnE) speakers
have when listening to English spoken by a native English
(nE) speaker.  Extensive use was made of the
Answer-in-Sentence Test (AIS) Irvine (1974) which consists
of 200 sentences arranged in blocks of 20 of equal
difficulty.  Each sentence contains the answer to the
sentence and usually requires a single word answer.  An
example is "What is the colour of a yellow motor car?"
Most of the data were collected by using a tape recording
of this author's voice, although experiments using four
other voices indicated no real differences in
intelligibility scores.  The other voices exhibited no
marked regional accents.

The study reported below is a first attempt at
investigating the intelligibility of English when spoken by
nnE speaking males and listened to by both nnE and nE
subjects.


METHOD

Five people were prevailed upon to make tape-
recordings of two blocks of twenty questions each of Test
AIS.  These helpful individuals spoke the following native
languages, Lunyule (Ugandan), Yoruba (Nigerian), Nepali
(Nepali) and two who spoke Arabic (both Iraqi).  They had
spent varying amounts of time in Britain, so this factor is
confounded with any differences in accent as such, and its
effect would need to be clarified in future research.  All
the speakers were students at the Faculty of Mangement
Studies, PCL.

The subjects consisted of two groups.  One group of N=11
were all nnE speakers who had, with one exception, lived in
England for two months only.  They were given the AIS
sub-tests, the block of 20 questions by each speaker, as a
group with the material played back on a high quality
reel-to-reel recorder.

The nE speaking subjects were N=14 Higher Technical
Diploma students and N=6 Management Students.  Thus each
subject answered a block of 20 questions from each of five

nnE speakers.  The results were treated by Analysis of
Variance using the SPSS, and means and standard deviations
by speaker and by subjects were calculated.  In addition
1 x n $\chi^2$ was calculated from scores on each voice for nE
and nnE subjects.


RESULTS

The analysis in Table 1 shows that native English
speaking subjects are much better at understanding the
speech of the five foreign voices than are the non-native
English.

Table 1.  ANOVA of differences between nE and nnE Speakers
for each voice.  Source = Between groups DF=1.

| Voice | Sum of Squares | F | Sig. |
|-------|---------------|--------|-----------|
| A | 164.130 | 25.769 | p < 0.0001 |
| B | 622.229 | 46.682 | p < 0.0001 |
| C | 267.229 | 26.951 | p < 0.0001 |
| D | 96.916 | 9.147 | p < 0.0052 |
| E | 372.557 | 23.960 | p < 0.0001 |

In fact only two subjects in the nnE group have scores
which, on the basis of the research by Haug and by Irvine
(1983), suggest that they have any reasonable understanding
of the English spoken by these five voices.  Table 2,
below, gives the means and standard deviations of this
group.

On the other hand the nE speaking subjects are
reasonably proficient at understanding the voices of the
nnE speakers.  Table 3, gives some partial norms from data
collected by Haug and by Irvine (1983).  (More detailed
comparison can be obtained on request)  The best comparison
would be the N=45 DMS group whose mean score is 70.1%.  It
would also appear that the nE subjects are better able to
deal with "foreign" accents than nnE subjects are at
understanding an nE accent.  But this conclusion is
tentative.

Table 2.  Means and Standard Deviations of correct answers
to sub-tests of AIS by voices and by subjects.
a) nnE Subjects N=11

| Subjects | Means | SDs | Voices | Means | SDs |
|---|---|---|---|---|---|
| 1 | 5.8 | 2.17 | A | 9.07 | 3.36 |
| 2 | 10.6 | 4.32 | B | 8.64 | 5.73 |
| 3 | 9.6 | 2.88 | C | 12.36 | 4.90 |
| 4 | 8.75 | 2.17 | D | 9.59 | 4.18 |
| 5 | 10.0 | 2.83 | E | 8.45 | 4.91 |
| 6 | 11.5 | 1.12 | All* | 9.62 | 4.73 |
| 7 | 4.0 | 3.54 | = | 48.1% | |
| 8 | 17.2 | 2.39 | | | |
| 9 | 5.6 | 1.82 | *Combined Voices & | | |
| 10 | 6.8 | 3.77 | Subjects | | |
| 11 | 15.8 | 2.49 | | | |

b) nE Subjects N=20

| Subjects | Means | SDs | Voices | Means | SDs |
|---|---|---|---|---|---|
| 12 | 17.4 | 2.41 | A | 13.9 | 1.94 |
| 13 | 14.8 | 3.70 | B | 18.0 | 1.75 |
| 14 | 16.4 | 1.82 | C | 18.5 | 1.57 |
| 15 | 17.2 | 3.42 | D | 13.15 | 2.30 |
| 16 | 16.4 | 3.78 | E | 15.7 | 3.33 |
| 17 | 11.6 | 2.51 | All* | 15.85 | 3.09 |
| 18 | 14.8 | 3.11 | = | 79.25% | |
| 19 | 13.8 | 3.49 | | | |
| 20 | 19.0 | 1.00 | *Combined Voices & | | |
| 21 | 14.2 | 3.96 | Subjects | | |
| 22 | 16.2 | 3.03 | | | |
| 23 | 15.0 | 3.81 | | | |
| 24 | 13.4 | 2.70 | | | |
| 25 | 17.2 | 2.17 | | | |
| 26 | 15.6 | 2.19 | | | |
| 27 | 17.0 | 2.92 | | | |
| 28 | 17.0 | 1.87 | | | |
| 29 | 16.6 | 3.45 | | | |
| 30 | 16.2 | 1.79 | | | |
| 31 | 17.2 | 2.17 | | | |

Table 3. Samples of nnE norms obtained using AIS.  Max=40.

| Course | N | Mean | SD |
|---|---|---|---|
| DMS (Applicants) | 628 | 23.33 | 9.24 |
| DMS (Poly) | 45 | 28.04 | 8.84 |
| Cambridge 1st Cert | 109 | 15.07 | 6.59 |
| Cambridge Proficiency | 86 | 24.28 | 7.90 |
| BSc Mech. Eng. | 60 | 23.69 | 8.30 |

DISCUSSION

For each voice, and taking the groups separately, $1 \times n \chi^2$ values were calculated. For the nnE group all values were significant at $p < 0.05$ or less except for voice A. For the nE group all values were significant at $p > 0.99$ or greater ecept for voice E at $0.90 > p > 0.50$. Thus for the nnE group the subject's responses vary significantly for all voices except one, while the nE subjects tend not to vary much in their responses to the voices except for voice E.

Theories of speech perception are in conflict. Moore (1977) discusses a motor-matching theory, but the fact that individuals who have been deprived from birth of the ability to generate the neural motor-matching which such a theory demands, seems sufficient to negate it. Broadbent (1967) and others have favoured a model based on signal-detection theory and the concept of an "internal dictionary".

English speech spoken by an nE speaker is usually recognisable by an nE listener over a wide range of intensities and distortions. Nor does it matter to the listener whether the speaker is a high soprano or a basso profundo. The fact that speech is intelligible at all is something of a mystery. Psychologists tend to take refuge in learning theory and concepts such as "Stimulus generalisation". However this has been strongly criticised by Chomsky (1959). In our experiments it would seem that the nnE subjects have problems in matching what they hear with what is stored in their "internal dictionaries". It is suggested here that there are two "internal dictionaries" for each language, one written and one for speech. The written language dictionary may contain words which are not in the spoken language dictionary and vice-versa, while the latter may also not "match" the former, and both may not match or generalise (if one dare use the term) to what someone else says. The problems become more acute if the speaker has an "outlandish" accent. The nE speaking, literate subject, on the other hand, has well-matched spoken and written dictionaries. One solution might be a common artificial language, Humblet (1984), but Ogden's (1968) Basic English might be better. Speech is replacing conventional displays. Witten (1982) discusses the advantages, but the cross-cultural problems have been neglected.

BIBLIOGRAPHY

Bansal, R.K., 1969, The Intelligibility of Indian-English
    (Monograph No.4, Central Institute of English),
    Hyderabad.
Broadbent, D.E., 1967, Psychological Review, 74, 1.
Cherry, C., 1957, On Human Communication. (Chapman & Hall)
    (M.I.T. 1966).
Chomsky, N., 1959, Language, 35, 26.
El Anani, M.I., 1968, An Assessment of the Intelligibility
    of Jordanian English to Educated Speakers of British
    English, Unpublished M.Phil Thesis, University of Leeds.
Haug, U., 1981, Student Selection with particular reference
    to Applicants whose native Language is not English,
    Unpublished PhD Thesis, PCL.
Haug,U. and Irvine, D.H., Assessment of Spoken English
    Language Problems of Non-Native English Speakers.  In
    Human Assessment and Cultural Factors edited by S.H.
    Irvine & J.W. Berry, (Plenum), p.617.
Humblet, J.-E., 1984, International Social Science Journal,
    26, 1.
Irvine, D.H., 1974, Ergonomics, 17, 6.
Irvine, D.H., 1977, Language & Speech, 20, 4.
Lane, H., 1963, The Journal of the Acoustical Society of
    America, 35, 4.
Miller, G.A., 1951, Language & Communication,
    (McGraw-Hill).
Moore, B.C.J., 1977, Introduction to the Psychology of
    Hearing, (Macmillan).
Ogden, C.K., 1968, Basic English.  International Second
    Language, (Harcourt, Brace and World).
Witten, I.H., 1982, Principles of Computer Speech,
    (Academic Press).

# SLIPS AND MISTAKES: TWO DISTINCT CLASSES OF HUMAN ERROR?

**James Reason**          Department of Psychology,
                          University of Manchester,
                          Manchester. M13 9PL.

## WHY THE ERROR BUSINESS IS BOOMING

Nearly sixty years ago, Spearman (1928) complained that "crammed as psychological writings are, and must needs be, with allusions to errors in an incidental manner, they hardly ever arrive at considering these profoundly, or even systematically." If he were around today, he would have little cause for complaint. The past decade has seen a rapid increase in what might loosely be called error studies. These have been both applied and theoretical in their orientation.

The recent upsurge in applied research has been fuelled by a growing public concern over the terrible cost of human error in high technology enterprises. Three Mile Island, though a near-disaster for     Pennsylvania, has since provided work for a large number of human factors specialists. Not entirely coincidentally, cognitive psychologists have also shown a renewed interest in human error. Aside from the obvious fact that more effective error prevention must be predicated upon better cognitive models, it has become increasingly apparent that adequate theories of mental control processes must explain not only correct performance but also more predictable varieties of human fallibility. And these have proved to be neither as extensive nor as varied as prior consideration of their enormous potential might suggest.

THE ORIGINS OF SYSTEMATIC ERRORS

Far from being rooted in irrational or maladaptive tendencies, these systematic error forms have their origins in fundamentally useful, even vital, psychological processes. Thus, a large proportion of action slips take the form of strong habit intrusions in which procedural knowledge structures, essential for handling our routine interactions with the world, continue to hold sway even though changes in present plans or circumstances render them inappropriate (Norman, 1981; Reason & Mycielska, 1982; Reason, 1983, 1984, 1985). Likewise, many of our errors of judgement can be traced to the over-utilisation of intuitive rules of thumb, or heuristics, which economize on cognitive effort and give good service most of the time (Tversky & Kahneman 1974; Kahneman et al., 1982; Nisbett & Ross, 1980). The same kinds of underlying error principles can also be detected in memory function (Bartlett, 1932; Neisser, 1982), and in inference, problem-solving and everyday thinking (Bruner et al., 1956; Wason & Johnson-Laird, 1972; Johnson-Laird & Wason, 1977; Evans, 1983).

My concerns here are more those of the cognitive theorist than the human reliability specialist, but they are offered in the belief that all of us can benefit from a closer interchange between the two sides of the error business. I want to examine the case for regarding slips (failures of execution) and mistakes (planning failures) as two distinct classes of human error. Although I, along with Donald Norman (1983), have been active in promoting this categorization over the past few years, I would like to share with you some doubts I have about whether this really is the most basic error dichotomy, or whether it actually obscures a more fundamental distinction: that between errors emerging from resource-limited conscious processes, and those associated with the apparently limitless capacity of the unconscious processors, or knowledge structures.

DEFINITIONS

Leaving aside the intervention of chance events, planned actions may fail to achieve their desired objective for one or both of the following reasons: because the actions did not go as planned (slips and lapses), and/or because the plan itself was inadequate (mistakes). On this basis, it is possible to define two kinds of error:

## 1. Slips and lapses

These are errors which result from some failure in the execution stage of an action sequence, regardless of whether or not the guiding plan was adequate to achieve its purpose. Slips and lapses differ in their manifestations rather than their origins. Whereas slips are potentially observable as externalized actions-not-as-planned (e.g. slips of the tongue, slips of the pen, slips of action, etc.), lapses are more covert error forms, largely involving failures of memory. The criterion for the existence of a slip (or lapse) is to ask whether or not the intended actions deviated from the plan governing the period in question. At a purely behavioural level of classification, these deviations can be readily allocated to one of four categories: omissions, repetitions, intrusions and misorderings (Reason, 1984). Such taxonomies, however, tell us very little about the underlying error mechanisms. Any one of these categories can be produced by a wide variety of error mechanisms. Likewise, the same basic error tendency can reveal itself in various surface forms.

## 2. Mistakes

These are deficiencies or failures in the judgemental and/or inferential processes involved in the selection of an abjective and/or in the specification of the means to achieve it, irrespective of whether or not the actions directed by this decision-scheme actually run according to plan. Clearly, these can take a wide variety of subtle and dangerous forms, and are identified most directly in the comprehensive classificatory schemes of Campbell (1958) and Altman (1967).

## THE CASE FOR DISTINGUISHING SLIPS AND MISTAKES

Norman (1983) spelled out the basic distinction between slips and mistakes very succinctly: "The division occurs at the level of intention: A person establishes an intention to act. If the intention is not appropriate, this is a mistake. If the action is not what was intended this is a slip". This implies a difference not only in the levels of cognitive operation, but also in aetiological complexity. Even when the actions go as intended, a particular plan may fail for many different reasons.

The most obvious psychological difference between slips and the mistakes is the readiness with which these respective errors forms can be detected by their makers. One of the functions of attention is to catch departures of action from intention (Mandler, 1975). This custodial role is most readily seen in the fine-grained analysis of speech errors (Baars, 1980), where editing occurs during the construction of the articulatory program and right up to the moment of output. A similar kind of selective self-awareness can be inferred from the pattern of shoppers' absent-minded slips and lapses (Reason & Lucas, 1984). But there appears to be no comparable regulatory mechanism to act against the occurrence of mistakes. In fact, quite the reverse is true. There are powerful psychological forces at work to preserve completed plans. Francis Bacon expressed this confirmation bias most eloquently in 1620: "The human understanding when it has once adopted an opinion...draws all things else to support and agree with it." (Anderson, 1960, p.50).

A completed plan is not only a set of directions for later action, it is also a theory concerning a future state of the world. It confers order and reduces anxiety. As such, it will be strongly resistant to change, even in the face of fresh evidence clearly indicating that the planned actions are unlikely to achieve their objective, or that the objective itself is realistic. Prisoners' reports revealing that deep chalk bunkers in the German front-line rendered their defenders largely immune to the preliminary bombardment did not halt the disastrous Battle of the Somme in July, 1916; nor did aerial photographs showing the presence of fresh panzer divisions in and around Arnhem stop the catastrophic air-borne assault in September, 1944. The list of such instances is very long.

## DOUBTS ABOUT THE FUNDAMENTAL NATURE OF THIS DISTINCTION

Given the distinctions between slips and mistakes outlined above, it is tempting to argue that these two error classes have their origins at different levels of the control system. Slips and lapses could be said to stem from the unintended activation of largely automatic procedural routines, invariably associated with some degree of attentional 'capture' (Reason, 1979; Norman, 1981). While mistakes could be said to result from failures of the higher-order cognitive processes involved in judgement, thinking and decision-making. But if this

distinction were valid, one would expect slips and
mistakes to take quite different forms.  However this
does not seem to be the case.

Over the past three years, I have been carrying out a
literature search in order to produce a handbook which
catalogues the predictable varieties of human error, both
slips and mistakes.  My aim was to summarise in a
probabilistic fashion the circumstances under which an
error is likely to occur, and, given those conditions,
the form that it is most likely to take.  In doing this,
I have become increasingly aware that while the surface
forms of errors vary considerably according to the task
being executed and the cognitive functions implicated,
they all appear to be governed by a relatively few
underlying principles.  The most obvious of these is
that changes of various kinds produce judgements,
decisions and actions that are more in keeping with past
circumstances than with the prevailing ones.  In short,
a large proportion of human errors involve the
inappropriate application of well tried solutions to
new problems.

THE ATTENTIONAL AND SCHEMATIC MODES OF CONTROL

In this concluding section, I wish to argue that
human errors of all kinds are best understood in relation
to the complex interaction between two distinct modes of
cognitive control:  the attentional and schematic modes.

1.  The attentional control mode

This is closely identified with consciousness (though
not necessarily co-extensive with it -- see Reason, 1983).
It has powerful, analytical, feedback-driven, information
processing capabilities, and is essential for coping
with novel or changed circumstances and for detecting
and recovering slips.  However, it is also severely
resource-limited, slow, laborious, sequential and
difficult to sustain.

2.  The schematic control mode

The schematic mode, in contrast, has no known limits
to its capacity.  It can process familiar information
rapidly, in parallel, and without conscious involvement
or effort.  But it is ineffective in the face of
unforeseen circumstances.  *The schematic knowledge-base*

comprises a vast number of experts, or specialised
theories, each one dealing with a particular aspect of
the world through one or a variety of cognitive domains
(e.g. perception, action, memory retrieval, language,
thought, etc.).  Together, they constitute a richly
interconnected, immensely powerful, labour-saving
apparatus for governing the routine and largely
predictable activities of life.  But they do not generally
come 'hard-wired'.  In their beginnings, they make heavy
demands upon the limited attentional resource, but then
acquire a large measure of autonomy.  Once this autonomy
has been achieved, they may be activated both by current
plans and by a number of non-intentional factors: recency
and frequency of previous employment, the prevailing need
state, the present environment, and features shared with
other schemata (Reason, 1984, 1985).

If individuals are operating within a largely
predictable environment, then their performance will be
predominantly under schematic control, and their errors
will arise from one or more of the following processes:
(a) from fitting the data to the wrong schema;  (b) from
employing the correct schema too enthusiastically so that
gaps in a stimulus configuration are filled with default
assignments (best guesses) rather than the available
data; (c) from relying too heavily upon readily available
or well-used schemata.  As Taylor and Crocker (1981)
pointed out, "...virtually any of the properties of
schematic function that are useful under some
circumstances will be liabilities under others.  Like all
gamblers, cognitive gamblers sometimes lose."

The schematic mode cuts across all the 'levels' of
cognitive control discussed earlier.  Routinization is
not confined to simple motor or perceptual skills;  it
pervades all cognitive operations.  The extent to which
even the most elevated of mental activities are largely
schema-drivenhas been elegantly demonstrated by de Groot
(1965), Chase and Simon (1973), Tversky and Kahneman
(1974), Johnson-Laird and Wason (1977), and many others.

The attentional control mode has two major functions:
to monitor ongoing activity, and to cope with change.
If these latter adaptations are successful, then they
too will eventually become schematized.  Only in
conditions of novelty, is its guidance likely to
predominate.  But even then, the basic urge to avoid
cognitive strain will result in unconscious switches into

the schematic mode, yielding well-used but currently
inappropriate solutions or heuristics.  Other, less
predictable, errors will arise as the result of resource
limitations associated primarily with working memory.

Implicit in these. arguments is the belief that the
differences between the errors of experts and those of
novices are more profound than those between slips and
mistakes.  Differences in task proficiency, perhaps more
than anything else, govern which of the two control
modes will dominate performance.  This, in turn, will
determine the nature of the errors produced.

REFERENCES

Anderson, F. H., 1960, Bacon: The New Organon (Bobs-
    Merrill Educational Publishing).
Altman, J. W., 1967, Classification of human error. In
    Symposium on Reliability of Human Performance, edited
    by W. B. Askren (Report AMRL-TR-67-88) (Aerospace
    Medical Research Laboratories, Wright-Patterson AFB,
    Ohio).
Baars, B. J., 1980, On eliciting predictable speech errors.
    In Errors in Linguistic Performance: Slips of the
    Tongue, Pen and Ear, edited by V. Fromkin (Academic
    Press).
Bartlett, F. C., 1932, Remembering (Cambridge University
    Press).
Bruner, J. S., Goodnow, J. J., & Austin, G.A., 1956,
    A Study of Thinking, (Wiley).
Campbell, D. T., 1958, Information and Control, 1, 334.
Chase, W. G. & Simon, H. A., 1973, The mind's eye in chess.
    In Visual Information Processing, edited by W. G. Chase
    (Academic Press).
De Groot, A. D., 1965, Thought and Choice in Chess
    (Mouton).
Evans, J. St. B. T., 1983, Thinking and Reasoning:
    Psychological Approaches (Routledge & Kegan Paul).
Johnson-Laird, P. N., & Wason, P.C., 1977, Thinking:
    Readings in Cognitive Science (Cambridge University
    Press).
Kahneman, D., Slovic, P., & Tversky, A., 1982, Judgement
    under Uncertainty: Heuristics and Biases (Cambridge
    University Press).
Mandler, G., 1975, Mind and Emotion (Wiley).
Neisser, U., 1982, Memory Observed: Remembering in Natural
    Contexts (W. H. Freeman and Company).

Nisbett, R., & Ross, L., 1980, Human Inference:Strategies and Shortcomings (Prentice-Hall).

Norman, D. A., 1981, Psychological Review, 88, 1.

Norman, D. A., 1983, Position paper on human error (NATO Conference on Human Error, Bellagio).

Reason, J. T., 1979, Actions not as planned: The price of automatization. In Aspects of Consciousness Vol.1, edited by G. Underwood & R. Stevens (Academic Press).

Reason, J. T., 1983, Absent-mindedness and cognitive control. In Everyday Memory, Actions and Absent-mindedness, edited by J. E. Harris & P. E. Morris (Academic Press).

Reason, J. T., 1984, Lapses of attention. In Varieties of Attention, edited by R. Parasuraman & D. R. Davies (Academic Press).

Reason, J. T., 1985, Absent-mindedness. In Psychology Survey No. 5, edited by J. Nicholson & H. Beloff British Psychological Society).

Reason, J. T. & Lucas, D., 1984, British Journal of Clinical Psychology, 23, 121.

Reason, J. T., & Mycielska, K., 1982, Absent-minded? The Psychology of Mental Lapses and Everyday Errors (Prentice-Hall).

Spearman, C., 1928, Journal of General Psychology, 1, 29.

Taylor, S. E., & Crocker, J. C., 1981, Schematic bases of social information processing. In Social Cognition: The Ontario Symposium, Vol. 1, edited by E. T. Higgins, P. Herman, & M. P. Zanna (Erlbaum).

Tversky, A., & Kahneman, D., 1974, Science, 185, 1124.

Wason, P. C., & Johnson-Laird, P. N., 1972, Psychology of Reasoning: Structure and Content (Batsford).

# ERGONOMICS AND POLICE DETECTION METHODS

**Walter R. Harper**        Anacapa Sciences Inc.,
Santa Barbara, California, U.S.A.

ABSTRACT

A classic problem in criminal information analysis
is how to counteract the natural human tendency to jump
to some kind of conclusion based on hunches, guesses,
and whims of the investigative team.  Police personnel
need to improve their analytical effectiveness and apply
more scientific techniques to their problems.

The ergonomic approach used to attack this problem
was to apply principles of cognition and use of inductive logic to organise and clarify collected information
resulting in precise and valid inferences (or theories).
The inferences usually must be made in the face of changing
and incomplete information, which is characteristic
of real-life dynamic data bases.

The practical application of this concept required
use of:  **link analysis/network diagramming** to show the
relationship among individuals, organisations, or other
entities, **event-flow analysis** to show specific events
related in time, **process analysis** to show the path of
a commodity through different stages of a process, **activity analysis** to show a pattern of activities, and **cor-
relational analysis** to show the effect of one variable
on another.

All of these information-processing techniques fitted
into a logical analytical framework that helped develop
an overall theory useful for prediction of a major participant, or reconstruction of a series of events.  A
systematic assessment method used subjective conditional
probability techniques to provide confidence levels
for the developed inferences.
This application of ergonomics research is now used
in major law enforcement agencies in Great Britain,
Australia, Canada, and the United States.

## INTRODUCTION

Results of ergonomics research are often criticised by observers from other disciplines because of a seeming difficulty in transforming results into practical applications that will benefit user populations.  The work described here involved adapting principles and methods from ergonomics research to a unique user population-- specifically, police personnel.  The adaptation, subsequently adapted and used in many operational settings, resulted in a logical model to help clarify and integrate information that could lead to development of hypotheses on criminal activities.

## THE PROBLEM

The problem was to identify what methods/techniques would be most useful to the police population, to adapt them for operational use, and to teach the techniques to police users.

Police require valid and reliable information to help them operate in a difficult, and often hostile, environment.  Unfortunately, the information is not always available.  And, if is is available, it is often invalid or unreliable.  Recent confrontations suggest the external environment has become more hostile.  The internal environment is exemplified by working conditions and training that nurture resistance to change, and opposition to nonclassic training precepts.

Even so, police in the early stages of an investigation must make decisions from an incomplete and changing data base, without a structured model of decision-making. These difficulties are exaggerated by repeated stress on the acquisition and use of wholly **factual** data during training.  This emphasis on facts has a firm and necessary basis, in that successful prosecutions of criminal acts do require factual substantiation, e.g., witnesses, weapons, photographs, copies of relevant documents, and the like.

But much police work goes on long before the case is presented to judge and jury.  In the intelligence phase of police work, information has to be collected, evaluated, clarified, and integrated to help develop projections of future targets, required resources, and hypotheses about nefarious activities.  This pre-prosecution phase is analogous to the strategic long-term collection and analysis of military intelligence mandatory for later tactical assaults.

## THE APPROACH

We define intelligence in the context of police work to mean information to which something has been added. The "something added" results from analyses; i.e., usually an explanation of what the information means. To further define the term, consider that one does **not** collect intelligence--one collects data and subjects them to analysis to assess their true meaning. Analysis is the process, then, of developing meaning from collected information. This analysis is critical to the assessment and investigation of complex criminal activities such as organized crime, narcotics trafficking, white-collar crime and other frauds, because all desired information is not available. The analyst, therefore, must apply appropriate and efficient methods to glean the most meaning from what he has. The objective of analysis is to develop the most precise and valid inferences possible from available information.

The model which guided the development of analytical methods is shown in Figure 1. The components of the model are described briefly in the following paragraphs.

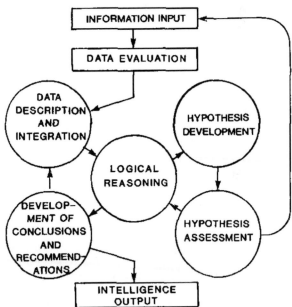

FIGURE 1. Analytical model.

The diagram portrays a classic system model and process, i.e., the components interact, there is linear flow, and provision for feedback is shown. The components of the model are described below.

**Information input** represents "raw" data resulting from initial investigation of an actual, or suspected criminal activity. **Data evaluation** embodies a filtering function whereby decisions are made to accept or reject collected data according to validity and reliability

criteria. **Data description and integration** requires that bits and pieces of information are organized into appropriate formats for clarification via charting and diagramming. Several techniques were developed to fulfill this function.
The use of each depends upon the nature of the data and the objective of the analysis. Figure 2 shows an example of **link charting,** sometimes described as network diagramming, which helps define the relationship among entities such as individuals and organizations. For a more detailed discussion, see Harper & Harris (1975).

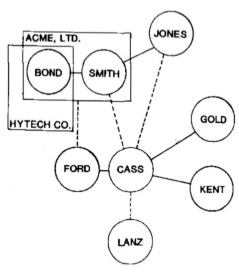

FIGURE 2.    A simple link chart.

**Commodity flow charting** (Figure 3) illustrates the flow of money, narcotics, stolen goods, or other commodities through the elements of a criminal network.

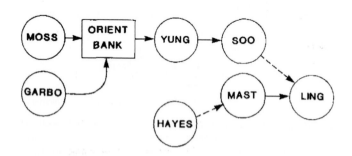

FIGURE 3.    Example of a commodity flow chart.

**Event charting** (Figure 4) shows the relationship among criminal events against a time line.

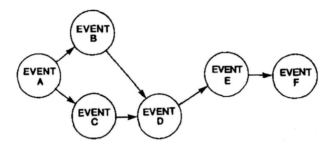

FIGURE 4. Example of an event chart. (Elapsed time is implied.)

**Activity charting** (Figure 5) defines patterns or sequence of criminal operations, including modus operandi.

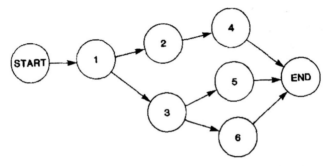

FIGURE 5.   Example of simple activity chart.

Descriptive statistics can be used to advantage to clarify masses of numerical data.  Frequency charting for example,  organizes, summarizes, and helps in the interpretation of quantitative information.  Data correlation, if portrayed by non-inferential statistics in a graphic format, is useful to help illustrate the relationship between two variables.  The results of financial analyses can also be presented in tabular form to highlight changes in net worth or suggest hidden sources of income.

**Logical reasoning,** at the heart of the analysis process, involves the application of inductive logic to develop inferences about criminal operations, key individuals, methods of operation and the extent of criminal scope or influence.  Inductive logic is the process of inferring broader meaning from specifics and details.  This implies that the analyst goes beyond the facts at hand and follows a path of item (data) selection, leading to premise/proposition development.  The premises, in turn, lead to an overall inference.

Police analysts have the unenviable task of assessing the truth through interpretation of incomplete and changing information. Inferences are made and hypotheses developed under a mantle of uncertainty. If the inference is valid, one could immediately go to a conclusion or recommended action. But more usually, in the scientific method analogy, one would develop a **hypothesis,** or theory, about the integrated information. The hypothesis is a structured form of the essential meaning derived from the analysis and only serves as a vehicle to permit confirmation or denial by additional assessment.

**Hypothesis assessment** or testing usually involves collection of additional data to fill information gaps, to supply confirmation of known data, and to follow promising leads.

## APPLICATIONS AND RESULTS

The users of these techniques were police personnel. Police personnel required the methods to be presented in simple, clear terms, since that audience had a high degree of expertise in the subject matter but low tolerance for classic didactic learning.

A paradigm of instruction was developed (Harper & Harris, 1971) that proved effective. The method included:

- Provision of a short background (of the item)
- Documentation and illustration
- Joint problem-solving under supervision and with assistance
- Individual unassisted practice
- Course materials and references described in police-oriented terminology

Several thousand analysts have been trained in these techniques, from agencies including Scotland Yard, Royal Canadian Mounted Police, Australian Bureau of Narcotics, the FBI (USA), and hundreds of local law-enforcement agencies from many countries. This new dimension in police work has contributed to understanding and solution of complex cases.

## REFERENCES

Harper, W.R., & Harris, D.H., 1975, The application of link analysis to police intelligence, **Human Factors**, 17, 157-164

Harper, W.R., & Harris, D.H., 1971, **Development of a training program for intelligence analysts** (Technical Report 140-1). Santa Barbara, California: Anacapa Sciences, Inc.

# 1986

| Durham University | 8th–11th April |
|---|---|
| **Conference Secretary** | Rachel Birnbaum |
| **Exhibitions Officer** | |
| **Honorary Meetings Secretary** | Heather Ward |
| **Chair of Meetings** | Ted Megaw |
| **Programme Secretary** | D.J. Oborne |

## Social Entertainment

The social activities included a trip to the Beamish Museum

| Chapter | Title | Author |
|---|---|---|
| **Keynote Addresses** | Phenomena, function and design: Does information make a difference? | P. Wright |
| **Keynote Addresses** | A model of mental health and work attitudes | P. Warr |
| **Organisational Ergonomics** | Attitude surveys: An authoritarian technique masquerading as a participative process | R. Sell |
| **Visual Processes** | The role of colour in object recognition, categorization and naming | J.B. Davidoff |

# PHENOMENA, FUNCTION AND DESIGN:
## DOES INFORMATION MAKE A DIFFERENCE?

Patricia Wright

MRC Applied Psychology Unit,
15 Chaucer Road,
CAMBRIDGE.

## ABSTRACT

Three phenomena relating to written information are:
(a) the pervasiveness and importance of written technical
information in working life, (b) the frequently poor design
of written materials, (c) the failure of tasks involving
written information to engage the research attention of
many ergonomists outside the subdomain of HCI. Among the
reasons for (c) may be the lack of an adequate research
methodology. The traditional ergonomic approach,
attempting functional mappings between information design
and performance, encounters difficulties because of the
context sensitivity of design solutions (i.e. "it all
depends"). The alternative psychological approach,
focussing on underlying cognitive functions, is also
limited in its ability to address the breadth of issues
relating to information design. Nevertheless a cognitive
skills analysis may function both to motivate research, and
also to map between research and application. Such an
approach has implications for the way that the knowledge
gained from research is most usefully made available to
information designers. In short, the answer suggested to
the title question is, "Yes".

SOME PHENOMENA

The focus of this talk will be tasks involving written information.  Twenty years ago Chapanis (1965) very eloquently complained about the amount of poor written information that we come across in daily life.  In spite of this, the Plain English Campaigns, both in the UK and in the USA, bear witness that the incidence of poorly designed information seems to continue unabated.  Among the common culprits are the manuals for computers and their peripherals.  But many other products enter the workplace similarly ill-equipped with respect to their accompanying documentation.  Nor is this phenomenon exclusively product related.  A survey carried out in 1977 of 48 people who were either councillors or council officers in 11 Scottish local authorities found that everyone working for the larger authorities thought that they had too much to read for committee meetings, and over half the interviewees reported finding it difficult to identify the main issues in the reports they had to read.  One of the main causes of difficulty was said to be the poor layout and organisation of the material.  Have things changed since 1977? Certainly the amount of reading necessary in some kinds of work seems to have increased rather than decreased. Estimates suggest that in the 1970s a professional had to read 30 documents a day to keep up with the field, but by the mid-80s this figure had doubled (Frase, Macdonald & Keenan, 1985).

If you ask, "What are ergonomists doing about this?" you will encounter another phenomenon.  My attention was drawn to it by a Civil Servant who commented that he used to find occasional articles of interest in journals such as Applied Ergonomics and Ergonomics, but not any longer. While the interest in information design shown by Ergonomists could never have been called excessive, it appears to be on the decline.  As evidence for this assertion consider the articles published in the Society's journal Ergonomics during 1984.  There were 105 papers, but not one was concerned with the design of non-computerised information.  A contributory factor has been mentioned. People interested in information design seem more likely to devote their attention to problems of computer-based information displays than to be concerned with technical materials more generally.  But many other kinds of written information will be essential parts of working life for many years to come (Glushko & Mashey, 1982).

As an indication that this overlooking of non-computerised information is not simply a national bias, a

more global perspective can be achieved by considering the
papers given at the International Congress held in
Bournemouth in September 1985.  261 titles of presented
papers and seminars were published in Ergonomics (1985).
Of these, only 2 were concerned with the design of non-
computerised written information:  one examined signposting
in hospitals, the other assessed the contribution made by
the journal Ergonomics to the development of the
profession.  If you check the review of Engineering
Psychology published last year you will find a similar
picture (Wickens & Kramer, 1985).  The work on presenting
information about bus routes is mentioned (Bartram, 1980)
as are four studies of diagrammatic adjuncts or
alternatives to prose.  This adds up to less than half a
dozen studies mentioned in a review of 399 publications.

   It should not be thought that psychologists are
showing any more interest than the ergonomic community.
Nickerson (1984) wrote, "Surprisingly little is known about
how to determine whether a particular representation of a
concept, principle, relationship or body of knowledge is
better or worse than another for purposes of communication
or exposition.  We lack an appropriate conceptual framework
for describing information apart from its representation."
(Nickerson, 1984, p.14).  But the reasons for this research
gap are probably quite different for psychology and
ergonomics.

   The limitations of the 24 hour day are very real.
Ergonomists must obviously concentrate on those issues
which seem to them most important.  So perhaps information
design just does not merit attention alongside other kinds
of workplace problems.  People may get lost because of
inadequate sign-posting, but these wasted man-hours may not
be serious.  The newcomer can always ask, and everyone else
eventually gets to know where things are.  But the economic
consequences of the poor design of some categories of
written information can be considerable.  For example,
Rayner (1982) suggested that one government department
could save over £2 million a year by a one per cent
improvement in the effectiveness of its forms.  Although
Rayner's costing were done for government departments, can
anyone doubt that similar savings could be made in
industry.  No business can run without administrative
procedures involving record keeping, filing and form-
filling.  Even more seriously, patients die because nurses
misread the information on drug labels.  The problems
extend outside the workplace but nevertheless cost industry
through the loss of work they cause.  Consumers become ill
because the warning information on the bottle is on the

back of a label that has to be read through the liquid in the bottle. There are reports of babies dying because mothers misunderstand the instructions for diluting baby foods. So how is it that the problems of information design seem to be so undeserving of the attention of ergonomics?

COMMON SENSE or RESEARCH?   You can probably recite many reasons why information design rates virtually zero coverage alongside workplace problems such as noise, temperature and vibration. Undoubtedly research occurs where the funds are, but let us consider just two other contributory factors:  the need for research and the kind of research needed. On the one hand the problem of poor information design seems to some people to be amenable to "common sense" solutions. Certainly this view can be found in professions such as engineering and law, where the intellectual challenge exists only in relation to substantive content, be it fluid mechanics or torts (Rutter, 1985).

The assumption is made that writers could communicate more successfully if only they tried. From this perspective there is no call for ergonomics expertise. No research is needed on information, in the way that studies are needed to discover what height a workbench should be. You couldn't just guess the height of a workbench. You would need data if you were going to get the design right. In contrast, anyone could express themselves clearly if they wanted to. There is no reason for research. Well perhaps a little bit, but only relating to sensory constraints which might influence legibility - factors like size and colour, that sort of thing. Nothing needs to be done about the form or content of the message. After all, getting that right is part of the competence of any native speaker of the language. Isn't it?

This assumption can be challenged from several standpoints. No organisations deliberately set out to make their written materials complicated, to confuse their employees or their customers. While it is true that in some instances an organisation's legal department may reduce the communicative efficiency of a document, there are numerous instances where something less than perfect was obtained without specialist legal help. So the ability to present written information clearly does not seem to follow automatically from being a native speaker of English. A little reflection would assure you that this is far from surprising. When did you last speak something that sounded like a railway timetable? Receiving

instructions by telephone is something that we soon
discover people are not as good at as we might have hoped
when we asked for the information. Even the literal
transcription of a committee meeting or conference
discussion suggests that what may have been readily
understood at the time does not preserve this
comprehensibility when written down. So I am suggesting
that, in spite of the common sense view of what a native
speaker should be able to do, our oral language skills are
not well-honed for the precision needed when communicating
technical information in writing.

SCHOOL SKILLS:   If written communication is a special
skill, does it fall outside the domain of ergonomics?
Perhaps it is an educational problem.  Certainly Sticht
(1985) has pointed out that school experience of reading
and writing leaves people very ill-equipped for generating
materials designed for "reading to do" rather than "reading
to learn". Maybe schools should be teaching children how
to write technical materials. But by the same argument
shouldn't you learn how to make satisfactory workbenches in
woodwork or metalwork classes at school? Ergonomists
consider themselves different from the teachers of basic
skills because of their emphasis on getting the designed
product tailored to the user's requirements. Because of
this, the workplace designer needs to know more about users
than the school can teach. On the face of it this would
seem to be as true for information design as for other
elements in the workplace.

RESEARCH METHODS:   A second factor which may have
contributed to Ergonomics focussing elsewhere than on
information design concerns the availability of a research
methodology for exploring the domain of written
information. Ergonomics general approach has been to start
with a "phenomenon" (e.g. a workplace problem such as back
pain) and relate this to the causes of the problem (e.g.
seating or lifting); then alternative workplace designs
which might circumvent the problem are examined. In doing
this examination, careful attention will be given to
physical and social elements of the phenomenon. But, at
least before computers were so invasive, relatively little
attention has been given to cognitive issues, and the
phenomena of interest to ergonomists have tended to have
relatively small cognitive involvement, being blue collar
rather than white collar jobs.
     Assuming a shift in interest, it is worth asking
whether cognitive ergonomists will measure cognitive

performance in much the same way as traditional ergonomists measured overt behaviour? There are reasons for thinking that this will not be the case. Traditional ergonomics has been relatively successful in explicating functional mappings between environmental manipulation and performance changes without there being any need for multiple layers of theory. A range of anthropometric indices can be related directly to design requirements in the workplace. This approach will not suffice for information design. The interpretation of written information is knowledge dependent, and people's knowledge is not a stable entity like arm-span. Moreover the processes of interpreting written materials are very context sensitive. As Nickerson (1981) pointed out, we know what is intended by the message "STOP CHILDREN CROSSING" when we see it on the signs held by school crossing patrols. The literal meanings of the words are only a part of the story.

This context sensitivity was again in evidence in a series of recent studies which Ann Lickorish and I carried out on the sequencing of information in multi-stage instructions. Common sense seemed to dictate that these instructions should be written in the same temporal order as that in which they would be carried out. Our first study confirmed this prediction. Our second showed that performance could sometimes be better if the instructions were given "back to front". The critical factor seems to be whether people have to formulate mental "action plans" in order to carry out the instructions. If they do, then the instructions must be sequenced in the correct temporal order. However when the environment provides cues to the order in which the steps must be undertaken, then linguistic factors may make other sequences of information easier to remember. There is nothing difficult about the instruction "Insert a coffee break after the second speaker on Tuesday" even though this information is "backwards" from the standpoint of the person making the alteration to the programme.


FUNCTION AND FUNCTIONAL LIMITATIONS:

Let us examine in more detail why research relating to information design is not amenable to a solely functional approach. Then we can explore some alternative approaches and perhaps see why several qualitatively different kinds of theory might be necessary for solving information design problems.

Consider the design of any work-related information. It scarcely matters what example we pick, so let us deliberately select one that is very simple. Consider the way aircraft departure information is displayed at airports. Typically this is a four column list which gives the carrier (e.g. British Airways), the flight number, the departure time, and instructions to passengers about embarcation. Although airports vary the way these elements are grouped, it is common in the U.K. for the columns to be arranged left to right in the order just described, with the list itself organised on the basis of the information in the third column, namely the departure time. You need to discover the rationale for the table's organisation because the physical location of the information will not be constant within the display as the time frame moves and you may need to search through the tables several times while awaiting to board your flight.

Seasoned travellers may have forgotten how bewildering this array of information can be for the newcomer. What could ergonomists do to improve matters? Working at a totally non-theoretical level and looking for changes to the display which reduce bewilderment, might produce a research program which starts with 96 display variants (the 24 possible column orderings times the 4 possible row orderings for the table as a whole). Of course the presentation options are much richer than these mere 96. All sorts of other cues (colour, flashing lights, etc.) can be introduced in order to make salient particular classes of information (e.g. "boarding now" vs "wait but on schedule" vs "delayed"). So a five year research program to sort all this out would have more than enough on its hands to keep a small team busy.

Even when the data were in it would not necessarily be clear where else in the universe of tabulated information these findings might have relevance. Would bus time-tables profit from the findings of such a project? Or company accounts? Last year APU were asked if the information presented to jurors in fraud trials could be improved. Numerous tables there, but could any of them be improved as a result of a study of airport departure information?

Lest it be thought that the parametric approach just outlined is very much a straw man, let me remind you that Hartley & Trueman (1985) were advocating exactly this approach to information design, and put their research money where their philosophy was by reporting 17 studies examining whether headings in 4-page texts should occur in dedicated margins or within the text itself, and whether they should be worded as questions or as statements. (For

those of you who think that information design is really
all common sense I won't need to summarise Hartley's
results.)   The motivation for Hartley's 17 studies was a
disquiet, which I fully share, about one-off studies which
explicitly ignore the context-sensitivity of design
factors.  Hartley is not alone.  Others have argued that
the only way of handling context-sensitivity is by means of
large multivariate designs (e.g. Willeges, 1985).  But are
there no alternatives?  Alternatives are urgently needed
because, no matter how many parameters are involved in a
series of experiments, the risk is that the project may not
have captured the critical factors determining cognitive
performance, at a level of abstraction where these factors
can be mapped across different categories of information.
Knowing what sorts of abstractions to work with requires
some underlying model that can drive the research process.

MODELS:  What is it that we need models of?  It would
clearly be helpful to have a model of how people interact
with information systems.  Such a model would motivate
enquiry not into whether the first column shown in a table
should be yellow or blue, but into what questions people
were trying to answer when scanning departure boards at
airports, and into where such people anticipated their
answers would be on the board.  (The first of these
questions may seem like a "cognitive task analysis").  Both
these issues, i.e. people's formulation of questions and
their anticipation of information structures, are heavily
dependent upon prior knowledge.  Who is supposed to know
about what sort of knowledge structures people operate
with?  Cognitive psychologists.  So why couldn't cognitive
ergonomics just crib some theoretical infrastructure from
cognitive psychology?

LIMITATIONS OF COGNITIVE PSYCHOLOGY:  To some extent the
ergonomics of information design can and must see itself in
relation to cognitive psychology.  But as the song says,
"It's not where you start but where you finish".  Starting
from cognitive psychology may be fine, but I want to
suggest that in principle psychological models may not
always be capable of getting you to the design point where
as an ergonomist you would want to be.  Cognitive
psychologists have developed a very well-defined mapping
between phenomena and function (Baddeley, 1977).  They see
it as their job to explain why particular phenomena occur
(e.g. why certain information is difficult to understand).
Unfortunately the usefulness of cognitive psychology is
limited in several ways.  All sorts of phenomena which lie

outside the scope of current theories are not necessarily
viewed as exciting challenges but as irrelevant to the core
development of the science.

One example of an overlooked phenomenon might be the
tendency of readers to stop reading. Whether it is a
newspaper, a journal article or a VDU "help" screen,
readers often take decisions to read no further. Cognitive
psychology has a large research literature on reading.
Numerous theories exist. None of them worries about this
common-place phenomenon. That it is an important
phenomenon was evident during the British Library project
on "electronic journals" (Shackel, 1982). Readers
selecting a journal were presented with a contents list of
available papers. This list functioned as a menu from
which readers could jump directly to the article of
interest. Page numbers therefore seemed irrelevant. But
page numbers are an important cue to the length of an
article, and length is part of the stopping rule that
readers apply (i.e., "That article is too long to read now;
I'll come back later").

Of course it is inevitable that in its relatively
short life there will be phenomena that cognitive
psychologists have not yet got round to. Apart from this,
need cognitive ergonomists do much more than apply the
available cognitive psychology in their own investigations?
One major problem is the breadth of the domain which
psychological theories cover. For example, many
researchers have been concerned with text and sentences.
There are models of cognitive processes on the basis of
which advice can be given about the sentence structures to
avoid and the kind of paragraph structures to employ. The
missing element that an ergonomist needs is consideration
of other forms of communication. Consequently advice about
replacing the text with an illustration, or a video, lies
outside the scope of the psychological theoretical
framework.

COGNITIVE SKILLS ANALYSIS: Although for the purposes of
cognitive ergonomics the theories being developed within
cognitive psychology may be limited, this is not all that
cognitive psychology has to offer. It also has an implicit
methodology, a way of approaching problems. Let us go back
to that large design space in which we were discussing
aircraft departure information. The cognitive psychologist
would seek to narrow the questions worth asking by trying
to analyse in cognitive terms what it is that people are
doing (or trying to do) when interacting with the
information, specifically what previous knowledge they are

drawing upon and what further information they are seeking.
The alternative design options are then mapped onto these
cognitive functions.  The great advantage of this approach
is that it enables generalizations to be made across
content domains.  Generalizability is gained at the level
of cognitive functioning, i.e. where similar cognitive
processes are recruited then similar design solutions will
be effective.

The potential advantages of this method, compared with
a more traditional task analysis, are several.  A task
analysis might define the task as being one of reading the
display, so leading the researchers to worry about whether
the display elements were legible or appropriately grouped.
No doubt you are all familiar with Stewart's (1976) summary
of this approach.  I am not contesting the usefulness of
that summary, but how can we use such a summary in the
design of our airport departure information?

There have been several recent illustrations of the
way a cognitive skills analysis can yield insights which
are very different from those which focus on performance.
Studies of the interpretation of x-rays have shown that the
experts' perceptual skills are a product of a detailed
mental model of human anatomy (Lesgold, 1984; Lesgold et
al. 1981).  Similarly, studies of how aircraft mechanics
diagnose what is wrong with an aircraft engine have shown
that experts conceptualise the task as inter-related sub-
units, whereas for novices these units are often considered
as discrete elements (Gitomer, 1984).

The critical cognitive functions were made visible to
the researchers by adopting a methodology which contrasted
the ways in which novices and experts did the task.
Whereas a performance-oriented approach would note which
task elements the novices failed on (and perhaps recommend
that more practice was needed on these elements), the
cognitive approach tried to specify what information was
being used by the expert, whether the novice had this
information available, what were the difficulties that
novices faced in trying to use the knowledge they had.
This approach led the researchers to propose new training
schemes to help novices become experts (e.g. Gott et al.
1985/6).  The approach was novel because it focussed on
creating the underlying representations of which the
performance will subsequently be a byproduct, rather than
trying to generate the performance more directly.  From
this perspective it becomes easier to understand how
experts might be worse at some tasks than novices (Adelson,
1984).

So what prior knowledge and underlying representations influence air travellers when consulting departure information?  I know of no research on the subject but it seems plausible that travellers are trying to answer a question such as "What is happening to my flight?" or "What gate does my flight leave from?"  The question is likely to be formulated with reference to time and destination, because this is an application of the traveller's prior knowledge of schedules for other transport systems such as trains and buses.  In the absence of research we do not know whether travellers characterise the concept "my flight" as "the 11.30 to New York" or as the system label "BA123".  Knowing about travellers' questions would have two direct implications for the design of the display.  One implication is that the information should be organised systematically in relation to this question.  If people are searching for flights to New York then an alphabetic listing of destinations is desirable.  If they are looking for BA123, then an alphabetic listing of carriers with flight numbers nested numerically within this might be one solution.  The second design implication is that this information, which forms the basis for the organisation of the table, needs to go in the leftmost column of the table. How do we know that?  Several studies at APU have looked at how people consult tables and what it is that makes tables easy or difficult to use.  These studies have shown that for a variety of tabular structures people in the U.K. expect the organising principle to be on the left of the table, at the beginning of the row, at the front of an alphanumeric string (Wright et al. 1984).

So from this standpoint of examining the knowledge used and the knowledge required there is no need for a large-scale parametric exploration.  Issues, both of content and presentation, can now be satisfactorily handled by adopting a different approach.  Moreover this alternative approach can provide an input to the design process long before the stages of prototyping are reached.

RESEARCH CHARACTERISTICS:  Research examining the relation beween cognitive processes and information design has some curious characteristics.  It does not lead to the development of the kinds of models which cognitive psychologists normally create.  If anything it tends to be parasitic upon those very models, taking concepts such as presuppositions and schemas into new territories.  It avoids being a broad-band, mutli-parametric approach.  But the necessary predictability and generality may have been achieved by the very fact of taking cognitive processes into account.  Research which enhances our knowledge of those cognitive processes is a different enterprise.  Such research is certainly no less valuable in its own right, but does not itself adequately address the practical problem of narrowing the design space to a viable subset of options for the designer.

What I am suggesting is that a variety of research enterprises may need to combine in order to fully address the issues of information design.  If this is true then it may help to explain why the problem of poor information has been with us for so long, and indeed why that state of affairs may continue.  It becomes easier to appreciate that the tools, both conceptual and procedural, for any ergonomist interested in information design will be different from those used by the cognitive psychologist and the traditional human factors practitioner.

FUNCTIONALITY:  You may have noticed that there has been a shift in the use of the word "function".  We started by examining functionalist research, then we considered cognitive functions, now we are thinking of the functions of different kinds of research.  There remains yet another facet of functionality that is of great importance for information design.  Whereas both the cognitive psychologist and the cognitive ergonomist ask what are the cognitive functions involved in processing specific information, we could also ask about the goal/task functions which that information is subserving.  The advantage of this shift in focus is that we would then be prompted to ask whether these goals/tasks could be achieved by other means.  This is one remaining thorny issue.  How do the alternative designs get generated?

This aspect of functionality becomes particularly important in the area of computer-based information displays.  The usual ergonomic approach to design has been to tailor the task to fit the person.  Within HCI this has sometimes been extended to mean tailoring the task to map onto the way people think about it.  There are two reasons why this approach may not always be appropriate.  One is that most systems, whether computerised or not, have multiple constraints.  It may be better to take these system characteristics into account by modifying the task structure rather than insist on continuing to do the task exactly as before (Wright & Bason, 1982).

The second reason is certainly not temporary. Computer-assisted information systems offer new forms of communication.  Exploiting this potential may result in considerable changes to the way a task is done.  This was

another of the findings which emerged very strongly from
the British Library's project on electronic journals
(Pullinger, 1984). Journal articles were no longer
constrained to being linear structures. The nature of the
interaction among the readership could become very
different from the current pattern with traditional
journals. It follows from a desire to fully exploit any
particular medium that when moving information across
media, e.g. from print to screen, it may not be desirable
to retain too closely the structure of the source materials
(e.g. the familiar characteristics of the printed page).
As one illustration of this point, many library catalogues
are gradually becoming available on-line. The display
characteristics which readers find easy to work with are
looking very different from traditional card catalogues
(Bryant, 1984).


DESIGN OBJECTIVES and PROCEDURES

Could we avoid the numerous information catastrophies
that we encounter daily? Here again there is a contrast
between the goals of those concerned with information
design and the goals of other ergonomists. Neither
optimisation nor perfection need be sought in information
design. Achieving the avoidance of disasters would
represent substantial progress. The saving of small
amounts of time studying instructions or completing a form
is overshadowed by the need to ensure that at least the
communication is essentially successful. So often this
seems not to be the case. The twin objectives are to have
the appropriate content (have you noticed in how many
instruction booklets the information is either missing or
wrong?) presented in an adequate way. In terms that would
be more familiar to those concerned with human-computer
interaction this corresponds to determining the
functionality required and deciding how to instantiate that
functionality. The issues of content and presentation are
often interrelated, as in the application form which wants
to know the age of the form-filler but asks instead for
date of birth. Nevertheless, in so far as they are
separable, decisions about what to present and how to
present it will often be best informed by different
research procedures.

What kinds of information might contribute to these
objectives? From the cognitive skills analysis outlined
previously it might be argued that designers would be
assisted by broad-brush sketches of the kinds of cognitive

functions that need to be supported by design decisions.
It is not being suggested that we yet have evidence of how
to deliver such sketches to designers in a form which they
can easily use.   Hammond et al. (1983) in their examination
of how decisions are reached in designing computer
interfaces, have shown that a given notion of users'
requirements can lead to wide variation in subsequent
design decision.   But it is being suggested that detailed
psychological theory is probably not necessary.   The two
critical factors are firstly the ability to conceive
alternative designs, and then secondly the ability to have
some means of selecting among these alternatives.   This can
be illustrated by considering alternative ways of asking
the same multiple-choice question.   Supposing that you
wanted to know whether people attending this year's
conference had also come during the previous two years.
One design solution would be to ask a "jump" question such
as:

"Apart from this conference, have you attended any other
Ergonomics conferences since 1983?   If Yes, please tick
whichever of the following applies: (a)  only 1983,
                                     (b)  only 1984,
                                     (c)  both 1983 and 1984."
A different solution, and one which avoids the jump, is to
rephrase the question as, "Which years have you attended
the Ergonomics conference?" and then list all four
combinations of years including "neither 1983 nor 1984" as
alternative (d).

A third design solution is to reduce the questioning
to just two items, each being a single Yes/No question
about 1983 and 1984, and allow the analysis of the answers
to generate the conjunctions between the years.   Would it
really require a research study to know which of these
three design solutions makes the fewest cognitive demands
on the form-filler?

Note how this approach contrasts with suggestions
elsewhere that human factors will only succeed if it
becomes a "hard science" being able to yield quantitative
estimates for designers to incorporate into their decision
making (Newell & Card, 1985).

In the long term, one might hope that any technical
writers who understand that their readers will be searching
for specific information in technical publications will
find that this influences their decisions about what
information to put in running heads, how to organise their
index(es), where to put the contents pages, etc.   But
although I have argued that such information is more useful

than traditional "guidelines", I am not convinced that too much optimism is justified. Experience with APU colleagues designing their own written information suggests that even a sophisticated knowledge of cognition may not be sufficient to ensure good design. The vision of alternatives is needed, as well as the means of deciding among them.

Another approach is to advocate design procedures which serve as "catastrophe detectors". The need for empirical procedures as part of design has been stressed by the director of the Document Design Centre in Washington (Redish et al. 1981) and the director of the Communications Design Centre at Carnegie-Mellon (Duffy, 1981). The need to evaluate draft documents is an inevitable consequence of the context sensitivity of the interpretation of written materials. When specifying the exclusion clauses in an insurance policy you may have grounds for thinking that most of the target audience will understand a short, familiar word like "riot"; but only when you have moved into the context of someone trying to make an insurance claim will it become apparent that the concept has fuzzy borders - how many people, doing what for how long?

There are undoubtedly instances where empirical, trouble-shooting procedures can produce cost-effective benefits. In one large forms processing organisation, Phil Barnard and I tried persuading those who were sending wrongly completed forms back to members of the public, to share their experience about the trouble spots on the forms with colleagues down the corridor who were responsible for designing the forms. This highly pertinent information was bound to be valuable, perhaps more so than any other kind of information that could have been given. Here we get back to the issue of common sense, and to that curious phenomenon - its uncommonness! Waller (1982) commented that procedures of the kind just suggested, involving monitoring the errors on forms, were very uncommon in most Government departments. It has to be recognised that evaluation techniques are not self-evident, nor do the findings of different procedures necessarily concur (Wright, 1985). So here is an opportunity for a different kind of input from ergonomists, a contribution which complements rather than surplants the kind of analysis which offers assistance earlier in the design process.

It seems likely that information on paper will continue to be used alongside information on VDU's for many years to come (Glushko & Mashey, 1982). In spite of the advent of the electronic office, or perhaps because of it, the amount of paper people must handle as part of their

working life continues to increase. If the only people who
worry about information design are those concerned with
computerised displays, then we may be surrounded by as many
information catastrophies in 20 years time as we are now.
Surely that's not necessary?

## SUMMARY

Let us finish with a speedy retracing of the ways we
have circled around and through some of the issues relating
to phenomena, function and design as they concern written
technical materials.

1.    The phenomena we have focussed on relate to the
prevalence of badly designed written information.  Why is
it there?  What could ergonomics do about it?

2.    The function of a cognitive skills analysis in
enabling the design space to be reduced was outlined and
discussed both in its relation to motivating research and
to motivating design decisions.

3.    The information from research, it was suggested,
needed to be made available to designers in a way that
differed from traditional ergonomics handbooks.  Instead it
would seek to give the designer the ability to analyse the
cognitive requirements of the task.  By this means the
problems of context sensitive design recommendations could
at least partially be met, and be met at early stages of
the design process.  This would not remove the need for
other kinds of evaluation, but it would allow pre-
prototyping decisions to appropriately restrict the design
space.

## REFERENCES

Adelson, B., 1984, When novices surpass experts:  the
    difficulty of a task may increase with expertise.
    Journal of Experimental Psychology:  Learning, Memory
    and Cognition, 10, 483-495.
Baddeley, A.D., 1977, Applied cognitive and cognitive
    applied psychology:  the case of face recognition.
    Paper presented at the Uppsala conference on Memory.
Bartram, J.D., 1980, Comprehending spatial information:
    The relative efficiency of different methods of
    presenting information about bus routes.  Journal of
    Applied Psychology, 65, 103-110.

Bryant, P., 1984, Reading library catalogues and indexes. Visible Language, 18, 142-153.

Chapanis, A., 1965, Words, Words, Words. Human Factors, 7, 1-17.

Duffy, T.M., 1981, Organising and utilising document design options. Information Design Journal, 2, 256-266.

Frase, L.T., Macdonald, N.H. & Keenan, S.A., 1985, Intuitions, algorithms and a science of text design. In T.M. Duffy & R. Waller (eds) Designing Usable Texts. Orlando, Fla:Academic Press.

Gitomer, D.H., 1984, Cognitive analysis of a trouble shooting task. Doctoral thesis, the University of Pittsburgh.

Glushko, R.J. & Mashey, J.R., 1982, Modelling document processing. Paper given at the Bell Labs Office Automation Conference, October.

Gott, S.P., Bennett, W. & Gillett, A., 1985/6, Deriving ideal student models to propel intelligent tutoring systems. Journal of Computer Based Instruction, in press.

Hammond, N., Jørgensen, A., MacLean, A., Barnard, P. and Long, J. 1983, Design practice and interface usability: evidence from interviews with designers. In A. Janda (ed) Human Factors in Computing Systems. Special issue of SIGCHI Bulletin, p.40-44.

Hartley, J. & Trueman, M., 1985, A research strategy for text designers: the role of headings. Instructional Science, 14, 99-155.

Lesgold, A.M., 1984, Human skills in computerized society: complex skills and their acquisition. Behavior Research Methods, Instruments and Computers, 16, 79-87.

Lesgold, A.M., Feltovitch, P.J., Glaser, R. & Wang, 1981, The acquisition of perceptual diagnostic skill in radiology. Technical Report No. PDS-1, Learning Research and Development Center, Pittsburgh, Pa.

Newell, A. & Card, S.K., 1985, The prospects for psychological science in human-computer interaction. Proceedings of CHI85, San Francisco.

Nickerson, R.S., 1981, Understanding signs: some examples of knowledge-dependent language processing. Information Design Journal, 2, 2-16.

Nickerson, R.S., 1984, Research needs on the interaction between information systems and their users: report of a workshop. Washington, D.C.: National Academy Press.

Pullinger, D., 1984, The design and presentation of the Computer Human Factors Journal on the BLEND system. Visible Language, 18, 171-185.

Rayner, D., 1982, Review of administrative forms:   report
    to the Prime Minister.  Obtainable from the Central
    Management Library, Management and Personnel Office, Old
    Admiralty Building, Whitehall, London, SW1A 2AZ.

Redish, J.C., Felker, D.B. & Rose, A.M., 1981, Evaluating
    the effects of document design principles.  Information
    Design Journal, 2, 236-243.

Rutter, R., 1985, Resources for teaching legal writing.  In
    M.G. Moran & D. Journet (eds) Research in technical
    communication:  a bibliographic source book.  Westport,
    Con:Greenwood Press.

Shackel, B., 1982, The BLEND system:  a program for the
    study of some "electronic journals".  Ergonomics, 25,
    269-284.

Stewart, T.F.M., 1976, Displays and the software interface.
    Applied Ergonomics, 7, 137-146.

Sticht, T.G., 1985, Understanding readers and their uses of
    texts.  In T.M. Duffy & R. Waller (eds) Designing Usable
    Texts, 315-340.

Waller, R., 1982, Review of Administrative Forms in
    Government, Forms Under Control and Review of
    Administrative Forms, Information Design Journal, 3,
    142-148.

Wickens, C.D. & Kramer, A., 1985, Engineering Psychology.
    Annual Review of Psychology, 36, 307-348.

Wright, P., 1985, Is evaluation a myth?  Assessing text
    assessment procedures.  In D.H. Jonassen (ed) The
    Technology of Text, vol. 2, Englewood Cliffs, N.J.:
    Educational Technology Publications.

Wright, P. & Bason, G., 1982, Detour routes to usability:
    a comparison of alternative approaches to multipurpose
    software design.  International Journal of Man-Machine
    Studies, 18, 391-400.

Wright, P., Hull, A.J. & Lickorish, A., 1984, Psychological
    factors in reading tables.  Proceedings of 23rd
    International Congress of Psychology, Mexico, p.194.

# A MODEL OF MENTAL HEALTH AND WORK ATTITUDES

Peter Warr

MRC/ERSC Social and Applied Psychology Unit,
University of Sheffield.

## ABSTRACT

A five-component model of mental health is outlined,
in which aspects of affective well-being, subjective
competence, subjective aspiration and subjective autonomy
are treated as overlapping with attitudes.  Attitudes and
mental health which are specifically job-related are then
considered, and some suggestions for measurement are made.

## INTRODUCTION

Occupational and social psychologists have been
studying work attitudes for more than half a century.
There is no doubt that reliable assessments can be made,
and that measures are often valid predictors of other
aspects of experience and behaviour (e.g. Warr, 1978;   Cook
et al., 1981).  However, what is lacking is an overall
perspective which sets particular job attitudes within a
framework of other job reactions and which places attitudes
themselves within a broader psychological context.

It is the aim of this paper to describe one such
general framework.  The approach will start from an
interest in "mental health", but other starting-points are
of course possible for different purposes.

Attitudes fall within that family of processes which
includes motives, desires, wants, values, satisfactions,
pleasures, interests, hopes, and so on.  These are held
together by evaluative bonds, with positive or negative
feelings at the heart of each process.  The emphasis upon

affective tone is usually linked in definitions with the
presence of a disposition or readiness to act in certain
ways, and with an implication that individual persons
differ consistently in their pattern of attitudes and
associated behaviours.


## A VIEW OF MENTAL HEALTH

No universally-accepted definition of the term "mental
health" is available.  In part this is because the concept
is to a considerable degree value-laden, reflecting the
views of particular societies or sub-groups within a
society.  Nevertheless, within Western countries most
theorists and practitioners seem to identify one or more of
five overlapping features.


### Affective well-being

The first of these, affective well-being, provides a
key indication of the level of someone's mental health.
Well-being is often viewed in overall terms along a single
dimension, roughly from feeling bad to feeling good.
However, it is preferable to identify two separate
dimensions, "pleasure" and "arousal", which may be treated
as orthogonal to each other.

We can view any affective state in terms of its
location on the two separate dimensions, so that the
specific quality of a particular affect derives from both
of them (e.g. Russell, 1980).  Four principal quadrants may
thus be identified, as shown in Table 1.

Although higher levels of affective well-being are in
general associated with the two right-hand quadrants,
mentally healthy people can also experience feelings
located in the left-hand sectors  This indicates the need
for a third dimension, concerned with time, and ranging
from relatively transient or momentary feelings to more
permanent, enduring states  In general, we can describe a
person's affective well-being over a given period in terms
of the proportion of time which he or she spends in each of
the four principal sectors.

Time spent in the top left-hand quadrant (experiencing
tension and related feelings) is particularly important in
this interpretation, since tension is often a necessary

Table 1.  Four types of affective well-being, defined in terms of pleasure and arousal.

| Low pleasure | High pleasure |
|---|---|
| High arousal | High arousal |
| Low pleasure | High pleasure |
| Low arousal | Low arousal |

prelude to affects located in the right-hand sections.    It would be quite wrong to infer from short-term negative feelings alone that a person is exhibiting poor mental health;  in many cases the reverse is true, as low affective well-being which is merely temporary is associated with high aspiration and high autonomy (components three and four within the model).

Within the framework of two orthogonal dimensions, three principal axes of measurement deserve special consideration.  These are shown in Table 2, with opposite poles indicated in terms of levels of pleasure and arousal.

I take these three axes to be the main ones we need to measure in respect of affective well-being.  And we can study well-being at two different levels.  First, we can measure it in general, without confining it to a particular setting;  let us call that "context-free" well-being.  Or we can measure "context-specific" well-being, in one limited setting.  We might do that, for example, in relation specifically to the family environment, or in respect of jobs.  In that case it is job-related affective well-being which is of research concern.

The measurement of affective well-being in terms of the three principal axes of Table 2 will be considered in the next section.  For the present, it should be noted that, in addition to characterizing three forms of well-being within a description of mental health, this account also depicts three parallel types of attitude.  The overlap arises from the fact that feelings of the kind addressed here are in practice focused upon particular issues or

Table 2.  Principal axes for the measurement of affective
well-being.

| Pleasure | Arousal | MEASURE-MENT AXIS | Pleasure | Arousal |
|---|---|---|---|---|
| Low | Medium | "Discon-tented" to Contented | High | Medium |
| Low | High | "Anxious" to "Com-fortable" | High | Low |
| Low | Low | "Depressed" to "Actively pleased" | High | High |

objects:  one has feelings about, or attitudes toward, some
aspect of one's environment or oneself.  Statements about
job-related affective well-being are thus also often
statements about job attitudes, and attitudes may be
construed in the same three terms as affective well-being.


Competence

Turning to the other components of mental health,
competence has been emphasized by many writers (e.g.
Jahoda, 1958;  Lazarus, 1975).  Terms used have included
effective coping, environmental mastery, self-efficacy,
effectance motivation, and so on.  It is widely held that a
key feature of mental health is an ability to handle life's
problems and in some ways successfully to act upon the
environment.

However, it would be wrong to view all types of low
competence as evidence of low mental health;  everyone is
incompetent in some respects.  The key factor seems to be a
link with affective well-being, in that low competence
which is not associated with negative affect would normally
be viewed as having no bearing on mental health.

We also need to include within any model of mental health both "subjective" and "objective" perspectives.  The former are as experienced by the person, the latter as viewed by an independent observer.  Thus, for example, we can look separately as "subjective competence", how well a person believes he or she is able to deal with the environment, and also at "objective competence", how successful the person is seen to be in practice. Subjective features of this kind may also be viewed as attitudes in the present context, containing beliefs and feelings about oneself in relation to the environment.

## Aspiration

The mentally healthy person is often viewed as someone who establishes realistic goals and makes active efforts to attain them.  Such people show an interest in the environment, they engage in motivated activity, and seek to extend themselves in ways that are personally significant. The converse is apathy and acceptance of the status quo, no matter how unsatisfactory.  This dimension is sometimes thought to be of greater concern in Western societies than in the East or in less developed countries, where alternative models of mental health might be more appropriate.

## Autonomy

The fourth component has also been emphasized more by Western than by Eastern writers.  It is widely held that mentally healthy people are able to resist environmental influences and determine their own opinions and actions. They feel and are autonomous and personally responsible for what they do.  However, a curvilinear pattern is usually assumed.  it is interdependence that is considered healthy rather than extreme independence or extreme dependence.

## Integrated functioning

The final component of mental health, here referred to as integrated functioning, is qualitatively different from the previous features.  Statements about integrated functioning refer to the person as a whole, often in respect of multiple interrelationships between the other four components.  The importance of this feature arises from the fact that people who are psychologically healthy exhibit several forms of balance, harmony, and inner relatedness.

Writers tend to develop their account of integrated functioning within a preferred theoretical approach. For example, some have written in psychoanalytic terms, examining the relationships between ego, superego and id. Assessments may again be made by an observer, or be "subjective", through reports from the focal person.

This feature is of particular importance in clinical settings, where complex assessments of stability and change in mental health are required. It is of less concern in discussions of job-related mental health or of job attitudes, and will not be considered further here.

## JOB-RELATED WELL-BEING AND JOB ATTITUDES

Developing the point that feelings are directed towards something (for example, aspects of the job situation), we may now consider job-related affective well-being and job attitudes from the same single perspective. Although the two concepts differ in certain ways (for instance, "well-being" is relatively more concerned with feelings about the self, and "job attitudes" are more concerned with feelings about job characteristics), they have a great deal in common.

This is illustrated in Table 3, where indices of affective well-being at three levels of specificity are set out in respect of the three measurement axes introduced above. In addition to "context-free" well-being, two levels of "job-related" affect are shown. These include relatively broad assessments, of overall job satisfaction for instance, but also a category of "facet-specific" feelings. These cover attitudes about particular elements within the job situation.

In overall terms, it may be suggested that studies of context-free well-being (concerned with life in general) have made quite good progress in measuring the three principal axes. There are several established inventories of distress, life satisfaction, anxiety, depression and so on, which can tap each of the three forms of affective well-being in an overall, context-free sense. But job-related well-being has very widely been measured along the first axis shown in Tables 2 and 3, usually in terms of

Table 3. Affective well-being: Three axes of
measurement and three levels of specificity, with
illustrative types of index.

| Context-free affective well-being | Job-related affective well-being | Facet-specific job-related well-being |
|---|---|---|
| **Axis 1: Discontented -- contented** | | |
| Happiness Life satis- faction General distress Negative affect | Overall job satisfaction Alienation from work Job attachment Organizational commitment | Specific sat- isfactions (with pay, amount of responsi- bility etc.) |
| **Axis 2: Anxious -- comfortable** | | |
| Anxiety Neuroticism | Job-related tension Resigned satisfaction | Specific feelings of job strain |
| **Axis 3: Depressed -- actively pleased** | | |
| Depression Tedium Self-denigra- tion Positive affect | Job-related depression Job-related burnout Job boredom Job-related pleasure Job involvement Morale | Specific aspects of job boredom |

scales of job satisfaction. Those can vary in several ways, but they all fail to consider the arousal axis.

A second problem with measures of job satisfaction is that they yield scores which tend to be grouped towards the positive pole. This is especially the case with overall assessments, where nearly 90% of workers pronounce themselves to some degree satisfied with their job (e.g. Weaver, 1980). In part this arises from self-selection, in that extremely dissatisfied employees are likely to have left in order to find different work, but the notion of "satisfaction" itself also tends to create relatively low variance in scores.

This is because judgements along this dimension are more likely than others to be moderated by comparisons with other jobs and with what a respondent considers possible or acceptable. Reported satisfaction is often an assertion that "all things considered this job is tolerable". Such a judgement is a very limited one, which does not come close to tapping the variety of job feelings suggested in Table 3. We should therefore take care not to treat and measure "job attitudes" merely in terms of job satisfaction.

The second main dimension of well-being (from "anxious" to "comfortable") has been studied in occupational settings through measures of job-related tension, anxiety or strain. There are several measures which cover low pleasure and high arousal (the top-left quadrant in Table 1), but the opposite pole appears to have been largely ignored by occupational psychologists. That is a pity, since that (bottom-right) sector seems to be particularly important in respect of low-arousal, resigned job satisfaction: workers in that quadrant are not complaining about their job, but they are apathetic and uninvolved.

The third axis includes job-related depression and also responses like job-related exhaustion, burnout, and boredom. These have been measured fairly successfully, but once again we have done less well at the opposite pole: high arousal and high pleasure. Here we might include "job involvement", which seems like high-arousal job satisfaction, and also "morale", covering feelings of active involvement as well as a pleasure in working. In general, I think that we should be more concerned with tapping job-related feelings of high pleasure at each of the levels of arousal indicated in Table 2.

Although statements about job-related well-being have as their reference some aspects of the job setting, they do of course primarily describe the job-holder's relationship to those job features. Other components of mental health which in a similar way may be viewed as "attitudes" are subjective competence, aspiration and autonomy. We are here particularly concerned with a person's perception of his or her job-related competence, aspiration or autonomy. Beliefs and feelings about one's relationship to one's job may thus be measured in these terms as well as through the three main axes of affective well-being.

## A QUESTION OF TIME: DEVIATIONS FROM THE BASELINE

The notions of both "mental health" and "attitude" suggest some stability over time. Although aspects of each may change in response to variations in the environment, we might expect to find considerable consistency across different measurement occasions. It thus becomes important to consider how much of this consistency arises from stability within the person, irrespective of external circumstances, and how much is due to continuity of environmental conditions.

This is of course a huge question, and I can merely touch upon it here. Although it seems clear that job characteristics do indeed have significant impact upon both job-related and context-free mental health (e.g. Warr, 1986), it is also the case that these outcomes (including job attitudes) are determined in part by characteristics of the person. For example, some people are generally more cheerful, less anxious or more tolerant of environmental constraint than others.

Ideally, one needs to take baseline measures of, say, anxiety level prior to exposure to particular job conditions, in order to assess deviations from that baseline which are induced by those conditions. Such a procedure is clearly very difficult, and we must probably accept that our assessments of attitude or mental health in a given setting are determined jointly, in unknown proportions, by attributes of both the person and the setting.

Another temporal issue has been touched upon several times throughout the paper. This concerns the time-frame within which mental health or attitudes should be measured.

Within particular sequences of actions or during specific
periods of time, a person´s affective well-being may be
temporarily impaired or enhanced to an atypical degree.  We
temporary strain or occasionally high positive affect do
not necessarily reflect characteristic levels of mental
health or continuing attitudes.  In seeking to determine
the impact of job conditions upon these outcomes,
investigators must be conceptually and operationally
explicit about their primary focus:  upon enduring or upon
temporary states.

REFERENCES

Cook, J.D., Hepworth, S.J., Wall, T.D., & Warr, P.B., 1981,
    The Experience of Work (Academic Press).
Jahoda, M., 1958, Current Concepts of Positive Mental
    Health (Basic Books).
Lazarus, R.S., 1975, The healthy personality.  In
    Society, Stress and Disease, Vol.2, edited by L. Levi
    (Oxford University Press).
Russell, J.A., 1980, Journal of Personality and Social
    Psychology, 39, 1281.
Warr, P.B., 1978, Attitudes, actions and motives.  In
    Psychology at Work, edited by P.B. Warr (Penguin).
Warr, P.B., 1986, Work, Unemployment and Mental Health,
    (Oxford University Press) (in the press).
Weaver, C.N., 1980, Journal of Applied Psychology,
    65, 364.

# ATTITUDE SURVEYS: AN AUTHORITARIAN TECHNIQUE MASQUERADING AS A PARTICIPATIVE PROCESS

## Reg Sell

Work Research Unit, St Vincent House,
30 Orange Street,
London.

## ABSTRACT

Surveying of attitudes is usually seen as a means of finding out what the employees of an organisation think of their job, their employers etc. Because it is based on finding out their views it is usually thought to be a part of a participative process. All too often, however, such a survey asks questions which only the originator thinks are important, fails to analyse the environment within which the survey is being taken and does not address the issues on which the respondents would really like their views considered. It is, therefore, often an authoritative technique applied in a participative guise. Only by managing and carrying out the survey itself in a participative way can it really be seen as part of a participitative process.

## Why survey attitudes?

Attitudes surveys are usually carried out for one of two reasons: either it is done for research purposes or else as a basis of diagnosis for taking future action.

Confusion of these two aims can cause problems. The main interest of the author and the emphasis in this paper is with the second aim.

The usual confusion is to take a research based survey and use it for diagnosis. Instruments designed for research purposes are usually intended to test hypotheses. These hypotheses are those relevant to the researcher but may not be to the organisation, and so the respondents are likely to see the questions asked as incidental to their own problems. They have their own order of priorities as regards work issues and would prefer to answer questions based on them rather than somebody else's view and thus may see the answering of irrelevant questions as an imposed chore which is not helpful to them.

Standard attitude surveys do not usually cover in great deal the environment in which the study is being made. Even when they cover some aspects it is often not possible to get a complete understanding of all the interacting areas. For instance, job rotation, a much quoted low level way of improving job design, might be seen as advantageous if carried out in a participative way but not so if it is imposed autocratically.

## How to survey

The two main choices for carrying out surveys are questionnaires or interviews but they can be combined with varying mixes of each.

Questionnaires can cover larger numbers for minimum cost, are easier to analyse and collate, give the pseudo-security of a numerical conclusion and are very easy to set up. They are also likely to be seen as the most authoritarian approach and to lead to the problems outlined above.

Interviews cost much more to cover the same numbers, are more difficult to collate and od not give conclusions with the same degree of numerical precision. They do, however, give respondents much more opportunity to express their own concerns and to cover the aspects which they believe to be important.

There is a risk with surveys which do have a detailed statistical analysis with breakdowns into different categories etc. that decision makers will come to conclusions which the real situation does not justify. Often the best that can be expected from a survey of a large organisation is the strength of feeling on a few key issues. Interview data across a whole organisation should identify the main issues whilst detailed action can be based on the items raised by the constitutent small groups.

## Who should survey?

Surveys an be carried out either by staff from inside the organisation or by outside independent people. There is, of course, no one answer as to who should do it and the following factors need to be considered:

aims of the study,
values of the organisation being surveyed,
values of those doing the surveying,
confidentiality,
competence,
credibility,
availability of inside staff,
finance available for external staff.

Ownership of the results by the members of the organisation is more likely to occur if it can be carried out by inside people.

## When should an organisation be surveyed

Some people think that they can get reliable comparisons by surveying at regular intervals and comparing the results across time.  If actions are taken as a result of earlier surveys this assumption may not be correct because the frame of reference within which the respondents are replying to the later ones will have changed.

For instance, if the survey is concerned with job satisfaction and changes are made as a result to improve this the later survey may not, in spite of an obvious improvement, confirm it.  It could be that the early one was carried out with low expectations of what could be done whilst the second one comes after changes had shown what opportunities there are for change and had, in turn, so increased the expectations of the respondents that they now could see even further opportunities for change.

There may also be risks in interpreting surveys which were carried out at times of high uncertainty or following traumatic experiences such as a redundancy.

## How can surveys by improved?

Probably the most accurate views of the work force could be attained if they could all be questioned individually in the local pub without knowing who was doing the survey, why they were bing surveyed or perhaps even that a survey was taking place.  Whilst this approach may be seen as authoritarian it is highly manipulative.

An important aim is that the work force on the whole must own the idea of the survey being carried out.  This can be achieved by having the process managed by a group representing the whole organisation.  This group will be able to identify the main areas of concern, carry out pilot runs and preliminary interviews.  This is the approach aimed at by ACAS when using employee questionnaires in the course of an Industrial Relations audit.

If multi-choice questionnaires are to be used they should be designed in such a way that the respondents are forced to think about why they are ticking particular boxes by asking them give examples of actual incidents which cause them to answer in such a way.

A recent study in which the Work Research Unit has beeen involved gives an example of an attitude survey carried out with a steering group using a questionnaire with the opportunity to give concrete examples. This survey was carried out as part of an overall programme to improve participation (Cuthbert, Smith & Sell 1984).

It goes without saying that whatever type of survey is carried out the results should be fed back to the participants. To fail to do so is arrogant and a betrayal of the trust they have given in putting forward their views.

It must be stressed that, if as a result of a study, there is a clear need for particular actions, these must be carred out if the credibility of the management and of the process is to be maintained. Unless the management is very lacking in perception it should know in advance some of the conclusions likely to arise. They should not go ahead with any survey which is likely to produce recommendations they will not or cannot follow up.

References

Cuthbert, D., Smith, A. and Sell, R.G., 1984, Forming the future through working together, *Employment Gazette*, Vol. 92, no. 1. Reprinted as Work Research Unit Occasional Paper no. 30.

# THE ROLE OF COLOUR IN OBJECT RECOGNITION, CATEGORIZATION AND NAMING

J.B. Davidoff

Department of Psychology,
University College of Swansea.

In general, successful performance at a visual task requires attention to be selective. Certain aspects of the input which are not essential to the task are therefore disregarded. But what of colour? Subjectively, colour seems so much a part of the visual world that it is hard to believe it plays no part in object identification and classification even if not strictly essential to the task. Yet, the addition of colour to shape information is remarkably unhelpful in improving identification for both arbitrary (Mial *et al*, 1974) and natural (Power, 1978) combinations. At the neurophysiological level, the independence of colour and shape information has been substantiated (Zeki, 1980) but at the psychological level it is surprising especially given the widespread use of colours in displays.

In a comprehensive review (Christ, 1975) of the role that colour plays in human factors research it is concluded that the "most clear-cut tendency is that if the colour of a target is unique for that target, and if that colour is known in advance, colour aids both identification and searching". This sweeping conclusion is modified elsewhere in the review and is in conflict with the above research but it still deserves further scrutiny. After all, if the colour is unique to a shape, then a shape identification task is transformed into a colour identication task.

Christ (1975) based his conclusions on the available studies, which were very few in number. In all there were eight studies for which colour was used as a redundant dimension in an identification task. Of these, five gave no advantage for the use of colour and for the other three the advantage can be seen on further analysis to be artefactual. Eriksen & Hake (1955) asked subjects to identify a partiuclar item from a series of twenty squares differing in size in small steps. Adding a unique colour to each of the squares markedly improved performance but no more than would be expected from the assumption that information was being transmitted via independent channels. Christ's graphical display of Eriksen and Hake's data is somewhat disingenuous; while there is an improvement in discrimination by adding colour, there is no improvement over that predicted from the information transmission rates of colour or shape alone. The Garner & Creelman (1964) task is identical and likewise affects discrimination rather than identification. The Markoff (1972) study (like most of the negative reports, the positive report of Markoff is not generally available) found an improvement for a different reason. The Markoff study used degraded displays. Christ (1975) says that this 'may' be significant but it is more important than that. Using a relatively restricted set of alternative real objects, colour is surely going to be useful for ruling out possibilities when shape information is uncertain.

While, visual search tasks are not the present concern, it is worthwhile pointing out that redundant colour is not as helpful to them as Christ (1975) suggests. It is useful but only under special circumstances. Colour is more useful in locating a target as the stimulus density increases (Smith, 1963; Smith *et al.*, 1965). The reason for this is that colour brings a stimulus to focal attention (Williams, 1966) and thus colour becomes important as the display becomes crowded. However, if the subjects are not informed that the colour is a cue for target identification, then adding colour does not improve performance in a visual search (Eriksen, 1953).

The analysis of the data presented by Christ (1975) leaves considerable doubt as to whether colour really does aid identification. It was therefore decided to investigate the matter further specially for the identification of natural objects. The investigation was also extended to consider object naming. While there is no evidence on the matter from normal subjects, the neuropsychological literature (Bisiach, 1966) did suggest the possibility of objects being easier to name from coloured rather than black and white versions.

No two natural objects have exactly the same colour; this produced a major problem for the proposed investigation. Colour alone, especially if using a restricted set of stimuli, can provide sufficient cues for identification. However, the problem can be overcome. Three objects (tomato, strawberry and radish) that were all of a similar hue of red were chosen. The three objects were spray painted the same red colour and photographed on colour and monochrome film from identical viewpoints and on a plain white background. The red colour was acceptable as a natural colour for all three objects. The objects were then sprayed blue and similarly photographed on colour film. Spray painting produced identically coloured pictures, without removing texture. Thus three sets of pictures were produced: a red set (the natural colour of the objects), a black-and-white set, and a blue set.

Twenty-four subjects participated in a naming task, and 27 subjects participated in a recognition task. In the naming task, subjects were told that they would be shown pictures of a strawberry, a tomato, and a radish, and they were allowed to familiarize themselves with the stimulus items. They were required to name the depicted objects as quickly as possible without making errors. Three blocks of 24 trials were run, where the stimuli within each block were all of the same colour: red, black-and-white, or blue. Before each block of trials, the subjects were informed what colour the stimuli would be and they were shown the three alternative items. The order of the red, black-and-white, and blue blocks was balanced between subjects.

In the recognition task, one of the objects was chosen as the target. Each of the three objects was the target for 9 subjects. The subject was required to respond *yes*, as quickly as possible, whenever he saw the target item and *no* when one of the non-target items was presented. Three blocks of 24 trials were run, and in each block the target was presented on 8 randomly selected trials. The stimuli within each block were all of the same colour (red, black-and-white, blue), and the order of the blocks was balanced between subjects. Before each block the subjects were always told what colour stimuli would be involved, and they were shown the three stimulus alternatives.

The object recognition data (Ostergaard & Davidoff, 1985) revealed that colour made no difference to recognition latencies; red, blue and black-and-white versions were all equally easy to recognize. The same was not true of naming. The correct colour facilitated naming latencies despite it giving no advantage for recognition. To make sure that it was a connection between the colour and the natural object that caused the naming advantage, rather than some peculiarity of the stimulus materials, a control experiment was run. Three geometrical shapes (square, circle and triangle) were cut out of the photographs of the tomato, strawberry and radish. With these stimuli an identical experiment to the one with natural objects was run except that it was shapes that were named or recognized. Colour now no longer influenced either recognition or naming.

A considerable amount of cognitive processing takes place between recognition and naming. It would therefore be instructive to know what role colour plays in the processing of the tasks which intervene. It appears that stimulus categorization is one such task (Potter & Faulconer, 1975). Therefore stimulus categorization was compared to naming with respect to the facilitation that colour provides. Sixteen undergraduates were given black-and-white and coloured line drawings to categorize as living or non-living and also to name. Balancing stimuli across subjects,ensured that the same object was not named or categorized twice. The data which resulted were then analysed by a 2 x 2 (object naming vs semantic classification x coloured vs black-and-white) ANOVA. There was an interaction between the two main factors. For naming, coloured pictures produced faster responses than black-and-white pictures ($p < .01$); this was the case for 15 of the 16 subjects. For semantic classification, colour produced no difference to the speed with which the decision was made. It therefore appears that colour information combines with shape information after semantic classification has been carried out.

The results of the present studies suggest a rather limited role for the use of colour in visual displays. If we ignore those occasions where colour may give better contrast, the main use for colour would appear to be in crowded displays. Here colour provides an additional modality to give meaning to displays with the added advantage of being 'attention grabbing'. However, for isolated items, colour does not affect either recognition or classification latencies. More positively, colour does, for reasons which are yet to be determined, help naming. In future years, when vocal responses to visual displays trigger computers, colour could therefore be valuable when responses are required to natural objects. Colour does also appear to be more pleasing than monochrome.  Even when colour does not help performance (Greenstein & Fleming, 1984), the subjects rated the task as more agreeable.  If very long intervals are spent waching displays, colour for as yet no known reason might therefore improve motivation.

REFERENCES

Bisiach, E., 1966, Cortex, 2, 90-95.

Christ, R.E., 1975, Human Factors, 17, 542-570.

Eriksen, C.W., 1953, Journal of Experimental Psychology, 45, 126-132.

Eriksen, C.W. & Hake, H.W., 1955, Journal of Experimental Psychology, 50, 153-160.

Garner, W.R. & Creelman, C.D., 1964, Journal of Experimental Psychology, 67, 168-172.

Greenstein, J.S. & Fleming, R.A., 1984, The use of color in command control electronic status boards.  In Proceedings of NATO Workshop on color coded vs monochrome electronic displays (H.M.S.O. London).

Mial, R.P., Smith, P.C., Doherty, M.E. & Smith, D.W. 1974, Perception and Psychophysics, 16, 1-3.

Ostergaard, A.L. & Davidoff, J.B., 1985, Journal of Experimental Learning, 11, 579-587.

Potter, M.C. & Faulconer, B.A., 1975, Nature, 253, 437-438.

Power, R.P., 1978, Perception, 7, 105-111.

Smith, S.L., 1963, Journal of Applied Psychology, 47, 358-364.

Smith, S.L., Farquhar, B.B. & Thomas, D., 1965, Journal of Applied Psychology, 49, 393-398.

Williams, L.G., 1966, Perception and Psychophysics, 1 315-318.

Zeki, S., 1980, Nature, 284, 412-4

# 1987

| Swansea University | 6th–10th April |
| --- | --- |
| **Conference Secretary** | Rachel Birnbaum |
| **Exhibitions Officer** | Jane Prince |
| **Honorary Meetings Treasurer** | Jane Alexander |
| **Chair of Meetings** | John Wilson |
| **Programme Secretary** | Ted Megaw |

## Social Entertainment

Delegates were able to network at the Pool Party in a swimming pool with a wave machine

| Chapter | Title | Author |
| --- | --- | --- |
| **Keynote Addresses** | The cognitive bases of predictable human error | J. Reason |
| **Human Reliability** | SHERPA: A systematic human error reduction and prediction approach | D.E. Embrey |
| **Human Performance** | Minor illnesses and performance efficiency | A. Smith and K. Coyle |
| **Design, Simulation and Evaluation** | The use of people to simulate machines: An ergonomic approach | M.A. Life and J. Long |

# THE COGNITIVE BASES OF PREDICTABLE HUMAN ERROR

## J. REASON

Department of Psychology
University of Manchester
Manchester M13 9PL

This paper identifies two pervasive error-shaping factors: similarity and frequency. These biases are evident in a wide range of error types (mistakes, lapses and slips), involving many different cognitive activities. A 'dual-architecture' model of human cognition is outlined. This comprises a serial, restricted, but computationally-powerful 'workspace' interacting with an effectively unlimited, parallel, distributed knowledge base. It is argued that similarity and frequency effects are rooted in the simple but universal heuristics by which the outputs of stored knowledge units are identified and elicited.

## ERRORS TAKE A LIMITED NUMBER OF FORMS

Human error is neither as abundant nor as varied as its vast potential might suggest. Not only are errors much rarer than correct actions, they also tend to take a suprisingly limited number of forms - surprising, that is, when measured against their possible variety. Moreover, errors appear in very similar guises across a wide range of mental activities. Thus, it is possible to identify comparable error forms in action, speech, perception, knowledge retrieval, judgement, problem solving, decision making, concept formation, and the like. The ubiquity of these systematic error forms forces us to formulate more global theories of cognitive control than are usually derived from laboratory experiments which, of necessity, focus upon very restricted aspects of mental function in artificial settings.

The purpose of this paper is to sketch out some of the causal relationships between these pervasive error forms and the more fundamental properties of human information processing. It will be argued that systematic errors are inextricably bound up with those things at which the cognitive system excels relative to other information-processing devices, and especially with the characteristic ways in which it simplifies complex information-handling tasks.

## DISTINGUISHING ERROR TYPES AND ERROR FORMS
### Error types
The term 'error type', as used here, relates to the presumed origin of an error within the stages involved in conceiving and then carrying out an action sequence. These stages can be described under three broad headings: planning, storage and execution. Planning refers to the processes concerned with identifying a goal and deciding upon the means to achieve it. Since plans are not usually acted upon immediately, it is likely that a storage phase of some variable duration will intervene between formulating the intended actions and running them off. The execution stage covers the processes involved in actually implementing the stored plan. The relationship between these three stages and the primary error types is shown in Table 1.

Table 1. The primary error types.

| COGNITIVE STAGE | PRIMARY ERROR TYPE |
|---|---|
| Planning | Mistakes |
| Storage | Lapses |
| Execution | Slips |

Actions may fail to achieve their desired consequences either because the plan was inadequate (mistakes), or because the actions did not proceed according to plan. In the latter case, it is useful to distinguish between those unintended actions which arise as a consequence of memory failures (lapses), and those due to the imperfections of attentional monitoring (slips).

For most practical purposes, however, the crucial distinction is between errors which occur at the level of intention (mistakes), and those which occur at some subsequent stage (lapses and slips). Mistakes can be further subdivided into (a) failures of expertise, where some pre-established plan or problem solution is applied inappropriately; and (b) a lack of expertise, where the individual , not having an appropriate 'off-the-shelf' routine, is forced to work out a plan of action from first principles, relying upon whatever relevant knowledge he  or she currently possesses. These two types of mistakes correspond closely to the rule-based and knowledge-based levels of performance, as described by Rasmussen (1982).

## Error forms

Whereas 'error types' are conceptually tied to underlying cognitive stages or mechanisms, 'error forms' are recurrent varieties of fallibility which appear in all kinds of cognitive activity, irrespective of error type. Thus, they are evident in mistakes, lapses and slips. So widespread are they that it is extremely unlikely that their occurrence is linked to the failure of any single cognitive entity. Rather, this omnipresence suggests that they are rooted in universal cognitive processes.

Two such error forms will be considered here: similarity-matching and frequency-gambling. Before describing them further, however, it is necessary to provide a preliminary sketch of the basic structural components of the cognitive system, and then to indicate how they might interact to specify a given action or thought sequence.

## COGNITIVE STRUCTURES
### The conscious workspace (Ws)

This 'sharp end' of the cognitive system receives input from both the outside world, via the senses, and from the knowledge base. It is related to - though not necessarily co-extensive with - conscious attention and working memory (Baddeley & Hitch, 1974;  Mandler, 1985). Its primary concerns are with the setting of gaols, with selecting the means to achieve them, with detecting deviations from current intention, and with monitoring progress towards these desired outcomes. The products of this high-level planning and monitoring activity are accessible to awareness; but since consciousness is a severely restricted 'window', it is usually the case that only one such high level activity can be 'viewed' and worked upon at any one time. These resource limitations confer the important benefit of selectivity, since several high-level activities are potentially available to the Ws.

The conscious Ws has powerful, analytical, feedback-driven computational processes at its disposal, and is essential for coping with novel or changed circumstances, and for detecting and recovering errors. But it is severely resource-limited, slow, laborious, serial and difficult to sustain.

## The knowledge base (Kb)

Human cognition is extremely proficient at modelling
the useful regularities of its previous dealings with
specific environments, and then using these stored
representations as a basis for the automatic control of
subsequent perception and action. The minutiae of mental
life are governed by a vast community of specialised
processors, each constituting a 'mini-theory' regarding
some particular aspect of the world, and each being
instantiated by highly specific triggers supplied by both
the conscious workspace (intentional 'calling conditions')
and by the environment (contextual or task-related 'calling
conditions'). These 'knowledge packets' or 'schemata' thus
possess two closely related elements: (a) they embody
generic or prototypical knowledge concerning specific
aspects of the world, and (b) they can generate pre-
programmed instructions for eliciting particular actions,
words, images, percepts, etc. Only these products of schema
activity are available to the higher-level workspace; the
processes themselves lie beyond the reach of consciousness.

In contrast to the conscious Ws, which can function
over both long time spans and a wide range of circumstances,
schemata within the Kb are tied to highly specific
triggering conditions. The Kb, however, has no known limits
on its capacity. It can process familiar information
rapidly, in parallel, and without conscious involvement or
effort. But it is relatively ineffective in the face of
novel or unforeseen circumstances.

## THE SPECIFICATION OF COGNITIVE ACTIVITY

Correct performance in any sphere of cognitive performance is achieved by activating the right schemata in the
right order at the right time. Schemata may be brought into
play (i.e. deliver their products either to the conscious
Ws or to the outside world in the form of actions) by both
specific and general activators. Specific activators are
those which trigger targeted schemata at a particular  time
and place. Among these, intentional activity is likely to
be paramount. Other specific activators include contextual
cues and 'descriptions' (see Bobrow & Norman, 1975; Norman
& Bobrow, 1979) passed on by other sequence-related schemata
General activators provide background activation to
schemata, regardless of the current intentional state. Of
these, frequency of prior use is probably the most
important.

The central thesis of this paper is that predictable
error forms are rooted in a tendency to over-utilise what

is probably the most conspicuous achievement of human
cognition: its ability to simplify complex informational
tasks by resorting to pre-established routines, heuristics
and shortcuts. It is believed that the two most fundamental
of these heuristics are (a) match like with like, and (b)
resolve conflicts (between schema candidates) in favour of
contextually-appropriate, high-frequency knowledge
structures. Both of these tendencies are brought into
particular prominence in conditions of cognitive under-
specification.

Precisely what is missing from a sufficient specification, or which cognitive level fails to provide it, will
vary with the nature of the task. But notwithstanding these
possible varieties of under-specification, their
consequences are remarkably uniform. The (often erroneuous)
responses selected in conditions of under-specification
tend to (a) show formal similarities to either the prevailing contextual features or to the currently intended (or
normatively and evidentially appropriate) responses, or
both; and (b) they are likely to be more familiar, more
conventional, more typical, more frequent-in-context than
those that would have been judged correct or appropriate.

## SIMILIARITY AND FREQUENCY: COGNITIVE 'PRIMITIVES'

Such fundamental aspects of experience as the degree of
likeness between events or objects and their frequency of
prior occurrence have been termed intuitive concepts.
Similarity and frequency information appear to be processed
automatically without conscious effort or perhaps even
awareness, regardless of age, ability, cultural background,
motivation, or task instructions (see Wason & Johnson-Laird
1972; Shweder, 1977; Tulving, 1983; Hasher & Zacks, 1984).
There is a strong case for regarding them as being pre-
eminent among the computational 'primitives' of the
cognitive system.

## SIMILARITY EFFECTS IN MISTAKES, LAPSES AND SLIPS

Error forms can resemble the properties of both the
current intentional specification and the prevailing
environmental cues in varying degrees. The most obvious
tendency in mistakes - particularly those in  which the
problem solver has been limited by an incomplete or
incorrect knowledge base - is for the error forms to be
shaped by salient features of the problem configuration.
In the case of both lapses and slips, however, there can be
matching to both intentional and contextualcues. The forms
of these execution failures can show close similarities to

the intended word or action, as well as, on occasions, being appropriately matched to the situation in which they occur. Exactly what is matched by the error form appears to be related to the extent to which the correct response is, or could be, specified at the outset of the thought or action sequence. Difficulties encountered at the level of formulating the intention or plan tend to create errors that are shaped primarily by immediate contextual considerations; those which occur at the level of storage or execution may reflect the influences of both intentional and environmental 'calling conditions' (specifiers). But irrespective of the precise nature of the matching, similarity effects are evident across all error types.

## Similarity effects in mistakes

Similarity effects have been most clearly demonstrated in (a) laboratory studies of reasoning and inference; and (b) investigations of human judgement under conditions of uncertainty.

Wason and Johnson-Laird (1972) employed a variety of techniques (the Wason Card Test, syllogisms, the '2-4-6' rule-discovery task, etc.) to investigate the ways in which people draw explicit conclusions from evidence. In the course of this work, they identified a number of 'pathologies' of problem solving (the reluctance to utilise negative statements, the corresponding ease of handling affirmative statements, confirmation bias, the 'thirst for confirming redundancy', illicit conversion, and so on); but they concluded that all of these factors were the consequence of one general, overriding principle: "..whenever two different items, or classes,can be matched in a one-to-one fashion, then the process is readily made, whether it be logically valid or invalid".(Wason & Johnson-Laird, 1972, p.241). Thus, affirmatives are easier to handle than negatives because they involve just the single step of making a one-to-one relation between a statement and a state of the world. And the fact that people will naturally establish such a match when dealing with a problem will inevitably bias them toward affirming rather than falsifying their beliefs.

In a series of studies, Tversky, Kahneman and their associates (see Khaneman, Slovic & Tversky, 1982) have shown that when people are asked to judge whether object A belongs to class B, or whether A originates from process B, they typically over-utilise the representativeness heuristic. That is, their probability judgements are heavily determined by the extent to which A resembles B, regardless of such critical factors as sample size, base rates, and the like.

## Similarity effects in lapses

Perhaps the commonest type of memory lapse is forgetting to remember to carry out intended actions at the appointed time and place (Reason & Mycielska, 1982). For the most part, these failures of prospective memory lead to the omission of isolated planned actions, and, as such, cannot readily manifest similarity effects. There are, however, other commonly occurring lapses, particularly failures to retrieve known items (names or words) from long-term memory, which show this bias in a variety of obvious ways. There is now a wealth of evidence (from Aristotle onwards) indicating that 'intermediate solutions' - wrong words dredged up in the course of an active memory search for a blocked word (in a tip-of-the-tongue or TOT state) - show close phonological, morphological and semantic similarities to the target item (see Reason & Lucas, 1984, for a discussion of the relevant literature).

## Similarity effects in slips

As with TOT state 'intermediates', slips of the tongue show marked similarity effects between the actual and target (intended) utterances (see Fromkin, 1973, 1980). "The more similar a given unit is to an intended unit, the more likely the given unit or a part of it will replace the intended unit or a corresponding part of it" (Dell & Reich, 1980, p.281).

Different kinds of similarity are likely to operate at the various stages of formulating and executing the articulatory program. Fromkin (1971) has suggested that semantically-related substitutions may occur because of under-specification of the semantic features. At the more detailed level of phoneme specification, however, substitutions are facilitated by phonological similarity between word segments (Nooteboom, 1969; MacKay, 1970).

Although less widely investigated, slips of action reveal comparable similarity effects (see Reason & Mycielska, 1982). These are most obvious in slips involving wrong objects, recognition failures in which another item is substituted for the correct one during the execution of a highly routinised action sequence. A necessary condition for these and other action slips is some diminution of attentional monitoring, either through preoccupation or distraction. Just as some investment of the limited attentional resource is necessary for checking that the correct actions have been selected, so it is also involved in verifying the accuracy of perceptions, particularly during oft-repeated action sequences when the relevant recognition schemata seem ready to accept rough approximations to the expected object configuration (see Reason, 1979).

Absent-minded misrecognitions most commonly involve the unintended substitution of physically similar items, as in the following actual instances (Reason & Mycielska, 1972): "I intended to pick up the deodorant, but picked up the air freshener instead." "When seasoning meat, I sprinkled it with sugar instead of salt". "I filled the washing machine with oatmeal". "I put shaving cream on my toothbrush".

Similarity effects are also evident in strong habit intrusions. In one diary study (Reason & Mycielska, 1982), subjects were required to complete a standarised set of ratings in relation to each slip recorded. In particular, they were asked whether or not their intended actions were recognisable as 'belonging to' some other task or activity, not then intended. Such a relationship was identified in 77 (40 per cent) of the 192 action slips netted. In the case of these slips, the diarists were further asked to rate the extent to which the intended actions and the 'other activity' shared common features. Strong similarity effects were found in regard to locations, movements and objects, and somewhat weaker ones in relation to timing and purpose (see Reason & Mycielska, 1982, p.257 for the actual data).

## FREQUENCY EFFECTS IN MISTAKES, LAPSES AND SLIPS

Errors in all types of cognitive activity tend to take the form of contextually-appropriate, high-frequency responses. The more often a sequence of perception, thought or action achieves a successful outcome, the more likely it is to appear unbidden in conditions of incomplete specification(see also the Law of Effect). The psychological literature is replete with terms to describe these high-frequency error forms: 'conventionalization' (Bartlett), 'sophisticated guessing' (Solomon & Postman), 'persistence forecasting' (Bruner, Goodnow & Austin), 'strong associate substitution' (Chapman & Chapman), 'inert stereotypes' (Luria), 'banalization' (Timpanaro), 'strong habit intrusions' (Reason), and 'capture errors' (Norman). Though it sometimes leads to incorrect responses, this tendency to gamble in favour of well-used knowledge structures is a highly adaptive strategy for dealing with a world that contains a great deal of regularity as well as a large measure of uncertainty.

Frequency effects in mistakes

Planning, decision making and problem solving all require the generation from the knowledge base of possible courses of action. The products of these cognitive activities emerge from a complex interaction between current perceptions of the world and the recall of previous states. But however these processes are initiated, the outcome tends to favour the selection of salient (vivid) or familiar (frequent) scenarios of future action (see Nisbett & Ross, 1980; Kahneman, Slovic & Tversky, 1982; Fischhoff, Lichtenstein, Slovic, Derby & Keeney, 1981).

Irrespective of what other kinds of under-specification may promote them, mistakes are almost always the result of incomplete or inaccurate knowledge. A recent study (Reason, Bailey & Horrocks, 1986), involving the identification of quotations from US presidents, demonstrated the greater effects of 'frequency-gambling' upon ignorant as opposed to educated guesses in the retrieval of declarative knowledge. The less subjects knew about US presidents (as measured by a recognition test comprising a jumbled mixture of both presidential names and those of their famous contemporaries) the more inclined they were to attribute quotations to contextually-appropriate, high-frequency presidents (frequency scores for all the 39 presidents were derived from ratings made by another group of comparable subjects). The results are summarised in Table 2.

Table 2. The relationship between domain knowledge and frequency-gambling.

| | QUARTILE GROUPINGS BY KNOWLEDGE SCORES | | |
|---|---|---|---|
| | Mn.Knowledge Scores | Mn.Frequency Scores | No.in group |
| Group I | 6.36 (1.37) | 4.89 (0.71) | 29 |
| Group 2 | 8.75 (0.60) | 4.84 (0.72) | 28 |
| Group 3 | 12.06 (1.31) | 4.56 (0.70) | 28 |
| Group 4 | 26.20 (7.73) | 3.80 (1.02) | 29 |

NOTES:
(a) Knowledge scores run from 1-39 adjusted recognitions.
(b) Frequency scores run from 1 (lowest) to 7 (highest)
(c) Figures in parentheses are standard deviations
(d) $F_{(3,110)} = 11.26$ (p  .0001).

## Frequency effects in lapses

In the memory block study, touched upon earlier
(Reason & Lucas, 1984), 16 volunteers kept 'extended'
diaries of their resolved TOT states over a period of one
month. Of the 40 resolved TOTs recorded, 28 involved the
presence of 'recurrent intruders' - recognisably wrong
names or words that continued to block access to the
target item during deliberate search periods. In 77 per
cent of these TOTs, the recurrent intruder was ranked
higher than the target for either contextual frequency or
recency, or both.

These data are consistent with the view that recurrent
intruders emerge in TOT states when the initial fragmentary
retrieval cues are sufficient to locate the 'ball park'
context of the sought-for item, but not to provide a unique
specification for it. Recurrent intruders tend to be high-
frequency items within this general context.

## Frequency effects in slips

It has already been mentioned that a large proportion
of action slips (40 per cent in one study) take the form of
strong habit intrusions: well-organized action sequences
that were judged as belonging to some other task or activity
In addition to assessing the similarity between this other
activity and the intended actions, the subjects were asked
to rate how often they engaged in the activity from which
the erroneous sequences had apparently 'slipped'. Over 70
per cent of these unintended actions were judged as
'belonging to' a task that was performed very frequently
indeed (see Reason & Mycielska, 1982, p.257).

There is also a considerable literature relating to
the word frequency effect, the repeated finding that
common words are more readily recognised than infrequent
ones when their presentation is brief or otherwise
attentuated (see Neisser, 1967). When a perceived word
fragment is common to many words, and the subject is asked
to guess the whole word from which it came, "..he will
respond with the word of the greatest frequency of
occurrence (response strength) which incorporates the
fragment" (Newbigging, 1961, quoted by Neisser, 1967,p.117).
An important corollary to this 'fragment theory' is that
seen (or recalled) bits of relatively rare words will tend

to be erroneously judged as belonging to common words sharing the same features (see also Gregg, 1976).

## THE ROOTS OF PREDICTABLE ERROR: KNOWLEDGE RETRIEVAL

The first part of the paper made a distinction between error forms and error types, and between the properties of two parts of the cognitive system: the limited workspace (Ws) and the knowledge base (Kb). Evidence was then presented for the prevalence of similarity and frequency effects (error forms) across mistakes, lapses and slips (error types). In this concluding section, it will be argued that these two basic error forms have their origins in the processes by which knowledge items (schema products) are retrieved from the Kb.

It is postulated that the cognitive system has three mechanisms for bringing the products of stored knowledge into the conscious Ws and/or into action. Two of them - similarity-matching and frequency-gambling - constitute the computational primitives of the system as a whole, and operate in a parallel, distributed and automatic fashion within the Kb. The third retrieval mechanism - directed or inferential search - derives from the sophisticated processing capabilities of the Ws itself. Within the Ws, through which information must be processed slowly and sequentially, the speed, effortlessness and unlimited capacity characteristic of the Kb have been sacrificed in favour of selectivity, coherence and computational power.

### Similarity-matching

The Ws has a cycle time of a few milliseconds, and each cycle contains between two and five discrete informational elements. During a run of consecutive cycles, these elements are transformed, extended or recombined by powerful operators which function only within this restricted conscious domain (see Mandler, 1985).

A useful image for the conscious Ws is that of a slicer. Information, comprising elements from both sense data and the Kb, is cut into slices which are then dropped into the buffer store of the Kb.

Once in the Kb buffer store, the informational features (the 'calling conditions') contained in these 'slices' are automatically matched to the attributes of stored knowledge items. No special fiat is required for this matching to ocur, and the process of relating calling conditions to corresponding attributes is both rapid and efficient.

All stored items possessing attributes which correspond to the elements of the most recent 'run' of Ws 'slices' (I.e. those held in the limited buffer) will increase their activation by an amount related to the goodness of match. The closer the match, the greater will be the received activation. When the activation level of a given schema exceeds a certain threshold value, its products may be delivered to the conscious Ws (images, words and feelings) and/or to the effectors (speech, actions). However, not all the knowledge units so matched will deliver their products to the conscious Ws. In some cases, only a partial correspondence will be achieved; in others, the products of activated schema may be pre-empted at the Ws by higher priority inputs.

## Frequency-gambling

In many situations, the calling conditions emerging from the Ws are insufficient to match uniquely a single knowledge item. This can occur because either the calling conditions, or the stored item attributes, are incomplete. These two kinds of under-specification are functionally equivalent. In under-specified searches, a number of partially-matched 'candidates' are likely to receive an increase in their activation levels. Where these contenders are equally matched to the current calling conditions, the conflict is resolved in favour of the most frequently encountered item. This occurs because an oft-triggered knowledge unit (i.e. one with a proven utility) will have a higher 'background' activation level than one less frequently employed.

Notice that these two automatic search processes, similarity-matching and frequency-gambling, though exceedingly simple and involving no logical principles, can together elicit a reasonably adaptive response (i.e. one that is contextually well-matched and previously useful) in any situation. Thus, the cognitive system is sensitive to both the formal and the statistical properties of the world it inhabits. And it is just these characteristics which give predictable shpaes to a wide variety of error types.

## Directed search

As is made apparent by such phenomena as TOT states, the conscious Ws has no direct access to the Kb. Its sole means of directing knowledge retrieval is through the manipulation of calling conditions. The actual search it-self is performed automatically by the similarity-matching and frequency-gambling heuristics. All that the Ws can do, therefore, is to deliver the initial calling conditions, assess whether the search-product is appropriate, and, if not, to reinstate the search with revised retrieval cues. The Ws has the power to reject the high-frequency candidates thrown up by under-specified matches, but only when sufficient processing resources are available. In conditions of high workload, environmental stressors, preoccupation and distraction, the Ws has often little choice but to accept these 'default' options. While these may serve adequately enough in relatively routine situations, they can and do lead to predictable error forms when their the goals or the circumstances of action have changed.

So far, we have considered a system in which actions are set in train and knowledge products delivered to the Ws as an automatic consequence of the interaction between (a) prior conscious processing, and (b) knowledge unit activation. But these properties alone are not sufficient to guarantee the successful execution of goal-directed behaviour. What gives cognition its intentional character? How does it initiate <u>deliberate</u> actions or knowledge searches?

As with many other difficult questions, William James (1890) provided a possible answer: "The essential achievement of the will.. is to attend to a difficult object and hold it fast before the mind. The so doing is the fiat; and it is a mere physiological incident that when the object is thus attended to, immediate motor consequences should ensue" (James, 1890, p.561).

This statement maps readily onto the simple 'dual architecture' model of cognition, previously described. The "holding-fast-before-the-mind" translates into a sustained run of same-element Ws 'slices'. Once in the Kb buffer, the consistency of these 'slices' will generate a high level of focused activation within a restricted set of knowledge units. This will automatically and advantageously release their products to the Ws and/or the effectors.

But such an 'act of will' places heavy demands upon the limited attentional resources available to the Ws. The continuation of specific informational elements within consciousness has to be amintained in the face of other strong claimants to the Ws. Such an effort can only be sustained for brief periods. As James puts it: "When we are studying an uninteresting subject, if our mind tends to wander, we have to bring back out attention every now and then by using distinct pulses of effort, which revivify the topic for a moment, the mind then running on for a certain number of seconds or minutes with spontaneous interest, until again some intercurrent idea captures it and takes it off" (James, 1899, p.101).

CONCLUSION

It has been argued that the pervasive similarity and frequency effects, apparent in a wide variety of error types, are rooted in the parallel and automatic processes by which knowledge is retrieved from long-term memory. In particular, they are shaped by the cognitive system's remarkable ability to match stored represenations to current Ws 'calling conditions', and to resolve conflicts between partially matched items using a simple "most used, most likely" heuristic. This bias towards selecting the ... more frequent of the partially matched candidates is dependent upon an automatic facility for keeping a running tally of roughly how often an event or object has been encountered in the past. The picture presented of human cognition is that of an informational system which, though adept at internalising the complexity of the world around it, is driven by a restricted number of simple computational procedures (see also McClelland & Rumelhart, 1985; Norman, 1986).

REFERENCES

Baddeley, A.D. & Hitch, G. 1974 Working memory. In The Psychology of Learning and Motivation edited by G.H. Bower, Vol.8(New York:Academic Press).

Battig, W.F. & Montague, W.E. 1969 Category norms for verbal items in 56 categories: A replication and extension of the Connecticut category norms, Journal of Experimental Psychology Monograph, 80, 1-46.

Bobrow, D.G. & Norman, D.A. 1975. Some principles of memory schemata. In Representation and Understanding: Studies in Cognitive Science, edited by D. Bobrow & A Collins (New York:Academic Press).

Dell, G.S. & Reich, P.A. 1980 Toward a unified model of slips of the tongue. In Errors and Linguistic Performance Slips of the Tongue,Ear, Pen, and Hand, edited by V. Fromkin (New York:Academic Press).

Fischhoff, B., et al. 1981, Acceptable Risk, (Cambridge: C.U.P.).

Fromkin, V. 1971 The non-anomolous nature of anomolous utterances, Language, 47, 27-52.

Fromkin, V. (Ed.) 1973 Speech Errors as Linguistic Evidence ,(The Haugue:Mouton).

Fromkin V. (Ed.) 1980, Errors in Linguistic Performance: Slips of the Tongue, Ear, Pen, and Hand. (New York: Academic Press).

Gregg, V. 1976, Word frequency, recognition and recall. In Recall and Recognition, edited by J. Brown (London: Wiley).

Hasher, L. & Zacks, R.T. 1984 Automatic processing of fundamental information: The case of frequency of occurrence, American Psychologist, 39, 1372-1388.

James, W. 1890 The Principles of Psychology, Vol.2. (New York: Holt).

James, W. 1899 Talks to Teachers on Psychology: And to Students on some of Life's Ideals. (London:Longmans)

Kahneman, D. et al. 1982 Judgement under Uncertainty: Heuristics and Biases (Cambridge:C.U.P.).

Mandler, G. 1985 Cognitive Psychology: An Essay in Cognitive Science (Hillsdale:Erlbaum).

MacKay, D.G. 1970 Spoonerisms: The structure of errors in the serial order of speech, Neuropsychologia, 8, 323-350.

Neisser, U. 1967 Cognitive Psychology, (New York:Appleton-Century-Crofts).

Nisbett, R. & Ross, L. 1980 Human Inference: Strategies and Shortcomings of Social Judgement (Englewood Cliffs: Prentice-Hall).

Norman, D.A. & Bobrow, D.G. 1979 Descriptions: An intermediate stage in memory retrieval, Cognitive Psychology, 11, 107-123.

Norman, D.A. 1985 New views of information processing: Implications for intelligent decision support systems. In Intelligent Decision Aids in Process Environments, edited by G. mancini & D. Woods. (San Miniato, Italy: NATO Advanced Study Institute).

Nooteboom, S.G. 1969 The tongue slips into patterns. In Leyden Studies in Linguistics and Phonetics, edited by A. Sciarone et al. (The Hague:Mouton).

McClelland, J. L. & Rumelhart, D.E. 1985 Distributed memory and the representation of general and specific information, Journal of Experimental Psychology:General, 114, 159-188.

Rasmussen, J. 1982 Human errors: A taxonomy for describing human malfunction in industrial installations, Journal of Occupational Accidents, 4, 311-335.

Reason, J.T. 1979 Actions not as planned: the price of automatization. In Aspects of Consciousness Vol.1. Psychological Issues, edited by G. Underwood & R.Stevens (London:Wiley).

Reason, J.T. et al. 1986 Multiple search processes in knowledge retrieval: Similarity-matching, frequency-gambling and inference (Unpublished report).

Reason, J.T. & Mycielska, K. 1982 Absent-Minded? The Psychology of Mental Lapses and Everyday Errors, (Englewood Cliffs, New Jersey:Prentice-Hall).

Reason, J.T. & Lucas, D. 1984 Using cognitive diaries to investigate naturally occurring memory blocks. In Everyday Memory, Actions and Absent-Mindedness, edited by J. Harris & P. Morris (London:Academic Press).

Shweder,R.A. 1977 Likeness and likelihood in everyday thought: magical thinking and everyday judgements about personality. In Thinking: Readings in Cognitive Science, edited by P.Johnson-Laird & P. Wason (Cambridge:C.U.P.).

Tulving, E. 1983 Elements of Episodic Memory (Oxford: Oxford University Press).

Wason, P.C. & Johnson-Laird, P.N. 1972 The Psychology of Reading (London:Batsford).

# SHERPA: A SYSTEMATIC HUMAN ERROR REDUCTION AND PREDICTION APPROACH

## D.E. EMBREY

Human Reliability Associates Ltd
1 School House, Higher Lane, Dalton
Parbold, Lancashire   WN8 7RP

ABSTRACT

This paper describes a Systematic Human Error Reduction and Prediction Approach (SHERPA) which is intended to provide guidelines for human error reduction and quantification in a wide range of human-machine systems. The approach utilises as its basis current cognitive models of human performance.

THE OBJECTIVES AND STRUCTURE OF SHERPA

The overall function of SHERPA is to provide a framework within which human reliability can be analysed and assessed, both quantitatively and qualitatively. It also generates specific error reduction recommendations in the areas of procedures, training and equipment design. The optimal way in which these recommendations should be implemented is evaluated using the quantification module in SHERPA, which allows cost effectiveness assessments to be carried out. These essentially consist of sensitivity analyses which indicate the changes in the system which will have the greatest effect in enhancing human reliability. Ideally SHERPA should be applied at the design stage of a new system. In practice it will generally be used to modify an existing system.

Although, as described above, the separate modules within SHERPA are relatively independent, they are linked by a common theoretical orientation. This is that human reliability optimisation and quantification cannot be effectively achieved by considering only the surface aspects of human error. The underlying cognitive processes that give rise to errors must be addressed as part of the approach.

SHERPA consists of a series of sequentially applied modules which are described in the following sections.

Task Analysis

Any error prediction and reduction approach must begin with a task analysis to provide a detailed description of the situation being evaluated, and to identify the characteristics of the task likely to give rise to error. The recommended task analysis technique in SHERPA is Hierarchical Task Analysis (HTA) described in detail in Shepherd (1984).

HTA is a systematic method for identifying the various goals that have to be achieved within a task and the way in which these goals are combined to achieve the overall objectives. One of the major advantages of the HTA method is that it explicitly identifies the plans that are used by operators to achieve goals at various levels. Because HTA captures the operator perception of the goal structure of the task, this structure can be reflected in the procedures and training methods derived from the analysis.

## Classification of Task Information Processing

The philosophy of SHERPA involves understanding the information processing involved in task performance, in order to facilitate error prediction and reduction. A number of different task classification schemes exist. For the purposes of analysing human reliability, SHERPA uses a classification scheme originally developed by Rasmussen, which has been adapted and extended for use in this context. The scheme is discussed in more detail in Rasmussen (1983). There are basically four types of mental processing represented in the classification.

Skill-Based (SB) Processing

This occurs in situations involving mainly simple physical operations of controls, or where equipment is manipulated from one position to another. In SB processing, the operator is sufficiently highly skilled or practised such that no conscious planning or monitoring is required to execute the actions.

## Rule-Based Diagnostic (RBD) Processing

RBD processing involves diagnosis where a pattern of symptoms is associated with a cause by the operator using an explicit "production rule" of the form IF <condition X> THEN <cause Y>. For example, if a car fails to start, RBD processing could require invoking the rule IF <starter motor is OK> AND <lights dim when operating ignition key> THEN <cause is flat battery>.

## Rule-Based Action (RBA) Processing

RBA is similar to RBD processing but the production rule directly connects a condition with an action or set of actions e.g. IF <situation X> THEN <do Y>. The situation that triggers the rule is often, but not invariably, made explicit as a result of prior RBD processing. Thus, RBD and RBA rules are often linked together in the following way:

        IF <Condition X> THEN <Situation Y>
        IF <Situation Y> THEN <do Z>

RBA processing generally occurs in situations where the operator is following an explicit procedure or rule of thumb, which may be written or memorised. SB processing is also associated in tasks with RBA processing, as it is required to actually execute the physical operations needed to carry out the task.

## Knowledge-Based (KB) Processing

KB processing is required if the operator cannot refer to any existing procedure or rule of thumb, and therefore has to go back to first principles and utilise his overall technical understanding of the situation.

The type of processing which is required for a particular task depends on the level of training and experience of the individual. Thus, for a beginner, all tasks involve a high proportion of KB processing, because he or she will frequently have to refer back to first principles. With practice, most tasks enter the RBD or RBA domain, and some very frequently encountered operations, e.g. manipulations of commonly used controls, will become highly automatised (i.e. will involve mainly SB processing).

## Application of the Information Processing Classification in SHERPA

The information processing classification is used in SHERPA in the following ways:

1. Defining the appropriate form of subsequent analyses. The qualitative error prediction technique used in the analysis module is applicable primarily to SB, RBD and RBA tasks. The KB dominated tasks must be dealt with using other approaches, which are currently under development.

2. Assisting in the identification of potential error modes during the error analysis. Characteristic error modes are associated with each of the information processing categories defined above. The prior classification of tasks therefore considerably facilitates the systematic postulation of possible error modes.

3. Defining the most effective error reduction strategy. Effective strategies can only be defined if the underlying root causes of errors are identified. These will depend on the types of information processing likely to be used in the task under consideration.

## Other Information Produced by the Task Analysis

1. Definition of the main task elements to be executed during the performance of the task. This is basically the primary path via which the operator is expected to achieve the task goals. Alternative paths are analysed in the error analysis.

2. Definition of the plans and the conditions which determine transfer of control between the levels in the goal hierarchy and within task elements at a particular level. This has particular relevance for the definition of training content, and the identification of certain types of errors.

3. Commentary on (1) and (2) with regard to their implications for procedures, training and equipment design.

This information is subsequently supplemented by insights gained during the error analysis, and used in the error reduction module.

## Human Error Analysis (HEA)

The HEA takes as its inputs the task elements identified in the HTA. For each task element the following analytical procedure is adopted:

1. Definition of external error modes.
   At task element it is assumed that one or more of 18 different error modes could occur. Examples of these are: action too early; right action on wrong object; information not obtained. These error modes are based on a taxonomy described in Pedersen (1985). A computer program is used to assist the analyst in deciding on which of the error modes is credible in the situation being examined.

2. Identification of psychological mechanisms.
   For each external error mode identified, the underlying psychological mechanism is identified. These mechanisms are also based on the error modelling work of Rasmussen. The derivation of psychological methods is important because it assists in prescribing appropriate error reduction strategies at later stages of the HEA. A computer program is again used to assist the analyst in this process.

3. Recovery Analysis.
   The ways in which error recovery can occur are analysed at this stage.

4. Consequence Analysis.
   The consequences of unrecovered errors are evaluated.

5. Development of error reduction strategies.
   Strategies to reduce the likelihood of the postulated errors are considered at this stage. These will include measures to reduce the initial likelihood of the error, together with methods for improving the probability of recovery.

The overall outcome of the HEA is a clearly documented evaluation of the errors likely to occur in the system, together with their expected consequences, and methods of control.

## THE QUANTIFICATION MODULE

The quantification approach used in SHERPA is SLIM-MAUD. SLIM-MAUD (Success Likelihood Index Methodology using Multi-Attribute Utility Decomposition) is a computer-based technique developed under the sponsorship of the U.S. Nuclear Regulatory Commission (NRC) and Brookhaven National Laboratory, to quantify human error probabilities (HEPs) (see Embrey et al, 1984).

Quantification within SHERPA is primarily used to evaluate the degree of error reduction that will be produced by the different error reduction strategies postulated in the HEA. This allows cost-benefit analyses to be performed to identify the most effective error reduction approach within given cost constraints.

## ERROR REDUCTION MODULE

The final module in SHERPA is concerned with applying the results of the previous two modules to obtain the most cost effective error reduction. As discussed previously, the outputs of the HTA and HEA provide very specific recommendations for the modifications to procedures, training and equipment design which will improve human reliability. After these recommendations are implemented, their effectiveness is monitored via a quality assurance phase.

## CONCLUSIONS

The SHERPA approach is a comprehensive methodology for human reliability prediction and optimisation in NPP. It also has the advantage of an integrated quantification and cost-benefit analysis capability.

SHERPA has been applied to several critical systems areas in nuclear power plants and the results so far indicate that it produces auditable results, that are perceived by operators, engineers and managers to be valid, useful and cost effective. In a recent test of the technique SHERPA successfully predicted between 78% and 100% of errors that had actually occurred on a range of procedures used in a large plant.

We are continuing to develop SHERPA, with particular emphasis on a more systematic consideration of the organisational and other socio-technical variables which have a considerable influence on reliability at all levels in a system. Nevertheless, even at its present stage of development, we believe that SHERPA has wide applicability and considerable promise.

REFERENCES

1. D.E. EMBREY, P.C. HUMPHREYS, E.A. ROSA, B. KIRWAN and K. REA, "SLIM-MAUD. An Approach to Assessing Human Error Probabilities using Structured Expert Judgement", Vol. I & II. NUREG/CR-3518. Brookhaven National Laboratory, Upton, New York, USA (1984).

2. O.M. PEDERSEN, "Human Risk Contributions in Process Industry", Report no. Risø-M-2513. Risø National Laboratory, DK-4000, Roskilde, Denmark (1985).

3. J. RASMUSSEN (1983), Skills, Rules and Knowledge: Signals, Signs and Symbols and Other Distinctions in Human Performance Models. IEEE Transactions on Systems, Man and Cybernetics, SMC-13 (3) 257-266.

4. A. SHEPHERD, "Hierarchical Task Analysis and Training Decisions", Programmed Learning and Educational Technology 22.2 162 (1984).

# MINOR ILLNESSES AND PERFORMANCE EFFICIENCY

## A. SMITH and K. COYLE

MRC Perceptual and Cognitive Performance Unit
University of Sussex, Brighton
Sussex  BN1 9QG

        Results from our research programme on the effects
of experimentally-induced respiratory virus infections
on performance have led to the following conclusions. First,
these minor illnesses have signficant effects on performance efficiency, although the exact nature of the impairment will depend on the type of virus and the task being
performed.   For instance, influenza impairs the ability
to attend and colds impair motor skills.   Second, performance impairments have also been found with sub-clinical
influenza infections and infections with certain cold
viruses.   These results have important implications for
occupational safety and efficiency.   Studies· in progress
are also outlined in the paper.

INTRODUCTION

        Recent evaluations of the importance of respiratory
disease show that acute infections and their consequences
account for a substantial proportion of all consultations
in general practice, and are the major cause of absence
from work and education.  Despite the frequency with which
such illnesses occur there has been no research on their
effects on the efficiency of performance.  This has, in
part, been due to the difficulties inherent in carrying
out such studies.  These difficulties have been overcome
in the present project by studying the effects of experimentally-induced infection and illness at the MRC Common
Cold Unit, Salisbury.  The aims of the present paper are
to describe the methodology used, report the results obtained so far, and to describe studies which are in progress or planned.
        Studies of naturally-occurring illnesses could provide
useful information but it is difficult to predict when
they will occur,and it is often unclear which virus pro-

duced the illness (there are over 200 viruses that produce colds). Such studies would also only enable one to examine the effects of clinical illnesses. It is possible that sub-clinical infections may also influence behaviour and these can only be identified by using the appropriate virological techniques. The methodology employed by the Common Cold Unit overcomes these problems. The crucial features of the routine may be summarised as follows:

(1) Volunteers stay at the Unit for a period of ten days, being housed in groups of two or three and isolated from outside contacts.

(2) There is a quarantine period of about 3 days prior to virus challenge. Any volunteers who develop colds in this period are excluded from the trial.

(3) Each volunteer is assessed daily by the Unit's clinician who assigns a score based on objective measures such as temperature and number of tissues used, and on the presence of other symptoms (see Beare & Reed 1977 for full details of the scoring system).

(4) Nasal washings are taken so that virus shedding can be measured. Pre- and post-challenge antibody status can be measured from blood samples taken at these times.

The behavioural effects of challenge with the following viruses have been studied (1) Rhinoviruses, (2) Coronavirus, (3) Respiratory syncytial viruses (the symptoms produced by these are, in adults, indistinguishable from a common cold), and (4) influenza A and influenza B viruses. The clinical pattern and pathogenesis of the illnesses produced by the different viruses is quite variable. However, the major distinction is between colds and influenza, with colds producing mainly local symptoms (an increase in nasal secretion) whereas influenza also gives rise to systemic effects,such as fever, myalgia and malaise.

Many people feel that they already know about the behavioural effects of colds and influenza, and argue that it is not necessary to carry out empirical research on this topic. One view is that, almost by definition, if you are ill then you will perform less efficiently than normal. Another is that the illnesses are very minor and any behavioural effects are probably too small or transitory to deserve consideration. Recent research on stress has shown that different types of stress produce different effects on performance. Different illnesses might, therefore, affect performance in different ways.

We have carried out experimental studies to test which of these alternative views is correct. Two main experimental methods have been used. The first has consisted

of the volunteers carrying out a battery of paper and pencil tests 4 times a day (at 8.00, 12.00, 17.00 and 22.00)
on nearly all the days of the trial. The following tests
have been used: (a) logical reasoning test, (b) search
and memory tests with low and high memory loads, (c) a
pegboard test, and (d) a semantic processing test. It
was important to examine performance at several times of
day for two reasons. First, it is well-established that
performance varies over the day (see Colquhoun 1971).
Second, we have found that there is diurnal variation in
the severity of symptoms, with increases in nasal secretion
and temperature being greatest in the early morning.

The second method of assessing performance consisted
of administering computerised tests once during the pre-
challenge period, once when symptoms were apparent in some
volunteers, and sometimes during the incubation period.
Subjects were always tested at the same time of day on
all occasions, although some were tested in the morning
and some in the afternoon. This meant that diurnal variation in the effects of the illnesses could be examined.
These computerised tests have all been widely used to study
the effects of stressors and drugs. They were selected
to enable us to assess a range of functions such as attention, memory and motor skills. The tests which have been
used are:-

(1) a variable fore-period simple reaction time task
(2) a five-choice serial reaction time task
(3) a numeric monitoring task (5's detection task)
(4) the Bakan vigilance task
(5) a pursuit tracking task
(6) a Sternberg memory scanning task
(7) stimulus and response set attention tasks
(8) a time estimation task
(9) a free recall task, involving both immediate and delayed recall.

RESULTS

(1) Effects of clinical illnesses

Smith et al. (in press) found that colds and influenza
have selective effects on performance. Influenza slowed
responses to stimuli appearing at irregular intervals but
had no effect on a tracking task. In contrast to this,
colds impaired tracking but had no effect on the attention
tasks (variable fore-period reaction time task and number
detection task). Smith et al. (submitted) confirmed these
results using different performance tests (the motor task
was the pegboard task and the attention task was the high
memory load version of the search task) and different cold

viruses.

Results from the five-choice serial reaction time task also show that activities involving movement of the hands are impaired by colds. In one study volunteers who developed colds following challenge with a coronavirus were significantly slower than uninfected volunteers. This result has been confirmed in a study using respiratory syncytial viruses.

Many types of behaviour involve both attentional and motor skills. The above results suggest that such activities may be vulnerable to the effects of colds and influenza. Studies using cognitive tasks have so far failed to show clear effects of colds or influenza. Further analyses are being carried out to examine individual differences in the effects of infection and illness on these tasks. Other moderating factors, such as the time of testing, are being taken into consideration. Indeed, we have some evidence that colds may amplify the diurnal changes in performance. In one study we examined the effects of changing the stimulus-response compatibility in a choice reaction time task (in the compatible condition stimuli presented on the left of the screen were responded to with the left hand, stimuli on the right with the right hand, whereas in the incompatible condition, stimuli on the left were responded to with the right hand and those on the right with the left hand). In the pre-challenge period subjects tested in the afternoon were faster but less accurate than those tested in the morning. In the post-challenge period, those with significant colds showed a much greater time of day effect than the uninfected subjects.

(2) Effects of sub-clinical illnesses

The effects of sub-clinical infection have been studied in two ways. First, performance has been examined during the incubation period, when by definition, no symptoms are apparent. Second, virological techniques have allowed us to sub-divide volunteers with no significant clinical illnesses into those who were infected with the virus and those who were not.

Smith et al. (submitted) found that volunteers with sub-clinical influenza infections were impaired on a search task. Performance was also impaired during the incubation period of this illness. However, we found no evidence of behavioural effects of sub-clinical colds on this task or on a pegboard task. However, a study using respiratory syncytial virus has shown impairments during the incubation period (the task was a five-choice serial reaction time task). Volunteers who had no significant clinical symptoms, but showed a significant antibody rise following virus challenge, were also slower on this task.

RESEARCH IN PROGRESS

(a) Mediators of the changes in performance

It is known that interferon is endogenously produced following infection with influenza virus. Interferon may produce systemic effects which could be responsible for the attentional deficits. If this is the case, then similar impairments should occur when volunteers are given an injection of interferon but no virus challenge. Preliminary studies have provided some support for this view.

(b) Drugs

A great deal of research is being carried out at the Common Cold Unit on prophylactic and therapeutic drugs. We are currently examining whether the drugs change the behavioural effects of the illnesses.

(c) Motor impairments

Studies are in progress to determine whether the motor impairments found with colds are due to peripheral effects, such as changes in muscle tone, or whether they reflect impairments in more central processes, such as response organisation.

(d) Prediction of illness from behavioural indicators

During the course of our research programme we have observed that volunteers who develop illnesses are sometimes worse in the pre-challenge period than those who are subsequently uninfected. This raises the interesting possibility that we may be able to predict illness from behavioural measures.

(e) New aspects of behaviour

We are currently examining the effects of infection and illness on cognitive functions such as comprehension and decision making.

PLANNED RESEARCH

(1) After-effects of illness

We intend to examine performance after the clinical symptoms have gone to see whether any impairments still exist, and if so, for how long. Changes in immunological indicators will also be taken and it will be possible to examine the relationship between performance and changes in immune system function.

CONCLUSIONS

Our research programme has shown that minor illnesses do have significant effects on performance efficiency. The exact effect will depend on the activity being performed and the type of virus. Viral infection, unaccompanied by clinical symptoms,can impair performance although these effects appear to be largely restricted to influenza infections and attention tasks. These results have strong implications for occupational safety and efficiency, and suggest that detailed studies of the impact of these illnesses in real-life should also be carried out.

REFERENCES

Beare,A. S. & Reed, S. E., 1977, The study of antiviral compounds in volunteers. In Chemoprophylaxis and Virus Infections, Vol. 2, edited by J. S. Oxford (Cleveland: CRC Press), p. 27 - 47.

Colquhoun, W. P., 1971, Circadian variation in mental efficiency. In Biological Rhythms and Human Performance, edited by W. P. Colquhoun (London: Academic Press), p. 39 - 107.

Smith, A. P., Tyrrell, D. A. J., Coyle, K. B. & Willman, J. S., Selective effects of minor illnesses on human performance. British Journal of Psychology (in the press).

Smith, A. P., Tyrrell, D. A. J., Al-Nakib, W., Coyle, K.B., Donovan, C. B., Higgins, P. G. & Willman, J. S. The effects of experimentally-induced respiratory virus infections on performance (paper submitted).

# THE USE OF PEOPLE TO SIMULATE MACHINES:
## AN ERGONOMIC APPROACH

**M.A. LIFE and J. LONG**

Ergonomics Unit
University College London, 26 Bedford Way
London   WC1H 0AP

People have been used successfully to simulate certain types of future machine in the absence of the technology required to produce them. This paper describes a simulation which demands a particularly close adherence by a human to a technological specification; it will be necessary to identify any limitations in the human's ability to simulate the target system and to provide aids to minimize them. An approach is described for applying ergonomic methods in optimizing performance of the task of the human simulator.

SIMULATION IN THE DEVELOPMENT OF TECHNOLOGY
   Experiments using simulations permit designers to evaluate products at early stages in development. Product development typically involves novel utilization of existing technologies, and so development simulations can normally be implemented using existing technologies. For example, a simulation for the development of the human interface for a manually guided missile system could be constructed using current computer technology (Evans & Scully, 1986).
   In the case of technology development, simulation offers a means of determining user requirements so that development is led by these, rather than by the evolution of an engineering concept. This feed-forward function is important because a failure to establish user requirements is likely to result in the technology producing unusable products. Unfortunately, it may be impossible or too expensive to simulate future technology using extant technology; however, for certain types of future system a solution lies in the use of people to simulate system behaviour. People are capable of adapting their behaviour to match a predefined specification, i.e. role-playing, and this paper describes an approach to

assessing the demands of simulation tasks and for identifying the need for modifications to ensure adequate fidelity.

USING PEOPLE TO SIMULATE MACHINES: PREVIOUS APPLICATIONS
   Human simulation experiments typically take the form
shown in Figure 1. A simulation is developed to study the
performance of users, a sample of which population will be
represented in an experiment by "user subjects" (USs). USs
operate the simulated system through a communication
interface, A, equivalent to that in the target system; for
example, A might consist of a keyboard terminal or a speech
interface. However, rather than operating a computer system,
the US is actually communicating with a person whose task is
to imitate the system (a "system subject", SS) via a second
interface, B. This paper is particularly concerned with the
requirements for interface B in supporting the task of SS.

Figure 1. A generalised human simulation.

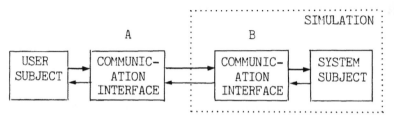

   Human simulation studies may be compared in terms of the
complexity of behaviour demanded of SS in order to represent
the target system. An example of a simulation requiring
relatively simple perceptual-motor behaviour of a person is
in the imitation of an isolated word speech recognizer.
Pullinger (1980) explored the user interface  requirements
for a videotex system (Prestel), comparing user performance
with a simulated speech interface and with a keypad. In the
former condition, SS listened for voiced commands from USs
and keyed them on their behalf. In this case, the behaviour
required to mimic the machine was recognition of spoken
numbers and translation of these to key-presses on a numeric
keypad. There was no physical instantiation of interface A in
this study, as the SS was hidden behind a screen close to US,
listening to commands directly. Interface B was a keypad.
   In contrast, the simulation of an expert system (ES) and
its user interface involves a number of complex cognitive
activities. Warren (1985) explored the interaction of users
with a simulated ES, and compared their performance using
natural language and command language front-ends. The target
ES provided a consultancy service relating to  computer
equipment, and the USs used the simulated machine to advise

them on a suitable system within a limited budget. The SS and USs communicated via keyboard terminals, and the SS simulated not only the language interface, but also the expert system itself. To perform the task, the SS therefore had to:

(a) interpret US communications in terms of the constraints imposed by the type of interface being simulated, (i.e. natural or command language)
(b) request further information to determine US requirements
(c) consult a knowledge base
(d) maintain a record of USs' intended purchases and budgets
(e) compose and key in replies constrained appropriately by the rules of the interface being simulated.

These two examples illustrate the range of applications in which human simulators have been utilized successfully, but there are many examples of implementations involving behaviour of intermediate levels of complexity; e.g. in speech transcription (Gould 1982); and speech interfaces for public information systems (Richards & Underwood, 1984).

In general, experiments using human simulation have been employed in studying the user behaviour elicited by general classes of technology and have not intended a close repres- entation of specific technology or actual products. Given these aims, fidelity is not a major issue, as claims are not made about performance with systems other than those which, coincidentally, have specifications generally similar to those presented in the simulations. However, fidelity becomes very important if the aim is to produce feed-forward inform- ation by determining the relationship between different implementations of future technology and user behaviour. The next section describes a study which imposes the requirement for such high fidelity human simulations.

DETERMINING USER REQUIREMENTS FOR FUTURE SPEECH INTERFACES

Although speech interfaces have been exploited in military applications (e.g. on aircraft flight-decks; Simpson et al., 1982) they have not been employed by the British Army for use on the land battlefield. One reason for this is that currently available devices do not meet the demands of operators. However, speech can offer unique benefits over other input/output methods (e.g. portability, operation in low ambient light and operation while the hands are otherwise occupied), and it is therefore likely that, with appropriate development, speech I/O could enhance the performance of certain classes of battlefield data communication task.

Research currently in progress at University College London (UCL) aims to develop methods for predicting the user requirements for speech interface devices in the context of specific tasks. This will enable those developing the tech- nology to direct their efforts most effectively. However, to achieve this it will be necessary to simulate a range of speech I/O devices in order to determine how various parameters of the specification, such as recognition error rate, vocabulary size and syntax constraints, influence performance. Experimentation using human simulation is a potentially useful approach, but the project demands high fidelity in the representation of target systems. It will therefore require a closer adherence of the human simulator to a technological specification than has been necessary in previous studies.

As the fidelity of the simulation will depend critically upon the performance of a human-machine system (the SS and interface B in Figure 1), it is appropriate that ergonomic methods be applied in its optimization. The next section describes an ergonomic approach to human simulation which will enable it to meet the rigorous demands of the study.

AN ERGONOMIC APPROACH TO HUMAN SIMULATION

Ergonomics seeks to optimize the relationship between people and work, and the ergonomist would typically achieve this by modification of tasks or by the provision of task aids: alternative measures might be modification of training or modification of personnel selection. Such intervention is initiated as a consequence of system evaluation against criteria such as operator peformance (e.g. speed or errors) and measures of operator comfort (both physical and social).

Ergonomic approaches have been applied in the optimization of human-computer interfaces. For example, Buckley & Long (1985) identified difficulties that users experienced in using a system, by observing the errors that they made. Errors were assumed to be an indication of incompatibility between users' existing knowledge representations and the "ideal knowledge" representations necessary to operate the system correctly. Buckley & Long showed that observations of errors may be used diagnostically (to identify the causes of incompatibility) and prescriptively (to suggest ways in which incompatibility might be reduced and hence performance improved). Ways to improve compatibility might be to change the system such that the "ideal knowledge" for error-free performance coincided with that held by the user, or to change the user's knowledge by means of training or documentation.

A similar approach may be applied to the case of a human simulating a machine: mismatches between the performance necessary for an adequate simulation and that actually achievable indicate a requirement for performance aids, training or selection of SSs against tighter criteria. In this instance, it would not be sufficient only to consider compatibility between the ideal knowledge required to simulate the target system and that held by the user: incompatibility might also be manifested in the dynamic behaviour of the simulation and in the ability of the SS to reproduce the physical signals (audible and visible) emitted by the target system.

To apply the approach would initially require specification of the parameters of the target system which are to be included in the simulation, and their values. A study would then be conducted to determine the difficulties encountered by the SS in meeting this specification unaided. These difficulties would be used to develop a model of the SS, expressed in terms of incompatibility with respect to knowledge, dynamic and physical representations. This would provide a basis for identifying appropriate interventions to achieve the performance demanded by the specification; these might take the form of aiding devices for SS, and/or training to reduce the incompatibility of SS representations, or alternatively procedures might be developed for selecting SSs with more compatible representations. The effect of changes would then be evaluated empirically to ensure adequate fidelity in the simulation.

Warren's study of an ES will be used to illustrate how SS's performance might be analysed using this model. In his simulation, incompatibility concerning knowledge might have occurred as a consequence of interference between the SS's knowledge of natural language and that necessary to simulate a command language interface. This might have been manifested as consistent errors in which SS generated responses appropriate to a natural language interaction rather than responses which were legal within the target interface. Incompatibility might have been reduced by means of training or by a representational aid such as a display of legal commands.

Incompatibility concerning dynamics might have occurred as a consequence of the SS's inability to respond at the same speed as the target expert system, perhaps due to inadequate keyboard operating skill. Again, incompatibility might have been reduced by means of training, by selecting SSs with fast performance or by providing the SS with a representational aid such as a high speed typist to transcribe SS spoken responses.

Incompatibility concerning physical signals was trivially present in the Warren study: the subjects had no means of communicating unaided in a way comparable with the target system, so the simulation had to be implemented with representative text displays and keyboards. A better, albeit hypothetical, illustration of incompatible physical representation would be a system intended to simulate a speech recognizer: incompatibility might occur due to the SS being a more effective decoder of speech signals than the target system, causing him not to generate representative word recognition errors. In this case the incompatibility could be reduced by introducing errors into the SS's transcription according to an appropriate set of rules of type and frequency.

DEVELOPMENT OF THE ERGONOMIC APPROACH
The ergonomic approach described here is currently being used at UCL in the development of simulations of speech interface devices, but, given this approach, the use of human simulations might be extended to new application domains. For example, many human tasks currently performed manually might in future be performed by autonomous or semi-autonomous robots, in which case the human operator will become a supervisor with responsibility for occasional manual intervention. Human simulation of autonomous elements in such systems provides a means of investigating experimentally appropriate allocations of function.

The title of this Conference is "Ergonomics Working for Society". If ergonomics is working for the society of the future it must provide feed-forward information to ensure that developing technologies produce products which are usable by people. Human simulation is a tool which could help ergonomics to achieve this goal.

ACNOWLEDGEMENT
This research is carried out for the Royal Signals and Radar Establishment under Contract No. 2047/127 (RSRE). Any views expressed in this paper are those of the authors and do not necessarily reflect those of the Ministry of Defence.

REFERENCES
Buckley, P. & Long, J.B., 1985. Identifying usability variables for teleshopping. In D. Oborne (ed.), Contemporary Ergonomics 1985. London: Taylor & Francis.
Evans, J.L. & Scully, D.C., 1986. The simulation of a ground to air guided weapon system: the requirement and a solution. IEE Second International Conference on Simulators, IEE Conference Publication no. 267, 192-196.
Gould, J.D., Conti, J. & Hovanyecz, T., 1982. Composing letters with a simulated listening typewriter. Proceedings: Human Factors in Computer Systems, Gaithersburg, Maryland 367-370.
Pullinger, D.J., 1980. Voice as a mode of instruction entry to viewdata systems. MSc dissertation, University of London.
Richards, M.A. & Underwood, K., Talking to machines: how are people naturally inclined to speak? In E.D. Megaw (ed.), Contemporary Ergonomics 1984. London: Taylor & Francis.
Simpson, C.A., Coler, C.R. & Huff, E.M., 1986. Human factors of voice I/O for aircraft cockpit controls and displays. Proceedings: Workshop on Standardization for Speech I/O Technology, Gaithersburg, Maryland. 159-166.
Warren, C.P., 1985. The Wizard of Oz Technique: a comparison between natural and command languages for communicating with expert systems. MSc dissertation, University of London.

# 1988

| Manchester University | 11th–15th April |
|---|---|
| **Conference Secretary** | Rachel Birnbaum |
| **Public Relations** | Carol Mason |
| **Honorary Meetings Secretary** | Jane Forsyth |
| **Chair of Meetings** | Rachel Benedyk |
| **Programme Secretary** | Ted Megaw |

| Chapter | Title | Author |
|---|---|---|
| **Structural Systems Development** | Soft systems approaches and their integration into the development process | P.B. Checkland |
| **Display Design** | Eye movements and the conspicuity of routing information | T. Boersema and H.J.G. Zwaga |
| **Impact of New Technology** | Introducing word processing to novice users: A study of 'procedural' and 'conceptual' approaches | M.A. Sasse, G.I. Johnson and P. Briggs |
| **Vehicle Ergonomics** | Sorry, can't talk ... just overtaking a lorry: The definition and experimental investigation of the problem of driving and handsfree carphone use | M. Boase, S. Hannigan and J.M. Porter |

# SOFT SYSTEMS APPROACHES AND THEIR INTEGRATION INTO THE DEVELOPMENT PROCESS

## P.B. CHECKLAND

Department of Systems
University of Lancaster
Bailrigg
Lancaster  LA1 4YX

Soft systems methodology, being neither primarily
technically oriented nor primarily 'human factor'
oriented, but being a means of treating the two
together, is a candidate for enriching structured
systems development.

## INTRODUCTION

Since computers used to be large pieces of apparatus
requiring their special air conditioned locations and a
new breed of expert to install and run them, it is not
surprising that "projects" to set up "data processing"
(later "information" or "decision support") systems should
take as given a model for project management taken casually
from the world of engineering.  This was always a simplified
representation of what sophisticated attention to
organisational information needs actually required, but it
was not a serious problem under the conditions of the 1960's.

Changes in computer technology and the development of
a cadre of computer professionals obsessed with technical
considerations have now created a serious problem, and the
question of information provision in organisations calls
for significant rethinking.

One approach would be that reflected in the title of
this Symposium : to add some consideration of human factors
to existing structured systems development.  Another
could be based upon the new soft systems approaches which
treat human and technical considerations together.  This
paper describes Soft Systems Methodology, one of the current
sources of the re-thinking of systems analysis and design
methodologies.

## SOFT SYSTEMS METHODOLOGY:   AN OVERVIEW

The intention here[*] is covered by the title: to give
a birds-eye view of soft systems methodology (SSM) as a
strategic framework guiding intervention in real-world
problem situations, neglecting all detailed technical
aspects in the interests of providing a simple shape and
rationale for the approach.   SSM's development and its
nature is described in more detail elsewhere (Checkland,1981).

Context
        SSM was developed in a programme of research based in
the postgraduate Department of Systems at Lancaster
University.   Systems studies have been carried out in many
organisations large and small in both public and private
sectors over the last seventeen years.   Many of the studies
were associated with a Masters Course in Systems in Management which involved many of the mature students (of average
age 30) in action projects.   The research strategy was to
use a defined systems methodology to try to bring about
changes in a real problem situation which people in the
situation regarded as 'improvements'.   Reflections upon
the experiences led to modification and refinement of the
methodology and to definition of new research themes.
        The starting point, methodologically, was to take 'hard'
systems engineering methodology as given (Checkland, 1981,
Chapter 5), to attempt to apply it in unsuitably 'soft'
(ill-structured problem situations, and to learn from the
ensuing experiences.   SSM is the redefined methodology
which emerged from this process.

Systems Thinking and the process of SSM
        The general stance behind the research was that
'problems' are endemic in human affairs, cannot be 'solved'
once-and-for-all, and call for a process-oriented rather than
a technique-oriented approach.
        Systems ideas were regarded as potentially useful
since our intuitive knowledge of the world suggests that it
is densely interconnected.   Both its stability and its
continual change are problematical, and systems ideas can
potentially cope with both aspects.   In the basic system
image or metaphor a whole entity, showing emergent properties
and layered, or hierachical structure, may survive in a

* A version of this paper was given at a plenary session
    of the 31st Annual Meeting of ISGSR, Budapest, June 1987

changing environment by processes depending upon communication and control (Checkland, 1981, Chapter 3).

In the SSM research programme the concepts 'natural system' which might be mapped onto wholes created by Nature, and 'designed system' which might map man-made wholes, whether physical or abstract, were found to be useful in providing general insights into systemic behaviour, but not tightly relevant to real problems in human affairs. To cope with the latter, and because every problem situation was characterised by human beings trying to take purposeful action, a concept of 'human activity system' was developed (Checkland, 1981, Chapter 4). A human activity system is a concept of a purposeful 'machine' consisting of activities so linked that they constitute a purposeful whole. Just as natural system models and designed system models might map real world wholes – at least to the extent of illuminating the latter – so the hope was that models of human activity systems might be useful in elucidating actual real-world purposeful activity. Techniques for defining human activity systems and building models of them were developed (Checkland 1981, Chapter 6).

It was discovered that in defining a human activity system as a pure purposeful machine to carry out a transformation process, it was always necessary to declare the Weltanschauung which makes that description meaningful. This is because there are always multiple possible descriptions of purposeful activity. If we were describing purposeful systems which might be used to explore the nature of a football match, for example, it might be regarded as a system for displaying skill, a system for providing entertainment, a system for making people tired, a system for churning up a piece of ground, a system for following rules ....etc.

A model of a human activity system thus consists of a structured set of activities linked together to create a purposeful whole within the declared Weltanschauung. Techniques for formulating the root definition of such a system and building the model have been developed (Checkland 1981, Chapter 6; Appendix 1). Real-world action is always much more complex than the structured activity in the model of a human activity system, and a number of models built according to different viewpoints will have to be constructed in a systems study if the richness of the real situation is to be embraced.

Having broadly explored a problem situation, with its issues and tasks, and selected and modelled some human activity systems relevant to exploring it more deeply, the cluster of models can now be used to structure a debate

about change.    The models are compared with perceptions
of real-world action, the comparison providing the
structure of a dialectical debate, a debate which will
change perceptions of the problem situation, suggest new
ideas for relevant systems (leading to iteration), and
concentrate thought on possible changes.

    What has been described constitutes activity in the
right hand side of Figure 1:    a problem situation is
entered and explored;    human activity systems relevant to
debating its improvement are named and modelled;
comparison between models and perceived reality leads to
definition of possible changes.    All of this is logical,
and the guidelines described can enable conservative or
radical changes to be defined depending upon the kind of
choices made of root definitions.    But for the approach
truly to engage with the realities of human affairs, a
second stream of exploration is needed, that which explores
the human and social aspects of the situation.    This is
shown on the left hand side of Figure 1.    It is less well
developed than the right hand side, but is the focus of
current work.

    Firstly, it has to be accepted that every human
problem situation is the product of a history, one which
will dictate perceptions, judgements and standards.    That
history has to be discussed.    Secondly, an analysis of
the intervention in the situation examines who caused the
intervention, who wishes to do the work (the would-be
'problem solvers') and who could be taken to be 'problem
owners'.    This last list is itself a useful source of
ideas for relevant systems(Checkland 1981, Chapter 7).
Finally the situation needs to be examined in its social
and political aspects.    Ways of doing this have recently
been defined and tried out (Checkland, 1986).

    This stream of exploration both aids choice of relevant
systems and powerfully feeds the debate about change.
What is culturally feasible is as important as what is
systemically desirable;    equally, the debate itself changes
perceptions and readinesses to contemplate actions of
particular kinds, which emerge and develop as the debate
proceeds.

    In one sense the process ends with action to improve
the situation emerging from the debate structured by the
comparison stage;    but of course the decision to act
provides a new problem situation - the task now being to
implement the action decided - and the whole cyclic
learning process can begin again.    SSM, in fact, is a
systems-based learning system, one providing guidelines for
coping with real world complexity.

SSM:   The overall learning

    After several hundred studies by its originators and by other groups in many countries who have taken it up, the overall learning from the development of SSM can be briefly summarized in a few propositions, as follows:

1.    In SSM 'the system' is not some part of the real world but is the organized process of inquiry itself.

2.    The concept 'system' is not a label-word for something in the perceived world, but is an abstract concept which can be used to help make sense of the world.

3.    Given its nature as a system of inquiry, the use of SSM has to be participative.   The role of the 'expert' in SSM is restricted to helping the people in the situation carry out their own study.

4.    Thus users of SSM should not try to create and preserve its status as a body of professional knowledge.  They should be trying to give it away to people in problematical situations.

REFERENCES

Checkland, Peter., Systems Thinking, Systems Practice
               (J. Wiley, 1981)
Checkland, Peter., 'The Politics of Practice', International
    Roundtable on The Art and Science of Systems Practice,
    IIASA, November, 1986

# EYE MOVEMENTS AND THE CONSPICUITY
## OF ROUTING INFORMATION

## T. BOERSEMA[*] and H.J.G. ZWAGA[**]

[*] Department of Industrial Design Engineering
Delft University of Technology
Jaffalaan 9, 2628 BX Delft

[**] Psychology Laboratory
University of Utrecht
Sorbonnelaan 16, 3584 CA Utrecht
The Netherlands

An experiment is described to measure the
distracting effects of advertisements on the
conspicuity of routing signs. Subjects were
presented slides of railway station scenes, their
task being to search for a target word in the
routing signs shown in the scene. Eye movements
were recorded to measure search time. Search time
appeared to increase significantly with the number
of advertisements in two of three experimental
scenes.

INTRODUCTION

In the presentation of routing information for
pedestrians in places such as airport terminals, railway
stations, and shopping centres many questions remain
unanswered concerning practical methods of ensuring the
conspicuity of that information. The conspicuity of an
object can be operationally defined in terms of the
probability that the object will be noticed by an observer
within a fixed time or, conversely, in terms of the time
required to locate the object. Numerous studies have shown
that the conspicuity of an object is determined by
properties of the object itself, by properties of
surrounding objects, and also by the cognitive state of the
observer.

Several studies on the distracting effects of
background stimuli have been carried out within a practical
context. Some examples are:
- directional traffic signs embedded in photographed street
scenes were responded to more slowly in more complex scenes
(Shoptaugh and Whitaker, 1984);

- specific planted targets were reported by car driving
subjects least often in the clutter of shopping areas, more
often in arterial road sections, and most often in
residential road sections (Cole and Hughes, 1984);
- an advertisement inserted in slides of railway station
scenes decreased locating performance for routing signs, an
effect which was stronger with more or larger advertisements
(Boersema and Zwaga, 1985).
The experiment reported here also concerns the distracting
effects of surrounding advertisements on the conspicuity of
routing information.

Although the locating performance score used by Boersema
and Zwaga (1985) proved to be a reliable and sensitive
measure of conspicuity, keeping the objectivity of the
scoring at an acceptable level required a laborious scoring
procedure. In an unpublished study an attempt was made to
avoid these problems by using a reaction time measure in
which the task was to locate a target word in the routing
information and to report on the direction of the associated
arrow. Though previously used with success in a comparable
setting by Dewar et al. (1976) this measure appeared
insufficiently sensitive to determine the distracting
effects of advertisements. To explain this negative result
it can be argued that the reaction time consisted of two
parts: the time a subject needed to find the target routing
sign (search time) and the time used for reading this
routing sign and deciding on the response (processing time).
This search time should systematically increase with the
number of advertisements in a scene, while the processing
time should remain constant. The finding that the reaction
time did not increase significantly with the number of
advertisements can only be explained by assuming that an
increase was obscured by the accumulated effects of variance
in search and processing times. The use of eye movement
recordings to distinguish between search time and processing
time will make it possible to verify these assumptions.

METHOD
In the experiment colour slides showing scenes in railway
stations and other public buildings were used as stimulus
material. All routing signs in the scenes were blue with
white lettering. Six of the slides were filler slides and
showed randomly chosen scenes containing routing information
as well as advertisements. The other were experimental
slides in which the number of advertisements was
systematically varied using photographic techniques. The
experimental slides were made from three scenes (A, B, and
C). From each of these, three types of experimental slides

Figure 1.   The three experimental slides from Scene A with zero (a), one (b), and three (c) advertisements.

a

b

c

were obtained with, respectively, zero, one, and three
advertisements. Figure 1 shows prints of the three
experimental slides from Scene A. The scene in Figure 1 is
representative of the kind of environments used with regard
to complexity, transparency, crowdedness, etc. In each of
the experimental scenes the advertisements were about twice
the size of a routing sign, which is smaller than
advertisements normally encountered in those circumstances.
In general, the arrangement of routing signs and
advertisements in Scenes B and C was the same as in Scene A
shown in Figure 1.

There were three experimental conditions, one for each of
the three types of experimental slides. After six practice
trials, each subject was shown nine slides: three
experimental slides, all of one type, and six filler slides.
Each scene slide was preceded by a slide with a target word
(a destination). By pressing a button the subject made the
target slide disappear and started a two second presentation
of a slide with a fixation cross. This slide was followed by
the presentation of the scene slide. The subject's task was
to find as quickly as possible both the target word and the
arrow associated with the target word in the scene's routing
information, then to press the button again to stop the
presentation of the slide, and finally to report verbally
the direction of the arrow. Reaction time and eye movements
were recorded. Search time was defined as the time from the
start of a scene slide till the beginning of the first
fixation on a routing sign. Processing time was the
difference between reaction time and search time. A Demel
Debic 84 eye marker with an accuracy of one degree was used.

Data were collected from three independent groups of
subjects (N=54). All subjects had uncorrected and at least
normal vision.

RESULTS
Prior to the analysis of the data from the experimental
scenes the three independent groups were compared on their
results from the non-experimental scenes. These data showed
no significant differences between the groups on any of the
three dependent variables.

For each of the three experimental scenes results and
statistical analyses are presented separately because of the
uncontrolled qualitative differences between the scenes. The
data were analysed according to the nonparametric approach
developed by Meddis (1984). Complete data were collected
from twelve to seventeen subjects depending on condition and
scene.

The results of Scenes A, B, and C are presented in

Table 1.  Median search times (st), processing times (pt), and reaction times (rt) for Scenes A, B, and C under conditions 0, 1, and 3 are presented (in seconds). The numbers of subjects with complete data (n) are indicated.

|  |  | Condition 0 | | Condition 1 | | Condition 3 | |
|---|---|---|---|---|---|---|---|
|  |  | Med. | n | Med. | n | Med. | n |
| Scene A |  |  | 12 |  | 15 |  | 15 |
|  | st | 0.63 |  | 0.91 |  | 1.13 |  |
|  | pt | 1.40 |  | 1.43 |  | 1.32 |  |
|  | rt | 2.01 |  | 2.29 |  | 2.57 |  |
| Scene B |  |  | 16 |  | 14 |  | 15 |
|  | st | 0.59 |  | 0.57 |  | 0.64 |  |
|  | pt | 0.79 |  | 0.88 |  | 0.74 |  |
|  | rt | 1.32 |  | 1.47 |  | 1.39 |  |
| Scene C |  |  | 17 |  | 15 |  | 15 |
|  | st | 0.59 |  | 0.73 |  | 0.76 |  |
|  | pt | 1.30 |  | 1.58 |  | 2.23 |  |
|  | rt | 2.17 |  | 2.31 |  | 3.02 |  |

Table 1. Analysis of the results shows that in Scenes A and C, as predicted, search times increase with the number of advertisements (p's<0.01), while processing times and reaction times remain constant (p's>0.05). In Scene B adding advertisements has no effect on search time (p>0.05). Again, processing time and reaction time do not change over conditions (p's>0.05). Typical characteristics of the subjects' scan paths during search are:
- fixations on advertisements occur only incidentally;
- for all scenes and all conditions the first saccade is always straight upwards from the fixation cross;
- within each scene the scan patterns are very similar.

DISCUSSION
    For two of the three constructed scenes (A and C) the results are as expected: search time varies systematically with the number of advertisements, whereas processing time and reaction time do not vary. Apparently, search time alone, and not search time in combination with processing time (i.e. reaction time), can be used effectively to measure the distracting effects of advertisements in a scene. It seems that the observed increase in search time

does not result from additional time spent actually fixating advertisements. The increase appears to be caused by a rise in the number or duration of fixations in the scene, not necessarily on the advertisements themselves. Introducing advertisements appears simply to result in a less efficient scanning of the scene.

It is not clear why there is no effect on the search time in Scene B. This scene was constructed using the same rules as for the other two. Even after the experiment, knowing the results, it is impossible to point to essential differences between Scene B and Scenes A and C. Therefore, accepting the validity of the procedure, it has to be concluded that there are scenes in which advertisements can be inserted without diminishing the conspicuity of routing signs. As mentioned in the Method the advertisements were small, only two times the size of the routing signs. Considering this point the effect on search time in Scenes A and C allows the conclusion that search time can be a sensitive measure of the conspicuity of routing information.

## REFERENCES

Boersema, T. and Zwaga, H.J.G., 1985, The influence of advertisements on the conspicuity of routing information, Applied Ergonomics, 16, 267-273.

Cole, B.L. and Hughes, P.K., 1984, A field trial of attention and search conspicuity, Human Factors, 26, 299-313.

Dewar, R.E., Ells, J.G., and Mundy, G., 1976, Reaction time as an index of traffic sign perception, Human Factors, 18, 381-392.

Meddis, R., 1984, Statistics using ranks (London: Basil Blackwell).

Shoptaugh, F.C. and Whitaker, L.A., 1984, Verbal responses to directional traffic signs embedded in photographic street scenes, Human Factors, 26, 235-244.

ACKNOWLEDGEMENT

The authors wish to thank Austin S. Adams for his helpful comments on the final draft of this paper.

# INTRODUCING WORD PROCESSING TO NOVICE USERS:
## A STUDY OF 'PROCEDURAL' AND 'CONCEPTUAL' APPROACHES

## M.A. SASSE[*], G.I. JOHNSON[**] and P. BRIGGS[***]

[*] Department of Computer Science
University of Birmingham, Birmingham  B15 2TT

[**] MRC/ESRC Social and Applied Psychology Unit
University of Sheffield, Sheffield  S10 2TN

[***] Department of Psychology
City of London Polytechnic, London  E1 7NT

An experimental study of word processor (WP) tutoring is described. Conceptual (˜top-down˜) and procedural (˜bottom-up˜) tutoring approaches were evaluated. These different approaches were compared with respect to time spent on training and performance tasks and verbal protocol data. Users˜ errors were also analysed in order to identify areas in training and system design for improvement. Generally, the results demonstrated that conceptual approaches to WP tutoring are more efficient than procedural approaches.

INTRODUCTION

It is generally agreed that the training requirements of novice users of computer-based systems differ greatly from those of computer professionals, although research in computer education has tended to concentrate on the latter (Carroll and Caruthers, 1984). Yet there are many issues in this area that still need to be addressed: How should novice users be introduced to the systems and packages they want to use; how should the necessary information be presented; and which kind of training or tutoring best supports the learning process? (see Allwood, 1986).

Word processors (WP) are ideal for investigating novices˜ learning and understanding of computer-based systems as they are a perfect example of systems that are merely ˜instrumental˜ to those who use them. People usually learn to use WPs because they want to produce letters, reports, etc. quickly, not because they have a great desire to understand computing. Also, it is significant that the number of jobs requiring WP ability outnumbers all other jobs that involve computer use.

The typical and most common way of learning to use a WP package is via handbooks and self-instruction manuals which place learners in a role which is effectively passive. However, learning is fundamentally an active process. Manual authors often seem to hold a ˉtabula rasaˉ view of the learner, and yet users can and do bring their own goals, experience and expectations to the learning process (Carroll and Rosson, 1987).

In recent years, alternative approaches to training have been investigated. These include computer based ˉlearning by doingˉ methods, such as the ˉtraining wheels systemˉ of Carroll and Caruthers (1984), and guided exploration (see Carroll et al., 1985).

In this study personal tutoring was adopted, and within this, two approaches to training design were contrasted. The advantages of personal tutoring include flexibility (tutors can adapt to users), and assistance in explanation building. As Lewis (1986) has noted, explanations are an important mediating structure in learning to use a system, and the incidence of undetected errors can be reduced if the workings of the explanation process can be controlled or anticipated.

The objective of this study was to investigate differences between two approaches to the teaching of WP skills, namely a conceptual and a procedural approach (for a full description of the study see Sasse, 1986). The conceptual or ˉtop-downˉ strategy seeks to provide the learner with an understanding of the overall structure of the system, whereas the procedural or ˉbottom-upˉ strategy concentrates on teaching the operations (or procedures) needed to use the WP package in a sequential manner. It is this latter approach which is often found in user guides and manuals.

Monitoring paradigms, of the kind described in this study, are not uncommon in the field of human-computer interaction research, since they provide an empirical basis for exploring relationships between user knowledge, learning and system use (Hammond et al., 1982).

METHOD
Subject
Twelve subjects (four male and eight female) agreed to participate in the experiment. They were all volunteer

graduate and post-graduate students at the University of
Sheffield.  None of the subjects had used a WP before.
Using a random selection procedure, half of the subjects
were allocated to the conceptual learning condition, and
half to the procedural learning condition.

## Materials and apparatus

The WP package selected for the investigation was
Superwriter Version 3.2, running on an Apricot dual disc
drive personal computer.

Fourteen training tasks, plus one final transfer task
formed a basis for subject training and evaluation.  These
tasks covered various editing procedures (cursor movement,
deletion, formatting) plus the essentials of creating,
saving and printing files.

Slightly different training materials were devised for
the two different training methods.  For the conceptual
method, subjects were given a menu tree which illustrated
the hierarchical command structure of the interface.  For
the procedural method, in which subjects were taught
specific command sequences with each task as the need arose,
subjects were given ˉmemory cardsˉ – brief summaries of the
information learned at each point.

## Procedure

Training was given in individual sessions.  At the
beginning of each session the trainer conducted a short
interview with each subject.  The subjects in the conceptual
group were then given the menu tree, and had the interface
command structure explained to them.  The subjects in the
procedural group were given the first instruction sheet
immediately.

At the beginning of each task the trainer handed out the
instruction sheet and answered any queries which arose.
Subjects were asked to signal to the trainer when they were
ready to start the task, and when they had finished.  Time
on task was then recorded.  Trainers also kept a record of
the subjectsˉ use of the on-line help function; the menu
tree (for the conceptual group) or the menu cards
(procedural group).  In addition, the entire session was
audio-taped.  Each subject was given a full dayˉs training,
with a break for lunch.

At the end of the day subjects in both conditions were
given an identical ˉtransfer taskˉ designed to assess their
skill in the use of the word-processing package.

RESULTS

The training sessions were analysed in terms of time spent on learning the system, performance on the tasks, use of the help facility and backup materials, and errors made.

(a) Training times.

Total training times for both groups were calculated by summing the times recorded for each of the individual tasks. Subjects in the conceptual group completed their training significantly faster than those in the procedural group (t=2.851, tcrit=2.228, p<0.05). However, their performance on the final transfer task was not shown to be significantly faster (see Table 1).

Table 1: Training times and time spent on final task

|  | Training | Final task |
|---|---|---|
| Conceptual | 1 hr  42 mins | 30 mins 24 secs |
| Procedural | 2 hrs 19 mins | 36 mins 30 secs |

(b) Performance on tasks.

Degree of task completion was assessed by the trainer, using a four point rating scale. There were no significant differences in performance between the two groups.

(c) Use of help and backup.

The procedural group made significantly more use of the on-line help facility during training (t=2.526, tcrit=2.228, p<0.05). This was not due to a preference for on-line help over off-line backup, since this group also tended to make more use of their memory cards (however, this difference was not significant). These results are shown below in Table 2.

Table 2: Use of help facility and memory aids

|  | Help facility | Memory aids |
|---|---|---|
| Conceptual | 55 | 42 (menu tree) |
| Procedural | 93 | 56 (memory cards) |

(no. of references made)

(d) Errors.

Four categories of error were identified: slips; cursor control; command-execution; and command selection. Forty-three percent of all errors fell into this last category, i.e. they were caused by the subject selecting the wrong command for the operation required. Although subjects in the bottom-up category tended to make more errors, the differences were not statistically significant (see Table 3).

Table 3:   Analysis of learner's errors

|  | procedural | conceptual | totals |
|---|---|---|---|
| Slips | 14 | 6 | 20 |
| Cursor Control | 10 | 13 | 23 |
| Mode | 36 | 22 | 58 |
| Command-Execution | 41 | 39 | 80 |
| Command-Selection | 71 | 67 | 138 |
| Totals | 172 | 147 | 319 |

DISCUSSION

Although there were no significant differences between groups in terms of task performance or recorded errors, subjects in the conceptual group did complete their training significantly faster than subjects in the procedural group. They also made fewer references to the on-line help system. This suggests that a top-down, or conceptual approach might be a more efficient, but not necessarily more effective means of word-processor training. Given the high cost of implementing training programmes within organisations, this may be a significant factor in deciding between the two strategies.

The analysis of errors revealed that both groups found it difficult to associate command names with the actions they were trying to perform. This suggests that, for this WP system command sequences were not meaningful for the users, which made the learning process more difficult. This conclusion is supported by an analysis of the protocols generated during the study. It was interesting to note that subjects would tend to make assumptions about the system which were reasonable, given the evidence available to them, but were nevertheless ˉwrongˉ, leading to errors which might have easily been avoided, given more thoughtful interface design.

REFERENCES

Allwood, C.M., 1986, Novices on the computer: a review of the literature, International Journal of Man Machine Studies, 25, 633–658.

Carroll, J.M., 1984, Minimalist Training, Datamation, 30, 125–136.

Carroll, J.M. and Mack, R.L., 1983, Actively learning to use a word processor. Cognitive Aspects of Skilled Typewriting, ed. W.E. Cooper (New York: Springer).

Carroll, J.M. and Caruthers, C., 1984, Blocking learner error states in a training-wheel system, Human Factors, 26, 377–389.

Carroll, J.M., Mack, R.L., Lewis, C.H., Grischkowsky, N.L. and Robertson, S.R., 1985, Exploring a word processor, Human–Computer Interaction, 1, 283–307.

Carroll, J.M. and Rosson, M.B., 1987, Paradox of the active user. In Interfacing thought: cognitive aspects of human–computer interaction, ed. J.M. Carroll (Cambridge Mass: MIT Press).

Hammond, N., Morton, J., Maclean, A., Barnard, P. and Long, J., 1982, Knowledge fragments and usersˉ models of systems. Hursley Human Factors Report HF071.

Lewis, C., 1986, Understanding whatˉs happening in system interactions. In User Centred Systems Design: New Perspectives in Human Computer Interaction, eds. D.A. Norman and S.W. Draper (Hillsdale, N.J.: Erlbaum).

Sasse, M.A., 1986, Design and evaluation of a training program for word processor use: an experimental study. M.Sc. Thesis (unpublished), MRC/ESRC SAPU, University of Sheffield.

# SORRY, CAN'T TALK ... JUST OVERTAKING A LORRY: THE DEFINITION AND EXPERIMENTAL INVESTIGATION OF THE PROBLEM OF DRIVING AND HANDSFREE CARPHONE USE

## M. BOASE, S. HANNIGAN and J.M. PORTER

Department of Human Sciences
University of Technology
Loughborough
Leicestershire  LE11 3TU

This study investigated the interactional effects of driving and talking using a handsfree carphone.  Nine handsfree users were interviewed to provide preliminary information for the design of a laboratory study.  The experimental data show that the quality of a complex dialogue suffered at higher simulated driving workloads.  Simulated driving performance was also adversely affected with both simple and, to a greater extent, complex dialogues.  These findings contrast with the comments from many users who state that their driving does not suffer because they have the option to reduce driving speed, to not answer a call or to stay silent during a conversation.

## INTRODUCTION

Information technology is now pervading all aspects of society for business and leisure purposes.  Car telephones have existed for many years, but the advent of cellular radio in 1985 massively increased their availability to business and private motorists.  New systems ensure that this trend will continue.  Webster (1987) suggests that 2.1 million workers could benefit from such a service.

At the same time European Governments are starting to increase the level of awareness about road safety.  The accident figures behind this are alarming; 5165 fatalities and 70,980 serious injuries in Britain in 1985.  Efforts to reduce these figures appear to conflict with the introduction of new technology, starting with carphones.  Accordingly the Highway Code has effectively banned the use

of telephone handsets while driving.  Telephone
conversation, however, can take place via 'handsfree'
telephones incorporating a microphone mounted near the
sun-visor and a loudspeaker in the handset.  Both hands are
then available for driving.  An investigation into
handsfree telephones (Crown 1987) points out that the act
of engaging in the conversation itself may be dangerous.

AIMS OF THE PROJECT
    This project aimed to investigate the interaction
between the primary and secondary tasks of driving and
handsfree (HF) conversation respectively, but omitting the
manual and visual tasks of call set-up.  This was achieved
by interviewing users about their driving and
conversational ability by interview in order to define the
problems and to provide a basis for the experimental study.

INTERVIEWS OF HF USERS
Introduction
    Nine semi-structured interviews of regular HF users
were undertaken to look into the task variability,
discover types of conversation held and build a profile of
user behaviour when using the HF carphone.

Results and Discussion
    The users fell into two groups:  5 executives and 4
salesmen.  The mean number of calls was 37 per week.
Outgoing calls were 80% and 45% of the total for
executives and salesmen respectively, 60% of calls
involved the users' workbase in order to answer queries,
brief staff etc., 35% of calls were to or from clients to
plan meetings, give requests for stock delivery etc., or
to negotiate sales.
    The sample size, although small, was representative of
the user population in terms of job roles, sex and age
range (Ward, 1987).  The main findings that were relevant
to the laboratory based experiment were as follows:-
1. Call length was about 2-3 minutes.  2. Conversations
fell into two categories:  the most common being simple
question and answer type, e.g. making appointments,
checking on dates etc., and the second being more complex,
involving deals and negotiations.  Most users did not
complete deals but just kept them 'on the boil'.  For this
report, the simple type was called Information Dialogue
(ID) after Beun (1985), and the complex type was called
Negotiation Dialogue (ND).  3. Drivers were happy to get
involved in ID but most felt that their ability to
negotiate was reduced.  4. Seven people said that the

carphone was now essential to their job and it gave them up to two hours extra per day. 5. Users stated that their driving performance did not suffer because they had the option of either slowing down, not answering or making a call, ringing off, or not talking in a difficult road situation. But driving can be considered to be a forced pace task (Johnston and Perry 1980), and thus we might predict that where the driving task becomes demanding there might well be decrements in performance of driving, talking or both.

EXPERIMENTAL STUDY
    A laboratory-based experiment was chosen for the following reasons:- (a) in order to remove the variation in the signal to noise ratio and interference which both affect reception, (b) safety: the study included telephoning at high driving workloads, (c) to eliminate environmental variables, e.g. weather and road conditions, (d) drivers may choose to slow down when answering a call and this can mask any deterioration in task performance. This is avoided by presenting a forced-paced task.
    The experiment was designed to investigate the interactions between the primary task of simulated driving and the secondary task of two types of handsfree telephone conversation, ID and ND. This was for three levels of difficulty, i.e. workloads (WLs), of the simulated driving task. 24 subjects were used:- 6 young male adults, 6 young female adults, 6 older male adults, 6 older female adults. Each experimental session involved two subjects and the experimenter. One subject (S1) would sit at a computer with a game simulating driving, and an HF telephone beside him. The other subject (S2) took the role of an office worker and telephoned S1 and involved him in five separate IDs: the first and last ID before and after S1 had played the computer game, and the middle three occurred at different game WLs of low, high and mixed. Presentation order was balanced carefully. Each conversation was taped and scores were obtained for the computer game when it was played on its own (single task) and while the ID was occurring (dual task). S1 also answered 4 questions at the end of each conversation about the ease of conversation and driving and whether he felt that his performance at either had been affected. This same pattern was repeated with the experimenter taking the role of the office worker and involving S1 in ND.

## Development of the Computer Game

A 'squash' type game was used, and once adapted this involved tracking, prediction and some decision making, which are all aspects of the driving task. The game ran for 18 minutes through low, high and mixed WLs which could be arranged in any order. The WLs were calibrated prior to the experiment according to the subject's ability, which was determined after their learning curve had flattened. These WLs represented different driving conditions; low: an open road with time to plan ahead, high: a busy 'A' road or motorway, mixed: an unpredictable situation like a busy shopping centre or a motorway in fog. The game was scored by looking at balls hit as a percentage of total hits possible in the 90 second periods before and after the start of the dual task situation, (omitting the 30 second band in which the telephone call began).

## Development of the Conversation

Simulated salesmen's and executives' conversations were not suitable for use by ordinary subjects, and no suitable telephone tasks were found in the literature. Six simple question and answer conversations were then devised for the ID using topics that would be familiar to all subjects, e.g. favourite foods or past education. The NDs were devised by informally asking colleagues about real situations where they had been involved in negotiations in which they had to come to a satisfactory solution. Six NDs were then devised from the most common of these, e.g. returning faulty goods to a shop or having a holiday booking altered by the company. All IDs and NDs were piloted prior to the experimental situation and improved accordingly.

The conversations were scored by measuring: (a) mean conversation length in seconds, (b) mean utterance length in words, (c) mean pause length in seconds. This analysis was carried out for the two control conversations (single task performance), and for those conversations in the low and mixed WLs (dual task performance).

## Results and Discussion

1. Simulated driving performance (SDP) deteriorates significantly at the high and mixed WLs when conversation is introduced (approx. by 11%) and non-significantly at the low WL (3.5%). 2. ID produced a greater decrement (14.5%) in SDP than did ND (6.5%). 3. Age differences occur in decrements in SDP in the dual task situation with older subjects performing significantly worse in the mixed

Figure 1.    Mean computer game scores (% balls missed) for each workload for the periods preceding and following the start of the telephone call (0 seconds).

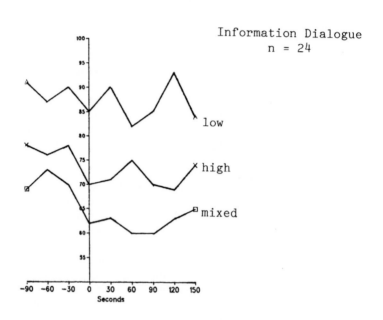

Information Dialogue
n = 24

low

high

mixed

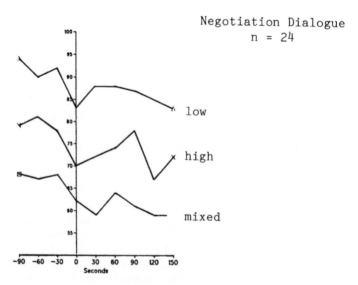

Negotiation Dialogue
n = 24

low

high

mixed

WL.  No consistent differences were found between males
and females.  4. Conversation and Pause Length increased
significantly in the dual task situation.  All alterations
in conversational performance are probably not highlighted
by these simple measures, but clues that ability to
negotiate is reduced are present in the tape recordings.
5. Subjects are good assessors of their game-playing
ability as indicated by game scores and questionnaire
responses.

DISCUSSION
    The main difference between the above two studies is
the way people rate their performance.  Drivers believe
that their driving does not deteriorate when using the HF
phone because they have the option of not participating in
a call and they also can reduce their WL by slowing down.
In the experimental situation, however, subjects did
perceive and state that their game-playing ability
deteriorated.  Drivers may well not receive such clear cut
feedback on their performance as the game-players who had
the feedback of lost balls.  Driving therefore may
deteriorate while using HF, and so drivers' opinions that
they are safe to drive and talk may be misguided.  Further
research is clearly required to investigate this important
safety issue.

REFERENCES

Beun, R.J., 1985, The Function of Repetitions in
    Information Dialogue, J.P.O. Annual Progress Report,
    20, 91-98.
Crown, J., Hannigan, S., Ward, C., 1987, An Experimental
    Investigation of Intelligibility and Acceptability of
    Speech Communication Using a Handsfree Car 'Phone,
    HUSAT Research Centre, Department of Human Sciences,
    Loughborough University of Technology.
Johnston, I.R., Perry, D.R., 1980, Driver Behaviour
    Research - Needs and Priorities, Australian Road
    Research Board Report, ARR No. 108.
Ward, C., 1987, Cellular Telephone Use:  A Questionnaire
    Survey, Issue 1 Doc. No. 45015.  HUSAT, Department of
    Human Sciences, Loughborough University of Technology.
Webster, P., 1987, Data Transmission over Cellular Radio.
    Cellular Radio and Mobile Communications, IBC.

# 1989

| Reading University | 3rd–7th April |
|---|---|
| **Conference Secretary** | External Conference Organiser |
| **Public Relations** | Carol Mason |
| **Finance Secretary** | |
| **Chair of Meetings** | Rachel Birnbaum |
| **Programme Secretary** | Ted Megaw |
| | |
| **Secretariat included** | Sandy Robertson |

## Social Entertainment

A Canal Trip was organised for the delegates and they were also able to entertain themselves when it snowed.

| Chapter | Title | Author |
|---|---|---|
| **Keynote Addresses** | Accident and intention: Attitudinal aspects of industrial safety | D.V. Canter |
| **Process Control** | Flow displays of complex plant processes for fault diagnosis | K.D. Duncan, N. Praetorius and A.B. Milne |
| **Process Control** | Reduction of action uncertainty in process control systems: The role of device knowledge | S.C. Duff |
| **Information Presentation and Acquisition** | Closed circuit television and user needs | A.J. Pethick and J. Wood |

# ACCIDENT AND INTENTION:
## ATTITUDINAL ASPECTS OF INDUSTRIAL SAFETY

**D.V. CANTER**

Department of Psychology
University of Surrey
Guildford
Surrey   GU2 5XH

Examination of a number of major fires shows that these became disastrous because inappropriate actions continued to be supported by conventional social processes. This is illustrated by behaviour in the King's Cross underground fire. It is hypothesised that similar phenomena maintain risky attitudes in industrial settings. This hypothesis was tested using an attitudinal survey of workers in 16 plants on the same site of a heavy process industry. The correlation coefficients between these attitudinal measures and recorded accident ratios are as high as the reliability of the measures involved, thus giving strong support to the hypothesis. The implications are that in safety training management should give as much emphasis to the consideration of social processes as to presenting technical knowledge.

ACCIDENT AND INTENTION

Many safety officers will insist that safety improves within an organisation after it has suffered a major accident, even if no immediate

changes are made to the industrial processes.
The simple conclusion from this anecdotal
evidence is that there is some degree of
voluntary control over what are regarded as
unexpected, unwanted events. Indeed, there is
some weak, but important evidence, that the
number of domestic fires that occurred during
the firemen's strike were fewer than for the
same period in previous years (Canter 1981),
fuelling the possibility that human
intentionality is not entirely absent from
events that are defined as unintentional.

Although it is logically inconsistent to
consider accidents as having an intentional
component, for if they are intended they are not
accidental, the whole practice of accident
prevention  is based on the assumption that the
pattern of accident occurrence can be predicted
and their causes identified. So that even though
the individual may not consciously carry out
actions that imply the possibility of an
accident, the context in which those actions
occur will carry implicit probabilities of
accidental events. Therefore the organisation
which carries responsibility for that context
can be thought of as intending some accidents in
so far as its members are aware of the possible
causes of accidents and does nothing to
eradicate them.

The enormous improvements in industrial
safety over the last century have been brought
about by recognising that it is possible to plan
for the unexpected.  Two general approaches have
been taken to this. One is by making the working
environment and industrial processes much safer.
Secondly by the introduction of safe working
practices.

Yet widely used and well understood systems
still give rise to catastrophic failures.
Flixborough, Seveso, Chernobyl, Bradford City,
King`s  Cross, and Clapham Junction, are now
names that represent places in which there was great

familiarity with the relevant technology, but human failings meant it was misused in a way that lead to disaster. In every case some person, or persons, carried out actions that gave rise to the disaster. Unlike the Lockerbie aeroplane crash, there is no evidence that anybody intended their actions to cause an accident. Yet in every case hindsight reveals that people could have known about the disastrous potential of their actions. The question that must therefore be asked is what framework of understanding and expectations supported these actions.

Major disasters are not the only accidents about which the questions of intentions must be asked. As reviewed by Powell and Canter (1985) there is considerable evidence that a majority of accidents, whether they be of the scale of a major catastrophe or involve just one individual, have their primary cause in the inappropriate actions of people. So, dismissing these actions as ill-informed or unintended would undermine the possibility of understanding the framework of objectives which support accidents and therefore of eradicating them.

Furthermore, it is not just in the occurrence of an accident or not that intention is relevant. Whether a dangerous situation turns into a disastrous one is usually a function of how people behave in that situation. Here again it cannot be claimed that people often, deliberately aggravate the dangers to which they are exposed, but there have been many incidents where they deliberately do things that turn out to make matters worse. The question in these cases, also, is what were they intending to do. Or more generally "what pattern of conceptions and expectations shaped the damaging behaviour?".

## THE ATTITUDINAL CONTEXT

The conundrum of how accidents can be shaped by intentions, especially in situations where people may be expected to understand what the safe acts should be, is often resolved within debates in industry by calling upon the notion of "attitudes". Yet like most apparent psychological explanations cherished by non-psychologists, professional behavioural scientists have so analysed, dissected and digested the terminology and the understanding the terms purport to provide that they are no longer fit for normal human consumption.

The notion of "safety attitudes" have certainly been through this process of metamorphosis. Whereas industrial managers tend to see "attitudes" as a bundle of unfathomable human foibles that in some ambiguous sense explain otherwise inexplicably silly behaviour, social scientists see these as socially structured, implicit intentions that are derived from a person's understanding of the nature of the situation in which they find themselves.

This rather convoluted form of words draws attention to the fact that all individuals are seen as developing some understanding of the situation in which they find themselves. Their actions are based upon their interpretation of the implications of those actions within the context as understood. It is only necessary to realise that in most situations, prior to those situations becoming dangerous, the cues available as to the potential dangerousness of the situation are usually ambiguous and often confused to see what significance personal interpretations can have.

The crucial point here is that there are usually cues available that could be taken to indicate that some actions, other than those actually taken, would be less dangerous, but that because of their understanding of what is demanded of them, people do not do the safest thing. Hale and Glendon (1987), in their review of behaviour in dangerous conditions, illustrate this phenomena in their emphasis on the need for people to recognise whether the situation in which they find themselves is different from a normal day to day situation. If it is different then different behaviour would be appropriate otherwise typical behaviour is maintained.

The maintenance of conventional behaviour despite evidence indicating that this is inappropriate is an illustration of how there are intentions at the core of accidents. Establishing the mechanisms that support these intentions is the research question that needs to be addressed.

This research question has relevance to a number of different groups of people in threatening circumstances:

a) non-specialists coping with a rapidly developing emergency such as a fire in a public building,

b) the activities of staff and other responsible officials during a potentially dangerous public emergency,

c) workers in a dangerous working environment.

It is proposed that in all three cases there are similar social psychological processes that can sustain dangerous behaviour, even though these processes will manifest themselves in very different ways. Management and training procedures need to take account of these processes.

In order to elaborate these issues and provide empirical evidence for the attitudinal components of accidents some recent research will be reviewed.

THE KING`S CROSS EXAMPLE

As is now well known, in November 1987 a fire erupted on an escalator in King`s Cross underground station, developing rapidly to engulf the ticket hall area and kill 31 people, badly injuring many others. In their detailed study of this incident, presented as evidence to the public enquiry, Donald and Canter (1988) showed how conventional behavioural patterns helped to make the loss of life greater than it might otherwise have been.

In examining the behaviour at King`s Cross the general model of behaviour in fires proposed by Canter (1985), deriving from a review of work extending back ten years (Canter 1980), was found to describe the unfolding incident quite well. This model proposes three broad stages to the development of a dangerous incident. In the first stage an interpretation process is gone through in which people try to make sense of whatever information is available. This eventually gives rise to the decision that some form of action different from the normal is needed and preparation for these actions occur. The final stage then evolves out of these preparations in which a variety of actions are actually taken.

The King`s Cross fire was a large, complex, interrelated set of incidents involving hundreds of people, each of whom in different ways can be seen to go through these three stages in relation to the actions of other people around them.  For the present, two different sub-sets of people are particularly interesting to consider, members of the public who were on the station platforms below the seat of the fire and the underground staff who were in the station.

People waiting on Piccadilly and Victoria line platforms had no knowledge of the fire developing on the Piccadilly line escalator above them until they got near that escalator, or were informed by police officers that there

was a fire and the station needed to be
cleared. Of special importance for understanding
the fatal circumstances of the fire were the
number of people who were waiting for trains and
would have got onto the next train if it had not
been for the police officers instructing them to
leave the station. A number of these people,
conscientiously obeying these instructions, were
guided up into the ticket hall at about the time
that the fire flashed over into the hall. Some
of them were killed and some barely escaped with
severe injuries.

These actions serve, horrifically, to
illustrate that what people see as being the
appropriate actions is shaped by a combination
of what they expect of the circumstances and
what figures of authority do and say to help re-
define those circumstances. The authority
figures themselves would appear to gain their
significance from the context. For although it
seems that London Underground staff appeared to
have attempted to instruct people, the evidence
indicates that very few passengers took any
notice of them. (Surveys carried out, did
indicate that this was in accordance with the
attitudes people held towards Underground
staff.) In contrast, as far as can be
ascertained, every one followed the instructions
given by the uniformed police officers.

The power of the intentions people had before
they were caught in the fire is also revealed
quite dramatically by Donald and Canter's (1988)
examination of the positions of the bodies found
by the fire brigade. This examination drew on
information made available by the police of the
journeys that the deceased had been making. It
shows very clearly that people were overcome by
the noxious gases and flames in positions that
related directly to their customary journey.
Even in the last moments as the smoke and flame
erupted into the ticket hall people, who must
have already had some idea of the danger they

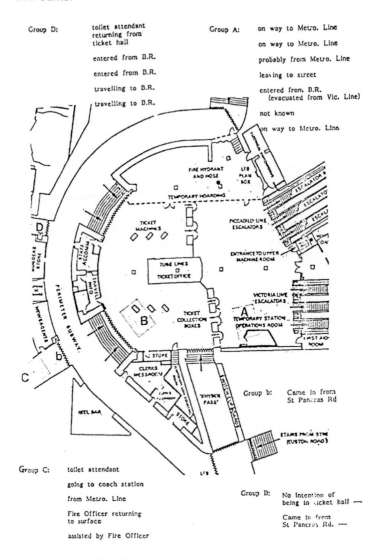

Group D:    toilet attendant
            returning from
            ticket hall

            entered from B.R.

            entered from B.R.

            travelling to B.R.

            travelling to B.R.

Group A:    on way to Metro. Line

            on way to Metro. Line

            probably from Metro. Line

            leaving to street

            entered from B.R.
            (evacuated from Vic. Line)

            not known

            on way to Metro. Line

Group b:    Came in from
            St Pancras Rd

Group C:    toilet attendant

            going to coach station

            from Metro. Line

            Fire Officer returning
            to surface

            assisted by Fire Officer

Group D:    No intention of
            being in ticket hall —

            Came in from
            St Pancras Rd. —

Plan of Ticket Hall at King's Cross Underground
Station showing directions people were going
grouped according to where their bodies were
found.

faced, appeared to have continued on their known route. Those fortunate few who saw the flames erupt in front of them, or were far enough beyond the conflagration not to be killed outright, then fell or crawled as best they could in the direction they perceived as leading to possible safety.

PLACE RULES

In the King`s Cross fire, then, the actions of people can be understood most readily by taking account of their normal behaviour in that situation. Their actions were modified when they either received instructions that came from people who appeared to be appropriate authority figures, or the evidence of their senses was so totally overwhelming that they knew different action was required for survival. The role of authority figures is of course most significant, because by the time that the influence of direct physical cues was strong enough the time left within which survival was possible was severely limited.

In other words, even unusual behaviour emerges from within a framework of socially understood rules. These rules are an integral part of the place within which they occur. In an underground station they derive in part from the regular, routine actions that people carry out on their daily journeys and in part from the understanding that people have of who carries what responsibilities for the control of people in those places. Interestingly, although officially it is the Underground Staff who are responsible for crowd control, the police are generally regarded as the figures who really carry that authority.

The impact of expected customs and practice in shaping interpretations and expectations of appropriate actions is also well illustrated in the study of the fire at Bradford City football ground (Canter et al., 1989). The television video film that was fortuitously made of that fire shows quite clearly that in the early stages people were  reluctant to climb out of the stands on to the football pitch. It was again when police officers started giving direct instructions to people to step onto the sacred turf that the crowd then began to move quite rapidly.

## ROLE RELATIONSHIPS

The King`s Cross fire also painfully illustrates how the system of expectations on which people draw are themselves a product of the positions people hold within an organisation.  For the patterns of rules, associated with places, not only have potentially dangerous consequences for the actions of staff who work in those places, but they also interrelate closely with the actual positions people have within the organisation. This significance of organisational role relationships serve to highlight the importance of management systems in preventing accidents and coping with emergencies.

The importance of organisational factors can be readily deduced from consideration of the initial interpretation stage of the  general model of behaviour in fires. People who feel some responsibility for dealing with a potentially threatening situation, and who become aware of cues they regard as ambiguous, may waste critical minutes and give unhelpful guidance by going to investigate the sights or sounds they think might be threatening.

At King`s Cross, for example, at least four different groups of people investigated the fire in its early stages before giving instructions or taking actions to deal with the threat. These investigations happened, broadly, by junior people making an initial examination, calling on more senior people who investigated for themselves and so on. The time lost by these investigations, prior to informing the fire brigade and closing off the ticket concourse area, undoubtedly contributed to the eventual loss of life.

From the point of view of our analysis here the point to emphasise is that intended actions of people were shaped directly by their position within the relevant organisation. Studies of many other fires (Canter 1985) show similar results. The very good record that hospitals have in dealing with fires, for example, is a reflection of the fact that in hospitals there is a disciplined, hierarchically organised group of people – the nursing and medical staff – who are used to dealing with emergencies in a quick and efficient way. By contrast fires in hotels are potentially much more lethal because the authority relationship between staff and customers is the reverse of what is essential for an effective response to an emergency.

AN EXAMPLE FROM A HEAVY PROCESS INDUSTRY

When it is members of the public and station staff responding to a rapidly changing and obviously threatening fire growth then, once the media myth of "panic" is removed, the significance of conventional expectations is apparent. But in an industrial context, such as a heavy process industry, with flowing molten metal, large perilous moving machinery, noise and heat, it is less obvious that social patterns, reflected in attitudes, can override the dangers clear to even an outsider and lead people to carry out the sort of dangerous actions that can lead to an accident.

Yet there is now clear evidence, not only from anecdotal accounts of many incidents from minor cuts and abrasions to fatalities, but from empirical research that those attitudes that reflect social processes do indeed relate closely to accident rates.

There are two stages to the argument here. The first is the existence of coherent, consistent accident patterns within any part of a plant. The second is the demonstration of attitudinal relationships to these patterns.

## CONSISTENCY OF PLANT ACCIDENT RATIOS

The argument that social processes sustain accidents is of course not a new one, but its implications have only begun to be fully appreciated in recent years. Yet the evidence that there must be some contribution from the organisational milieu to accidents is not hard to find. The big differences in accident figures between the same industries in different countries, or the large variations in accident rates between different industries in one country, cannot be explained solely in terms of the relative dangerousness of the industrial processes. The processes used in the building industry and farming industry are inherently far less dangerous than those in the petrochemical industry or the steel industry yet the ratios of serious accidents to numbers employed are far higher in building and farming than in petrochemicals and steel.

When different plants on the same site of one industry are examined it is even more obvious that local sub-cultures develop that maintain accident patterns, providing a fruitful context in which to explore these processes more closely. The starting point is to consider whether different plants within a site do indeed have consistently different accident rates. It is certainly true that if they do not any hypothesis linking accidents to local social processes will be unsupportable.

As part of a large study of attitudes towards safety commissioned by the Industry (a preliminary report is given by Olearnik and Canter, 1988) we compared the companies own ratios of accidents to number employed for each of 16 plants over two years. The

product-moment correlation between these ratios
from one year to the next was 0.74 for the lost
time injuries and 0.87 for all injuries.  Given
that the recording of accident figures is not
totally reliable and that the company was
actively trying to reduce accidents these
correlations may be regarded as being close to
the reliability of the recording procedure. They
therefore lend some support to the possibility
that plant level processes of management and
social interaction maintain the stability of the
accident patterns.

It is also possible, however, that the
differences simply reflect differences in the
hazardousness of the different plant, against an
overall similar  approach to safety. This is a
difficult hypothesis to test because there is no
direct, objective way of establishing how
hazardous a plant is, based on objective
information say of its engineering. Some attempt
at estimating this can nonetheless be obtained
by asking experts who know the plants to rate
them for their dangerousness.  The site safety
committee was therefore asked to rate each plant
for its dangerousness.  Correlating the average
of these ratings with the accident ratios gives
a value of 0.58, showing that there is some
link to perceived dangerousness but that this is
not high enough to account for the year on year
consistency in the accident rates.

RELATIONSHIP TO PLANT ATTITUDE SCORES
As has been argued attitudes are posited as a
reflection of the social process of which people
are a part. It therefore follows that if the
accident ratios are a reflection of social
processes that there will be correlations
between the accident ratios and the declared
attitudes of different plants. This is a
hypothesis eminently open to conventional test.

An attitude questionnaire was therefore
developed and distributed to a carefully

selected random sample of employees on one site.    564 questionnaires were returned (a response rate of 51%) from 16 plants.

The hypothesis is concerned with the relationships between attitudes for a plant, so Average scores in response to each question were calculated for each plant. This is, of course, rather a crude way of establishing the attitudinal climate for a plant, but as has been mentioned the safety figures themselves can only be regarded as an estimate.    This direct test would as a consequence seem to be reasonable rather than developing a more precise, but possibly spurious assessment.

Product-moment correlations were calculated between the mean scores for each question for each plant and the lost time accident ratios, as provided by the company, for each plant. The highest of these correlations were close to the value of the product moment correlations between the accident figures over two years. The particular questions that have the high correlations are also instructive.

---

Correlations between Lost time Accident Ratios and Mean Scores on four attitude questions across 16 plants.

| Question<br>"How satisfied are<br>you with: | Correlation<br>with Lost time<br>Accident Ratios. |
|---|---|
| Commitment to safety<br>of workmates | − 0.78 |
| Rules about safe<br>working | − 0.76 |
| Recommended precautions | − 0.67 |
| Accident Inquiries | − 0.65 |

(for 15 degrees of freedom r < 0.61 is p < 005)

---

All these questions show a clear awareness of the interpersonal processes of which the worker is a part. The answers reflect their understanding of what people do in relation to potential dangers, lending clear support to the hypothesis that accidents develop within a network of intentions.

HARNESSING THE HAWTHORNE EFFECT

The questionnaire was administered in the early Summer of 1987. In the 18 weeks before it was administered there had been 85 lost time accidents on the whole site. In the 17 weeks after it was administered the number had dropped noticeably to 57. This can be compared with 1985. During the same early period there had been 62 lost time accidents and in the subsequent period the rate had increased markedly to 87. A comparison for 1986 gives figures of 85 and 89, a slight increase through the year. Examination of the total number of accidents give similar trends.

So, although these figures cannot be regarded as conclusive proof they do accord with what was said in later interviews with the workforce. A number of respondents, quite unprompted, said that the completion of the questionnaire had raised people's awareness of the need to work more safely and they had felt that an improvement in safe working had resulted. So, even if indirectly, this supports the contention that people can intend to work more safely with positive results.

The recent tragic frequency of accidents within man-made systems and their horrific consequences, has increased public awareness of the need to improve procedures that will reduce accidents. The indications are that by making people more self conscious about the social processes that shape their risky behaviour it may be possible to reduce accidents. If this heightened awareness is combined with improved management effectiveness then it can also reduce the probability of emergencies turning into disasters.

The historic Hawthorne investigations revealed the power of the researcher's involvement in the social processes he was studying for changing those processes. This has often been presented as a fundamental weakness in field based social science research. The management of safety may well be a context in which the Hawthorne effect is far from being a weakness. It could actually save lives.

REFERENCES

Canter, D., (ed.) 1980, Fires and Human Behaviour (Chichester: Wiley).

Canter, D., 1981, Fires and Human Behaviour: Emerging Issues Fire and Safety Journal, 3, 41-46.

Canter, D., 1985, Studies of Human Behaviour in Fires: Empirical results and their implications for education and design. (Borehamwood: Building Research Establishment).

Canter, D., Comber, M. and Uzzell, D., 1989, Football in Its Place (London: Routledge).

Donald, I. and Canter, D., 1988, Behavioural Continuity under Fatal Circumstances. paper presented to the British Psychological Society London Conference 19-20 December 1988.

Hale, A.R. and Glendon, A.I., 1987, Individual Behaviour in the Control of Danger (Amsterdam: Elsevier).

Olearnik. H. and Canter,D. 1988, Empirical Validation of the Relationship between Safety Attitudes and Industrial Accidents. paper presented to the British Psychological Society London Conference 19-20 December 1988.

Powell, J. and Canter, D., 1985, "Quantifying the Human Contribution to Losses in the Chemical Industry" <u>Journal of Environmental Psychology</u>, 5, 37-53.

# FLOW DISPLAYS OF COMPLEX PLANT PROCESSES FOR FAULT DIAGNOSIS

**K.D. DUNCAN***, **N. PRAETORIUS**** and **A.B. MILNE***

* School of Psychology
University of Wales College of Cardiff
P.O. Box 901, Cardiff   CF1 3YG

** University of Copenhagen
Frue Plads, 1168 Copenhagen K, Denmark

All the information needed for fault
diagnosis is rarely  displayed at once, but
must be retrieved, typically from a paged
hierarchy. A hierarchy is described in which
all pages  use the same small set of abstract
flow symbols. Loss of flow is often seen to be
the central problem in fault diagnosis. It is an
open question whether abstract representations
help operators to abstract from previous
experience to solve novel problems.

## INTRODUCTION

Any complex system may be too large for all its
functions to be represented on one display.  As systems
become more complex the number of displays or "pages"
needed increases.  If the popular hierarchical page
structure is adopted the problem of the relation between
the super- and sub-ordinate pages arises.  The hierarchy
employed in our work is a hybrid:  the relationship
between super- and sub-ordinate pages is either one of
redescription or of dependence.

All our pages  use the same small set of symbols
corresponding to functions which we hope to demonstrate
experimentally are functions and symbols which naive
subjects use when reasoning in a natural way, or perhaps
in a natural language, (Figure 1).

Conventional displays represent pressures, levels, or
temperatures, frequently against a time base, usually
supported by a plant diagram which will often be an
intuitive mixture of function, geography, and morphology.
A fair example is the power plant representation in Figure
2, which is a part of the plant used in our research.

Figure 1. Flow symbols and functions.

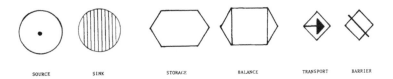

Figure 2. Conventional representation of a turbine system, (Larsen,1987).

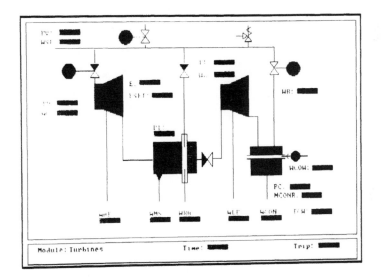

Our pages represent only functional relationships. The relationship between functions which we are depicting is flow. Flow and its interruption or degradation is seen by many people in industry to be the essential problem in fault diagnosis. Fault diagnosis is the task we are addressing.

## FLOW REPRESENTATION

Each of our pages represents a modular sub–system or flow unit which, in the absence of plant failure, is in thermo–dynamic balance in terms of mass or energy distribution. Each sub–system is also itself a condition for the availability of other flow units in the system, (Lind, 1983).

One effect of such a flow representation is to make transparent how imbalance in any unit affects

Figure 3.   Energy flow through the plant.

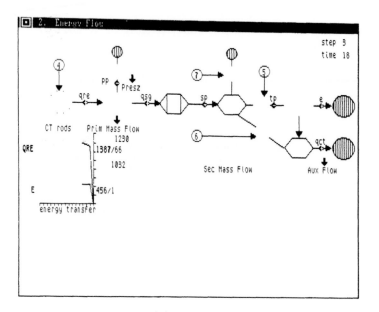

the balance of other units and the system as a whole. This representation should convey to the operator, not only which component has been lost during a plant disturbance, but also what function it serves for the balance of the system, thus making transparent the set of appropriate interventions to restore the system to normal.

Figures 3 and 4 show example pages of the abstract functional representations of energy and mass flow in the system. Figure 3 shows the energy flow through the plant from primary to secondary systems. Figure 4 shows the flow of mass, (water and steam) in the secondary system.

The pages are arranged in the hierarchy shown in Figure 5, which is displayed on an adjacent touch-screen. As well as enabling access to pages, the adjacent touch-screen shows which page is currently being addressed and which pages have "alarms".

Figure 4.   Secondary system mass flow.

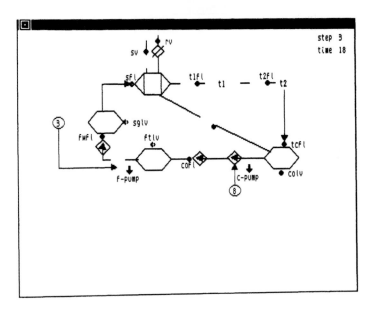

To illustrate the emphasis on function, take the steam generator which has three functions:  barrier to nuclear activity in the primary system, distribution of energy between primary and secondary systems, and storage of water and steam.  Note that Figure 4, the storage and transport of mass in the secondary system, is connected to both higher and lower representations.  These functions are conditions for the achievement of the two top goals, safety (Figure 5, box 1), and production and distribution of energy (Figure 5, box 2). And the same functions of storage and transport of mass in the secondary system are dependent on two "lower" pump functions represented in boxes 10 and 11 in Figure 5.

Figure 5.  Page hierarchy and plant overview.

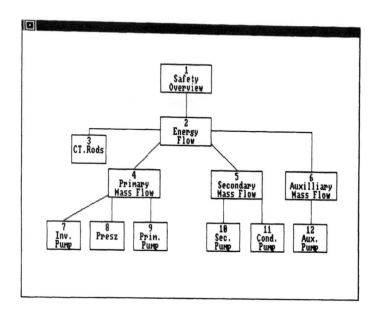

In our pages, alarms are represented as an integral part of the flow function units. This is quite unlike many control room displays where alarms are "added on", so to say, to the temperature, pressure, and level instruments, typically in the form of annunciator panel blocks which light up or flash when the sensed parameter is outside tolerance limits. These limits, usually decided when the plant is first instrumented, determine whether or not alarms occur, rather than the diagnostic significance of annunciator arrays.  As a result the operator may be confronted with a bewildering number of alarms, many of which taken singly are of no diagnostic significance. And any diagnostic pattern, if it exists, may be swamped by the profusion of out-of-tolerance signals which typically follow a fault in a complex continuous process.

Our flow representations alert operators, not to temperature, pressure and level dangers per se, but rather to whether a function is available, threatened or lost. If a function is threatened or lost the corresponding symbol is edged, flooded, coloured, or otherwise changed according to the seriousness of the threat.  Alarms thus indicate the status of a function, and in this sense are an integral part of the display. Moreover, the alarm distinguishes between a function "threatened" by component failure, or by consequent compensatory control-loop regulation.

## THE POWER PLANT SIMULATION

Flow representation has previously been applied to a nuclear power plant (Goodstein,, et al., 1983, Hollnagel et al., 1984). Results from these experiments were encouraging, but the power plant simulated in these experiments was simple and too unrealistic to permit generalization to real fault diagnosis problems. Moreover, it was clear that major revisions were needed of the flow representation. the flow units and the multi-level structure.

We have now made these revisions. Moreover, we employ a rather complex simulation of a process plant in which we are able to "inject" a very large number and variety of possible failures, (Larsen, 1987). For example leaks in kilograms per second can be specified in or between the primary, secondary and auxiliary systems, in either the liquid or gas phase. Pumps can slow or stop. Valves can stick. And spurious control system signals can be precisely specified.

## CRITERIA

One of the most important criteria in testing diagnostic performance is the ability to distinguish between difficult, highly confusable arrays corresponding to very different failures. This is easier said than done, but the power plant simulation now available has enabled us to generate a large number and considerable variety of faults from which to select a much smaller representative set of such different, highly confusable and difficult to diagnose failures.

We are thus able to address in a thorough-going fashion the intricate, too often entirely neglected, problem of a valid fault set and a non-trivial task.

In addition to making sure that the diagnostic task is non-trivial, indeed is of the kind which has so tragically proved to defeat operators in situations such as that of Three Mile Island, we have also applied rather rigourous criteria to the performance of our diagnosticians. These go beyond correct diagnoses, or diagnosis-time to attempting to uncover the diagnostic strategies which people employ - see the discussion which follows..

DISCUSSION

Abstract functions are represented. The safety and integrity of many high capital cost installations depends on the operator's capability to generalise and abstract from previous experience to a novel problem. As technologies advance, operators have frequently to learn to cope with new plant or with the same plant which has been, perhaps drastically, reconfigured. It is an interesting theoretical question to what extent abstractions may facilitate generalisation or transfer of the high order these tasks changes demand.

The choice of hierarchy, if apposite, should enable obscure relationships between functions to become transparent in at least one respect - the classic difficulty in diagnosis of recognizing symptom referral should be reduced.

A general problem with any, usually VDU, multi-level display is that all the information needed for a diagnosis is almost never displayed at once. Industrial operator difficulties have been reported, some of which almost certainly stem from the need for skills of information retrieval, (Marshall et al., 1981a).

At a theoretical level retrieval skills need research as much as interpretation skills once information is retrieved. For example, heuristics are powerful in control panel symptom interpretation, (Duncan, 1986) But will heuristics and the same techniques of eliciting them prove effective when it comes to retrieval skills?

If we wish to identify whether strategies provided by the experimenter or generated by the subject are being applied, we will gain only limited information from measures such as number of correct diagnoses, time to diagnose, number of pages inspected, or time spent inspecting a page. These "product" measures are limited unless supplemented by the sort of "process" measure which may be possible if withheld information is a feature of the diagnostic task, (Duncan 1981, Duncan and Gray, 1975).

However, in the very nature of multi-level displays, information about plant states is withheld until retrieved by the subject.

Withholding information is a rather powerful methodological tool. It may enable an observer to improve the trainee's acquisition of diagnostic skill by intervening with comment about what the subject is apparently trying to do, (Duncan, 1986, Marshall,et al.,1981a). Moreover, it may enable the application of more fine-grained measures of problem solving which, in turn, enable richer feedback of information during the course of problem solving, (Duncan 1981, Duncan and Gray, 1975). Lastly, withholding information has been shown to provide opportunities to elicit and refine useful verbal reports about diagnostic strategy and mental models of the plant from both novices and experts, (Duncan, 1986, Duncan and Praetorius, 1987).

## ACKNOWLEDGEMENTS

This work is supported by the Economic and Social Research Council and the University of Copenhagen

## REFERENCES

Duncan, .D., 1981, Training for Fault Diagnosis in Industrial Process Plant. In Human Detection and Diagnosis of System Failure, eds. J. Rasmussen and W.B. Rouse (New York: Plenum).

Duncan, K.D., 1986, Reflections on Fault Diagnostic Expertise. In New Technology and Human Error, eds. J. Rasmussen, K.D. Duncan and J. Leplat (Chichester: Wiley)

Duncan, K.D. and Gray, M.J. 1975, An evaluation of a fault finding training course for refinery process operators, Journal of Occupational Psychology, 48, 199-218.

Duncan, K.D. and Praetorius, N., 1987, Knowledge capture for fault diagnosis training, Advances in Man-Machine Systems Research, 3.

Goodstein, L.P., Hedegaard, J., Hojbjerg, K.S. and Lind, M. 1983. The GNP testbed for operator support evaluation. (Roskilde:RISO N-16-83).

Hollnagel, E., Hunt, G., Praetorius, N. and Yoshimura, S., 1984. Reports from the pilot experiment on Multi-level Flow Modelling displays using the GNP Simulator.(Halden, HWR 114).

Larsen, N.,1987. Simulation Model of a PWR Power Plant (Roskilde: RISO-M-2640).

Lind, M., 1982, Multi-level Flow Modelling of Process Plant for Diagnosis and Control. Paper presented at the International Meeting on Thermal Nuclear Research Safety.

Marshall, E.C., Duncan, K.D. and Baker, S.M., 1981, The role of withheld information in the training of process plant fault diagnosis, Ergonomics, 24, 711-724.

# REDUCTION OF ACTION UNCERTAINTY IN PROCESS CONTROL SYSTEMS: THE ROLE OF DEVICE KNOWLEDGE

## S.C. DUFF

MRC Applied Psychology Unit
15 Chaucer Road
Cambridge   CB2 2EF

Controlling any relatively complex device requires users to know or infer action sequences that are appropriate in a variety of situations, including novel ones. Obviously information provided about system states is crucial, but also of importance is the user's underlying mental representation of the system itself. A series of experiments has been carried out in the context of an industrial plant, the control of which requires resolution of uncertainty of action specification. They examine what kinds of prior knowledge best support the development of a representation and thus, user learning and problem solving. The data suggests that the type of knowledge users have available will strongly influence their ability to effectively and efficiently control a complex device.

## INTRODUCTION

The process controller (PC) is required to make decisions, and on the basis of these decisions, to carry out actions in order to maintain the functionality of the system. To achieve this there are certain basic ergonomic requirements of the system which are, for the most part, recognised and implemented during the design phase. However, even when these basic design requirements are met, the PC's knowledge of the system will be an important determinant of task performance.

It is assumed that knowledge embodied in the possession of a coherent representation of the system and its operations will facilitate both learning and problem solving behaviour on the part of the PC. In the literature such mental representations have variously been referred to as 'mental models', 'device models' or 'conceptual models' (for example, Kieras & Bovair 1984; Young 1981 ). Such models are viewed as providing a supportive body of knowledge from which procedural skills can develop, or which provide a structure onto which new knowledge can be collected and current knowledge restructured and organised (Ausubel 1968; Glaser et al. 1986). There is evidence in the literature, both theoretical and empirical, which suggests that providing an individual with such a representation of a system will improve the use of that system under learning and problem solving (Halasz & Moran 1983; Kieras & Bovair 1984; Gentner & Gentner 1983; Williams, Hollan & Stevens 1983; Young 1981), but such evidence is not unequivocal. There are a number of negative and unclear findings from similar research (Alexander 1982; Foss Smith & Rosson 1982) and other such negative findings possibly remain unreported.

The present paper will focus specifically on the use of device knowledge in assisting decisions about a sequence of actions. In moving from a goal to the actual performance of actions Norman (1986) argues for a mental stage of action specification. During initial learning or during problem solving users may have an appropriate goal but be uncertain as to the precise actions to be performed. It is at this stage that knowledge of the system is likely to play a major role in resolving that uncertainty. Two forms of uncertainty of action specification will be considered; Order uncertainty where a set of controls must be used in a predetermined sequence, and item uncertainty where a set of controls must be chosen from a larger selection.

The PC interacts with devices which impose such problems of action specification every day. This being the case, the resolution of such uncertainty appears to be a candidate factor which the possession of a mental representation might influence and so affect the process controller's behaviour. Given an appropriate paradigm it should be possible to look at the knowledge most useful for developing representations, the way in which this knowledge is best acquired, and the way in which representations may affect performance.

## METHOD
### Brief Summary of Design and Procedure

Two experiments were carried out to investigate the influence of prior task knowledge on the resolution of the two forms of uncertainty of action specification detailed above. Both experiments employed a paradigm requiring subjects to learn how to control a computer based, on line, representation of an industrial manufacturing plant, thus taking on the role of process controller. The plants used in each experiment were similar, consisting of five sub-units or components in which reactions take place. Subjects were required to correctly use controls which would bring about successful manufacture of the hypothetical product. Each of the components needed to be initialised and then subsequently closed down and it was these processes which the subjects controlled. The experiments were carried out on a BBC B microcomputer, input being via the keyboard and the attached monitor providing necessary output information. The screen display provided a 2 dimensional diagram of the particular component under control, a vertically ordered menu of controls available and, associated with each control, a one word description of current status (On, Closed etc.) and the letter key which corresponded to each control (always the initial letter of the control eg. control - Pump, corresponding key - P).

The tasks subjects had to perform were as follows; firstly an initial learning task (L) where, depending on experimental group membership, they received some information for each component in turn and then controlled that component three times. This was repeated for the five components. Following this there were three trials of problem solving (T), involving controlling the whole plant where subjects were warned in advance that only two controls were required for each component process (initialise, close down). The remaining statuses changed as and when they would do under normal operating conditions. After a two week break, subjects returned and performed a number of relearning trials (R) controlling the whole plant as originally learned. Subsequently, the problem task required subjects to manipulate a variable number of controls for each component process (V), again remaining statuses changing as and when they would in normal operating conditions. This time there was no advance knowledge of the number of changes required to operate each component process. During these four tasks the computer programs stored key press and time data. Finally, subjects answered a questionnaire looking at their ability to generate certain aspects of components and control orders/sets.

Under conditions of order uncertainty (Experiment 1) the on screen menu comprised of four controls for each component, all of which had to be used in predetermined orders during initial learning. Under conditions of item uncertainty, (Experiment 2) the on screen menu comprised of eight controls, four of which had to be used during initial learning, but in any order.

When a control was used correctly, its associated status changed. Incorrect use of a control caused an error message from the system which the subject had to respond to before continuing.

The experimental groups in both experiments consisted of a group receiving explicit statements of control/control (for Order - Experiment 1) control/system (for Item - Experiment 2) relationships and explanations (referred to as Declarative + information) and a group receiving a list of the necessary correct actions and the appropriate new statuses for the controls (referred to as Procedural information). In addition, Experiment 1 included an exploratory group receiving no prior component specific information. Examples of information types are given in Table 1. Item information was similar to this but was provided on hard copies of component diagrams so as to reduce order effects produced by written text. Control orders and control selections were counter balanced across subjects, and menu orders for each component were constructed in arbitrary orders.

Table 1.  Examples of prior information given to subjects.

Declarative +. Pumps should not be on with the Gates closed. This would
              cause a large increase in pressure on the Gates and would
              damage the plant.

Procedural     ....... Gate - Open,  Pump - On, ..........

### Subjects

Subjects for both of the experiments were drawn from the APU subject panel and were randomly assigned to their respective groups. Each independent group consisted of ten subjects with a mean age over the two experiments of 37 years. All subjects were new to these tasks and only participated in one or other of the experiments.

RESULTS

Experiment  One  -  Order  Uncertainty

The data  shown in Figure 1 (key presses per required action)  indicates that the  declarative group performs more accurately during the latter three tasks (phasesT, R and V). Statistical analysis by ANOVA shows that these differences are reliable ($F=3.08$, $df=3,36$ $p<.05$). During relearning trials, all the groups perform similarly to when they first learned the system except the procedural group who perform considerably less accurately. This situation arises because during initial learning  the procedural group make no errors, which is not surprising as they have rote learned a list of required actions, but later this proves to be a less useful source of information. Whilst subjects are learning and then relearning the system, their performance accuracy improves significantly, ($F=12.63$, $df=2,72$ $p<.0001$) indicating that under all conditions the subjects are not inhibited from learning about the system by the experimental condition they are experiencing.

Time data in Figure 2 (time per key press measured in seconds) shows no reliable differences between groups when analysed by ANOVA except during the initial learning phase where the procedural group are performing significantly faster ($F=8.93$, $df=3,36$ $p<.0001$) than the other two groups. There are however significant differences between trials during the two learning tasks (Initial learning - $F=108.52$, $df=2,72$ $p<.0001$;  Relearning - $F=61.34$, $df=1,36$ $p<.0001$), the speed of action execution increasing across trials, indicating that the subjects are learning something during these tasks.  Total time to carry out trials produces identical significant results but the overall pattern of the data suggests that the declarative groups are speeding up more quickly on subsequent trials of each task. This result suggests that subjects are not merely learning about the positions of keys on the keyboard as this should be a constant factor for all groups.

Experiment Two - Item Uncertainty

This data is displayed in Figures 3 and 4. Key press per required action data (Figure 3) is similar to that found in the previous experiment. During the latter three tasks (phases T,R and V) the declarative group consistently perform more accurately, and these differences prove to be statistically significant when analysed by ANOVA ($F=22.46$, $df=1,20$ $p<.001$). Here however, during initial learning the two experimental groups perform almost identically, thus the initial benefits of procedural knowledge found earlier do not occur under item uncertainty. There are significant differences between trials during learning tasks ($F=8.49$, $df=2,40$ $p<.001$) and the declarative group tend to become more accurate across the latter three tasks. Thus again, the experimental conditions do not inhibit either group from being able to learn something about the system, but the group receiving the declarative, relational knowledge appear to benefit more during on line experience.

Time per key press data (Figure 4) demonstrates that initially the two groups perform similarly, the time benefit the procedural group showed during conditions of order uncertainty being lost.  In the latter three tasks, the declarative group take more time to carry out actions, significant by ANOVA ($F=8.49$, $df=1,20$ $p<.01$). There are reliable effects of training during both learning tasks, ($F=9.49$, $df=2,40$ $p<.001$), speed of action execution increasing and again this is unlikely to be due merely to familiarity with the computer keyboard.

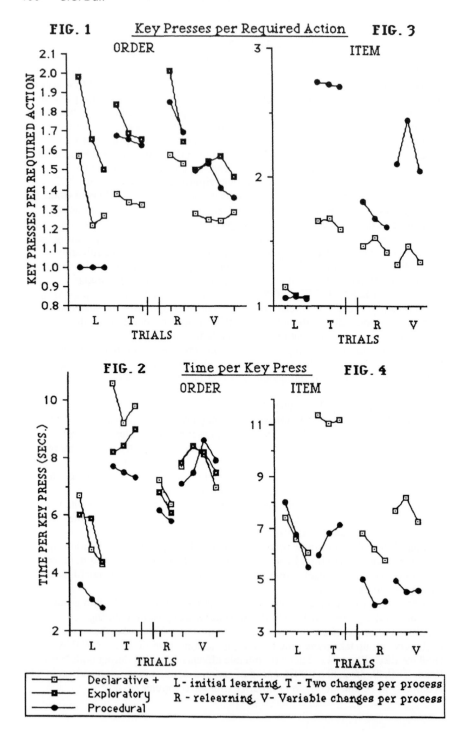

**FIG. 1**    Key Presses per Required Action    **FIG. 3**

ORDER

ITEM

KEY PRESSES PER REQUIRED ACTION

TRIALS

**FIG. 2**    Time per Key Press    **FIG. 4**

ORDER

ITEM

TIME PER KEY PRESS (SECS.)

TRIALS

—□—  Declarative +    L - initial learning. T - Two changes per process
—■—  Exploratory      R - relearning. V - Variable changes per process
—●—  Procedural

Supplementary Experiments

Briefly, the questionnaire data from both of the experiments indicated that the declarative groups were more able to produce accurate answers which required recall and recognition of component menus, control orders and control names. However, the data also showed that there were no between group differences for the number of questions attempted, suggesting that the differences found in the experiments are not merely a result of the declarative groups possessing better memories.

During both of these experiments, a major part of the task involved interacting with the system and this may have influenced the behaviours of the various groups. To investigate this, a series of experiments were carried out to determine to what extent the menu order and diagrammatic presentation of the components may have effected subjects' action judgments and to check if prior, non experimental knowledge was influencing subjects. By asking for action judgments from exploratory, procedural and declarative sub groups (Order) and procedural and declarative sub groups (Item) under four menu/diagram compositions (made up from combinations of original menu-original diagram and new menu-new diagram) it was possible to look at the effects of these factors. Very briefly, the results demonstrate no reliable differences between sub groups, but reliable differences between groups, suggesting that in these experiments the system representation was not responsible for producing the behavioral variations across groups.

DISCUSSION

The results from these two experiments suggest that the type of knowledge which a user possesses is crucial in determining ability to use a device. The evidence shows that although during initial learning there may be benefits of rote learning under some conditions, but certainly no detrimental effects, such benefits will not carry over to later problem tasks, and relearning after disuse. It appears that prior knowledge which consists of explicit assertions of relationships within the device representation (referred to as "declarative" in the experiments), from which inferences about the system can be made, is most beneficial for enhancing on task performance, though any prior knowledge will be useful. However, knowledge which subserves the same functions as the explicit relational statements does not appear to be naturally abstracted in learning through exploration or rote procedures.

Both order and item uncertainty are resolved to a certain extent under learning and problem solving conditions when users are provided with relational knowledge. This advantage appears to be mediated by such knowledge enhancing the internal mental representations, the explicit relational assertions becoming internally represented. It is likely that this is due to enhancement, rather than declarative information being the sole type of information which will support the development of such a representation. As Young (1981) suggests, whenever users interact with a device they are developing some form of representation of it in order that they can learn and make inferences about it, and all the groups in both experiments demonstrate an ability to learn. These assertions furnish the raw material from which inferences can be drawn when the user requires help with resolving action uncertainty. The need to draw such inferences may well reduce with increased system experience as the behaviours become proceduralised, but as the data indicates, when such behavioral repertoires are inappropriate, for example in novel situations, declarative knowledge is more flexible and so more widely applicable. Also, representations which embody such knowledge appear to be more supportive of accurate recall of items of information within the representation, as well as supporting more generally applicable processing.

For the process controller, the evidence strongly suggests that ensuring the individual possesses a representation of the system, supported by explicit relational assertions, should provide the controller with a useful aid for learning parts of the system and operating the system under novel conditions through the resolution of uncertainty of action specification.

## REFERENCES

Alexander, J.H. 1982. Computer text-editing; The development of a cognitive skill. Unpublished doctoral thesis, Dept. of Psychology, University of Colorado, Boulder

Ausubel, D.P. 1968 Meaningful reception learning and the acquisition of concepts. Analysis of Concept Learning. In H.J. Klausmeier & C.W. Harris (eds.)

Foss, D.J., Smith, P.L & Rosson, M.B. 1982. The novice at the terminal: Variables affecting understanding and performance. Paper presented at the Psychonomic Society Meeting, Minneapolis.

Gentner, D & Gentner, D.R. 1983 Flowing waters or Teeming Crowds. Mental Models of Electricity. In Mental Models. D. Gentner & A.L. Stevens (eds.). Erlbaum, Hillsdale, N.J.

Glaser, R., Lesgold, A & Lajoie, S. 1986 Toward a cognitive theory for the measurement of achievement. In The influence of cognitive psychology on testing and measurement. R. Ronning (ed.) LEA, Hillsdale, N.J.

Halasz, F.G & Moran, T.P. 1983. Mental models and problem solving in using a calculator. Proceedings of ACM, SIGCHI.

Kieras, D.E. & Bovair, S. 1984 The role of a mental model in learning to operate a device. Cognitive Science, 8, 255-273

Norman, D.A. 1986. Cognitive Engineering. In User Centered System Design; New perspectives on human-computer interaction. D.A. Norman & S.W. Draper (eds.) Erlbaum, Hillsdale, N.J.

Williams, M.D., Hollan, J.D., & Stevens, A.L. 1983 Human reasoning about a simple physical system. In Mental Models. D. Gentner & A.L. Stevens (Eds.), Erlbaum Hillsdale, NJ;

Young, R.M. 1981 The machine inside the machine; user's models of pocket calculators. International Journal of Man-Machine Studies, 15, 51-85

# CLOSED CIRCUIT TELEVISION AND USER NEEDS

## A.J. PETHICK and J. WOOD

CCD Limited
76 Church Street
Weybridge
Surrey  KT13 8DL

Recent improvements in camera, transmission and control equipment now enable large concentrations of CCTV into remote monitoring centres. Typical of such centres are those responsible for security, motorways, banks, railways etc. In each a wide geographical area is observed from a common observation point. CCD's experience in such environments shows that CCTV is a widely misunderstood resource. Even where large sums are invested in these systems, there is all too often insufficient consideration of how to present, interpret and respond to CCTV information. It is also clear that ergonomic principles can be applied in the design of such installations, greatly facilitating their operation and increasing their cost-effectiveness. There is a need for a user-oriented holistic approach during the design of information handling networks that include CCTV. A framework, developed from a number of projects in different application areas, is proposed that may help the design of such systems and encourage technical developments.

## INTRODUCTION

The use of CCTV (Closed Circuit Television) for security purposes alone was a market estimated to be worth £55 million in the UK during 1986 and growing steadily (NEDO 1986). Readily available video recording has encouraged the growth of CCTV as a deterrent to criminal activities by reducing the dependence upon eye-witness testimony and increasing the likelihood of getting caught.

However, our experiences in a range of CCTV applications demonstrate that the methods employed in its installation and operation could benefit greatly from the use of an ergonomics approach. The selection of technology can benefit from an holistic assessment of the system. The optimal solution will be related to achieving performance objectives and these are largely dependant upon the human operators. This paper sets out to describe the basics of CCTV operations for ergonomists and the basics of the ergonomics approach for CCTV engineers. Many of the items mentioned have been found in security installations in the UK. (Unfortunately, this prevents us from citing our references in the normal way.)

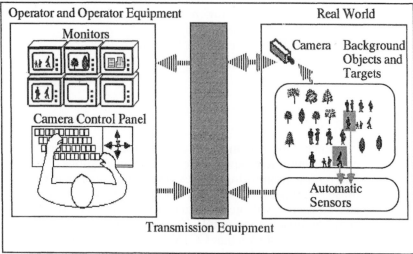

Figure 1. Simplified Description of a Typical CCTV System

## CCTV SYSTEM BASICS

Figure 1 shows a simplified representation of CCTV systems. The operators responsibilities can be broadly categorised as follows:-

• to respond to complex deviations from complex normal conditions, for example concealed bombs in luggage.
• to respond to incidents within a complex background, for example during crowd control in public transport.
• making value judgements of character on the basis of CCTV pictures, for example in courtrooms
• to respond to subtle changes of state, for example anticipating violent behaviour in a prison yard.
• to respond to simple changes of state, for example traffic jams.

All of these are real time tasks and in the majority of applications rapidity and accuracy of response are paramount. When discussing CCTV systems we must therefore address ourselves to the ergonomics of complex images in real time tasks.

For the ergonomist good system design should start with definitions of the performance objectives and allocation of function between the operator and the machine. In the case of CCTV, performance objectives are difficult to define. In many cases it's deterrent potential outweighs any likely performance. In some the deterrent potential will only exist if 100%

accuracy can be achieved. But even with modest performance objectives, the ergonomist can often demonstrate that without due care in the system design, those objectives cannot be met.

## ALLOCATION OF FUNCTION AND EQUIPMENT SELECTION.

When designing a CCTV system the important ergonomic features are related in the first part to the operators' ability to watch CCTV and see the picture and in the second part their ability to identify events/targets that need a response. Control centres are the termination point for many other systems such as radio, telephone, computers and remote control devices. In the most common uses of CCTV these other tasks are dominant and the time available for monitoring is low. The use of CCTV is usually triggered by eyewitness reports on the telephone or emergency communications systems. Where it is the main task, the sheer monotony of CCTV monitoring has a considerable effect upon operator performance. Experimental work (Tickner and Poulton, 1973, Tickner et al, 1972) has demonstrated this and other effects. Although their studies were specific to one application and not intended to provide a general model of operator performance our experience shows that their conclusions are useful guidelines for system design.

Our knowledge of CCTV monitoring centres shows that in general:-
• the number of interruptions is high
• the number of cameras for each monitor is high
• monitors are usually small
• monitor sizes are not properly related to viewing distances
• operators capabilities are not reflected in the performance objectives.

In one example of this 42 cameras were to be concentrated onto 10 monitors to be viewed continually 24 hours a day and particularly at night. The concentration of cameras was to be achieved by autocycling, a method by which the pictures from a number of cameras are shown sequentially on the monitor. The similarity in the pictures on the monitors would tend to give rise to extreme monotony especially in the middle of the night.

The proposed method of CCTV presentation was not suited to the operators capabilities so two basic methods were applied to resolve the ergonomic conflicts. From the outset it was obvious that the monitoring task needed to be changed. CCTV would have to be supported by automatic sensors linked to the CCTV system to show the relevant views of any area in which an event had been detected. Also a job design change was used to introduce a system of CCTV patrols using a check list built into the normal security procedures of the control.

Both of these moves involved a departure from the normal practices of monitor use. We feel that it is necessary to maximise the attention gaining potential from the monitors when they are triggered by sensors. Therefore some monitors need to be left blank. The automated display and camera selection also removes a time consuming task from the operators at moments when speed of response is important. The second departure is the need to group cameras dealing with related areas to be shown during the patrol period. It is unnecessary to fill all the monitors all the time and counterproductive when automated sensors are in use. We used a presentation structured to give a standard group of pictures for the gaps between patrols using six cameras and monitors. The remaining 4 monitors would be off. The other 36 cameras were grouped into two blocks of the perimeter and two of the building interior, related to a geographical mimic of the camera locations and their fields of view of the site as an aide memoir for the operators.

By this simple means an overall improvement in the reliability of intruder detection and response to alarm situations could be achieved.

Another common allocation of function problem is caused by autocycling where the picture from each camera in turn is shown for 2-3 seconds. Not only will transient events be missed completely if the operator is the sole detector but the rapidly changing pictures make long term changes difficult to detect and subtle behavioural changes mentioned earlier almost impossible to detect. Autocycling monitors are extremely distracting when in the visual periphery. Given the choice most operators will opt to switch off autocycling but in most control centres there is a practical limit to the number of monitors that can be viewed and housed. Where automatic detection methods are impractical, the structuring of the cameras on the monitors becomes more important and the use of representational mimics to supplement any on screen camera identification needs to be considered.

Compressions of in excess of 100 cameras into one control centre will not be uncommon in the near future and 10 to 1 or greater compressions of cameras to one monitor are already common. Given the unacceptability of autocycling, it may be necessary to adopt multiple picture presentations. At the moment, vertically split screen picturing with one picture in each half of the screen is found but is only suitable for particular shapes of target/event. Multiple pictures of up to 16 on each monitor are also possible with a variety of interim combinations. However the individual picture size will be greatly reduced and the technical quality may suffer accordingly. In most instances the operators' vision is not the limiting factor and anything smaller than two lines of the screen/camera will only be presented as a dot (Kirkpatrick et al 1976).

## STRUCTURING THE CAMERA INFORMATION

Priorities are sometimes self evident, but normally there are different priorities at different times of day or weather conditions. This means that a good CCTV system will need to have a large number of configurations. Locating cameras and designing monitor presentation configurations requires information about the distribution of events within the monitored areas. However in many instances this information is not known at the time of installation. The use of historical data and assessment by skilled operators can help and in conjunction with a properly conducted CCTV survey, the camera end of the system can be near optimal from the start. Simulations of the CCTV monitor structuring can be performed using

camera survey material. Another method is the instrumentation of the
camera selections made by operators. As a long term management
information practice there are obvious benefits from knowing the
distribution of camera use. By collecting this information, the operators
can help to configure the system as they use it but provisions for this must
be made from the
outset.

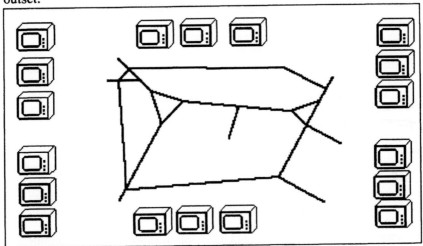

Figure 2. Monitor Distribution Around Motorway Mimic Diagram.

Having collated data on the priority and frequency of use of the
system, the operators' "mental model" of the world as seen through the
CCTV can be enhanced by careful structuring of the monitor-camera-mimic
relationships. Taking a motorway ring-road as an example there are
distinct relationships between cameras and the road (Figure 2). Traffic
flows from one field of view into another and this relationship needs to be
maintained. Major blocks of monitors can be reserved for showing the
overall picture of the motorway, but a few monitors can be left blank for
transient phenomena when they arise, thereby retaining the view of the
whole picture.

## ERGONOMIC FRAMEWORK FOR CCTV SYSTEM DESIGN
An holistic approach allows equal consideration of the human and
technical aspects of the system, relating them to performance objectives in
the light of their relative strengths and weaknesses. The use of an
ergonomic analysis of the operators' tasks is essential in this process. To
sum up, the framework for CCTV system design would:-
• Carefully assess the system objectives.
• Consider the role the operator must play in detecting incidents/targets.
• Attempt to reduce the visual workload of the operators
• Provide the most appropriate technology for the visual and perceptual
  tasks.

- Structure the information using geographical and temporal references
- Provide ancillary tasks where boredom may be a problem
- Provide manning levels and distribution of workload to allow proper relief breaks
- Assess the frequency and priority of incidents when allocating cameras to monitors
- Provide easy to use job aids and mimics for complex systems
- Assess the background in which targets may appear when locating cameras
- Provide a good quality working environment.
- Design a system that plays to the strong features of the operators.
- Remember that infra-red, x-ray or thermal imaging systems are not as easy to view as normal video pictures.

In conclusion, much more benefit can be derived from CCTV when the operators are considered early in the system design and are given the right tasks, facilities and equipment.

REFERENCES AND BIBLIOGRAPHY

Kirkpatrick M, Shields NL and Malone TB 1976 A Method and Data for Video Monitor Sizing. Proceedings of the Sixth Congress of the International Ergonomics Association and Technical Programme of the 20th Annual Meeting of the Human Factors Society, July 1976 pp218-221.

National Economic Development Organisation 1986 Security Equipment a Growth Market for British Industry. NEDO:London.

Tickner AH, Poulton EC, Copeman AK and Simmonds DCV 1972 Monitoring 16 Television Screens Showing Little Movement. Ergonomics 15: 279-291.

Tickner AH and Poulton EC 1973 Monitoring up to 16 Synthetic Television Pictures Showing a Great Deal of Movement. Ergonomics 16: 381-401.

CCD Ltd. 1988 Greater Manchester control upgrading: Report on CCTV monitoring and interface design. CCD Report No. CCD/234.4(B)/88

CCD Ltd. 1988 Pilot assessment of CCTV monitor sizes for motorway monitoring: Working paper. CCD Report No. CCD/233.1/88

CCD Ltd. 1982 Advice on design of CCTV control panels for the Central Command Complex at New Scotland Yard. CCD Report No. CCD/146.1/82

CCD Ltd. 1981 Summary report on workstation design and room layout for traffic control. CCD Report No. CCD/106.14/81

# 1990

| Leeds  University | 3rd–6th April |
|---|---|
| Conference Secretary | Carol Mason |
| Exhibitions Officer | Jackie Baseley |
| Finance Secretary | Sandy Robertson |
| Chair of Meetings | Rachel Birnbaum |
| Programme Secretary | Ted Lovesey |
| | |
| Secretariat Included | Carol Mason |

## Social Entertainment

Delegates took a trip to Wigan Pier.

| Chapter | Title | Author |
|---|---|---|
| Systems Integration | Eye-pointing in the cockpit | M.R. Hicks |
| Seating Posture and the Spine | Ergonomic evaluation of aircraft seating | A.D.J. Pinder |
| Alarms | Altering the urgency of auditory warnings: An experimental study | E. Hellier and J. Edworthy |
| Introducing Ergonomics | The ergonomics audit | C.G. Drury |

# EYE-POINTING IN THE COCKPIT

## M R HICKS

Human Factors Department, FPC 267
British Aerospace plc, (Sowerby Research Centre)
P O BOX 5, Filton
Bristol  BS12 7QW

This paper describes a laboratory based study which
investigated the effects of eye-pointing behaviour upon
certain cognitive mechanisms. The research represents the
first step in an ongoing programme designed to evaluate the
cognitive impact of integrating eye-pointing technology in
the cockpit. Results indicate that certain of the critical
cognitive operations which the pilot performs may suffer
interference during eye-pointing behaviour.

## INTRODUCTION

A top priority of the fighter pilot is to maintain a high level
of situation awareness, the level achieved largely determines his
ability to perform his assigned role adequately. A factor
contributing to situation awareness is the pilot's mental image of
the combat arena. This is a three dimensional space often
enclosing varied topographical features that may contain static
features such as ground targets and threat zones and dynamic
'implied' features such as threat envelopes, flight envelopes and
aircraft positions. Static and particularly dynamic ' implied'
feature locations must be continually calculated and updated in
respect to own and other movements. This and similar tasks place
heavy demands upon the pilots information processing system,
particularly those mechanisms responsible for the processing of
spatial information.

Eye-pointing has recently been mooted as a promising interface
technology for integration in advanced fighter cockpits. The
technology development is driven by the belief that eye-pointing
will provide a means of reducing pilot workload by opening up
another communications channel for the pilot to interface with the
aircraft systems. Whilst a considerable body of work is
addressing the problems of visual performance requirements and the
ergonomics associated with integrating this technology, little, if
any research has addressed the possible impact of this interface
upon the human's limited capacity information processing

capabilities. The aim of this study was to perform a preliminary investigation of the impact of eye-pointing upon certain concurrent cognitive tasks and particularly those associated with spatial information processing.

## Human limitations upon system performance

The 'driver' behind this research is straightforward; it is believed that the pilot's capabilities to fly and employ an aircraft as a weapon delivery platform are not yet fully exploited and that enhancing these capabilities will lead to an overall increase in system effectiveness. The probable future combat role of fighter aircraft is such that novel man machine interface technologies must be developed that will enable the pilot to perform with greater speed and flexibility than is possible in existing systems. Nonetheless, it is important to recognise that when new or 'improved' cockpit interface technologies are proposed it is important to assess the human impact of these technologies or it cannot be assured that the new technology will in any way enhance the man-machine system.

## Cognitive & eye-pointing interactions

The present study was designed to address an hypothesised relationship between voluntary eye movements and the mechanisms of a memory process termed the 'visuo-spatial scratch pad' (Farmer et al.1986), and specifically to identify whether the type of voluntary eye movements that may be made during an eye-pointing task actively interface with the operation of the VSSP during a visualisation task. Since Baddeley & Hitch (1974) proposed their model of Working Memory a considerable research effort has been focused upon two sub-systems termed the visuo-spatial scratch-pad (VSSP) and articulatory loop (AL) (Baddeley et al.1984). The VSSP is considered to input, store, manipulate and retrieve visuo-spatial material whilst the AL performs a similar function for verbal material. Of the two sub-systems, the AL has been examined in greater detail.

There is a body of evidence supporting the notion that the AL and VSSP are functionally independent limited capacity systems that appear able to operate in parallel upon different tasks without interference (eg Atwood 1971; Healy 1975; Baddeley & Lieberman 1980), but which due to their limited capacity may be unable to individually perform a number of similar tasks. This view is inherent to the multiple resource theory of attentional resources (Wickens 1980).

It has become clear that active suppression of the rehearsal processes implicit in either the VSSP or AL dramatically reduces the capability of these systems to store, consolidate or recall information. It has been found possible to suppress spatial imagery in the VSSP by asking subjects to perform concurrent, spatially based motor tasks such as pointing (eg Byrne 1974) and continuous tapping (Farmer et al.1986 op cit). However, the mechanisms underlying this suppression are not fully understood.

This study was designed to evaluate whether the types of eye-pointing task interference reported in the literature could be observed in the laboratory under dual task conditions and to determine whether interference (if any) was a general or task-specific effect. To this end both verbal and spatial tasks were run concurrently with eye-pointing tasks.

## METHOD
### Subjects
These were ten unpaid volunteer participants, (seven male, three female). Their ages ranged from 26 to 35 years. All were members of the British Aerospace Human Factors Department. All had normal or corrected to normal visual acuity.

### Procedure
A spatial primary task, similar to a spatial working memory task used by Brooks (1968), which required the memorisation and processing of a mental image was used to evaluate the effects of eye movements on the processing of spatial information. This task was designed to selectively load spatial processing mechanisms. A verbal primary task was used to assess the effects of eye movements on the processing of verbal information. This involved the serial presentation of a sequence of four 'phonetically spelt syllables that could be reorganised to form a word. This task was designed to selectively load verbal processing mechanisms. A secondary visual step-tracking pursuit task was employed, this was designed to keep the eye moving whilst the subject performed one of the primary tasks.

### Equipment
Eye tracking accuracy was recorded using an NAC Inc. Eyemark 5 (corneal reflection).

### Measures
Eye tracking accuracy was video recorded and monitored in real time. The experimenter provided general performance feedback after each trial. Response latencies and error rates were recorded under each condition. Subject's comments regarding task difficulty and subjective opinion concerning interferences were recorded.

## RESULTS
### Descriptive statistics
Mean response latency between tasks and conditions clearly indicated that eye-pointing strongly interfered with the spatial task but hardly affected verbal task performance (See Table 1 & Figure 1). Mean error rate data confirmed this finding (See Table 2 & Figure 2).

### Unvariate statistics
Response Latency Data : ANOVA revealed highly significant {$F(1,9)=17.1$, $p<0.005$} differences between eye movement and non

eye movement conditions and less significant {F(1,9)=5.33, p<0.05}, differences between spatial and verbal conditions as shown in Table 3. Interactions between these variables {F(1,9)=7.489, p<0.025} was evaluated using a Tukey's test of unconfounded means revealing that the main effect of task type can be attributed to the differential increase in response latency found between task types under eye movement conditions{p<0.01}. It was also found that the main effect of eye movements could be attributed to the increased response latency found between the spatial tasks under eye movement conditions {p<0.01}.

Error Data : ANOVA performed on error data revealed highly significant differences between eye movement and non eye movement conditions {F(1,9)=42.04, p<0.0001} and a less significant difference {F(1,9)=11.445, p<0.01} between spatial and verbal conditions as shown in Table 4. Interactions between these variables {F(1,9)=34.41, p<0.001} were evaluated using a Tukey's test of unconfounded means. Results showed that as with the response latency data, the effect of task type can be attributed to the differential increase in error rate found between task type and under eye movement conditions {p<0.01}. It was also found that the main effect of eye movement can be attributed to the increased error rate found between the spatial tasks under eye movement conditions {p<0.01}. It should be noted that the majority of errors incurred with the spatial task/non eye movement condition can be attributed to two subjects who between them scored 15 of a total of 20 errors (See Fig 2).

## Eye movement data

Eye tracking accuracy was inspected during and after the experiment; no formal analysis procedures were applied to these data. Nonetheless, visual inspection was sufficient to determine that eye-tracking performance deteriorated significantly during the dual task spatial condition and not in other conditions. The deterioration in eye movement performance within the spatial dual task condition was associated with lag and not accuracy. Subjects tended to dwell upon a particular target location for a long period after the target had 'moved on' (thereby missing other target locations), but when ready to move, moved to the next target location with a normal degree of accuracy.

## Subjective report

All subjects found the verbal task to be marginally harder than the spatial task when these tasks were run under single task conditions, conversely, all subjects found the spatial task to be far harder than the verbal task when both were run under eye movement conditions. All subject's reported that the spatial task under eye movement conditions was 'very difficult'.

## DISCUSSION

All data clearly indicate that eye movements significantly interfered with the performance of the spatial task. It is also clear that eye movements selectively interfered with the spatial

Table 1. Mean Response Latencies Over all Conditions
(Seconds)

|  | No Eye-Movement | Eye-Movement |
|---|---|---|
| Verbal Task | 3.34 | 4.18 |
| Spatial Task | 2.88 | 7.14 |

Table 2. Mean Errors Over all Conditions
(Maximum Possible 10)

|  | No Eye-Movement | Eye-Movement |
|---|---|---|
| Verbal Task | 2.1 | 2.22 |
| Spatial Task | 2 | 8.6 |

Table 3. ANOVA Sumary Table, between Task Type (A)
and Eye Movement Conditions (B) : Response Latency Data

| Source | DF | SS | F | P< |
|---|---|---|---|---|
| A | 1 | 15.337 | 5.33 (1,9) | 0.05 |
| AS | 9 | 25.893 | | |
| B | 1 | 64.823 | 17.10 (1,9) | 0.005 |
| BS | 9 | 34.117 | | |
| AB | 1 | 29.492 | 7.489 (1,9) | 0.025 |
| ABS | 9 | 35.440 | | |
| S | 9 | 67.384 | 1.901 (9,9) | - |

Table 4. ANOVA Summary Table between Task Type (A)
and Eye Movement Conditions (B) : Error Data

| Source | DF | SS | F | P< |
|---|---|---|---|---|
| A | 1 | 99.225 | 11.445 (1,9) | 0.01 |
| AS | 9 | 78.02 | | |
| B | 1 | 112.225 | 42.04 (1,9) | 0.001 |
| BS | 9 | 24.021 | | |
| AB | 1 | 105.625 | 34.41 (1,9) | 0.001 |
| ABS | 9 | 27.621 | | |
| S | 9 | 171.225 | 6.199 (9,9) | 0.01 |

Figure 1. Mean Response Latency Data

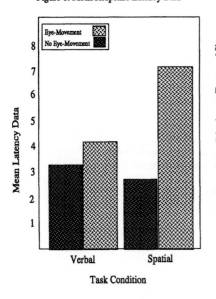

Figure 2. Mean Error Data

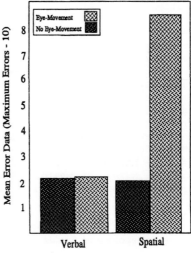

task and do not appear to have interfered significantly with the verbal task.  These results are very similar to those found by Baddeley et al.(1974 op cit) who observed the same patterns of interference under verbal and spatial task conditions during a pursuit-rotor tracking task.

Because the experimental method eliminated the possibility that eye movements may have interfered with the input of spatial or verbal information and as subjects had no trouble recalling any stimuli after the trials, it appears that the locus of interference lies with the processing of the spatial task.

The high error rate found in the spatial eye movement condition implies that eye movements actively interfered with the spatial task, and that interference cannot be attributed solely to a form of task incompatibility that could. be solved by time-sharing procedures.  The latter case could have been argued had merely response latencies been increased.

That the physical directions of eye movement and movement through mental imagery appears to interact suggests that interference cannot be solely attributed to attentional capacity limitations rather, that some interference must arise due to a process conflict associated with incompatible eye movements or a process related to the performance of these eye movements.

The effects observed within this experiment may add a new dimension to the multiple resource pool (MRP) model proposed by Wickens (1980 op cit).  The model only ascribes interference with spatial information processing (the type of activities ongoing in the VSSF) as due to a resource pool overload caused by the concurrent central processing or encoding of similar spatial activities.  This study has produced results that can be interpreted in two ways.  Either (i) central processing can be interfered with by a mechanism common to eye movements and spatial processing, in which case the MRP model may perhaps be usefully extended to accommodate the effects of a response modality[1] upon stages of processing or (ii) the process of eye movement in some way utilises aspects of spatial memory.  In the latter case the eye movement interference can be ascribed to simple overloading of the spatial code at the central processing stage.  There is no a priori reason to consider that the mechanisms of eye movement alone generate spatial memory codes, as there does not appear to be any obvious interaction between eye movements per se and memory processes.  However, if it is important for the cognitive system to keep a record of where the eyes have moved, the code interference hypothesis may be valid.

---

1.Whether or not the movement of the eyes can be described in all cases as 'responsive' needs to be considered.

CONCLUSION

This laboratory based study raises important questions for the future utility of eye-pointing technology in the cockpit. Although the experiment addresses very fundamental cognitive issues, the results have a clear relevance for tasks that involve concurrent eye-pointing and the processing of spatial information. Given the nature of the study it is not yet reasonable to make recommendations concerning the installation of eye-pointing technology in the cockpit. However, it is possible to identify aspects of the pilots task which may prove to be problematic when accompanied by eye-pointing behaviour, and to highlight these areas for future research.

REFERENCES

Atwood, GE 1971 "An Experimental Study of Visual Imagination and Memory" Cognitive Psychology, 2,290-9

Baddeley,AD; Hitch, GJ 1974 'Working Memory' In Recent Advances in Learning and Motivation VIII (G Bower,ed) pp.47-90. Academic Press, New York

Baddeley,AD; Lewis, VJ; Valler,G; 1984 'Exploring the Articulatory Loop' Q.J.Exp.Psych,36,233-52

Baddeley, AD; Lieberman, K 1980 'Spatial Working Memory' In Attention and Performance VIII (R.Nickerson, Ed) pp 521-39, Erlbaum, Hillsdale, NJ

Byrne B, 1974 "Item Concreteness vs. Spatial Orgnisation as Predictors of Visual Imagery" Memory & Cognition 2,53-59

Farmer EW, Berman JVF, & Fletcher YL 1986 'Evidence for a Visuo-Spatial Scratch-Pad in Working Memory' Q.J.Exp. Psych, 38A, 675-688

Healy, AF 1975 'Temporal-Spatial Patterns in Short-Term Memory' J.Verb.Learn.Verb.Behav, 14,481-495

Kosslyn SM & Scwartz SP 1981 'Empirical Constraints on Theories of Visual Mental Imagery' In Attention & Performance IX J Long and AD Baddeley, Eds) pp.241-260. Erlbaum, Hillsdale, NJ

Neisser, U 1976 'Cognition and Reality' WH Freemen, San Francisco

Wickens 1980 'The structure of attentional resources' In R Nickersen and R Pew (Eds), Attention and performance VIII. Hillsdale, NJ Erlbaum Assoc.

# ERGONOMIC EVALUATION OF AIRCRAFT SEATING

**A.D.J. PINDER**

Biomechanics Laboratory, Anatomy Department
Royal Free Hospital School of Medicine
Rowland Hill Street
London, NW3 2PF.

A prototype economy-class aircraft seat that incorporated ergonomic principles and the results of ergonomic tests was compared with two other economy-class aircraft seats using short-duration sitting trials. Each of eighteen subjects sat in each seat for ten minutes and gave ratings of overall comfort and body part comfort, using a seven point scale and a fifteen area body map. They also completed Chair Features Checklists of seventy items. The seats were found to be equally comfortable under the conditions of the test, but the new seat offered far more living space than the other seats. It was concluded that a better seat had been designed.

## INTRODUCTION
### Background

The field of seating has been one of perennial interest to ergonomists. Corlett (1989) has provided a comprehensive overview. However, passenger aircraft seating has been one of the less studied areas.

Many aircraft seat manufacturers are relatively small and therefore do not employ ergonomics specialists to advise on seat design, with the result that the area has tended to be ignored by professional ergonomists. Even so, comfortable seats have been produced in the past, but much depends upon the skill and knowledge of the individual designer. Mason (1965a, 1965b) described a particularly successful seat that Aircraft Furnishing Ltd. designed for the BOAC VC10.

### Requirements for aircraft seating

Aircraft seating design, especially for economy-class seats has tended to be driven by the economics of flight, structural considerations, and airworthiness regulations. Major concerns are seat flammability and the reduction of fuel costs by making the seat as light as possible, but still strong enough to pass the requisite CAA / FAA strength tests.

However, as far as the passenger is concerned, due to the long durations of some flights, and the economic need to fit as many passengers as possible into the aircraft cabin it is important that aircraft seat are designed to provide:

    a) the maximum living space per passenger
    b) the least discomfort initially and over the flight
    c) correct support for the various activities that passengers may reasonably expect to engage in during a flight.

## DESIGN METHODOLOGY ADOPTED

A collaborative project was set up under the auspices of the Teaching Company Scheme between the Faculty of Art and Design of the University of Ulster and Aircraft Furnishing Ltd., a specialist manufacturer of aircraft seating. The project concentrated upon the design of a new economy-class seat.

The design of the new seat was approached from an ergonomic rather than a structural perspective, with the structure being determined by the needs of the seat occupant rather than the occupant being forced to adapt to the seat.

The design of the seat proceeded in an iterative manner, with ergonomic concepts being identified from the literature and from first principles, then mocked up on test rigs and systematically tested. Thus suitable seat contours were identified and used as a basis for testing further seat features such as armrests and headrests.

As a result of a series of tests, a prototype economy-class double seat was constructed. This had a slimline shaped back designed to provide good lumbar support and to maximise leg-room for the person in the seat behind. Keegan (1953) recommended that a pivoted backrest should rotate about the hip joint axis of the occupant. A means of approximating this was found which not only gives the desired motion but also reduces the amount that the seat encroaches into the personal space of the rearward passenger as the seat is reclined. Because of the impossibility of providing a fixed headrest suitable for more than a very small proportion of the population (Branton, 1984), provision was made for the headrest to slide vertically upon straps.

## EVALUATION METHODOLOGY ADOPTED

The prototype seat was compared to two other existing prototype aircraft seats by means of short duration sitting trials since these have been found to produce rankings of seats very similar to those produced by long duration trials (Shackel et al., 1969).

The seats were set up in rows to simulate the situation in an aircraft cabin and to allow subjects to evaluate the rear of the seat in front. Up to three subjects participated at once. Each of eighteen subjects sat in each seat for ten minutes, the order being controlled with a Latin square. The subject then remained seated to give comfort ratings and to fill in a comprehensive

Chair Features Checklist (Shackel et al., 1969).  A seven point
scale with anchors of 1 = 'Perfectly Comfortable' and 7 =
'Intolerably Uncomfortable' was used to obtain overall comfort and
body part comfort ratings on a fifteen area body map (Bennett,
1963).  The Chair Features Checklist consisted of seventy items
covering the range of features of the seat (see Table 1.), where
subjects marked a 50mm line to show where the seat feature in
question fell between the descriptors at the ends of the line.

Table 1. Seat areas examined using the Chair Features Checklist.

| | |
|---|---|
| Entering Seat | Head Cushion |
| Passenger Space | Armrests |
| Recline Mechanism | Tables |
| Seat Width | Appearance |
| Seat Cushion | Covers |
| Upper Backrest | General |
| Lumbar Support | Leaving Seat |

    The height, sitting height, weight, age and sex of each subject
were also recorded (Table 2).  Using data from Pheasant (1986), it
was found that they ranged in stature from a 3rd percentile female
to a 92nd percentile male; in weight from a 6th percentile female
to a 99.5th percentile male; and in sitting height from an 8th
percentile female to a 99.4th percentile male.  The minimum age
was 24, the maximum age was 59.

Table 2. Subject data.

| | Males (n=9) | | Females (n=9) | | All (n=18) | |
|---|---|---|---|---|---|---|
| | Mean | SD | Mean | SD | Mean | SD |
| Height (mm) | 1739 | 70 | 1568 | 49 | 1652 | 103 |
| Sitting Height (mm) | 924 | 35 | 869 | 31 | 897 | 43 |
| Weight (Kg) | 83.0 | 14.6 | 69.0 | 16.5 | 76.0 | 17.1 |
| Age (years) | 41.3 | 10.2 | 48.3 | 10.1 | 44.8 | 10.8 |

RESULTS
Comfort ratings
    Two-way analysis of variance of the ratings of overall comfort
and body part comfort showed that there were no significant
differences between seats.  Very highly significant (P < 0.0001)
differences were found between body parts.  Examination of the
mean ratings for body parts showed that there was a tendency for
discomfort to decrease towards the feet.  Using the Tukey post-
hoc comparison of means, overall comfort and head and neck comfort
were found to be significantly worse (P < 0.01) than the region
towards the feet.  In particular, the neck was significantly less
comfortable (P < 0.05) than ten of the other body parts.  The
lower arms were also significantly less comfortable (P < 0.01)

than the lower legs and feet. The buttocks and low back were two
of the most comfortable areas.

The interaction between seats and body parts was significant at
the 1% level, but the Tukey test showed this to be due to
differences between body parts. In other words, within each body
part no significant differences were found between seats.

Table 3. Mean comfort ratings (7-point scales, 1 = 'Perfectly
Comfortable', 7 = 'Intolerably Uncomfortable').

| Body Part | Seat 1 | Seat 2 (new seat) | Seat 3 | Mean of seats |
|---|---|---|---|---|
| Overall Comfort | 3.67 | 4.06 | 2.83 | 3.52 |
| Head | 4.33 | 3.22 | 3.39 | 3.65 |
| Neck | 4.56 | 3.50 | 3.67 | 3.91 |
| Shoulders | 3.39 | 3.72 | 2.94 | 3.35 |
| Upper Back | 2.94 | 3.33 | 2.89 | 3.06 |
| Upper Arms | 3.39 | 3.44 | 3.44 | 3.43 |
| Sides | 2.83 | 2.78 | 2.89 | 2.83 |
| Mid Back | 3.28 | 2.67 | 2.72 | 2.89 |
| Lower Arms | 3.50 | 3.22 | 4.17 | 3.63 |
| Low Back | 3.17 | 2.72 | 2.56 | 2.81 |
| Buttocks | 3.11 | 2.50 | 2.33 | 2.65 |
| Upper Thighs | 3.11 | 2.78 | 2.39 | 2.76 |
| Lower Thighs | 2.94 | 3.39 | 2.67 | 3.00 |
| Back of Knee | 3.00 | 3.00 | 2.78 | 2.93 |
| Legs | 2.67 | 2.72 | 2.56 | 2.65 |
| Feet | 2.56 | 2.72 | 2.39 | 2.56 |
| Mean of body parts | 3.28 | 3.11 | 2.91 | |

Table 4. Significance levels for differences between mean comfort
ratings (Tukey post-hoc comparison of means).

| Source of Variance | P = 0.05 | P = 0.01 |
|---|---|---|
| Seats | 0.74 | |
| Body Parts | 0.76 | 0.86 |
| Interaction | 1.28 | 1.44 |

Chair feature ratings

Analysis of the data from the Chair Features Checklists to
investigate differences between seats within each seat feature was
carried out using both parametric and non-parametric methods. It
was found that the different statistical methods gave almost
identical results (Table 5). Gregg & Corlett (1989) had argued
that parametric analysis of such results is not valid, but these
results show that the error rate is negligible.

Table 5. Comparison of parametric / non-parametric analyses of
Chair Features Checklist ratings.

| Two-way Anova | One-way Anova | Friedman test | No. of ratings |
|---------------|---------------|---------------|----------------|
| SIG | SIG | SIG | 15 |
| SIG | SIG | NS | 1 |
| SIG | NS | SIG | 0 |
| SIG | NS | NS | 0 |
| NS | SIG | SIG | 3 |
| NS | SIG | NS | 2 |
| NS | NS | SIG | 4 |
| NS | NS | NS | 45 |

The general finding was that subjects could reliably rate
physical dimensions and living space, and that there were
significant differences found between seats where the actual
dimension (e.g. armrest height) differed between seats.  It was
also found that where the scale was 'Unipolar' (with the 'ideal'
point at one end) subjects were unwilling to use the ideal point,
but where the scale was 'bipolar' (with the 'ideal' point in the
middle between bad extremes) subjects were willing to give an
ideal rating.

Pearson correlation coefficients were calculated between the
seat feature ratings and the anthropometric measures taken.  Very
few significant correlations were found.  This is disappointing
since it means that these dimensions cannot be used as predictors
of ratings of the seat features measured.

The new seat was found to receive better ratings than the other
two seats in respect to factors related to the space available to
the passenger.  This can be attributed to its slim, shaped back,
and good shin clearance behind.

The adjustable headrest on the seat was generally seen as being
easy to adjust, which would encourage passengers to make use of
this facility.

It was noticeable that the most obese subjects could not fully
deploy the tables fixed to the backs of the other two seats, but
the table on the back of the new seat, while it could be used by
all individuals, was seen, on average, as being too far away.
This highlights the desirability of having horizontal adjustment
on economy-class meal-tables.

CONCLUSIONS

It had been hoped that the new seat would prove to be the most
comfortable seat of the three tested.  The tests carried out did
not demonstrate this.  Instead, they showed that, for sitting in
for ten minutes on the ground, the seats were equally comfortable,
but the new seat provided much more living space for the passenger
through the use of a slimline shaped back and a recline mechanism
that eliminated hard, sharp protrusions from the shin contact

area. It was therefore concluded that a better seat had been designed and also that judgements of seat comfort are not necessarily dependent upon the living space available to the occupant.

It is disappointing that the provision of an adjustable headrest did not affect the comfort ratings in the head / neck region. This may be due to the short duration of the trials, but suggests that further development of the headrest is needed.

The finding that subjects use 'unipolar' scales differently to 'bipolar' scales is important in the interpretation of comfort / discomfort ratings, since comfort is usually interpreted as being the absence of discomfort, making the use of unipolar scales inevitable. This means that even if all the subjects who sit in a seat are suffering no discomfort, they won't say so using a unipolar scale, but will give answers around a false zero offset from the no discomfort point. This problem will remain until we can conceptualise 'comfort' as being the good point between two equally and oppositely bad extremes.

REFERENCES

Bennett, E., 1963, Product and design evaluation through the multiple forced-choice ranking of subjective feelings. In Human Factors in Technology, ed. E. Bennett, J. Degan & J. Spiegel (New York: McGraw-Hill), Chapter 33, pp. 521-525.

Branton, P., 1984, Backshapes of seated persons - How close can the interface be designed?, Applied Ergonomics, 15, 2, 105-107.

Corlett, E.N., 1989, Aspects of the evaluation of industrial seating, Ergonomics, 32, 3, 257-270.

Gregg, H. & Corlett, E.N., 1989, Subjective assessment techniques: where to draw the line? In Contemporary Ergonomics 1989, ed. E.D. Megaw (London: Taylor & Francis), pp. 344-348.

Keegan, J.J., 1953, Alterations of the lumbar curve related to posture and seating, The Journal of Bone and Joint Surgery, 35-A, 3, 589-603.

Mason, R.V., 1965a, Passenger seats for the jet age: Part 1, Journal of the Society of Licenced Aircraft Engineers and Technologists, 3, 1, 11-17.

Mason, R.V., 1965b, Passenger seats for the jet age: Part 2, Journal of the Society of Licenced Aircraft Engineers and Technologists, 3, 2, 9-14.

Pheasant, S.T., 1986, Bodyspace (London: Taylor & Francis).

Shackel, B., Chidsey, K.D., & Shipley, P., 1969, The assessment of chair comfort. In Sitting Posture, ed. E. Grandjean (London: Taylor & Francis), pp. 155-192.

# ALTERING THE URGENCY OF AUDITORY WARNINGS:
## AN EXPERIMENTAL STUDY.

### Elizabeth Hellier & Judy Edworthy

Department of Psychology
Polytechnic South West
Plymouth, Devon.

Patterson (1982) said that auditory warnings could be improved by designing them so that they were prioritized in terms of perceived urgency. To do this we need to know how variations in spectral and temporal parameters affect urgency. Experiments are reported which investigated the effects of number of repetitions, warning speed and length upon perceived urgency. Increases in all parameters increased the perceived urgency of the stimulus. Steven's Power Law (1957) was applied to the data to quantify the relationship between the experimental parameters and perceived urgency.

INTRODUCTION

The auditory warnings in many working environments are indistinctive, too loud and too numerous. Patterson (1982) suggested ways in which warnings could be improved. He said that temporal characteristics could be varied to make them more distinctive and thus less easily confused; that they should be quieter to allow operator communication and thought; and less numerous so that they were easier to learn, and again, less easily confused. It was also recommended that warnings be constructed at different levels of urgency. If this were done warnings could come on at a moderate level of urgency for a short time, they could then play at a low moderate level of urgency to allow operator communication. If the fault was not remedied the warning would play at its most urgent level to interrupt the operator and demand attention.

By manipulating perceived urgency it would be possible to create low, moderate and high urgency warnings, which could them-selves play at low, moderate or high levels of urgency. Such prioritisation would allow warnings of different urgency to be matched to different events and would mean that when two alarms sounded simultaneously the operator could respond to the more urgent one first.

In order to construct such auditory warnings we need to know how changes in different sound parameters affect perceived urgency. Four experiments are reported which investigate the relationship between changes in three temporal parameters and perceived urgency. These parameters are speed, units of repetition and length. It was hypothesised that increases in these parameters would result in increases in perceived urgency. Subjects were played different levels of the three parameters and asked to rank and rate the urgency of the stimuli. This allowed us to confirm the predicted order of the stimuli, and to quantify the relationship between objective parameter changes and subjective urgency within the framework of Stevens (1957) Power Law,

$$S=ko^{\wedge}m. \tag{Equation 1}$$

S is the subjective value, 0 the objective value and k and m are the intercept and the slope of the line of best fit. In this application, m describes how powerfully the parameter under study produces changes in perceived urgency.

## METHOD
### Subjects and apparatus

The twelve subjects, (two male, ten female), were under-graduate students between 18-27 years. A Tandon microcomputer linked to a Cambridge Electronic Design 1401 interface and 1701 low-pass filters set at a cut off of 4 kHz was used to produce the experimental pulses and bursts.

### Stimuli

A pulse lasting 200ms with a 20ms onset and offset envelope, fundamental frequency of 300Hz and 15 regular harmonics was used in all experiments to construct experimental bursts. These bursts lasted approximately 2 seconds and were played at the same loudness level. Seven bursts were generated for each experiment. They were constructed in the following way: EXPERIMENT ONE (Units of repetition): Two pulses, the first at 300Hz, the second at 200Hz, represented one unit of repetition. The seven bursts contained 6, 5, 4, 3.5, 3, 2 and 1 such units. The stimuli were thus different total lengths. EXPERIMENT TWO (Speed): The inter-pulse intervals of the seven stimuli were as follows: 9, 29, 59, 118, 237, 475, and 950ms. All stimuli were approximately the same total length, with each pulse played at a fixed fundamental of 300Hz. EXPERIMENT THREE (Length): The number of pulses in the seven bursts was 8, 7, 6, 5, 4, 3 and 2, resulting in stimuli which varied in length. All pulses were played at a fixed frequency of 300Hz. EXPERIMENT FOUR (Number of repetitions and speed): The seven bursts varied along two parameters, speed and number of repetitions. The number of repetitions/speed combinations were as follows: 6/9ms, 5/50, 4.5/59, 4/118, 3/237, 2.5/300 and 2/475. All stimuli were approximately the same length, and the basic unit of construction was the two-pulse 300Hz/200Hz unit described in Experiment One. Stimuli were labelled A(most) to G(least) in order of predicted urgency, which is described above.

Procedure
    The procedure was identical for all four experiments, subjects
had to rank and rate the bursts according to their urgency.

RESULTS
    As shown in Table 1 the stimuli were ranked in the predicted
order – stimuli which contained more units of repetition, and/or
were faster or were longer were assigned higher rank orders, that
is, they were perceived as being more urgent.

Table 1:   Mean rankings of stimuli, Experiments One to Four

| Expt. | Stimulus. | | | | | | |
|---|---|---|---|---|---|---|---|
| | A | B | C | D | E | F | G |
| | Rank order | | | | | | |
| 1 | 1 | 2 | 3 | 4 | 5 | 6 | 7 |
| 2 | 1 | 2 | 3 | 4 | 5 | 6 | 7 |
| 3 | 1 | 2 | 3 | 4 | 5 | 6 | 7 |
| 4 | 1 | 2 | 3 | 4 | 6 | 5 | 7 |

    In all four experiments Kendall's coefficient of concordance
was high and very significant.  Experiment One, $W'$ =0.849
($F=61.744$, df=5,64, p <.001);  Experiment Two, $W'$ = 0.843
($F=59.212$, df=5,64, p<.001); Experiment Three, $W'$ =0.890
($F=88.941$, df=5,64, p<.001); and Experiment Four, $W'$ =0.939
($F=167.927$, df=5,64, p<.001).  There was thus much agreement
between subjects as to the rank orderings of the stimuli.  Each
subject's ranking of each stimulus was correlated with their
magnitude estimation (rating).  The means of these correlations
were, Experiment One, 0.841; Experiment Two, 0.855; Experiment
Three, 0.790; and Experiment Four, 0.940.  Such high correlations
indicated that in the ratings, as in the rankings task stimuli
that contained more units of repetition, and/or were faster or
were longer, were perceived as being more urgent.
    The ratings data from Experiments 1-3 was fitted to Steven's
Power Law which quantified the relationship between the objective
values of the stimuli (number of repetitions, speed, or length)
and the subjective values (perceived urgency).  Thus for
Experiments One, Two and Three:
    Perceived urgency = 22.13xunits of repetitions 0.69 (Eq.2 )
    Perceived urgency = 18.32x(2500/pulse-to-pulse time) 0.61(Eq.3 )
    Perceived urgency = 1.65 x length 0.49 (Equation 4 )
    A linear regression showed that the data from all Experiments
was well fitted by a straight line when plotted in log-log
co-ordinates, as predicted by Steven's Power Law.  The percentages
of variance accounted for by the straight line were 99.8% (Expt.1);
99.7% (Expt.2) and 99.0% (Expt.3).

DISCUSSION
    Our Experiments show that increasing the units of repetition
(and necessarily increasing the length) of a warning increases its

ugency, as does increasing warning speed or length.  The high values of W' for the ranking tasks suggests that subjects are very sure about the direction of change in perceived urgency for all three parameters.  The power functions derived for the three parameters suggest that the number of repetitions and the speed of the warning are more powerful in producing changes in perceived urgency than is a simple length change for these parameters had higher exponents. When all three parameters are combined (Experiment Four), subjects are even more sure about the direction of the change in perceived urgency as shown by W' of .939, the highest for the experimental series.

The results show that the perceived urgency of an auditory warning can be readily manipulated, and supports the work of Edworthy and Loxley (this volume).  The power functions suggest the contributions of individual sound parameters to the overall urgency of a warning.  The parameters with higher exponents are more economical to use for they require smaller changes in themselves to produce a unit of change in perceived urgency.  This may be of use to the manufacturer with only one or two parameters at his or her disposal.  Furthermore, use of the power function allows a manufacturer to quantify how more or less urgent one warning is to another.  Our results will eventually lead to a set of design principles for use in advanced auditory warnings work.

REFERENCES

Edworthy, J. & Loxley, S. (1990) Auditory Warning Design:  The Ergonomics of Perceived Urgency.  Contemporary Ergonomics.
Patterson, R., 1982, Guidelines for Auditory Warning Systems on Civil Aircraft. CAA Paper 82017.
Stevens, S., 1957, On the Psychophysical Law.  Psychological Review 7, 64, 153-181.

# THE ERGONOMICS AUDIT

Colin G. Drury

The Center for Industrial Effectiveness
Baird Research Park
1576 Sweet Home Road
Amherst, New York 14221-2029

If ergonomics is to have its full impact upon a company's competitiveness, it needs to emerge from its project-oriented mode and address overall company needs. As with any other discipline, it requires a methodology for evaluating the overall state or ergonomics within an organization. Such a methodology was developed at the request of a large multi-national company and has since been used in three cases. This paper summarizes the methodology as originally developed, and the lessons learned in applying it in diverse industries.

An audit program was initially developed from process and outcome measures, by comparing existing workplaces against ergonomics standards and by using accident and performance measures to locate sub-areas within the plant as foci for ergonomics effort. The workplace survey procedure covers many different aspects of the task (environment, manual materials handling, controls and displays, physical workload, anthropometry) but allows the analyst to branch past areas considered irrelevant to the workplace at hand. This survey is computer-based but can be performed by hand when required.

## THE NEED TO AUDIT

As ergonomics moves from a project-oriented mode of solving specific user-defined problems into a more strategic mode where it can help define what problems need to be solved (e.g. Drury, Kleiner & Zahorjan, 1989), there is an urgent need for strategic-level measurement tools. In particular, we need to develop methods for assessing the state of ergonomics within an organization for two reasons:

1. To provide objective feedback on effectiveness of ergonomics, so that the company can allocate resources based on true needs.
2. To allow the company to focus on areas which need ergonomics most if they are to meet company strategic objectives. Thus the company may need to know which plants, departments or even lines require ergonomics effort.

Several years ago, a multi-national corporation required just such a procedure. It operated over twenty plants with products as diverse as shoes, perfumes, bulk chemicals, pharmaceuticals and sports equipment. To structure such a procedure, called by the company an Ergonomics Audit program, the levels of measurement possible had to be considered. Any organization has three broad classes of measures it uses: Input, Process and Outcome. Input measures specify the structure of the company, the nature of its competitive environment and, the resources available in the company. Process measures are internal performance measures, such as how resources are used. Outcome measures are "bottom-line" measures of how well the company is achieving its objectives. Applied to the ergonomics functions these give:

| | |
|---|---|
| Input: | Structure or ergonomics effort. |
| | Size and nature of company. |
| Process: | How well are workplaces designed? |
| | How does the ergonomics effort function? |
| | How well is ergonomics received by users? |
| Outcome: | Performance: productivity, quality and delivery. |
| | Well-Being: accidents, absence, turnover. |

While all are possible to incorporate into the Audit program, the main concentration was on Outcome measures (to define how effective the company was in ergonomics) and Process measures to give prescriptive and diagnostic advice of value to the organization. Since developing the Audit program, one part of it has been used in other companies and other parts have evolved around it. This paper describes the original program development, and subsequent modifications.

AUDIT PROGRAM - VERSION 1.0

The original program, detailed in Mir (1982) used outcome measures in Phase I to provide an overall context of a plant, followed by a Workplace Survey (Phase II) of departments selected in Phase 1. The Workplace Survey consisted of mainly process measures.

Phase I measured the following variables, either from historical records or detailed interviews with managers, for each organizational unit or department.

> First Aid Reports/Medical records
> OSHA Reports of Accidents/Injuries
> Workers Compensation Payments
> Turnover Rate
> Absenteeism Frequency
> Lateness Reports
> Labor/management friction
> Productivity or Effectiveness

The accident/injury reports were further classified according to Hazard Pattern (Drury & Brill, 1983) to help understand the nature of the ergonomics problems, and their pattern across departments. On the basis of departmental differences in all of the above variables, particular representative departments were chosen for the Workplace Survey of Phase II, using a ranking technique similar to that of Brown (1976).

As a first step in developing the Workplace Survey, objectives were as follows:

1.  Must identify violations of ergonomics principles in a quantitative manner.
2.  Must be useable in field conditions, with data collected either by hand-held computer (e.g. Drury, 1987) or by pencil and paper.

3.    Must give unambiguous, directive output to alert the user to redesign opportunities and to the consequences of making changes.

A checklist was the obvious design solution to these objectives, but most checklists were either too narrowly directed (e.g. Cakir, Hart & Stewart, 1980), or lacking in quantitative data (e.g. review by Easterby, 1967). Perhaps the most extensively developed checklist is the IEA's Ergonomics Systems Analysis Checklist, last revised in Dirken (1969). Even that is not entirely quantitative, and does not provide the directive information required. In addition, it was recognized that any checklist must be modular, as in any one workplace only a subset of the ergonomics factors would be at issue.

The Workplace Survey was thus designed to have the following sections:

Visual Aspects
Auditory Aspects
Thermal Aspects
Instruments/Controls/Displays
Design of workplaces
Manual Materials Handling
Energy Expenditure
Assembly/Repetitive Tasks
Inspection Tasks

Each section, and many subsections, can be omitted if not required, although some are naturally linked. Thus the quantitative analysis of Thermal Aspects requires a calculation of the energy expenditure as well as measures of the thermal environment and clothing. In each section, quantitative questions are asked, based on handbook recommendations (e.g. Grandjean, 1980; for seating) or physical models (e.g. NIOSH, 1981; for manual materials handling or Givoni & Goldman, 1973 for predicting heart rate as a function of thermal aspects).

The computer program which assimilated this data, performed the calculations, and provided output was AUDIT1, written in CDC BASIC for portability. It has since been transferred to an IBM-PC as AUDIT.BAS. A rule-based logic is used to provide messages to the user, based on either a single input, e.g.:

MESSAGE:
Seats should be padded, covered with non-slip
materials and have front edge rounded.

or on the integration of several inputs:

MESSAGE:
The total metabolic workload is 174 watts.
Intrinsic clothing insulation is 0.56 CLO.
Initial rectal temperature is predicted to be 36.0 °C
Final rectal temperature is predicted to be 37.1 °C

If a section contains no violations of ergonomics principles, the message is:

Results from analysis of auditory aspects:
Everything OK in this section.

In use, each selected workplace is analyzed and the output used both for counting purposes, i.e. how many messages in each section, and to alert the company personnel to needed design changes.

This Audit program was tested in the company and found to provide the data required. In practice, Phase I was omitted and the Workplace Survey applied directly to specific chosen jobs within each plant. The ergonomics group within the head office organization used the Audit program as part of their training course for personnel in each plant, but plant personnel did not use it regularly. Since its inception with Mir (1982), the Workplace Survey has been used in several organizations, both as an Audit tool and as what would now be termed an expert system to diagnose ergonomics problems with particular workplaces. An example of this use is given.

## AUDIT OF A LARGE COMPANY

Recently, the Workplace Survey was used as part of a broader audit of an automotive components manufacturing company, with three plants and a corporate structure. The complete ergonomics audit was part of a broader evaluation of the company's competitiveness, performed at the joint request of the company, its unions and a state economic development agency concerned with maintaining employment within the region. Within the audit, the structure of ergonomics activities at corporate and plant levels was examined, the Workplace Survey was used to assess the effectiveness of ergonomics efforts in design of jobs, and interviews with providers and consumers of ergonomics services were interviewed to diagnose the changes required to the system.

At the three plants, a total of 76 workplaces were chosen for the Workplace Survey. The selection was based on observation and photographs of all jobs in the plants. True randomness was not used, but rather workplaces were chosen to represent the various activities (e.g. assembly, machine operating, materials handling, inspection) within clusters, typically different production lines within each plant. In this way questions of interest to the plant could be answered from survey data, e.g. "has new technology led to better ergonomics?" or "is line 15 different from line 27, which is the model for future lines?" No attempt was made to include or exclude jobs suspected of high injury rates. Surveys were conducted using the current workplace incumbents, and included operators from all three shifts.

The results can only be summarized here, and will be presented by section of the Survey. Overall, there were few significant differences between the three plants, or between lines within the plants, indicating both a uniform level of ergonomics application ( or lack of it!) and no measurable improvement in new lines.

Visual Aspects: The distribution of lighting levels at the task, at midfield, and outer field showed satisfactory mean values, but high variance. Level of luminance was too low in 37% of workplaces and only 36% workplaces had progressively decreasing luminance from task to mid to outer field.

Auditory Aspects: While the mean noise level was less than the OSHA limit, 42% of workstations were above the 8 hour limit of 85 dBA. All were above 75 dBA so that communications interference could be expected.

Thermal Aspects: For the level of energy expenditure in the tasks, even for sedentary work, the thermal environment was too hot and too humid for comfort, but below the levels of potential heat stress.

Instruments/Controls/Displays: Apart from a few workplaces lacking control shape coding and adequate label sizes, the survey showed that human/machine communication was well up to ergonomics standards.

Design of Workplaces: Very few workplaces met ergonomics standards. Chairs were of poor design, footrests absent, and kneeroom under most conveyors was nonexistent. The results were seen in excessive bending, twisting, reaching, and awkward hand/arm postures. In the Repetitive Tasks section, most jobs were found to have high repetition rates associated with these awkward postures, thus predisposing operators to repetitive trauma injuries (Putz-Anderson, 1988).

Manual Materials Handling (MMH) and Energy Expenditure: Energy expenditures were within typical recommendations, but forces on the body, estimated from static biomechanical models, were high. Ten percent of the 21 jobs where MMH was evaluated gave disc compressive forces exceeding 2500N and 85% exceeded 1000N. In terms of the NIOSH Action Limit concept (NIOSH, 1981), nineteen percent had actual loads exceeding the Action Limit.

Inspection Tasks: Training for inspection was found to be inadequate in all plants.

Conclusions from this study were that relatively simple changes to the visual environment were possible and desirable, but less noisy equipment would have to wait for new designs meeting ergonomics criteria. A major effort in workplace design was recommended, and demonstration projects were started in all plants as a final part of the assessment. MMH analyses were recommended as standard procedure for all new jobs and equipment.

Diagnosis of organizational changes to support these workplace changes was the function of the providers and consumer interviews. Many of the frustrations of implementing ergonomics in a large, multi-factory environment were the basis for recommendations for organizational changes made to corporate and plant management. As a result of the overall competitiveness evaluation, many of these organizational changes have been implemented. The company has the Workplace Survey methodology, so that the future status of ergonomics within the plant can be tracked.

CONCLUSIONS

This paper has concentrated on the Workplace Survey as the central instrument in an ergonomics audit program. It has proven relatively simple to use, and a very effective mechanism to provide feedback to a company on the effectiveness of its ergonomics efforts. Around this Survey, other audit techniques have evolved, generally replacing outcome measures with more process measures. The point has now been reached where the Workplace Survey itself needs to be reviewed and updated in the light of increased ergonomics knowledge and standards activities.

REFERENCES

Brown, D.B., 1976. System Analysis and Design for Safety, (New Jersey: Prentice-Hall, Inc).

Cakir, A., Hart, D.M., and Stewart, T.F.M., 1980. Visual Display Terminals, (John Wiley & Sons) pp. 144-152, 159-190, App. I.

Dirken, J.M., 1969, An ergonomics checklist analysis of printing machines, ILO, Geneve, 2, pp. 903-913.

Drury, C.D., 1987, Hand-held computers for ergonomics data collection. Applied Ergonomics, Vol. 18.2, pp. 90-94.

Drury, C.G. and Brill, M., 1983, Human factors in consumer product accident investigation, Human Factors, 25.3, pp. 329-342.

Drury, C.G., Kleiner, B.M. and Zahorjan, J., 1989, How can manufacturing human factors help save a company: Intervention at high and low levels., Proceedings of the Human Factors Society 31st Annual Meeting, pp. 687-689.

Easterby, R.S., 1967, Ergonomics checklist: An appraisal. Paper to 3rd International Congress on Ergonomics, Birmingham; Ergonomics, Vol. 10, No. 5, pp. 549-556.

Givoni, B. and Goldman, R.F., 1972, Predicting Rectal Temperature response to work, environment, and clothing, Journal of Applied Physiology, 32(6), pp. 812-822.

Grandjean, E., 1980, Fitting the Task to the Man, Chapter 15, (London: Taylor & Francis).

Mir, A.H., 1982, Development of an Ergonomics Audit System and Training Scheme. Unpublished MS thesis, University at Buffalo.

NIOSH, 1981, Publication No. 81-122, (Washington, DC: US Government Printing Office)

Putz-Anderson, V., 1989, Cumulative Trauma Disorders: A Manual for Musculo-skeletal Diseases of the Upper Limbs, (London: Taylor & Francis).

# 1991

| Southampton University | 16th–19th April |
| --- | --- |
| **Conference Manager** | David Girdler/Jonathon Sherlock |
| **Honorary Meetings Finance Secretary** | Sandy Robertson |
| **Annual Conference Chair** | Rachel Birnbaum |
| **Chair of Meetings** | Dave O'Neil |
| **Programme Secretary** | Ted Lovesey |

| **Secretariat** | E. So | C. Meller | A. Messenger |
| --- | --- | --- | --- |
| C. Kirkbride | J. Fowler | M. Gobel | S. Jeffs |
| P. Joice | K. Kingston-Howlett | J. Voit | A. Walker |

## Social Entertainment

A diversion was available in the form of a trip to the Aircraft Museum.

| Chapter | Title | Author |
| --- | --- | --- |
| **Speech Input and Synthesis** | Voice versus manual techniques for airborne data entry correction | P. Enterkin |
| **Noisy and Hot Environments** | Human thermal responses in crowds | T.L. Braun and K.C. Parsons |
| **Job and Workplace Design** | Office lighting for VDT work: Comparative surveys of reactions | A. Hedge |
| **Musculoskeletal Studies** | Use of wrist rests by data input VDU operators | C.A. Parsons |

# VOICE VERSUS MANUAL TECHNIQUES FOR AIRBORNE DATA ENTRY CORRECTION

## P. ENTERKIN

FS-9, F131 bldg, Royal Aerospace Establishment,
Farnborough, Hants GU14 6TD

A comparative study assessed voice versus manual error correction techniques for airborne voice data entry tasks. Subjects performed a continuous visual tracking task while simultaneously completing corrections on DVI feedback (simulated to contain errors) either by voice or keypad. The results showed voice correction took consistently longer than keypad correction. The expected head-up monitoring and faster transaction advantages of DVI were not evident. Subjective measures, from the NASA Task Load Index indicated voice correction did not reduce overall workload ratings comparative to using the keypad but shifted the workload across the dimensions that were assessed. The results were interpreted in terms of human attentional resources and in the context users' experience with automatic speech recognition systems.

## INTRODUCTION

Direct Voice Input (DVI), or "voice control" is viewed as a means of assisting pilots in gaining optimum flight and mission performance in the "head-up" and "hands on the throttle and stick" (HOTAS) environment of the military fast jet. DVI systems could provide a means of control some of the mission management tasks requiring head-down, manual operations (Bell et al, 1982; Berman, 1986).

### Simulator and ground based studies

Compared to conventional data input methods (e.g., keyboards) DVI has been shown to be useful in reducing data input transaction times (Welch, 1977; Laycock & Peckham, 1980; Beckett, 1986), improved accuracy of data input (Connolly, 1979; Laycock & Peckham, 1980), improving attention capabilities on a concurrent task (Poock, 1980; ) and improving head-up attention capabilities on a concurrent flight task (Laycock & Peckham, 1980; Beckett, 1986; White and Beckett, 1983). There was evidence that DVI improved performance particularly in higher workload conditions. Error rates have generally been quoted in the region of 0 - 5%, however some studies have reported error rates of up to 22% (Beckett, 1986).

## Airborne research

Airborne research has generally quoted error rates in the range of 2 to 13%, even up to 40% (Smith and Bowen, 1986; Little, 1986; England, Harlow & Cooke, 1986). The airborne environment's unique factors of noise, vibration, g-forces, the pilots face mask, and mission phase all interfere with DVI performance (Little, 1986; Smith and Bowen, 1986; Montague, 1977; Lea, 1979). Despite these factors DVI has been shown to be a useful asset in many aircraft types (Smith and Bowen, 1986; England, Harlow & Cooke, 1986; Little, 1986) especially in situations where pilot's hands, or both hands and eyes were busy with flight tasks.

Particular applications for DVI were evident in navigation and communications control, especially in low-level flight, high speed flight, setting up on navigation approach parameters and formation flying as they are essentially verbal tasks (or can be transformed into verbal tasks). They would therefore not interfere with the spatial task of visual tracking and allow a more consistent head-up posture. White (1987) suggested that time savings in transactions would be lost in time needed to correct error intrusions. DVI must therefore demonstrate itself to be operationally beneficial and acceptable for application in this area.

## Feedback and Error intrusion

Even with experienced users of conventional systems make errors (Shaffer and Hardwick, 1969). In DVI error intrusions are of three basic types: substitution (where something has been altered), rejection (where something has been deleted), and insertion (where something has been added). The reduction of errors can be achieved to a large extent from an engineering perspectives such as vibration reduction, noise abatement, and g suits. From the behavioural aspect, training in the use of ASRs (Bell et al, 1982), syntax structures and combined visual and voice feedback (McGuinness, 1987) would help reduce error rates further.

Errors are inevitable. ASR technology is far from being able to achieve the same level of speech recognition, semantics and understanding that we are capable of. Indeed we ourselves have slips of the tongue, mispronounce and muddle words, and mishear and misunderstand words and sentences. It would be unreasonable to expect ASR technology to perform any better.

## Dual task performance

Cognitive ergonomics concentrates on the human information processing characteristics of the interface and particularly relevant in the context of pilots' high levels of mental workload. The role of a fast jet pilot is characterised by the requirement to perform several concurrent tasks.

When one examines the nature of the tasks in terms of the resources they use it is evident that concurrent tasks interfere with each other more if they are presented in the same resource modality (visual or auditory). This concept has been addressed and supported in many cases (e.g., Treisman & Davies, 1973; Wickens & Kessel, 1980; McLeod, 1977; Wickens et al, 1983; Kantowitz and Knight, 1976). Models on human attentional resources consider attention to have a multiple resource structure with distinct modes of resources (Wickens, 1984; Baddeley & Hitch, 1974; Hitch & Baddeley, 1976: cited in Eysenck,M.W, 1984).

Without going into detail the modalities basically centre on auditory versus visual encoding and spatial versus verbal processing components. When any two tasks demand separate rather than common resources on any of the dimensions time sharing will be more efficient, changes in difficulty of one task will be less likely to influence performance on the other, the performance operating characteristic towards its limits as resources withdrawn from one task cannot be used to advantage in another due to its dependence on different resources.

Cognitive automaticity has been proposed as a factor involved improving dual task efficiency (Schneider & Fisk, 1982; Schneider & Shiffrin, 1977). As a task becomes more rehearsed it demands less conscious attention and can be completed more quickly, so concurrent tasks can be attended to more effectively.

How useful will voice as a text editor given the behavioural aspects of error correction and a concurrent spatial tracking? Resource compatibility in multiple tasks with respect to the efficiency time sharing activities may be dependent on the central processing code used in the task (Wickens, Sandry and Vidulich; 1983). Hence, verbal tasks are most effective with auditory input (to the subject) and vocal outputs (from the subject), while spatial tasks were more effective with visual inputs and manual outputs. Dual tasks performance would therefore be most efficient when taking the nature of the tasks and their feedback into consideration (McLeod, 1977; Wickens et al, 1983). Further support for resource compatibility came from Wickens and Yili Liu (1988) who found interference was greatest when tracking tasks involving spatial decision track were performed with a manual response rather than a verbal response.

Turning to workload in DVI, Laycock and Peckham (1980) suggested that whereas voice and keypad input tasks may have equivalent cognitive loadings, the time advantage gained with voice input would reduce workload. Mental workload may be reduced to some extent from the memory cues provided by the keypad legends (Poock, 1980; Long, 1976). It has been noted, however, that people revert to manual control functions under high stress situations (Wickens, 1984; Little, 1986) possibly due to the familiarity factor of conventional systems.

### In summary

From past research and cognitive models it was hypothesized that: voice correction would allow for better head-up attention on a concurrent visual tracking task compared to keypad correction, voice correction times would be quicker than keypad correction, and voice correction would reduce workload rating relative to keypad correction.

## METHODOLOGY

The primary independent variable was the input mode assessed across independent groups, the modes being manual (keypad) and voice (DVI). Twenty subjects were randomly assigned to either voice or keypad correction mode conditions (10 in each). The feedback in the data input task was simulated to include the error types of substitution, rejection, insertion & non-error feedback.

Each experimental run consisted of training, quantitative data collection, and subjective workload rating measures. Subjects followed a continuous visual tracking task on a head-up display (HUD) in a cockpit simulator. The tracking task was controlled manually using the flight stick. Concurrently subjects were required to repeat digit strings, presented on HUD, into their microphone and assess the visual feedback presented on an LCD readout next to the HUD. Corrections were to be made to the feedback by voice (using a Marconi "Macrospeak" ASR with a set vocabulary) or by keypad (directly in front of them below the HUD), depending on the assigned condition. Twenty-four digit strings in total were presented during the experiment.

Each session consisted of a training/practice session, the experimental run, and subjective ratings. The objective measures were as follows: tracking accuracy in terms of on-target accuracy (seconds) and average tracking error (RMS pixels), correction times (seconds), and post correction error frequency. The subjective measures were taken using the NASA Task Load Index (NASA TLX). An overall workload score and the weighted ratings of workload on the dimension defined in the TLX were assessed (i.e., mental demand, physical demand, temporal demand, performance,effort and frustration).

# RESULTS
## The Objective Data

A between groups multivariate analysis of covariance assessed the variables of on-target tracking accuracy, average tracking error, final error frequency and total 'mission' duration across voice and keypad correction modes. The multivariate statistic showed a difference in performance between voice and keypad correction techniques on the combined dependent variables (p=.007). The univariate statistics, however, suggested the main difference occurred in the overall "mission" times (p=.006) with keypad correction proving to be faster. The univariate statistics were not adjusted for the fact that four separate test rather than one were done. Therefore, no difference in tracking error across correction modes was evident suggesting a more favourable picture for keypad correct on the measures. The canonical correlation indicated that 64% of the variability was attributable to the between-groups differences on the multiple dependent variables.

A repeated measures analysis of covariance was performed on the variables of correction times to assess in more detail the effects of error type and digit string length across correction mode. The descriptive statistics showed a consistent trend for keypad correct being the shorter in duration which proved to be a significant main effect (p=.002). Other main effects were found in that longer digit strings take longer to correct (p=.000) and correction times were different for the different error types (p=.000) with insertion and rejection errors taking longer to correct. A significant interaction of DVI*error type illustrated that keypad correction was faster on rejection, substitution and insertion errors (p=.000)

On the basis of these results the hypothesis that voice correction would produce faster reaction times was rejected. In addition, the null hypothesis that there would be no difference between correction modes was also rejected due to the significantly larger correction times experienced under voice correction.

## The Subjective Workload Data

An independent t-test on the overall workload scores across correction mode showed no statistically significant difference (p=.326; $w^2$=.001). An analysis of variance was submitted on the individual workload dimensions as an exploratory measure. Mental demand for both correction modes was relatively high while voice correction was rated significantly lower on the dimension of **effort** (p=.026). Voice correction was rated more favourably on the **performance** dimension, but slightly more frustrating to use. These two results were not statistically significant. The canonical correlation indicated the percentage of variance attributable to correction mode was 36%.

## DISCUSSION

The first point to come from the results was that DVI at least equivalent to a conventional editing method in some respects. Head-up attention and post correction error intrusion data were comparable. However, when focussing on correction times it was evident that the case for DVI deteriorated as voice correction took consistently twice as long to complete than keypad correction. Two possible sources of difficulty in DVI correction were recognition difficulties and human resource incompatibility. Misrecognitions on the part of the speech recogniser system, typically between words like "eight" and "delete", and "insert" and "enter" introduced further errors in the digit strings which required additional correction times. Nonrecognitions only added additional time through the subject having to repeat a command. Further thought will therefore need to be given to appropriate vocabulary for error correction.

Human resource compatibility theory may be involved when considering cursor movements required for correction. It could be argued that tracking cursor movement is essentially a spatial tracking task, so resource compatibility theory suggests that this would be more suited as a manual task whereas correction commands themselves would be more effectively done by voice. This outlines that voice systems may probably need to be an interacting rather than a parallel avionics system (i.e. combining voice and keypad systems on some tasks rather than giving the pilot the option of using one or the other) hence maximum the effectiveness of the interface.

Turning to the question of subjective workload, the correction strategies compared here were equivalent on overall workload scores (i.e., no significant difference was evident across correction modes). Therefore DVI did not reduce subjective workload as one might have expect. On the individual dimensions the only significant difference manifest keypad correction as being more of an effort, possibly due to subjects having to monitor not only the head up display and feedback but the keypad as well. Mental demand was high for both correction modes. Small, though non-significant differences occurred where voice correction was viewed slightly more positively on performance but was contrastingly rated as more frustrating to use. It is possible that subjects felt overall performance was better through using a system they felt kept them in a head-up posture even though it was novel and slightly more frustrating to use.

As the interface stands at present, DVI is at least comparable to keypads in allowing the maintenance of a head-up posture. Whereas there is promise in the use of speech recognition and synthesis technology in the cockpit there are some outstanding human factors problems relating to speech recognition errors. A subsequent study has been undertaken to assess different error correction techniques taking into account aspect of human resource compatibility and subjects experience in using DVI systems (talking to machines).

## REFERENCES

Beckett, P. (1986), Voice Control in Cockpit Systems. Proceedings form AGARD AMP Symposium, Information Management and Decision Making in Advanced Airborne Weapons Systems.
Bell, R., Bennett, M.E. & Brown, W.E. (1982). Direct Voice Input for the Cockpit. In: AGARD Advanced Avionics and the Military Aircraft Man/Machine Interface.
Berman, J.V.F. (1986), Speech Recognition Systems in High Performance Aircraft: Some Human Factors Considerations. Institute of Aviation Medicine Report No. 646.

England, P., Harlow, R. & Cooke, N. (1986), Some Experiences of Integrating Avionics Systems on the Civil Flight Deck. R.A.E. Technical Memorandum FS(B) 653.

Eysenck, M.W. (1984), A Handbook of Cognitive Psychology. LEA.

Hays, W.L. (1988), Statistics (4th edition). Holt, Rinehart & Winston.

Laycock, J. and Peckham, J.B. (1980), Improving Pilot Performance Whilst Using Direct Voice Input. R.A.E. Technical Report 80019.

Lea, W.A. (1979), Critical Issues in Airborne Applications of Speech Recognition. Naval Air Development Centre, N62269-78-M-3770.

Little, R. (1986), Flight Evaluation of a Speech Recognition and Speech Output System in an Advanced Cockpit Display and Flight Management System for Helicopters. Proceedings of AGARD-AMP Symposium on Information Management and Decision Making in Advanced Airborne Weapons Systems, Toronto.

McGuinness, B. (1987), An Assessment of Feedback Modes for DVI, MSc thesis, Cranfield Institute of Technology, Bedford.

Norusis, M.J. (1988), SPSS-X Advanced Statistics Guide (2nd edition). SPSS Inc.

Poock, G.K. (1980), Experiments with Voice Input for Command and Control: Using Voice Input to Operate a Distributed Computer Network. Naval Postgraduate School, NPS 55-80-016.

Shaffer, L.H. and Hardwick, J. (1969), Errors and Error Detection in Typing. Quarterly Journal of Experimental Psychology, 21, 209-213.

Smith, K. and Bowen, W.J.C. (1986), The Development of Speech Recognition for Military Jet Aircraft. Proceedings from Military Speech Technology Conference, Arlington, Virginia.

Warwick, G. (1986), Eurofighter avionics: How advanced? Flight International, 4 October 1986, 28-35.

Welch, J.R. (1977), Automatic Data Entry Analysis. Rome Air Development Report No. RADC-TR-77-306.

White, R.G. (1987), Speaking to Military Cockpits. Technical Memorandum FS(B) 671 [RAE Farnborough].

White, R.G. and Beckett, P. (1983), Increased Aircraft Survivability Using Direct Voice Input. Proceedings from AGARD-FMP Symposium, Flight Mechanics and Systems Design Lessons for Operational Experiences, Athens.

Wickens, C.D. (1984), Engineering Psychology and Human Performance. Merrill, Ohio, USA.

Wickens, C.D., Sandry, D.L. and Vidulich, M. (1983), Compatibility and Resources Competition between Modalities of Input, Central Processing and Output. Human Factors, 25, 227-248.

Wickens, C.D. and Yili Liu. (1988), Codes and Modalities in Multiple Resources: A Success and a Qualification. Human Factors, 30(5), 599-616.

# Human thermal responses in crowds.

T. L. Braun and K. C. Parsons

Human Modelling Group
Department of Human Sciences,
University of Loughborough
Leicestershire LE11 3TU
United Kingdom.

There are many situations where humans are in crowded environments e.g. in a lift. Previous research has concentrated mainly on the psychological effects, this experiment investigated the effects on human thermal response. Climatic chamber experiments were carried out on four human subjects, under 3 levels of crowding. These were standing freely, loosely packed and tightly packed. Thermal physiological responses were measured in terms of mean skin temperatures, internal body temperatures and sweat loss. Subjective and behavioural measures were also taken. The results showed a significant affect of crowd density. A computer based model of human thermoregulation was modified to allow for crowding. A comparison with experimental data found that further modifications would be required to predict allowable exposure times before collapse due to heat stress.

## INTRODUCTION

It is generally accepted that air temperature, radiant    temperature, relative humidity and air movement across the body are the environmental factors which affect the human thermal response. In addition, the clothing worn and activity of the individuals are also important.

Emphasis is usually on man 'experiencing' the thermal environment in isolation from other individuals. Yet in the many situations the environment often involves the interaction, sometimes close, with other individuals. For example commuters on the London Underground at peak time, marathon runners, the audience at a 'pop' concert etc. Some of the incidents are much less pleasant than others and members of a group may differ in their experience, depending on their specific needs i.e. the individual will 'feel crowded' when their demand for space exceeds the available space.

In this investigation the term crowding refers to high occupant density. At this level of crowding , density alone would be sufficient to promote the experience of crowding and the relative effects of social and personal determinants will decrease Stokols (1972). Humans in close proximity may restrict heat loss from the body by convection, radiation and in particular by evaporation of sweat in hot conditions. Previous research on crowding has concentrated on subjective responses and behavioural observation. While measures of both were taken and used for comparison, the primary purpose of the experiment was to measure the human thermal response to a crowded environment, i.e the physiological strain imposed on the thermoregulatory system.

233

Since it is of great benefit to be able to accurately predict the effects of human exposure to different thermal environments, a computer based model of thermoregulation (the 2 node model - Nishi and Gagge 1977) was modified to allow for crowding and predictions were compared with experimental data.

## THE 2-NODE MODEL

The 2-node model is the simplest mathematical model to take into account the body's thermoregulation process. The body is considered to consist of two regions, the body core and the skin shell (see fig. 1)

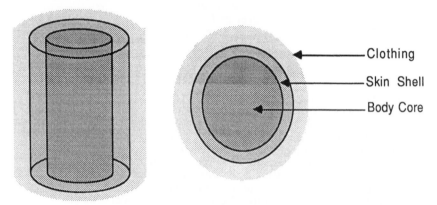

Figure 1. Diagrammatic representation of the 2-Node Model of Thermoregulation.

Metabolic heat is produced in the central core. Some of this heat is lost directly to the surroundings by respiration and the rest is exchanged at the skin surface. At the skin a proportion is lost by evaporation and the remainder is conducted through the clothing to be lost by radiation and convection. The physiological response of the model provides the reaction of the clothed person, through time, to exposure in hot, moderate and cold environments.

The modification to the model assumed that heat loss by evaporation would be the major physiological response affected by the 'crowded' environment. A coefficient was introduced into the program, ranging from '0', which represents minimal heat loss due to sweating (expected from total restriction), to '1' which represents normal conditions i.e. no crowding. This coefficient was a value estimated by the experimenter. Given that significant heat lost by the body is lost from the head, it would be reasonable to assume that the value for any crowded condition will not be less than 0.2 since heat loss from the head will not be restricted by a crowded condition. It should however be noted that no account was taken of increased temperature and humidity which would occur if the crowd were in an enclosed space with low ventilation.

## EXPERIMENTAL STUDY

Climatic chamber experiments were carried out on four male subjects under three conditions of crowd density, (see fig. 2, 3, 4). In all experimental sessions the subjects and other 'crowd' members were required to stand; this allowed a higher occupant density for crowding that could not have been achieved if seated. All participants in the crowded situation were male, since, pilot experiments with mixed groups lead to an uneasy atmosphere causing a high level of fidgeting.

Experimental conditions

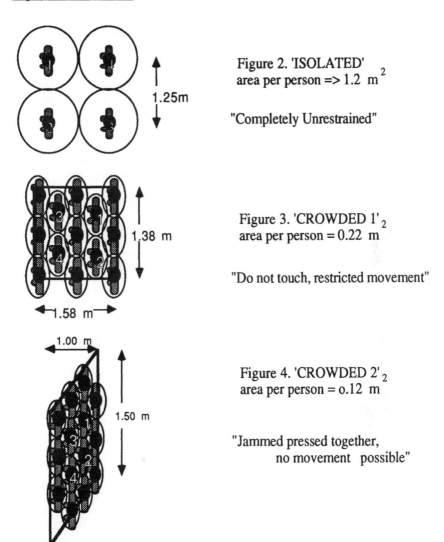

Figure 2. 'ISOLATED'
area per person => 1.2 $m^2$

"Completely Unrestrained"

Figure 3. 'CROWDED 1'$_2$
area per person = 0.22 $m$

"Do not touch, restricted movement"

Figure 4. 'CROWDED 2'$_2$
area per person = o.12 $m$

"Jammed pressed together,
            no movement  possible"

The exposure period in the thermal chamber was 60 minutes for each experimental condition and in the 'crowded' situations the subjects stood in 'isolation' for a preliminary 10 minutes so that a steady thermal state could be achieved. The time period chosen was based on the results from Kogi (1979) who suggested that a journey of 60 minutes represented an approximate tolerance limit for a physiological overcrowded vehicle.

It was preferred that the thermal environment chosen was within a comfort range. Relative humidity was chosen at 50% and air Temperature 25$^O$C.

Physiological Measures

The physiological response to the environment was measured in terms of the mean skin temperature (Ramanathan Weighting, see Mitchell, d. et al. 1969) and core temperature (using aural thermistors insulated from the external environment )

Metabolic rate , was measured using a douglas bag collection half way through each of the three sessions.

One of the aims of the study was to measure the reduction in evaporative heat loss that may be caused by the effects of crowding.  By determining the weight of perspiration absorbed in the clothing an approximation of 'sweat not evaporated' was calculated.

Subjective Measures

Thermal sensations are not dependent solely on the temperature of the environment.  The opinion of what constitutes a comfortable environment is sometimes very wide.  In this investigation it is necessary to associate the physiological response with the subjective sensations in both the crowded and isolated experiments so that conclusions, about discomfort caused by crowding, can be drawn.

The questionnaires included two types of question.  Firstly there were 3 rating scales; thermal sensation, thermal comfort and 'stickiness' (dampness felt by unevaporated sweat).  The rater was required to rate how each of the body segments felt i.e. head trunk, arms, hands, legs, and feet plus the overall feeling.  The final question was based on a series of bi-polar (opposite) adjectives e.g relaxed-aggitated and designed in a semantic differential format.

Due to the close proximity of subjects within the crowded conditions the questionnaire was kept brief and easy to answer so that any disruption,  would be minimised.  The four subjects would complete a questionnaire every 15 mins during the 3 conditions and in those experiments where there was a preliminary 10 minute period an additional questionnaire would be completed.

Observational  measures

Although it was stressed that all crowd members should remain on their marked squares, it would be unrealistic to assume that all individuals will maintain their position in such close proximity for the whole experimental period.  A behavioural response should be expected and therefore taken into account.

The movement of all members of the group would be recorded over 5 minute intervals throughout the session by an observer.

A note should be made here of the applicability of the results.  The study of the thermal effects of crowding in this investigation was within the context of a laboratory situation.  The dissimilarities between the experimental situation and the urban environment limit the ability to generalise and other contextual factors will need to be considered.

RESULTS

The following page displays 3 figures gained from the physiological measurements.  Figure 5 and 6 represent the experimental data gained from the readings of the skin and aural thermistors on the four subjects .  Figure 7 represents the difference of weight measurements taken before and after the exposure period.  A comparison of predicted and experimental data showed that the modified model in the present form did not make a good approximation to either internal body temperature or mean skin temperature.

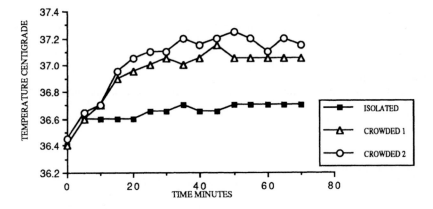

Figure 5. Mean values for 'core
temperature' in the 3 experimental conditions.

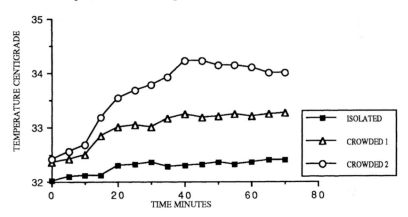

Figure 6. Average of mean skin temperature
calculation for 4 subjects in the 3 conditions.

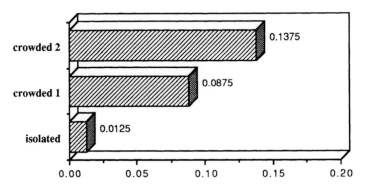

Figure 7. Average 'sweat not evaporated' (kg)

DISCUSSION

The subjects body temperature was hotter in CROWDED 2. Yet the difference between the two crowded situations does not truely reflect the substantial decrease in area per person. From the observational recordings it seems that crowd members had moved slightly away from the subjects to avoid contact, so reducing the crowd density and subsequently the thermal effects. Further investigation would require a physical restriction to maintain a fixed position in the high density condition.

Comparing the 3 different environments, the initial 30 mins of both crowded conditions shows a significant increase in 'arousal' (from behavioural measures). From the subjective responses the greatest intolerance is also during this period. This can be explained by the subjects interpretation of the situation i.e tolerance depends on the amount of time that the individual expects to remain within the crowded environment.

The amount of 'sweat not evaporated' was significantly greater in the crowded conditions. It was expected that the denser, 'touching' crowd would have shown a more substantial increase. However as mentioned above the high density was not strictly maintained and therefore the result obtained is a accurate measurement of the high density situation. Alternatively we should perhaps consider that the 'microclimate' within the crowd was significantly saturated so that a further increase in density would not restrict evaporation much more but instead increase the body temperature.

Since the front and back of the trunk are one of the most effective areas for heat loss by evaporation, and in addition these areas represent those restricted by crowding, it is not surprising that the sensation of "stickiness" was predominant in this area.

CONCLUSIONS

For conditions normally within the comfort range low density crowds can cause significant thermal strain. This can be attributed to restrictive evaporative loss in the thermal microclimate, as well as to other heat exchange mechanisms.

REFERENCES

FANGER, P.O. 1970 Thermal Comfort (Danish Technical Press.)

FREEDMAN, J,L. 1975 Crowding and Behaviour(Freeman & Company)

GAGGE, A.P., NISHI, Y. 1977. Heat exchange between human skin surface and thermal environment. In: Handbook of Physiology, Section 9 Chapter 5, p69-91.

HARDY, J.D. , GAGGE, A.D. and STOLWIJK, J.A.J. 1970 Physiological and Behavioural Temperature Regulation (Charles c. Thomas)

KOGI, K. 1970 , Passenger Requirements and Ergonomics in Public Transport. Ergonomics, , vol.22 p631-639

McINTYRE, D. 1980. Indoor Climate (Applied Science Publishers)

PINNER, B., RALL, M., SCHOPLER, J. and STOKOLS, D. 1973, Physical, Social and Personal Determinants of the Perception of crowding. Environment and Behaviour,, p87-115

STOKOLS, D. 1972, On the Distinction Between Density and Crowding. Psychological Review, , vol. 79, p275-277

TEICHNER, W.H. 1967, The Subjective Response to the Thermal Environment. Human Factors, , vol.9, p497-510

# OFFICE LIGHTING FOR VDT WORK: COMPARATIVE SURVEYS OF REACTIONS TO PARABOLIC AND LENSED-INDIRECT SYSTEMS.

A. HEDGE

Department of Design and Environmental Analysis,
Cornell University,
Ithaca, NY 14853-4401.

A study of the effects of lensed-indirect uplighting and parabolic downlighting on computer workers is described. Results from a pre-installation survey, and 3 month and 15 month post-installation surveys showed that both installed systems improved the office lighting. However, workers with lensed-indirect uplighting reported fewer complaints of eyestrain and eye focusing problems. Workers with parabolic downlighting reported more lighting-related problems and productivity losses, even though almost half of these fixtures had been modified by users. Overall, workers expressed a strong preference for lensed-indirect uplighting.

## INTRODUCTION

The increase in computer use in offices over the past decade has been accompanied by increasing concerns about the relationships of office lighting to the satisfaction, visual health, and productivity of workers. A recent national survey found that eyestrain tops the list of health complaints among U.S. office workers (Harris, 1989). Although a number of lighting solutions are now available for computer offices (I.E.S., 1989), there has been little comparative research on the effects of alternative systems. The present study was a field experiment designed to test the effects of two "state-of-the-art" types of lighting believed to effectively reduce glare and other lighting problems for computer workers: lensed-indirect uplighting, and parabolic downlighting.

## METHOD
### Experimental design

Three surveys were conducted in a virtually windowless 7,000 $m^2$ U.S. office building occupied by the Xerox corporation. Prior to renovation, the office was lit by direct light

distribution luminaires with prismatic diffusers. There were
four types of offices in the building: 18 3x4.5m enclosed
offices with windows; 27 3x4.5m enclosed offices without
windows; 26 3x3m enclosed offices without windows; and 82 3x3m
open-office cubicles without windows. In June 1988 a pre-
installation (PI) worker survey was conducted. As part of a
scheduled renovation, lensed-indirect uplighting was installed
in one-half of the building and parabolic downlighting was
installed in the other half. Lensed-indirect uplighting
(LIL) has ceiling suspended luminaires which provide upward
light which is then reflected down to the workplane by the
ceiling and walls. Parabolic downlighting (PBL) has ceiling
recessed luminaires which are shielded with a grid of
parabolic louvers to provide direct lighting to the workplane.
Both systems were installed according to standard practice:
the LIL system provided 500 lux, and the PBL system provided
750 lux at the workplane. The numbers of offices with each
type of lighting were balanced within reason. During the
renovation, all walls were re-painted light grey, discoloured
ceiling tiles were replaced, and new office furniture,
carpets, and fluorescent lamps were installed. In June 1989, a
3 month post-installation (3PI) survey was conducted, and in
June 1990, a 15 month post-installation (15PI) survey was
conducted. All employees in this building used computers for
their professional and technical work.

Survey Questionnaire

Separate versions of a self-administered questionnaire, all
with common core questions, were developed for use in each of
the surveys. Questionnaires collected information on
environment conditions and health complaints, including the
frequency of, disruption to and the amount of productive work
time lost because of environmental problems and health
complaints over the previous three months. Questionnaires also
asked about general worker information and lighting
preferences, and job satisfaction and job stress. The 15PI
questionnaire asked about how users had adapted their
workspaces and/or their lighting.

Procedure

The procedure was standardized for all surveys. A researcher
distributed and collected a self-administered questionnaire on
the same day. A questionnaire and reply envelope were left
where a worker was away from his/her desk. Illumination levels
at the center of each worker's normal working surface, e.g.
desk, computer table, and any modifications to the office
lighting, were recorded at the same time as the questionnaire
was distributed.

Survey samples

One hundred and forty seven workers completed the PI survey

(92% return rate), 90 workers completed the 3PI survey (61% return rate), and 121 workers completed the 15 PI survey (82% return rate).

## Data Analysis
Survey data were analyzed using the Statistical Package for Social Sciences (SPSSXv.4.0).

## RESULTS
Some results from the PI and 3PI surveys previously have been reported(Hedge, Sims, and Becker, 1989), however, this paper presents results from all three surveys. There were no significant differences in job satisfaction and job stress between LIL and PBL workers in the surveys. Only responses to questions on lighting and visual health showed any consistent significant differences, and given the minimal researcher involvement with workers and the duration of the study it is unlikely that the results are attributable to non-lighting effects, such as the "Hawthorne effect".

### Visual Health Complaints
Figure 1 shows the prevalence of daily complaints of eye focusing problems and figure 2 shows the prevalence of daily complaints of tired, strained eyes for all respondents in the PI survey, and then separately for LIL and PBL workers in the 3PI and 15PI surveys. Daily complaints of these visual health problems were significantly lower for workers with the LIL system compared with prismatic lighting, and the PBL system. However, after 15 months workers with the PBL system had extensively modified their lighting, and there was no significant difference between daily visual health complaints for the LIL and PBL systems, although the overall prevalence of eye focusing problems still was significantly lower for the LIL workers ($\chi^2$ = 9.33, p<0.02).

### Lighting Modifications
When the 3PI survey was conducted, lighting fixtures had been modified at the request of workers in 28% of the PBL offices, but none of the LIL fixtures had been modified. When the 15PI survey was conducted lighting fixtures had been modified in 48% of the PBL offices, and these modifications included disconnecting one, two, or all three fluorescent lamps in the fixture, and repositioning fixtures in two of the offices. In one LIL office the fluorescent lamps in the fixture had been disconnected because the worker reported sensitivity to flicker from any type of fluorescent lighting.

### Illumination Levels
The average illumination levels in the surveys were 593 lux (PI); 526 lux (3PI-LIL); 671 lux (3PI-PBL); 475 lux (15PI-LIL); and 603 lux (15PI-PBL).
In the PI and 3PI surveys respondents were asked to rate

Figure 1. Daily complaints of eye focusing problems

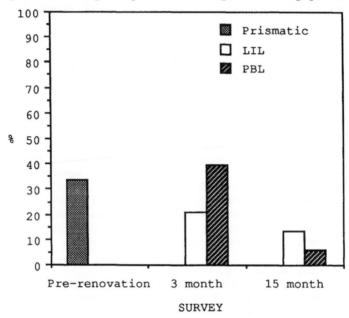

Figure 2. Daily complaints of tired, strained eyes

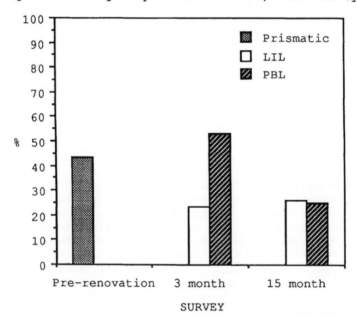

their lighting on a 7 point scale (1=too bright, through 4= comfortable, through 7=too dim) at the time that illumination was being measured. Illumination levels did not correlate with ratings of lighting comfort for either the prismatic or PBL systems, but there was a significant negative relationship for the LIL system ($r= -0.33$, $p<0.05$).

In the PI survey, ratings of lighting comfort were significantly correlated with eye focusing problems ($r= -0.31$, $p<0.05$) and eyestrain complaints ($r= -0.30$, $p<0.05$), but illumination levels were not correlated with these complaints.

In the 3PI survey, illumination levels were correlated with eye focusing problems ($r= 0.46$, $p<0.01$) and eyestrain complaints ($r= 0.44$, $p<0.01$), but ratings of lighting comfort were not correlated with complaints. For the PBL workers there were no correlations of visual health symptoms and illumination, however, eyestrain complaints were correlated with ratings of lighting comfort ($r= -0.39$, $p<0.05$). In the 15PI survey, illumination levels did not correlate with any visual health complaint.

### Lighting and Work Function

In the PI and 3PI surveys respondents were asked to rate, on a 4 point scale ( 1 not at all bothersome through 7 very bothersome) how bothersome their lighting was for the work which they were doing. Mean ratings for glare and illumination problems are shown in Table 1. The LIL system consistently was rated as the least bothersome.

Table 1.    Mean "bothersome" ratings for each type of lighting for the PI and 3PI surveys.

| Complaints | Prismatic | PBL | LIL |
|---|---|---|---|
| Worksurface glare | 2.7 | 2.6 | 2.3 |
| Screen glare | 3.0 | 3.1 | 2.6 |
| Office lighting level | 2.6 | 2.6 | 2.2 |

In the 15PI survey workers were asked which lighting problems they had experienced daily over the previous 3 months. Again the LIL system had fewest reported problems (Table 2). Workers were asked to estimate productive time losses caused by lighting problems (too bright, reflected glare, direct glare). For the LIL workers 1%-4% reported losing 1-15 minutes/day, and 1%-3% reported losing >15 minutes/day. For the PBL workers 2%-10% reported losing 1-15 minutes/day, and 8%-18% reported losing >15 minutes/day.

### Lighting preferences

In the 3PI survey 80% of LIL workers said that they preferred the LIL system to any other lighting they had experienced, and almost 50% of PBL workers also expressed this preference. In the 15PI survey, 70% of LIL workers said that they still preferred the LIL system to any other they had experienced, and over 75% of PBL workers expressed a preference for the LIL system.

Table 2.    Daily complaints about each type of office lighting in the 15PI survey.

| Item | LIL | PBL | $\chi^2$ | P |
|------|-----|-----|----------|---|
| unsatisfactory lighting | 26% | 54% | 10.6 | <0.001 |
| direct glare from lighting | 26% | 56% | 10.9 | <0.001 |
| lighting too bright | 15% | 55% | 22.0 | <0.001 |
| lighting is unpleasant | 17% | 50% | 14.1 | <0.001 |
| glare on paper | 17% | 43% | 9.2 | <0.01 |
| uncomfortable lighting | 17% | 50% | 14.5 | <0.001 |
| harsh shadows from lighting | 6% | 18% | 4.7 | <0.05 |
| glare on computer screen | 40% | 57% | 3.5 | <0.06 |

## DISCUSSION

This study showed that office lighting significantly effects the visual health and related problems experienced by computer users. Illumination generally was not a good indicator of these problems. Both the LIL and PBL systems created better lighting for computer work than that from prismatic diffusers. The LIL workers consistently reported fewer visual health problems, fewer lighting-related problems, and better productivity than the PBL workers. Overall, reactions to the LIL system were significantly more favorable than those to the PBL system, and workers said that they preferred working under the LIL system.

## ACKNOWLEDGEMENTS

This research was supported by a grant from the Peerless Lighting Corporation, Berkeley, CA, and was conducted with the permission of the Xerox Corporation, Rochester, NY.

## REFERENCES

Harris, L., and Associates, 1989 Office Environment Index, (Grand Rapids: Steelcase, Inc.).

Hedge, A., Sims, W.R. Jr., and Becker, F.D., 1989 Lighting the computerized office: a comparative field study of parabolic and lensed-indirect lighting systems. In Proceedings of the 33rd Annual Meeting of the Human Factors Society, Vol. 1, (Santa Monica: Human Factors Society), pp. 521-525.

I.E.S., 1990 VDT Lighting: RP-24-1989, (New York: Illuminating Engineering Society of North America).

# USE OF WRIST RESTS BY DATA INPUT VDU OPERATORS

DR C. A. PARSONS

Post Office Research Centre
Swindon, Wilts.
SN3 4RD.

Forty full time data input VDU operators tried nine prototype wrist rests differing in height, profile and consistency. Information about existing upper limb discomfort and wrist rest preferences, suitability and possible improvements was collected using a self administered questionnaire. Four subjects suffered from discomfort in the wrists alone, nine in neck/shoulder region and thirteen in both wrist and neck/shoulders. Only four subjects found the rests useful, three in reducing wrist discomfort alone and one both wrist and neck/shoulder aches. None of the other subjects found them useful and seven commented that discomfort increased when using a wrist rest.

INTRODUCTION
Many studies have shown a high incidence of neck and upper limb discomfort in VDU operators due to the constrained posture and resultant high static muscle load, Hunting et al (1981), Parsons and Thompson (1989), Hagberg and Sundelin (1986). One factor influencing this was found to be inadequate support for wrists and forearms, (Grandjean, 1988). Wrist rests are frequently recommended to reduce both the static muscle load on the neck and shoulders and disorders of the wrists, (Darby, 1984). Occipinto and Colombini (1985) also found a reduction in lumbar spinal pressure. Despite finding an increase in EMG activity in the trapezius and no difference in forearm EMG activity when using a wrist rest Bendix and Jessen (1986) found that most of their subjects preferred to type with their wrists supported. Sauter et al (1987) stress the importance of smooth contours and padding on wrist rests to prevent trauma by compression at the wrist.

As the literature suggested that wrist rests were useful to VDU operators a trial was carried out in a large data input unit to determine the most appropriate design of wrist rest for the type of keyboard in use.

METHOD

The trial took place using forty full time Data Input VDU Operators over a period of four weeks. The work was mainly one handed using a numerical keypad at the edge of the keyboard. The key depression rates were in excess of 10 thousand per hour. Almost all the subjects were female. Nine prototype wrist rests were formed from styrofoam with a vinyl cover. Six were padded using a thin layer of open cell foam. They differed in height, profile and padding according to the table 1 below. All wrist rests were 15 cm across and 9 cm from front to back. The sloped profile had the same angle of slope as the keyboard. The keyboards in use were 3.4 cm high at their front edge.

Table 1. Design of the wrist rests used.

|  |  | MAXIMUM HEIGHT (when compressed) | | |
|  |  | 3.4cm | 2.4cm | 3.4cm |
| PROFILE | Flat | padded | padded | no padding |
|  | Sloped | padded | padded | no padding |
|  | Curved | padded | padded | no padding |

Each subject was asked to try all the wrist rests over a period of a week and to use their prefered rest for at least half a day. Using a self administered questionnaire information was gained about; wrist rest preferences, upper limb discomfort, (and whether it was helped by using the wrist rest), and possible improvements to the wrist rests. Other comments were also invited.

RESULTS

The distribution of reported discomfort is shown in Table 2. Only four subjects found the wrist rests useful. Three stated that they reduced wrist discomfort, one both wrist and neck/shoulder aches. Three preferred the wrist rest with the same slope as the keyboard, one the curved option. Two preferred 3.4 cm height, two 2.4 cm, all preferred the padded options.

The other 36 operators did not find any of the wrist rests useful and seven commented that discomfort was worse when using a wrist rest because in order to rest their wrists they needed to change their posture. Several suggested that a padded chair arm would be preferable so that their arm was supported further back allowing free movement of the wrist and hand.

Other problems were that the vinyl covers tended to cause sweating and the rests sometimes fell off the tables.

Table 2. The distribution of discomfort suffered by the VDU
Operators

| Position | Number of Subjects |
| --- | --- |
| none | 14 |
| neck/shoulders alone | 9 |
| wrists alone | 4 |
| wrists and neck/shoulders | 13 |

DISCUSSION

It was suprising to find that only 10% of subjects found using
wrist rests beneficial and that 18% stated that they increased
discomfort. The main reason for their unsuitability may have
been that during data entry work keying is continuous and the
wrist and hand tend to move forward and backwards at a high rate
with infrequent pauses. Having the wrist supported in a fixed
position would tend to reduce the movement of the arm requiring
more movement of the fingers and thus increased activity of the
forearm muscles. Provision of support for the arm allowing free
movement of the wrist would seem more suitable, and this is
currently being investigated.

The results of this study do not support the theory that
wrist rests are of use to data input VDU operators, and may have
disadvantages.

REFERENCES

Bendix, T. and Jessen, F., 1986, Wrist support during typing.
    Applied Ergonomics 1986, 17.3, pp 162-168.
Darby, F.W., 1984, Visual Display Units, A Review of the
    Guidelines. Department of Health Division of Public Health.
    New Zealand, pp.114-124
Grandjean, E., 1988, Fitting the Task to the Man 4th edn
    (London: Taylor & Francis).
Hagberg, M. and Sundelin, G., 1986, Discomfort and load on the
    upper trapezius when operating a wordprocessor. Ergonomics 29
    12 pp.1637-1645.
Hunting, W. et al, 1981, Postural and visual loads at VDT
    workplaces. Ergonomics 24 12 pp. 917-931.

Occipinto, E. and Colmbino, D., 1985, Sitting posture: analysis of lumbar stresses with upper limbs supported. Ergonomics 1985, 28 9 pp 1333-1346.

Parsons, C.A. and Thompson, D., 1989, Comparison of cervical flexion in shop assistants and data input VDT operators. Contemporary Ergonomics 1990 pp 299-304

Sauter, S.L. et al, 1987, Case of wrist trauma in keyboard use. Applied Ergonomics 18 3 pp. 183-186.

# 1992

| Aston University, Birmingham | 7th–10th April |
|---|---|
| **Honorary Meetings Finance Secretary** | Sandy Robertson |
| **Conference Manager** | David Girdler/Jonathon Sherlock |
| **Chair of Meetings** | Dave O'Neil |
| **Programme Secretary** | Ted Lovesey |

## Social Entertainment

An evening's entertainment was provided with a trip to see some greyhound racing.

| Chapter | Title | Author |
|---|---|---|
| **Keynote Address** | Causes of motion sickness | M.J. Griffin |
| **Drivers and Driving** | A survey of car driver discomfort | J.M. Porter, C.S. Porter and V.J.A. Lee |
| **Selection and Workstress** | The occupational well-being of train drivers – An overview | R.A. Haslam |
| **Ergonomics Applications** | The importance of ergonomics in plastic surgery | D. Falcao and A. McGrath |

# CAUSES OF MOTION SICKNESS

## MICHAEL J. GRIFFIN

Human Factors Research Unit
Institute of Sound and Vibration Research
University of Southampton
Southampton SO9 5NH
England

Motion sickness is not an illness but a normal response to motion
which adversely affects many fit and healthy people during leisure
activities and at work. A variety of different motions can cause sickness
and reduce the comfort, impede the activities and degrade the well-being
of both those directly affected and those associated with the motion sick.
Studies conducted in laboratory and field environments have provided
information on the causes of sickness. This paper summarises the data
underlying current understanding of motion sickness; some of the
information can be applied to design by ergonomists and by others.

## INTRODUCTION

Ergonomics has been defined as *an applied science concerned with the characteristics of
people that need to be considered in designing and arranging things that they use in order
that people and things will interact most effectively and safely*. The 'people' who use the
'things' will likely have opinions on the 'design' and 'arrangement' produced by this
applied science. Even 'applied scientists' (e.g. ergonomists) may find it difficult to
distinguish their 'opinions' of designs and arrangements from that body of knowledge
which has been accrued by scientific study.

Motion sickness is just one of many areas of ergonomics on which 'people' often
express 'opinions' based more on their experience than on scientific study. People have
are said to have the ability to 'reason from the particular to the general'. Opinions on
motion sickness (and other subjects) are frequently constructed from fragmented
experience rather than comprehensive study. Opinions which are derived without a study
of the available evidence may be condemned. Yet scientists attempt to construct an
understanding of the subject from fragmented evidence: ideally, they develop theories
and test hypotheses. However, scientists are also prone to jump to conclusions without
testing hypotheses or checking their observations.

Some ergonomists approach ergonomics as a qualitative subject in which the
identification of the relevant variables is a sufficient conclusion. Others seek to discover,
or apply, quantitative information. Those concerned with motion sickness are similarly
divided: some evolve mathematical relationships between variables while others scoff at
the attempt.

Some ergonomists believe that ergonomics is primarily concerned with the application

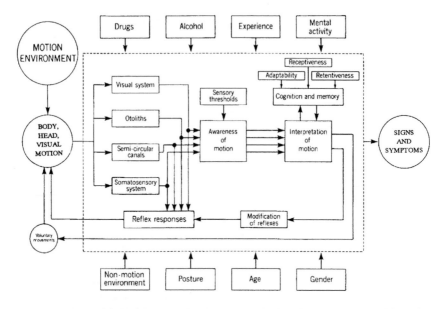

**Figure 1** Model of factors involved in causation of motion sickness (from Griffin, 1990).

of knowledge to design. Others feel that ergonomics is also concerned with discovering the characteristics of people that need to be considered in design. Most scientists concerned with motion sickness do not classify themselves as ergonomists: relevant characteristics of people are determined by vestibular physiologists while the data may be applied by naval architects. Neither all vestibular physiologists nor all naval architects are ergonomists, but the application of their combined knowledge may be within the province of ergonomics.

This paper attempts to provide a summary of some *science concerned with the characteristics of people that need to be considered* in situations in which motion sickness may occur. The paper also addresses the application of the data to *designing and arranging things* that people use.

## DEPENDENT VARIABLES: SIGNS AND SYMPTOMS OF MOTION SICKNESS

The study of the way that people and things interact requires a 'dependent variable'. In the study of motion sickness the dependent variables are the signs and symptoms of sickness. As with other areas of ergonomics, some of the dependent variables shown in Table I are open to different interpretations - both by subjects participating in experiments and by experimenters. Unlike some other areas of ergonomics, there is one easily observed dependent variable: vomiting. Although this is not the only effect, or always the most important effect, it has become the most commonly used indication of motion sickness. Many studies have therefore reported the severity of motion exposures in terms of the percentage of exposed persons who vomit.

## INDEPENDENT VARIABLES

The variables associated with causing, or modifying, the incidence of motion sickness are great in number and varied in nature. Figure 1 provides a diagrammatic illustration of some of the variables and their possible interrelationships.

### Individual variability

Some variables are related to subject characteristics and might be classified as 'physiological' or 'psychological'. Studies have found significant correlations between sickness susceptibility and physiological function, but the only well recognised conclusion is that persons without a functioning vestibular system (i.e. semi-circular canals and otoliths) are immune to motion sickness.

Psychological attributes may be associated with susceptibility to sickness, but the evidence is far from conclusive. It is clear that sickness is not confined to those with abnormal psychological attributes. Behaviour can result in differing susceptibility but this may often be due to differing excitation of either the vestibular system (due to head movements) or the visual system (due to differing field of view) or the proprioceptive system (due to postural variations).

**Table I**    Signs and Symptoms of Motion Sickness

Vomiting
Retching
Nausea
Epigastric symptoms
Colour changes
Cold sweating
Irregular breathing
Yawning
Drowsiness
Dizziness
Headaches

Studies have consistently shown an increased susceptibility to motion sickness among females (see Griffin, 1991b; Lawther and Griffin, 1988b). Age also affects susceptibility with immunity in the very young (less than about 1 year old) and an apparent, but gradual, decrease in sensitivity through adult life. Posture has widely been considered important with many recommendations to adopt a horizontal position on ships. However the full understanding of why this posture may be beneficial and how the optimum posture depends on the direction of the motion stimulus awaits further study.

### LABORATORY INVESTIGATIONS

A rigid object has six possible axes of motion (see Figure 2). In each axis the motion may be continuous or oscillatory. The orientation of the motion with respect to body posture and with respect to gravity may also vary. The human body is often not rigid (e.g. the head may rotate relative to other parts of the body) and motions may occur in more than one axis. Additionally, motion of the visual field can cause sickness and the nature of the visual field can be important even when it is not moving. Even within the laboratory it is therefore an exhaustive task to even identify the motions which are nauseogenic. It is not surprising that the characteristics of those motions which do, and those motions which do not, cause motion sickness are only partially established (see Griffin, 1991a).

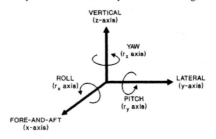

**Figure 2** Axes of motion.

### Continuous rotation

Continuous rotation of the body about the vertical axis causes little sickness if there are no head movements and the eyes are closed. However, small amounts of head movement cause considerable sickness (by the 'cross-coupled' or 'Coriolis' effect), see below.

Continuous rotation about an off-vertical axis produces a 'rotating acceleration vector' from the force of gravity: this can be very nauseogenic. When the rotation occurs about the horizontal axis it is sometimes called 'barbecue spit rotation' and can produce sickness in less than five minutes (e.g. Benson and Bodin, 1966). It appears to make little difference whether rotation about the horizontal axis is about the x-axis, the y-axis, or the z-axis of the subject but there tends to be less sickness when there is an external visual frame of reference (Leger *et al*, 1981).

The rate of development of sickness with off-vertical rotation depends on the angle of rotation to the vertical and on the rate of rotation. Greatest susceptibility to motion sickness may occur when the rate of rotation is in the range 15 to 20 r.p.m. The oscillation frequency caused by the rotating linear acceleration vector is then similar to that giving greatest susceptibility to vertical oscillation (see below). There appears to be little sickness with less than about 5 degrees tilt but, as the angle of tilt increases, so sickness increases (Miller and Graybiel, 1973). In the 'tilted-axis rotation test' (TART), blindfolded standing persons are rotated about their z-axes while they (and the axis of rotation) are tilted at various angles (Lentz and Guedry, 1978).

## Rotational oscillation about a vertical axis

Oscillation in yaw about a vertical axis is not very nauseogenic, but symptoms of sickness can arise if a visual search task is presented so as to require eye (or head) movements. This is the basis of the 'visual-vestibular interaction test' (VVIT) in which seated subjects are oscillated in yaw sinusoidally at 0.02 Hz with a peak angular velocity of $\pm 155$ deg.s$^{-1}$. During rotation, they identify the co-ordinates of numbers shown on a visual display within an enclosed cabin. Lentz and Guedry (1978) say that the motion itself, when experienced with simple visual displays or in darkness is not disturbing. A problem arises when motion-induced nystagmus is superimposed on the saccadic eye movements required during shifts of visual fixation. Guedry *et al* (1982) found that oscillation in yaw at 2.5 Hz produced no convincing signs of sickness, but oscillation at 0.02 Hz was highly nauseogenic with a head-fixed visual search task.

## Rotational oscillation about a horizontal axis

Moderate angles of roll or pitch oscillation about a horizontal axis are also not very nauseogenic, although often these motions also cause (or are closely associated with) motions in other axes which are nauseogenic. Morton *et al* (1947) exposed subjects simultaneously to roll motion (through 25.5 degrees) and a combined pitch and vertical motion through 3.6 metres at the end of a 4.9 metre arm of a seesaw. Oscillation on the seesaw alone at 0.125 Hz resulted in 40% of subjects vomiting, whereas with this motion combined with roll motion at 0.08 Hz, 33% of subjects vomited. The illness rates were similar in both conditions and the authors concluded that vertical motion from the seesaw was the cause of the sickness.

McCauley *et al* (1976) investigated response to pitch and roll motion at 0.115, 0.230 or 0.345 Hz. When the rotational motions were added to a vertical motion of 0.25 Hz at 1.1 ms$^{-2}$ r.m.s., the incidence of sickness was not significantly different from the incidence of sickness with the vertical motion alone. No subject vomited with roll motion alone (33.3 deg.s$^{-2}$ at 0.345 Hz) but two subjects vomited with this magnitude of pitch. The authors concluded that vertical oscillation was a greater cause of sickness.

## Rotational acceleration

Severe, or sudden, rotational acceleration about a vertical axis has been reported to cause symptoms of motion sickness, but the extent to which sickness will occur without a visual task or head movements is less clear. In the 'sudden-stop vestibular-visual interaction test' subjects are accelerated at 20 deg s$^{-2}$ to 300 deg s$^{-1}$ (50 rpm), maintained at this velocity for 30 seconds and then brought to a stop in 1.5 seconds. For 10 subjects receiving their first exposures with eyes open during the test, an average of about 14 stops is required before slight nausea occurs; with eyes closed an average of 38 stops is required (Lackner and Graybiel, 1979).

## Coriolis, or cross-coupled, stimulation

If, while the body undergoes constant speed rotation, the head is rotated about an axis other than the axis of constant speed rotation, nausea and other classic symptoms of motion sickness soon appear in most subjects. The test is usually conducted with seated subjects who make pitch and roll movements of their head while they are rotated about their vertical axis (i.e. during yaw rotation about the z-axis).

The problem is said to arise from a 'false' indication of head motion produced by the semicircular canals when a canal is rotated so as to change the extent to which it is within the plane of the constant speed rotation. Consider, for example, a vertical canal during constant speed rotation of the body about the vertical axis (i.e. rotation in yaw) so that the canal is not in the plane of the rotation. Assume the canal is then rotated forward in pitch, so as to become horizontal and in the plane of rotation. The rotational velocity of the canal will have changed from zero to the velocity of the constant speed rotation. The fluid within the canal (i.e. endolymph) will need to accelerate to reach the speed of rotation. Until this speed is reached the endolymph will deflect the cupula and so indicate that there has been a change of rotational velocity in the plane of the canal (i.e. in the yaw axis of the body but in the roll axis of the head, for pitch head movements). However, the only movement of the head containing the canal has been in the pitch axis. So, while the semicircular canals indicate pitch and roll rotation of the head, the otoliths and proprioceptive information indicate only a pitch motion of the head. In summary, when a person is exposed to continuous rotation about one axis and makes a rotary movement of the head about a second axis, there is a feeling of being rotated in a plane approximately orthogonal to the two true motions.

The above illustration is only a simplified introduction to the cross-coupled effect. It is also necessary to consider the signals that arise from canals that leave the plane of rotation and, since the semicircular canals are not truly orthogonal and aligned in the pitch, roll and yaw axes of the head the analysis for any simple motion can be complex.

The Coriolis Sickness Susceptibility Index (CSSI) provides a means of quantifying individual susceptibility to sickness by determining the number of head movements required during rotation at any rotation rate between 2.5 and 30 rpm to cause a defined degree of malaise (e.g. Miller and Graybiel, 1970) . The CSSI index implies that a doubling of the revolution rate will have a greater effect than doubling the number of head motions. For example, a three fold increase in the rate of rotation is approximately equivalent to a six-fold increase in the number of head movements.

The Brief Vestibular Disorientation Test (BVDT) uses a fixed rotation rate of 15 rpm; after 30 s at constant velocity, a seated subject with closed eyes makes 45 degree head movements every 30 s in the order: right, upright, left, upright, right, upright, left, upright, forward, upright. On completion of the sequence (after 330 s) the motion is stopped and the subject opens his eyes after sensations of movement have ceased. Guedry (1968) says that the results of testing 500 student pilots with this test showed that observers judged that only 5% of subjects remained unaffected after six head movements.

Studies have been conducted with subjects exposed to rotation for several days within 'slow rotating rooms'. Graybiel (1969) said that rotation at 1 rpm gives little or no disturbance of vestibular origin but "*at 10.0 rpm it is comparable to exposure on rough seas*".

Vertical oscillation

Low frequency vertical oscillation can cause sickness in both people and animals. Only two series of systematic laboratory studies of the production of motion sickness by vertical oscillation have been conducted.

In the 'Wesleyan University studies' an elevator, or lift, capable of 5.5 metres of displacement (peak to peak) was used to impart 20 minute exposures to alternating periods of constant acceleration and constant velocity so that the resulting displacement waveforms were, very approximately, sinusoidal. Various experiments were undertaken, mainly with four different frequencies (0.22, 0.27, 0.37 and 0.53 Hz). The higher frequencies resulted in less sickness (Alexander et al, 1945a) and increasing the magnitude of the motion from low to moderate magnitudes increased the incidence of sickness (Alexander et al, 1945b). At higher magnitudes, sickness rates decreased (Alexander et al, 1945c). The results have been reanalysed so as to express the vomiting incidence as a function of r.m.s. acceleration for each frequency (see Lawther and Griffin, 1987).

Investigations by Human Factors Research Inc. used a 2.4 metre square closed cabin supported by an hydraulic motion system capable of 6.1 metres of vertical displacement, ± 15 degrees of roll and ± 15 degrees of pitch (the axes of rotation were approximately 0.4 metres below the floor of the cabin). The first study exposed groups of 20 to 33 subjects to each of 14 experimental conditions involving various magnitudes of vertical sinusoidal motion at 0.083, 0.167, 0.333 and 0.500 Hz (O'Hanlon and McCauley, 1974). Seated subjects sat with heads on a headrest during exposures of up to 2 hours with no external view. Vomiting incidence ranged from 0% to 60% with higher magnitudes of acceleration being required to generate sickness at the higher frequencies; at each frequency the vomiting incidence increased with increasing acceleration magnitude. The frequency of maximum sensitivity to motion sickness was found to be 0.167 Hz. The manner in which vomiting incidence varied with both the magnitude and the frequency of vertical sinusoidal motion in the series of studies is shown in three-dimensional form in Figure 3.

Experimental studies of habituation showed that vomiting incidence fell during five repeated exposures but rose again after about seven days without motion exposure (McCauley et al, 1976). Guignard and McCauley (1982) investigated the effect of adding harmonics (at 0.33 or 0.50 Hz) to a fundamental frequency of 0.17 Hz vertical sinusoidal oscillation. The motion sickness incidence varied from 50 to 78% in the five conditions and was reasonably consistent with the predictions of the motion sickness dose value procedure defined below.

Formulae for predicting motion sickness incidence (MSI) were proposed by O'Hanlon and McCauley (1974), McCauley and Kennedy (1976) and McCauley et al (1976). The method applies to motion in the frequency range 0.08 to 0.63 Hz with maximum sensitivity to acceleration at about 0.16 Hz. The assumption that MSI will vary with acceleration and with time in ogival form (i.e. a cumulative normal distribution) resulted in somewhat complex mathematical operations. Motion sickness incidence (MSI), expressed as a percentage, is the product of a term representing the influence of motion magnitude and frequency, $P_A$, and a term expressing the effect of motion duration, $P_T$:

$$MSI = 100 \, P_A \, P_T$$

The term $P_A$ is calculated from a term, $z_A$, quantifying the effect of magnitude and frequency and a term describing the form of a cumulative normal distribution (in practice this may be obtained from statistical tables). The effect of motion magnitude and motion frequency was determined from a curve describing the acceleration required to produce vomiting at various frequencies in 50% of persons during two hour exposures:

$$z_A = 2.13\log_{10} a - 9.28\log_{10} f - 5.81(\log_{10} f)^2 - 1.85$$

Where $a$ is the r.m.s. acceleration in g; $f$ is the frequency in Hz. The term $P_T$ is calculated similarly to give a value for $z_T$:

$$z_T = 2\log_{10} t + 1.13z_A - 2.90$$

where $t$ is the exposure time in minutes.

Values of $P_A$ and $P_T$ are obtained by consulting a table of the normal deviate z at the values of $z_A$ and $z_T$ respectively. For example, with a 60 min. exposure to an acceleration of 2.1 ms$^{-2}$ r.m.s. (i.e. 0.21 g r.m.s.) vertical sinusoidal motion at 0.25 Hz, $z_A$ = 0.19 and $z_T$ = 0.87 so $P_A$ = 0.57 and $P_T$ = 0.81 giving a predicted motion sickness incidence of 46%.

## Horizontal oscillation

Only one study has explored the sickness produced by horizontal motion. Golding and Kerguelen (1991) investigated the sickness produced by 0.3 Hz sinusoidal motion at a magnitude of 1.8 ms$^{-2}$ r.m.s. The motion occurred along the z-axes of subjects who were either lying on their backs (i.e. horizontal motion) or seated upright (i.e. vertical motion). For these postures, the horizontal motion was less nauseogenic than vertical motion. When subjects were asked to perform a visual search task their sickness was greater than when their eyes were closed.

## International Standard 2631 Part 3 (1985)

Part 3 of ISO 2631 suggests magnitudes of vertical oscillation in the range 0.1 to 0.63 Hz expected to produce a 10% incidence of sickness in sitting or standing fit young men over 30 min, 2 h and, tentatively, 8 h exposures (International Organization for Standardization, 1985). The magnitudes and durations are in an inverse-square relationship so that the magnitudes for 2 hours are double those for 30 min. The magnitudes required for 10% vomiting in 2 hours are shown in Figure 3. Sensitivity to acceleration is greatest from 0.1 to 0.315 Hz but falls at higher frequencies.

## British Standard 6841 (1987)

British Standard 6841 (1987) defines a 'motion sickness dose value', MSDV, based on a frequency weighting ,$W_f$, (i.e. a filter) and a time dependency. The standard fully defines the realisable weighting for implementation by analogue or digital filters. The greatest sensitivity to acceleration is in the range 0.125 to 0.25 Hz, with a rapid reduction in sensitivity at higher frequencies. The exposure duration, $t$ (seconds), and the frequency-weighted r.m.s. acceleration, $a_{rms}$ (ms$^{-2}$ r.m.s.),

**Table II** Filter gains for $W_f$ defined in BS 6841 (1987).

| Frequency Hz | Gain |
|---|---|
| 0.100 | 0.800 |
| 0.125 | 1.000 |
| 0.160 | 1.000 |
| 0.200 | 1.000 |
| 0.250 | 1.000 |
| 0.315 | 0.630 |
| 0.400 | 0.391 |
| 0.500 | 0.250 |

may be used to compute the motion sickness dose value:

$$MSDV_z = (a_{rms}^2 . t)^{1/2}$$

The percentage of unadapted adults who may vomit is then given by:

$$percentage\ vomiting = \frac{1}{3}\ MSDV_z$$

This relation is based on exposures lasting from about 20 min to 6 h with a prevalence of vomiting up to 70% (see Figure 3).

Lawther and Griffin (1987) compared the motion sickness dose value ($MSDV_z$) procedure with the motion sickness incidence (MSI) method. It was shown that the frequency weightings were similar and that the dependence of vomiting on the magnitude and duration of motion were similar for magnitudes up to about 2.5 ms$^{-2}$ r.m.s. and for durations up to about 6 hours.

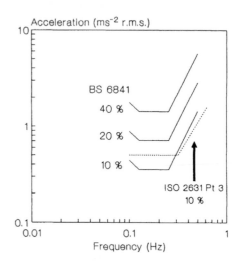

**Figure 3** Vertical z-axis oscillation expected to cause 10%, 20% and 40% incidence of vomiting during 2 h exposures according to British Standard 6841. Incidence will double if magnitude is doubled or exposure duration is increased by a factor of four.

Oscillation on swings

Studies on swings have been conducted for two main purposes: to assess the effectiveness of drugs and to devise a procedure for eliminating susceptible persons during selection for military service. Only a few studies have sought to determine the influence of the type and direction of swing motion on motion sickness.

Two different types of swing have been used: two-pole (like the garden swing) and four-pole (in which the platform remains horizontal). The motions experienced by observers on the two swings are different and neither imparts a unidirectional translational acceleration. Not all studies describe the form of swing used and the motions of exposed persons (especially their heads) are often not the same as the motion of the swing.

Swings with radii of about 4 m have often been used; 30 minute exposures are then sufficient to make many subjects ill and vomit. Sickness increases with increasing angles of swing and with increasing swing radii, at least as the swing frequency falls to about 0.25 Hz. It appears that sickness is not primarily due to rotational motion, but neither is sickness adequately explained by current knowledge of susceptibility to translational oscillation. Sickness also depends on body posture and visual conditions.

Visually-induced motion sickness

An illusion of self-motion may be produced by moving a visual scene past an observer. If the scene rotates around the observer, the illusion of body rotation is called circular vection. The Coriolis effect occurs if, while the body is rotating about one axis, the head is rotated about another axis (see above); the term 'pseudo-Coriolis effect' describes the similar consequences that arise if head movements are made during circular vection. Sickness can be produced by a particular pattern of head movements but may also occur without deliberate head movement. In some experiments with unrestrained animals, optokinetic stimulation has been as nauseogenic as a similar rotation of their bodies. With human subjects it appears that sickness may increase with increasing rates of drum rotation up to about 10 r.p.m. and decline with faster rotation.

The wearing of left-right or up-down reversing spectacles while walking or making other voluntary or involuntary head movements can induce symptoms of motion sickness. Ambulation while wearing left-right reversing goggles appears more nauseogenic than ambulation with up-down reversing goggles.

There are many reports of motion sickness in simulators, some of which may be attributed to distortion or other deficiencies of the visual simulation. The presence of inappropriate lags between subject responses and movements of the visual scene and the motion of the simulator may also be responsible.

## FIELD STUDIES
### Sea sickness

It seems possible that scientific study of sea sickness has been impeded by its very commonplace occurrence: why study a phenomenon which is so well known? Although sea sickness is commonly observed, the cause-effect relationships have rarely been the subject of scientific study. Consequently, there are many historical accounts of seasickness and many suggestions for reducing its effects, but most claims are not founded in reproducible study and there is little useful application of knowledge to design.

Studies among troops on vessels crossing the Atlantic Ocean and studies on smaller vessels have shown that a variety of different types of drug have a more beneficial effect in reducing seasickness than a placebo (see Griffin, 1991b). Amongst passengers on ferries the incidence of sickness among those choosing to take anti-motion sickness drugs was found to be higher than amongst those not taking drugs! (Lawther and Griffin, 1988b). Presumably, those most likely to be seasick were those most likely to take drugs, but the beneficial effect of the drugs was not sufficient to reduce the susceptibility of the group of passengers who took drugs to below that of those choosing not to take drugs. In the same study there was a tendency towards less sickness amongst those who consumed more than two alcoholic drinks. This, and a decrease in sickness susceptibility among those who travelled at sea more frequently, might also arise from a self-selection process.

Studies by Kanda *et al* (1977) and Goto and Kanda (1977) in two sail training ships, by Wiker *et al* (1979, 1980) comparing three craft steaming side by side, by Applebee *et al* (1980) in a seakeeping trial in a Coast Guard Cutter, and by Lawther and Griffin (1986, 1987, 1988a,b) with 20,029 passengers on 114 voyages on six ships, two hovercraft and a hydrofoil, are the best known investigations in which the relation between ship motion and vomiting incidence have been studied. A more detailed review of relevant studies is provided in Griffin (1991b).

Notwithstanding popular expression of a variety of motions being the cause of sickness (e.g. roll motion, corkscrew motions etc.), the evidence is strongly in favour of much sea sickness being attributable to the vertical oscillation at frequencies below about 0.6 Hz. It has been possible to combine the findings at sea with the results of laboratory studies with vertical oscillation to define a 'motion sickness dose value' (see above, Lawther and Griffin, 1987, British Standards Institution, 1987). Figure 4 shows that when presented on a graph of vomiting as a function of the motion sickness dose value for vertical motion alone, there is close agreement between laboratory data and sea data. It should not be concluded, however, that there exists a simple mechanistic relationship between

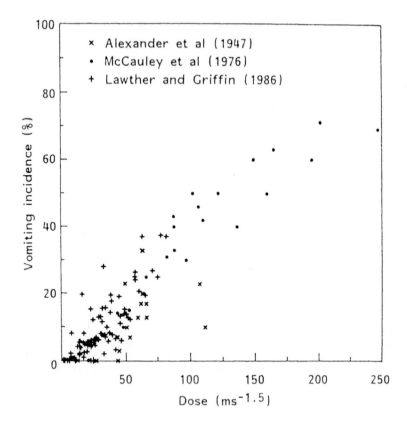

**Figure 4** Relation between motion sickness dose value and vomiting incidence for laboratory studies with vertical motion reported by Alexander et al (1947) and McCauley et al (1976) and sea studies reported by Lawther and Griffin (1986).

these two variables or that other motions are never contributing factors. It may be concluded that changes which reduce the motion sickness dose value (i.e. reductions in the vertical motion, changes in motion frequency, reductions in exposure duration) will tend to reduce the incidence of sea sickness.

### Car sickness

The systematic study of sickness in road vehicles has been even less evident than the study of sea sickness. It seems unlikely that such sickness is primarily due to vertical oscillation (Griffin, 1990). It is known that repeated acceleration and deceleration can cause sickness (Probst et al., 1982; Vogel et al., 1982) but the degree to which sickness can be attributed to acceleration measured in the fore-and-aft or lateral directions of vehicles is not known. One problem is that as a vehicle rolls, for example, a part of the acceleration measured in the lateral direction is due to a gravitional component ($g \sin \theta$). A useful understanding of the motions causing sickness in road vehicles requires a knowledge of whether the relevant non-vertical acceleration is that in the horizontal direction or that measured within the axes of the vehicle. It also requires a consideration of postural, vestibular and visual responses to such motion. Studies to determine dose-effect relationships in road vehicles which are in progress should provide useful design guidance.

### Air sickness

While airsickness has declined as a problem in large passenger aircraft it remains a problem in flying training, especially among pilots of military fast jet aircraft. Vertical oscillation may be a cause of sickness in some aircraft but this is probably not a sufficient explanation of sickness in aerial aerobatic manoeuvres. Useful dose-effect data for sickness in aircraft do not exist. The approaches to sickness in flying training vary between countries: in some there are highly successful programmes of desensitization (by a progressive program of ground and air exposures to nauseogenic stimuli), while in other countries little help is offered to the trainee and only those not impeded by motion sickness are allowed to proceed through training.

### Spacecraft

Up to about 70% of astronauts in larger spacecraft have developed a form of motion sickness: clearly oscillatory motion is not the only cause of the problem. Head movements in a 'weightless environment' are generally the provocative stimulus. While much research has been devoted to the problem there has been little success in predicting individual susceptibility to 'space sickness' (Money, 1991).

### Simulators

Sickness has been a problem in various simulators (e.g. car, tank and aircraft simulators). Although the problem may be ascribed to 'deficiencies' in the cues provided by the simulated visual, somatosensory and auditory cues, the problem is not simply cured by increasing the 'fidelity' of the simulation. Indeed, the problem can be small with very crude simulators and greater with those providing more 'realism'. Since, by definition, complete realism is not the objective of a simulator, the knowledge required is the extent to which the simulator may depart from reality without causing problems. The lags between component parts of the simulation may be important in some cases.

### DISCUSSION

Sadly, for the sufferer, proven causes and cures of sickness are more illusive than the consequences of motion sickness! Guidance to minimise sickness is often unproven, highly varied, and sometimes a source of disagreement between experts; for example, some advise trying to comprehend the true motion while other advise ignoring the motion. It is proven that some drugs can reduce the incidence of vomiting but also apparent that current drugs do not eliminate the problem. Indeed, among passengers on ferries at sea it appears that self-selection processes result in more sickness among those who decide to take anti-motion sickness drugs! While some individuals report benefit from remedies based on 'alternative medicine', scientific proof of their value has not yet been shown in controlled studies.

For marine environments there may be some confidence in the application of the various procedures for predicting motion sickness caused by vertical oscillation. These allow designers and operators to optimise the combination of physical variables, especially motion magnitude, motion frequency and exposure duration. These can be calculated from the vessel design and the sea conditions so it is possible to predict the degree of sickness at the design stage. The standards suggest that reductions in motion magnitude will have a more beneficial effect than a similar reduction in exposure duration: increased speeds which also increase the motion will not reduce the numbers of people vomiting. The oscillation frequency of the motion may also depend on the speed: on some craft the motion is more nauseogenic when the vessel stops at sea. Operators of ships may note that the vertical motion is far greater at the bows and the stern, so sickness will be least amidships. On hovercraft the motion is least at the stern.

In non-marine environments there is a less satisfactory understanding of the physical causes of sickness. In road vehicles a minimisation of acceleration, deceleration and cornering forces may be expected to be beneficial. In most environments an external view which reveals the true body motion is usually beneficial (i.e. not nearby waves or moving vehicles). Tasks should involve the minimum of head and eye movements and should not involve the use of optical devices which magnify or otherwise distort the visual field. A recumbent posture may be beneficial, otherwise it may be preferable to stand and attend to somatosensory cues to movement rather than respond to visual and vestibular sensory information. If the unwanted effects of drugs are acceptable, they may be taken to advantage but the effects (wanted and unwanted) are highly dependent on the individual. Continued exposure to a particular motion usually brings habituation, but this may take many exposures, it is specific to each type of motion, and may be lost after a period of time away from the motion.

There is a very large variability between individuals yet little explanation of these differences. Females have been found more susceptible than males but the reasons are not known. Any simple physical, physiological or psychological variable indicating susceptibility to sickness has yet to be discovered.

Theories of motion sickness suggest that there is not one physical cause, but that the problem arises from a 'conflict' between various sources of sensory information. It may be helpful to think in terms of reflex responses to motion which occur in response to sensory information from the vestibular system (semicircular canals and otoliths), from the visual system, and from the somatosensory system. Problems arise when one, or more, reflex response to motion developed for daily life becomes inappropriate in a new motion environment (see Griffin, 1990). The problem disappears when new reflex responses are developed (i.e. habituation). This appears compatible with the 'sensory conflict theory' of motion sickness (see Reason and Brand, 1975; Benson, 1984).

For the non-sufferer, the occurrence of motion sickness in others can be a source of fun, a sign of some indadequacy, or an inconvenience. For the sufferer, motion sickness can be embarrassing or humilitating, impede activities, and demotivate to the point of removing a desire to survive. For some, motion sickness is something that happens to others or only to themselves in their leisure time. For others it is a part of the job. Few other aspects of ergonomics have such a dramatic effect on comfort, performance and health yet remain laughable.

There is a need for more systematic study so as to advance the understanding of motion sickness. There is also a need to increase the application of existing knowledge to design. In summary, by both scientific investigation and by the application of ergonomics, motion sickness merits more attention to ensure that *people and things will interact most effectively and safely.*

## REFERENCES

Alexander,S.J., Cotzin,M., Hill,C.J., Ricciuti,E.A., Wendt,G.R. (1945a). Wesleyan University studies of motion sickness: I. The effects of variation of time intervals between accelerations upon sickness rates. The Journal of Psychology, 19, 49-62.

Alexander,S.J., Cotzin,M., Hill,C.J., Ricciuti,E.A., Wendt,G.R. (1945b). Wesleyan University studies of motion sickness: II. A second approach to the problem of the effects of variation of time intervals between accelerations upon sickness rates. The Journal of Psychology, 19, 63-68.

Alexander,S.J., Cotzin,M., Hill,C.J., Ricciuti,E.A., Wendt,G.R. (1945c). Wesleyan

University studies of motion sickness: III. The effects of various accelerations upon sickness rates. The Journal of Psychology, 20, 3-8.

Applebee,T.R., McNamara,T.M., Baitis,A.E. (1980). Investigation into the seakeeping characteristics of the U.S. coast guard 140-ft WTGB class cutters: sea trial aboard the USCGC mobile bay. Report DTNSRDC/SPD-0938-01, David W. Taylor Naval Ship Research and Development Center, Ship Performance Department.

Benson,A.J. (1984). Motion sickness. Chapter 19 in: Vertigo, Editors: M.R. Dix, J.S. Hood, Published: John Wiley & Sons Ltd., 391-426.

Benson,A.J., Bodin,M.A. (1966). Interaction of linear and angular accelerations on vestibular receptors in man. Aerospace Medicine, 37, (2), 144-154.

British Standards Institution (1987). Measurement and evaluation of human exposure to whole-body mechanical vibration and repeated shock. British Standards Institution BS 6841.

Golding,J.F. and Kerguelen,M. (1991) A comparison of the nauseogenic potential of low frequency vertical versus horizontal linear oscillation. Paper presented at the United Kingdom Informal Group Meeting on Human Response to Vibration, HSE, Buxton, 25th to 27th September.

Goto,D., Kanda,H. (1977). Motion sickness incidence in the actual environment. Proceedings of U.K. Informal Group Meeting on Human Response to Vibration at UOP Bostrom, Northampton, September 7th -9th.

Graybiel,A. (1969). Structural elements in the concept of motion sickness. Aerospace Medicine, 40,(4), 351-367.

Griffin,M.J. (1990). Handbook of human vibration. Published: Academic Press, London, ISBN: 0-12-303040-4.

Griffin,M.J. (1991a). Physical characteristics of stimuli provoking motion sickness. In, Motion Sickness: significance in aerospace operations and prophylaxis. Advisory Group for Aerospace Research and Development (AGARD) Lecture series 175.

Griffin,M.J. (1991b). Sea sickness. In, Motion sickness: significance in aerospace operations and prophylaxis. Advisory Group for Aerospace Research and Development (AGARD) Lecture Series 175.

Guedry,F.E. (1968). Conflicting sensory orientation cues as a factor in motion sickness. 4th Symposium on the Role of the Vestibular Organs in Space Exploration, Florida, 24-27 September, NASA SP-187, 45-51.

Guedry,F.E., Benson,A.J., Moore, H.J. (1982). Influence of a visual display and frequency of whole-body angular oscillation on incidence of motion sickness. Aviation, Space and Environmental Medicine, 53,(6), 564-569.

Guignard,J.C., McCauley,M.E. (1982). Motion sickness incidence induced by complex periodic waveforms. Aviation, Space and Environmental Medicine, 53, (6), 554-563.

International Organization for Standardization (1985). Evaluation of human exposure to whole-body vibration - Part 3: Evaluation of exposure to whole-body z-axis vertical vibration in the frequency range 0.1 to 0.63 Hz. International Standard ISO 2631/3.

Kanda,H., Goto,D., Tanabe,Y. (1977). Ultra-low frequency ship vibrations and motion sickness incidence. Industrial Health, 15,(1), 1-12.

Lackner,J.R., Graybiel,A. (1979). Some influences of vision on susceptibility to motion sickness. Aviation, Space and Environmental Medicine, 50, (11), 1122-1125.

Lawther,A., Griffin,M.J. (1986). The motion of a ship at sea and the consequent motion sickness amongst passengers. Ergonomics, 29, (4), 535-552.

Lawther,A., Griffin,M.J. (1987). Prediction of the incidence of motion sickness from the magnitude, frequency, and duration of vertical oscillation. The Journal of the Acoustical Society of America, 82, (3), 957-966.

Lawther,A., Griffin,M.J. (1988a). Motion sickness and motion characteristics of vessels at sea. Ergonomics, 31, (10), 1373-1394.

Lawther,A., Griffin,M.J. (1988b). A survey of the occurrence of motion sickness amongst passengers at sea. Aviation, Space and Environmental Medicine, 59, (5), 399-406.

Leger,A., Money,K.E., Landolt,J.P., Cheung,B.S., Rodden,B.E. (1981). Motion sickness caused by rotations about earth-horizontal and earth-vertical axes. Applied Physiology, 50, (3), 469-477.

Lentz,J.M., Guedry,F.E. (1978). Motion sickness susceptibility: A retrospective comparison of laboratory tests. Aviation, Space, and Environmental Medicine, 49, (11), 1281-1288.

McCauley,M.E., Kennedy,R.S. (1976). Recommended human exposure limits for very-low-frequency vibration. Pacific Missile Test Centre, Point Mugu, California, TP-76-36.

McCauley,M.E., Royal,J.W., Wylie,C.D. O'Hanlon,J.F., Mackie,R.R. (1976). Motion sickness incidence: Exploratory studies of habituation, pitch and roll and the refinement of a mathematical model. Human Factors Research Inc., Goleta, California, AD-A024 709.

Miller,E.F., Graybiel,A. (1970). Comparison of five levels of motion sickness severity as the basis for grading susceptibility. Naval Aerospace Medical Institute, Pensacola, Florida, Report No. NAMI-1098, AD 708 040.

Miller,E.F., Graybiel,A. (1973). Perception of body position and susceptibility to motion sickness as functions of angle of tilt and angular velocity in off-vertical rotation. Naval Aerospace Medical Research Laboratory, Pensacola, Florida, Report NAMRL-1182, AD-772-702.

Money,K.E. (1991) Space sickness. In, Motion sickness: significance in aerospace operations and prophylaxis. Advisory Group for Aerospace Research and Development (AGARD) Lecture Series 175.

Morton,G., Cipriani,A., McEachern,D. (1947). Mechanism of motion sickness. Archives of Neurology and Psychiatry, (57), 58-70.

O'Hanlon,J.F., McCauley,M.E. (1974). Motion sickness incidence as a function of the frequency and acceleration of vertical sinusoidal motion. Aerospace Medicine, 45,(4),366-369.

Probst,T., Krafczyk,S., Buchele,W., Brandt,T. (1982) Visuelle praevention der bewegungskrankheit im auto. Archives fur Psychiatrie und Nervenkranheiten 231: DRIC-T-6823.

Reason,J.T., Brand,J.J. (1975). Motion sickness. Academic Press, London. ISBN 0-12-584050-0.

Vogel,H., Kohlhaas,R., Baumgarten,R.J. von (1982) Dependence of motion sickness in automobiles on the direction of linear acceleration. European Journal of Applied Physiology, 48: 399-405.

Wiker,S.F., Kennedy,R.S., McCauley,M.E., Pepper,R.L. (1979). Susceptibility to seasickness: Influence of hull design and steaming direction. Aviation, Space and Environmental Medicine, 50, (10), 1046-1051.

Wiker,S.F., Pepper,R.L., McCauley,M.E. (1980). A vessel class comparison of physiological, affective state and psychomotor performance changes in men at sea. Report CG-D-07-81, U.S. Department of Transportation, U.S. Coast Guard, Office of Research and Development, Washington.

# A SURVEY OF CAR DRIVER DISCOMFORT

J.M. PORTER[1], C.S. PORTER[2] & V.J.A. LEE[1]

1  Vehicle Ergonomics Group, Department of Human Sciences, Loughborough
University of Technology
2  School of Art & Design, Coventry Polytechnic

1000 drivers were interviewed at 3 motorway service stations in
England. The major areas of reported discomfort were the lower back
and the neck. Drivers were found to be more uncomfortable if they
drove for long periods of time and/or drove a car with a manual
gearbox or a seat without adjustment provided for the height, tilt or
lumbar support. These findings show the benefit of providing
flexibility within the driving package so that drivers of various sizes
can all adopt healthy and comfortable postures.

## INTRODUCTION

In a case-control study of the epidemiology of acute herniated lumbar
intervertebral discs (i.e. 'slipped' discs), it was found that men who spend half or
more of their working day driving a motor vehicle are approximately three times
as likely to develop a 'slipped' disc compared to those who do not hold such jobs
(Kelsey & Hardy, 1975). In addition, they found that people, of both sexes, who
said that they drove a car were found to be more likely to develop such disc
problems than those who did not drive. These findings clearly warrant further
research on the discomfort and health aspects of driving. The aims of this project
were (a) to conduct a discomfort survey of car drivers identifying the body area and
extent of any reported discomfort; and (b) to identify factors, either driver
dependent (e.g. age, sex, driving time) or car dependant (e.g. provision of
adjustable seat, lumbar support) which are related to the reported discomfort.

## METHOD

One thousand drivers were interviewed by 5 trained experimenters at 3
motorway service stations (Scratchwood M1, Leicester Forest East M1, Corley M6)
during July 1990. The different locations were chosen in an attempt to capture
drivers who had a wide variation in the length of time driving and the type of
driving (i.e. city/town, country, motorway). The comfort/discomfort data were
recorded using two types of rating scale. Firstly, levels of discomfort were assessed
for 20 individual body areas using a 4 point discomfort scale: no discomfort (0),
slight (1), moderate (2) and considerable (3). Secondly, the drivers were asked to
make an overall assessment of a) their body comfort and b) their seat, using a 5

point comfort/discomfort scale: very comfortable (1), comfortable (2), neutral (3), uncomfortable (4), very uncomfortable (5). The remainder of the questionnaire covered the drivers' personal details, including the amount of time they spend driving and whether they suffer from any chronic aches, pains or stiffness, and the specification of the cars (i.e. adjustable features, gearbox type).

## RESULTS

88% of drivers were male, the mean age was 41 years (range 17-80 years, s.d. 13 years) and 66% were travelling on business. The sample of cars included those from 36 manufacturers of which 74% were first registered in 1987 or later. 89% of the cars studied had a manual gearbox, 36% of the cars studied had seat height adjustment, 37% had cushion tilt adjustment, 32% had steering wheel adjustment (in/out 3%, tilt 24%, both 5%) and 28% had lumbar support adjustment

17% of drivers were not able to easily reach the pedals in their car, 0.4% were not able to easily reach the steering wheel (although many drivers will be unaware of how their driving position has been compromised) and 9% only just had sufficient headroom and 4% complained of insufficient headroom. 68% of all cars were fitted with a sunroof and 90% of those drivers with 'only just' or 'no' headroom were in cars with sunroofs.

11% of drivers reported suffering from physical discomfort before driving that day. For these drivers the most frequent locations were the back (55%) and the neck (9%). 43% of drivers had suffered in the past from aches, pains or stiffness that had lasted for a few weeks. For these drivers, the most frequent locations were the back (37%), knees (16%) and neck (5%). 23% of drivers were still suffering 'often' or 'sometimes' from these chronic aches, pains or stiffness. For these drivers the most frequent locations were the back (total 44%; low back 27%, other areas on the back 17%), knees (22%) and neck (6%). The most frequent reported cause for this chronic discomfort was sport (total 46%; football 13%, rugby 9%, cricket, squash, skiing, golf, tennis, all 3-4%). Other frequently reported causes were work (8%), car, motorbike or bicycle accidents (10%), falls (4%) and gardening, decorating, DIY (4%). Of these chronic sufferers, 11% had been suffering for over 20 years, 21% for between 10 and 20 years and 68% in the last 10 years. 66% reported that driving exacerbated their problem.

53% of drivers reported discomfort in at least one body area. Figure 1 shows that the most frequently reported body areas were the lower back (25%) and the neck (10%). The chest, stomach, arms and the left calf showed the least number of reports (all below 2%) whilst the remainder of the body areas were reported by 3-7% of drivers.

Female drivers reported more discomfort than the males in the neck (reported by 16% and 9%, respectively; $p < 0.05$) and upper back (6% and 2%, respectively; $p < 0.05$). No significant differences were found for the other body areas or for the overall body comfort or seat assessment.

Age was found to show a negative correlation with the ratings for overall body comfort (-0.18, $p < 0.001$) and seat comfort (-0.23, $p < 0.001$), and the discomfort ratings for the upper back (-0.10, $p < 0.001$), mid back (-0.11, $p < 0.001$), lower back (-0.10, $p < 0.01$) and the right foot and ankle (-0.07, $p < 0.05$). The mean age for those drivers describing their overall body comfort as 'very comfortable' and 'very uncomfortable' were 46 and 33 years, respectively.

The effect of road type was investigated by allocating drivers to either 'motorway', 'town' or 'open' roads if they had spent greater than 60% of their driving time in that category of road. No significant differences were found.

The number of minutes driven today (mean 172, s.d. 102) was positively correlated with the ratings for overall body comfort (0.09, $p < 0.01$) and the

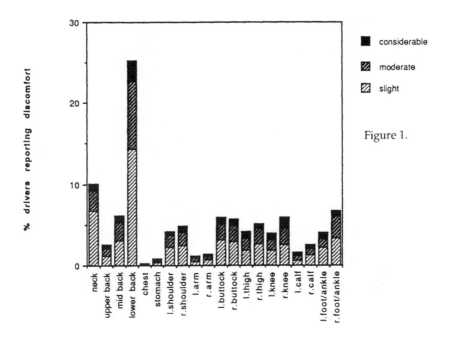

Figure 1.

discomfort ratings for the mid back (0.10, p<0.001) and lower back (0.07, p<0.05). The discomfort ratings for the chest were negatively correlated (-0.07, p<0.05). The mean number of minutes for those drivers describing their overall body comfort as 'very comfortable' and 'very uncomfortable' were 168 and 232 minutes, respectively. The number of minutes driving since the last break (mean 104, s.d. 59) was positively correlated with the ratings for overall body comfort (0.07, p<0.05) and the discomfort ratings for the neck (0.09, p<0.01), lower back (0.06, p<0.05) and the left buttock (0.07, p<0.05). The number of minutes driving per week (mean 1124, s.d. 797) showed a positive correlation with the ratings for seat comfort (0.11, p<0.001) as did the annual mileage (mean 26,436, s.d. 17074; 0.08, p<0.05).

The price of the car when new was negatively correlated with the ratings for overall body comfort (-0.13, p<0.001),and seat comfort (-0.26, p<0.001) and the discomfort ratings for the neck (-0.07, p<0.05), upper back (-0.08, p<0.05) and right knee (-0.11, p<0.001). The mean price of the cars driven by those drivers describing their overall body comfort as 'very comfortable' and 'very uncomfortable' were £14,139 and £11,392, respectively.

Drivers using a manual gearbox rated their overall body comfort and seat to be less comfortable than those using an automatic gearbox (p<0.001). 72% with manual gearboxes rated their seat as 'comfortable' or 'very comfortable' compared to 88% with automatic gearboxes. Drivers with manual gearboxes reported more discomfort in the mid back (p<0.05; manual 7%, automatic 2%) and lower back (p<0.05; manual 27%, 15% automatic).

Drivers who found their pedals difficult to reach rated their overall body comfort and seat to be less comfortable than those who found it easy (p<0.001). 41% of those who had difficult reach rated their seat as 'comfortable' or 'very comfortable' compared to 74% who had easy reach. Drivers with difficult reach reported more discomfort in the left calf (p<0.001; 12% difficult, 1% easy), right calf (p<0.05; 12% difficult, 2% easy), left foot and ankle (p<0.001; 24% difficult, 4% easy) and right foot and ankle (p<0.001; 35% difficult, 6% easy).

Drivers with a fixed seat height rated their overall body comfort and seat to be less comfortable than those with an adjustable seat height (p<0.001). 70% with a fixed height rated their seat as 'comfortable' or 'very comfortable' compared to 80% with an adjustable height. Drivers with a fixed seat height reported more discomfort in the left buttock (p<0.001; fixed 7%, adjustable 3%) and the right buttock (p<0.05; fixed 7%, adjustable 3%).

Drivers with a fixed cushion tilt angle rated their overall body comfort and seat to be less comfortable than those with an adjustable tilt angle (p<0.001). 69% with a fixed angle rated their seat as 'comfortable' or 'very comfortable' compared to 82% with an adjustable height. Drivers with a fixed cushion tilt angle reported more discomfort in the mid back (p<0.01; fixed 8%, adjustable 4%), lower back (p<0.001; fixed 29%, adjustable 20%), left shoulder (p<0.05; fixed 5%, adjustable 2%) and right knee (p<0.01; fixed 8%, adjustable 3%).

Drivers with a fixed lumbar support rated their overall body comfort and seat to be less comfortable than those with an adjustable support (p<0.001). 69% with a fixed support rated their seat as 'comfortable' or 'very comfortable' compared to 86% with an adjustable support. Drivers with a fixed lumbar support reported more discomfort in the neck (p<0.05; fixed 11%, adjustable 7%), lower back (p<0.05; fixed 27%, adjustable 21%), left shoulder (p<0.01; fixed 5%, adjustable 2%) and right shoulder (p<0.05; fixed 6%, adjustable 3%). However, these drivers reported less discomfort in the stomach (p<0.05; fixed 0%, adjustable 2%).

DISCUSSION

Figure 1 clearly shows that the most frequently reported discomfort area was the lower back with a quarter of all drivers commenting upon it. This is clearly a cause for concern particularly in the light of the study described earlier by Kelsey & Hardy (1975)  The second worst area of discomfort was the neck which was reported by 10% of the sample, although the female drivers reported more discomfort than the males both in this body area and in the upper back. No other sex differences for areas of reported discomfort were identified and the assessment of overall body comfort and seat comfort were not significantly different. It is suspected that the females experienced more neck discomfort primarily because of their shorter shoulder height, when compared to the males, causing the seat belt to come into contact with the neck more frequently. Previous work investigating static seat comfort, without the provision of seat belts, found no significant differences in the extent or location of discomfort between males and females (Porter & Sharp, 1984). The factors which were found to be related to the drivers' perception of overall body comfort are listed below:

Age of the driver

The older drivers reported less discomfort than the younger drivers in all areas of the back and the right foot and ankle. This resulted in a more comfortable assessment of their overall body comfort and seat comfort. However, previous work did not identify any significant differences in reported discomfort for groups aged 18-24, 30-40 and 50+ years (Porter and Sharp, 1984). This latter study used identical seats in a laboratory setting whereas the present study assessed drivers in their own vehicles and as a function of their personal driving profiles.

However the price of the car and the driver's age were found to be positively correlated (0.113, p<0.001) which suggests that the age effect may be secondary to the price, and hence the specification, of the car. Analysis showed that the people driving cars with automatic gearboxes or adjustable cushion tilt or lumbar supports were significantly older. As the provision of these features were all associated with fewer reports of discomfort and more comfortable assessments of the overall body comfort and seat comfort then this analysis also suggests that the older drivers may experience less discomfort because they have more 'luxury' features fitted to their cars. The age of the driver was also found to be negatively correlated with the time spent driving on the day of the interview (-0.09, $p<0.01$), the weekly driving time (-0.09, $p<0.01$) and the annual mileage (-0.11, $p<0.001$). This is another likely explanation for why the older drivers reported less discomfort than the younger drivers.

*Driving time*

As the driving time increased, either in total or since the last break, the assessment of overall body comfort became more uncomfortable. Interestingly, the assessment of seat comfort was not related to the driving time on the day of the interview but it was to the weekly driving time and the annual mileage. This indicates that the assessment of seat comfort is made over a period of time whereas the assessment of body comfort is made at a particular moment in time.

*Specification of the car*

The drivers of cars which had an automatic gearbox or a seat with adjustment provided for the height, tilt or lumbar support were found to rate their overall body comfort and seat comfort to be more comfortable than those whose cars did not offer such features. This difference is likely to be a result of the improved posture permitted if these features, either singly or combined, are available. The adoption of a good posture whilst driving (e.g. as recommended by Rebiffé, 1969, or Grandjean, 1980) is essential to delay the onset not only of discomfort but also of any potential long term impairment.

The benefit of the automatic gearbox was not a direct result of the reduced work load for the left arm and leg as shown by the lack of any significant differences for the discomfort reported in these areas. The only significant differences were found in the mid and lower back where the drivers with a manual gearbox experienced more discomfort. This was possibly due to the increased need for forward leaning to engage the gears with a manual gearbox. Also it is likely that most of the automatic cars had power steering which helps promote the likelihood of the driver adopting a reclined posture as the effort levels are low. A reclined posture is beneficial for the lower back because it reduces the muscle activity and the pressure in the intervertebral discs of the lumbar spine (i.e. lower back) compared to forward leaning (Andersson, 1980; Nachemson, 1975). A further factor is that the cars fitted with an automatic gearbox often have a high specification including good seat adjustability to help reduce back discomfort.

The provision of cushion tilt or an adjustable lumbar support were both associated with a lower incidence of discomfort in the back, again due to an improved posture. The tilt angle can be adjusted to suit the individual's requirements resulting in a more reclined posture (by reclining the backrest as well as the cushion) or, by lowering the front of the cushion, improving the ease of reach to the pedals (right knee discomfort was lower for drivers with tilt adjustment). This latter option allows a greater trunk/thigh angle for a given backrest angle which also helps to reduce back discomfort (Mandal, 1981). A further explanation is that drivers with short legs often find the seat cushion to be too long causing discomfort behind the knee. Many of these people may slide their buttocks forward on the cushion to avoid this but, in doing so, they induce slouching and subsequently discomfort in the back because the backrest is too far away.

The adjustable lumbar support typically provides in/out adjustment which enables the driver to select more support in the lower back if desired. This extra support is known to help reduce disc pressure. Few cars were supplied with a height adjustable lumbar support even though research has shown that the requirement for height adjustment is greater than for in/out adjustment (Porter & Norris, 1987)

The provision of seat height adjustment was associated with fewer reports of discomfort in the buttocks. It was also seen that the provision of adjustable cushion tilt and/or lumbar support were associated with fewer reports of shoulder discomfort.

The drivers of the more expensive cars reported more comfortable assessments compared to the drivers of the cheaper cars. This is not surprising given that the more expensive cars are more likely to offer an automatic gearbox and/or more comprehensive seat adjustments, all of which have been shown to improve ratings of overall body comfort and seat comfort.

CONCLUSIONS

The results of this study show clearly that drivers were more uncomfortable if they drove for long periods of time and/or drove a car with a manual gearbox or a seat without adjustment provided for the height, tilt or lumbar support.

A clear recommendation arising from this survey is for individuals to select their cars carefully to ensure that they can adopt a good posture whilst driving. This will often require a highly adjustable driving package. The major problems for car manufacturers, employers purchasing company cars and individuals buying private cars is that a highly adjustable driving package will cost more than a basic, fixed package. More attention should be paid to the costs that may be incurred if any adjustable package is not provided. For the individual driver who drives for several hours a day the personal cost may well be comparatively high levels of discomfort and possible chronic musculo-skeletal complaints. The employer will experience hidden costs because of the number of days lost through sick leave. Eventually the manufacturer may possibly find his market share falling as the more aware employers and public purchase cars from the manufacturers who offer suitably adjustable driving packages.

REFERENCES

Andersson,. G.B.J., 1980, The load on the lumbar spine in sitting postures. In: Human Factors in Transport Research, edited by D. Oborne and J.A. Levis, Volume 2, pp. 231-239, (Academic Press).

Grandjean, E., 1980, Sitting posture of car drivers from the point of view of Ergonomics. In: Human Factors in Transport Research, Volume 2, edited by D.J. Oborne & J.A. Levis, pp 205-213, (Academic Press).

Kelsey, J.L., Hardy, R.J., 1975, Driving of motor vehicles as a risk factor for acute herniated lumbar intervertebral disc. American Journal of Epidemiology, 102, 1, pp. 63-73.

Mandal, A.C., 1981, The seated man (Homo Sedens) - the seat work position, theory and practice. Applied Ergonomics, 12, 1, pp. 19-26.

Nachemson, A., 1975, Towards a better understanding of low-back pain: A review of the mechanics of the lumbar disc. Rheumatology and Rehabilitation, 14, pp. 129-143.

Porter, J.M. and Sharp, J.C., 1984, The influence of age, sex and musculo-skeletal health upon the subjective evaluation of vehicle seating. In: Contemporary Ergonomics 1984, edited by E.D. Megaw, pp. 148-154, (Taylor & Francis Ltd).

Porter, J.M. and Norris, B., 1987, The Effects of posture and seat design on lumbar lordosis. In: Contemporary Ergonomics 1987, edited by E.D. Megaw, pp. 191-196, Taylor & Francis Ltd.

Rebiffé, R., 1969, Le siège du conducteur: son adaptation aux exigence fonctionneles et anthropométriques. In: Sitting Posture, edited by E. Grandjean, (Taylor & Francis Ltd).

# THE OCCUPATIONAL WELL-BEING OF TRAIN DRIVERS - AN OVERVIEW

## R.A. HASLAM

Department of Human Sciences
University of Technology
Loughborough
Leicestershire
LE11 3TU

Survey investigations into the work of train drivers
have identified potential health risks, traumatic
incidents, shiftworking, the physical working
environment, cab design, and organizational climate as
areas for concern. Shiftwork, the driving task, driver
vigilance, and train cab design have received further
experimental and analytical attention from researchers.
An overview is given of this work. The paper concludes
by suggesting that much could be achieved towards
overcoming problems with the occupation through fully
considering train drivers' needs and limitations when
taking decisions that affect their work.

## INTRODUCTION

The work of train drivers has changed significantly over the
160 years of the occupation's existence. The high level of
proficiency required to drive steam traction commanded respect
from both those within and outside the profession. The
introduction of diesel locomotion brought with it marked changes
to the skills required, the associated prestige and social
relationships. These changes have continued with the steady
advance of technology, to the point where there is far greater
scope today for railway managers and designers to give thought to
drivers' needs when taking decisions that affect their work.
However, trade unions in countries around the world continue to
voice concern regarding aspects of the modern train driver's
occupation.

## TRAIN DRIVER SURVEYS

The working conditions and attitudes of train drivers around
the world have been examined through many survey investigations
(e.g. Cox and Haslam, 1985). Areas of concern that have been
identified include: implications that the drivers' work might

have for their health, the effects of traumatic incidents such vandalism and suicides, shiftworking, inadequacies of the physical working environment, poor cab design and the organizational climate within which drivers work.

Certain aspects of the work of train drivers have attracted further investigation. These include shiftwork, the train driving task, driver vigilance, and cab design. These will be considered in turn.

## SHIFTWORK

Irregular shiftworking is a particular problem for the railways. The nature of railway operation results in train drivers having to work hours varying from day to day and from week to week. It is clear from studies (e.g. Akerstedt et al., 1983) that the type of shiftwork involved in railway operation can cause disturbance to eating and sleeping patterns and to home and social life. It is important that railway managers pay attention to developing working rosters that minimize as far as possible the need for night work, especially during the early hours of the morning. Rosters should be consistent over as long a period as is practical. It is possible that benefit might be gained by applying selection techniques to identify individuals suited to shiftwork; this should be investigated further.

Where irregular shiftworking is unavoidable it might be useful to provide counselling to advise drivers how to cope with irregular working hours and the problems that it may cause. The costs of selection and counselling should be weighed against the benefits of having a more alert workforce.

## THE TRAIN DRIVING TASK

The primary task of the train driver is to adhere to a timetable by controlling the speed of the train, while observing speed restrictions, and stopping at red signals and stations. To be able to stop a train at appropriate stations, the driver must have a thorough knowledge of the characteristics of the route so that braking can commence at the correct time. The behaviour of a train's deceleration depends on among other things its speed and load, the gradient of the track and the weather. Thus, the stopping distance of a train may vary from place to place and from day to day, and will often necessitate that the driver commences braking before the station is in sight. Branton (1978) considered the skills of train driving in some detail and interested readers are referred to this paper for further details.

## DRIVER VIGILANCE

While trains on some railway systems are now automatic, requiring minimal human intervention, networks such as British Rail still rely on drivers obeying trackside signals and speed limits for their safe operation. In these cases driver vigilance is of obvious importance. Studies of lapses in driver vigilance have implicated automation in information processing, with this exacerbated by fatigue (e.g. Sharp Grant, 1971).

Signals are still being passed when set at danger and speed limits exceeded through driver errors. It appears that the nature of the task ensures that on occasions errors will occur. This would seem to be an argument for increased automation. However, while automatic systems may overcome shortcomings in human vigilance, their reliability has been questioned and they are poor at dealing with unexpected events. Given that public confidence is likely to require a driver's presence for the foreseeable future, the driver's work should be arranged so that vigilance is maintained at as high a level as possible. Fail-safe mechanisms should then protect against the lapses that will inevitably occur.

Figure 1. The driver's cab of the LNER A4 class locomotive, introduced mid 1930s (National Railway Museum, York).

## Monitoring and maintaining driver vigilance

Most of the attempts to combat the problem of lapses in driver vigilance have employed some form of "vigilance device". Devices that have been used include the dead-man pedal or handle, cyclic devices that require manual resetting, and cyclic devices that reset automatically. With each of these devices, if the correct response does not occur an emergency brake application results.

As monitors of driver vigilance the first two are far from ideal (e.g. Powell and Cartwright, 1977). The action of most of these devices is to monitor vigilance rather than attempting to maintain it. Where devices are intended to maintain vigilance and alleviate monotony they should either form a meaningful component of the driving task or be automatic and noninvasive.

Concern has been expressed that the problems of monotony and its effects on vigilance may have been made worse by the introduction of single manning (e.g. Sen and Ganguli, 1982). Single driver operation is an emotive political issue for the railways, with unions tending to oppose and managers supporting such working arrangements. Single driver operation has existed now in some countries for many years, with no apparent reduction in safety. There are, however, disadvantages associated with single manning of trains. Branton (1978) recognizes that drivers operate in a socially isolated environment and that there may be a link between this and their motivation. This in turn may have an influence upon the driver's "anticipation" or vigilance. Thus, while the presence of a second person makes no direct contribution to the performance of the task, there may be an indirect contribution in terms of motivation.

A technological solution to the problem of lapses in driver vigilance is employed on some European railways, in Japan, and has been on trial on the British Rail network. These systems compare the speed of a train with information provided by trackside transmitters and instruct the driver when to slow down. If the train exceeds the safe speed the brakes are applied automatically, and in time to prevent the passing of red signals. However, these systems are expensive and care is needed with their implementation to ensure that inappropriate deskilling does not occur.

TRAIN CAB DESIGN

A recurrent problem identified by studies has been unsatisfactory cab design. It is clear that the design of the driver's cab has received attention and the implications that it may have for a driver's health acknowledged (e.g. Hedberg et al., 1981). It is interesting to compare a locomotive cab from the era of steam with the design of the modern equivalent, figures 1 and 2.

There have been improvements made to the design of train cabs but these have for the most part been limited to the design of the immediate physical environment. Only limited attempts have been made to examine the driver's actual information requirements and alternative means of presenting this. These requirements should be investigated.

The design process should involve more direct consultation with drivers and the extensive use of simulations and prototypes. It is important to remember that the design of the driving cab greatly influences the nature of the drivers' work, in turn having implications for job satisfaction and general well-being. Train designers should be fully aware of this relationship.

Figure 2. The driver's cab of the British Rail HST 125, as in current service (National Railway Museum, York).

DISCUSSION AND CONCLUSIONS

Investigations into drivers' working conditions have highlighted possible health risks, traumatic incidents, irregular shiftworking, inadequacies of the physical working environment, train cab design, and general organizational climate as unsatisfactory aspects of train drivers' work.

Many of the health risks can be related to either postural problems caused by inadequate equipment design, or irregular shiftworking. Other possible health effects, for example the relatively high incidence of cardiovascular disorders, require further investigation to identify the causes and the drivers at risk.

Steps could be taken to minimize vandalism and suicides by clearing debris from the side of tracks and ensuring appropriate security. Given that complete prevention is impossible, thought should be given to preparing drivers for incidents and providing counselling for those who have the misfortune to have such an experience.

Shiftworking is a serious problem for many drivers but unfortunately an unavoidable component of an efficient railway network. It has been suggested that improvements could me made through careful scheduling and from the use of selection techniques.

Improvements have been made to the design of the driving cab but these have been mainly limited to the physical environment. More attention should be paid to identifying the driver's actual

informational requirements and how best to present this
information.  It is important that the design process should
include the participation of train drivers.

Complaints from drivers of having poor relations and
communication with their managers and unions should clearly be a
matter of concern for the organizations involved.  While problems
of this nature are usually deeply rooted in the culture of an
organization, steps could and should be made towards improving
this situation.

Railway transportation is important for both social and
environmental reasons.  Movement of people and goods by rail
helps reduce pressure on road transport, conserving oil reserves
and reducing pollution.  With the appropriate will it should be
possible to provide the people employed to drive the trains with
a satisfactory working environment, with benefits for the safety
of passengers and the operation of the system.

ACKNOWLEDGEMENT
    The collection and collation of much of the literature used
for this work was undertaken while the author was working for
Stress Research, Department of Psychology, University of
Nottingham, on a project sponsored by ASLEF, British Rail and the
Health and Safety Executive.  Please note, however, that the
views expressed here are solely those of the author.

REFERENCES
Akerstedt, T., Torsvall, L., and Froberg, J., 1983.  A
    questionnaire study of sleep/wake disturbances and irregular
    work hours.  Paper presented at the 4th International Congress
    of Sleep Research, Bologna, 18-22 July (Abstract).
Branton, P., 1978.  The train driver.  In: The Analysis of
    Practical Skills (Edited by: W. T. Singleton)(MTP Press:
    Lancaster), pp. 169-188.
Cox, T. R., and Haslam, R. A., 1985.  Occupational stress in
    train drivers (Stress Research: University of Nottingham).
Hedberg, G., Bjorksten, M., Ouchterlony-Jonsson, E., and Jonsson,
    B., 1981.  Rheumatic complaints among Swedish engine drivers
    in relation to the dimensions of the driver's cab in the Rc
    engine.  Applied Ergonomics, 12, 93-97.
Powell, A. J., and Cartwright, A., 1977.  The design of drivers'
    cabs.  Proceedings of the Institute of Mechanical Engineers,
    191, 195-205.
Sharp Grant, J., 1971.  Concepts of fatigue and Vigilance in
    relation to railway operation.  In: Methodology in human
    fatigue assessment (Edited by: K. Hashimoto, K. Kogi, and E.
    Grandjean)(Taylor & Francis: London), pp 111-117.
Sen, R. N., and Ganguli, A. K., 1982.  Preliminary investigations
    into the loco-man factor on the Indian railways.  Applied
    Ergonomics, 13, 107-117.

# THE IMPORTANCE OF ERGONOMICS IN PLASTIC SURGERY

DELFINA FALCAO* and ANDREW McGRATH**

* Institute of Advanced Studies, Manchester Polytechnic
All Saints Building, All Saints
Manchester M15 6BH, UK

** Department of Three Dimensional Design, Art School,
Manchester Polytechnic, Cavendish Street,
Manchester M15 6BH, UK.

This research relates to an ergonomic analysis of the
comfort and compatibility with the hand of Plastic Surgery
instruments. This analysis led to a design of a standard
hand grip system adaptable to a variety of tasks. Some
prototypes were tested in the operating theatre by the
Plastic Surgery team at Withington Hospital, Manchester.

## INTRODUCTION

In surgery generally large advances have been made in such areas
as sterilisation, surgical techniques and equipment; but there has
been little innovation in instrument design. According to Curutchet
(1964) our surgical instruments have developed from two pre-historic
ones, the shears and the tongs. The surgeons took over the scissors
from the barbers without changing the design, though the method of
use is completely different. This is an important point that was
considered in the analysis of the instruments. Also, the
justification for this analysis lies in the fact that the design of
the instruments has not changed in almost 300 years. There is no
organised pattern in their design and they are poorly compatible
with the shape and dimensions of the hand. Their users are
accustomed to the instrument, and not consciously aware of the
problems. Further, they have little choice of instruments because of
the limited range available.

In seeking a solution to these problems, a study was made of the
movements of the hands in relation to the use of the instruments
comparing the actions of five surgeons.

## METHOD OF INDIVIDUAL AND COMPARATIVE ANALYSIS

A considerable number of instruments is used in plastic surgery.
Thirty six basic instruments were selected, which are used
throughout the range of operations, from large-scale to micro-
surgery. They include forceps, retractors, suction tubes, skin
hooks, volkman scoops, pens, towel clips, needle holders, artery

forceps, scissors, rulers and diathermy instruments, as showed in
figure 1.

All 36 instruments were analysed individually to identify
problems in handling, their causes and consequences. The kinds of
grip were classified into five groups: pinch, tips, lateral, squeeze
and digital.

Also analysed was the range of movements of the hands in relation
to the correct position of the forearm. A specific recording table
was developed.

The frequency of use of the instrument was registered, along with
the length of time they were in use. This study was carried out by
direct observation of 40 operations and by indirect observation
(Chapanis, 1965) from video tape with subsequent analysis of the
pictures captured by computer. From the analysis it was possible to
detect which instruments gave more difficulty. Of 5595 frames, 325
showed awkward situations relating to hand movements.

A selection was made of 18 instruments that gave special
difficulty, and these were analysed comparatively. They were divided
into five groups according to their functions and characteristics.
The groups are: forceps, forceps with racks, needle holders,
scissors, and a group of some instruments that all perform static
tasks.

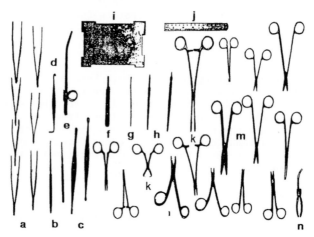

Figure 1. Instrumental System

RESULTS AND DISCUSSION

The analyses showed that the basic problem is that most of the
instruments possess the wrong shape. Scissors and forceps with racks
are much used by surgeons, but both are uncomfortable; the rings
restrict the fingers and the rack is hard to open and close.

They do not allow the neutral position of the wrist to be
maintained, which is when the axis of the grips makes an angle of
100-110 degrees with the axis of the forearm (Pheasant, 1986),
figure 2. Most of the time the wrist is in adduction, putting strain
on the muscles in most movements.

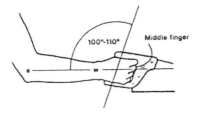

Figure 2. Neutral position of the wrist (Pheasant, 1986)

This study led to a redesign of the instruments. Each instrument can be characterised by an operating member and a handle. The operating member that performs the actions such as cutting, pressing, holding, hooking, etc, was kept in its traditional design. A standard hand grip system was devised, adaptable to a variety of tasks, figure 3.

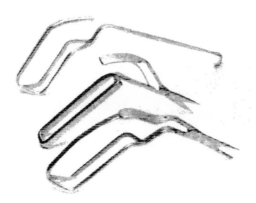

Figure 3. Samples of the standard hand grip system.

The traditional rings in scissors, needle holders and forceps have been eliminated and substituted by a pair of handles. The upper handle is concave in shape to form a thumb rest. By pressing it the instruments can be opened or closed. The lower handle is a elongate member formed in the front into two parallel members to improve the feel of the instrument joining in the back in the shape of a semi-ring. The handle will accommodate 3 or 4 fingers and can be used to hang the instrument up. Instruments for static tasks, such as the retractor, skin hook and suction tube have only the lower handle. Instruments performing active tasks, such as clamps, include a spring and a soft rack. This halves the number of actions the instrument needs to be either opened or closed, due to the counteraction of the spring. The main result is when in use the handle is gripped with the users wrist in straight position, figure 4 (a,b).

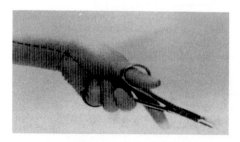

Figure 4a. Traditional needle holder,
wrist in adduction (wrong position).

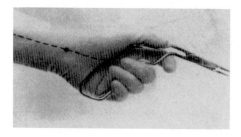

Figure 4b. New needle holder,
wrist in straight position.

The production process of this new instruments is simplified and the cost of production reduced compared with the current drop forging method of manufacture, where the instruments tend to be two-dimensional and thus less than perfectly comfortable in the hand. This causes strain and injury when the instruments are used constantly for long periods of time.

This project utilises the bending properties of steel (already used to an extent in some medical instruments) to create a more ergonomic three-dimensional instrument shape that is nevertheless of low weight and material volume. The instruments can all be made out of steel strip, thus reducing price and waste (from trimming drop forged blanks), storage, transport and finishing (rods and smooth edges can be inherent to the original strip) while increasing the ergonomic handle of the instruments.

Scissors, needle holders and retractors prototypes have been in the theatre at Withington Hospital in Manchester used by the plastic surgery team, figure 5. During the experimental stage the electromyographic audio feedback technique has been applied to detect the muscle strain in comparison with the current instruments, figure 6. All prototypes tested proved the merit of this concept what could be extended to the most usual surgical instruments.

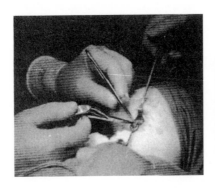

Figure 5. Scissors and needle holder prototypes used in the theatre.

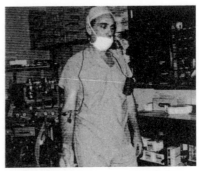

Figure 6. Audio feedback electromyographic technique.

REFERENCES

Chapanis,A., 1965, Research Techniques in Human Engineering (London: John Hopkins Press).

Curutchet,D.P., 1964, The Origin, Evolution and Modification of Surgical Instruments.

Pheasant,P., 1986, Bodyspace: Anthropometry, Ergonomics and Design (London: Taylor and Francis), pp.160-161.

ACKNOWLEDGEMENTS

This research has been supported financially by CNPq and Federal University of Rio de Janeiro, Brazil. The author wish to thank to Mr. Luis Lozano for his collaboration in the early stage of this research; to Mr. Paul Callaham, Head of Department, and Dr. John Langrish, Dean, for the facilities provided at Manchester Polytechnic; to the Plastic Surgery team at Withington Hospital, Manchester, specially to Mr. P.Davenport and Mr. C.Orton for their cooperation during the data collection and suggestions during the experimental stage.

# 1993

| Heriot-Watt University, Edinburgh | 13<sup>th</sup>–16<sup>th</sup> April |
|---|---|

| | |
|---|---|
| **Conference Organiser** | Janis Hayward |
| **Chair of Meetings** | Dave O'Neil |
| **Programme Secretary** | Ted Lovesey |
| **Untitled Position** | Sandy Robertson |

| **Secretariat Included** | Margaret Hanson | Janette Edmonds |
|---|---|---|

## Social Entertainment

A river trip was included in the entertainments package.

| Chapter | Title | Author |
|---|---|---|
| **Auditory Warnings** | Alarms in a coronary care unit | N. Stanton |
| **Health and Safety** | Safe surface temperatures of domestic products | K.C. Parsons |
| **Health and Safety** | An ergonomics appraisal of the Piper Alpha disaster | W.H. Gibson and E.D. Megaw |
| **Drivers and Driving** | Carphone use and motorway driving | A.M. Parkes, S.H. Fairclough and M.C. Ashby |

# ALARMS IN A CORONARY CARE UNIT

**Neville Stanton**

**Human Reliability Associates Ltd**
**1 School House**
**Higher Lane**
**Dalton**
**Wigan**
**Lancs**
**WN8 7RP**

This paper presents the findings of a study that
investigates alarm systems in a coronary care unit.
Data collected over a three days of observations
suggest that the design of such systems have severe
shortcomings in ergonomic terms. Explanations of
the shortcomings and proposed solutions are given.

## Alarm handling in the CCU

Alarm handling forms a small part of the nursing staff duties which are mainly
connected with patient care. These duties include assistance with washing and
dressing, administration of drugs, arranging meals to be delivered, preparing
patients for theatre, keeping patients and family informed of their progress, and
generally seeing to the patients' needs. Although the staff/patient ratio is quite high
(normally three nursing staff in the Coronary Care Unit (CCU) for each shift) the
patients' status means that they are kept busy, particularly if one of the patient's
condition deteriorates. The monitoring systems mean that the nursing staff are
relieved of monitoring each patient continuously. The alarm systems acts as an
interruption and call for attention if any of the parameters falls outside tolerance.

## Alarm systems in the CCU

This paper considers the design of alarms in a CCU. Alarms take two forms in
the CCU: auditory alarms and auditory-visual alarms. The syringe pumps and
blood pressure meter are alarmed by auditory media, whereas the patient's heart rate
is alarmed by auditory and visual media via the ECG monitor. In addition there is a
panic button, to call for assistance, which initiates an audible alarm and a red light
outside the CCU.

Different types of equipment are used within the CCU, some of it performing
the same type of function, i.e. the different types of syringe pump. The ECG
monitoring system, manufactured by Hewlett Packard, displays each patients ECG
with a 30 second trace via a VDU. This information is accompanied by the patient
ward and bed number, heart rate (beats per minute), and alarm status. There were

two VDUs in the CCU and four beds, two either side of the VDUs. The VDUs also displayed the ECGs of some patients in the cardiac wards.

### Brief description of the study

The study was undertaken over three working days, from 9.00 a.m. to 4.18 p.m. This was to determine if the data collected had been representative of a 'typical' day in the CCU. Whilst it is accepted that all three of these days may be atypical, it could be viewed as a useful exercise to see if they were similar. The approach taken and detailed findings are presented in Stanton (1992).

### Main findings

The three days of observation could be considered homogeneous as there were no statistical differences in the number of alarms that were presented. This provides a basis for confidence in interpreting the other data. It was reported that the ECG alarms could be linked to patient activity which accounted for 37.6% of all the alarms. This figure would have undoubtedly been higher if the facility of suspending alarms and changing thresholds had not been available. Therefore this degree of control needs to be regarded as an important feature of the alarm system and should be included in future systems design.

Very few of the signals were identified as 'alarms', i.e. satisfying all of the criteria proposed in that they:

- attract attention;
- were not predicted;
- called for intervention, were inaction would have been detrimental patient.

Between two and five percent of the signal satisfied all three criteria. All of the signals attracted attention. Most of the signal were not predicted 'a priori', but they were not surprise events at the same time in that their presentation could be explained by the circumstance of their onset, for example the patient moving. To have predicted the alarm, one would have to predict the degree of patient movement that would be sufficient to trigger the alarm. Approximately twenty seven percent of the signals resulted in some level of intervention, but most of these would not have led to any deterioration of the patient if the intervention had not taken place. Two thirds of the interventions were checking activities. In this respect the alarms may be seen as useful in prompting the nursing staff in checking the patients and the monitoring system. However, this needs to be carefully balanced as the alarm system should not falsely call for attention too frequently if it is to be trusted.

Alarm presentation is characterised by busy periods followed by intervals of relative calm. During the busy periods the nursing staff are put into a 'fire-fighting' role, dealing with the high priority demands. Whereas in the calmer periods they are able to carry out their normal nursing duties.

### Design considerations

The ECG monitor has some shortcomings in ergonomic terms and could benefit from redesign. There were five main problems that were recognised in the observational studies.

### (i.) Multiple Alarms

There could potentially be problems if there was more than one alarm at a time on the same VDU. This is because the audible call and the 'red/amber/green' light system may only draw attention initially to one patient. However, it is recognised that the likelihood of this ever occurring is very small.

### (ii.) Suspending Alarms

It was also noticed that if the staff suspended alarms on a particular patient, the "ALARMS SUSPENDED" tag covered the patients heart rate. This is of particular concern because of the recognition that the nursing staff are often able to use the heart rate as an indication of potential problems, which may be the very reason that the alarms are suspended.

### (iii.) Identifying Patient

The nursing staff had obviously had difficulties in the past in identifying the bed of the patient that related to the ECG presented on the VDU. Evidence of this lay in the adhesive label stuck to the bottom of the VDU that indicted the bed position of patients. A possible means of resolving this is to configure the screen layout in a more meaningful way and to present some means of identifying the bed of the patient the alarm refers to.

### (iv.) Monitoring Patients in Other Wards

Another problem relates to the difficulties of being unable to see the patients in other wards that are being monitored in the CCU. In one example the nurse noticed that a patient had 'flat lined'. This could have been a very urgent situation. The first course of action was to telephone the ward and ask the nurse to check the patient. Given the serious nature and possible consequences of this incident, this kind of procedure could waste valuable seconds in recovery time. The provision of CCTV (closed circuit television) could significantly reduce the time taken to ascertain the status of the patient as well as reducing the incidence of false alarms.

### (v.) Prioritisation

Finally the system attempted to provide a prioritised auditory warning, yet there was no evidence from the observations that this was warranted. The purpose of the auditory system is to attract the attention of the nurses. There did not appear to be any useful additional information provided by the urgency rating. Very few of the alarms that were categorised as high urgency actually result in any action. Therefore the merit of this auditory distinction must be questioned. Once attention has been drawn to the event, the maintenance of a visual distinction between the different types of event may help in determining the appropriate type of response, if any is to be taken.

### Syringe pumps

The syringe pumps also have shortcomings in ergonomic terms as was highlighted in the introduction. The Welmed goes a long way to resolving these problems. However two problems still remain: identifying the source of the alarm and integrating the alarms. During the observations it was noticed that the nurses did not always immediately recognise which bed the syringe alarm was coming from. This is a problem of an omnidirectional signal in the absence of a clear visual indication. Although there was a red LED on every syringe pump this was far too small to be seen from a position at any distance away from the bed. It is clear that the auditory signal needs to be accompanied by some clear visual reference to the bed that it is emitting from. As was also indicated in table 1 in the introduction there are a variety of syringe pumps which appear to be largely set up for the patient based on availability than any other criteria. This results in a variety of different audible signals, all which essentially carry the same message, i.e. "the syringe is running out". Human factors principles call for consistency in the transmission of information, clearly then this elementary guideline is broken.

### Blood pressure

The blood pressure monitor has the same shortcomings in ergonomic terms as the syringe pumps in identifying the source of the auditory signal. However, there may only ever be one patient in the ward whose blood pressure is being monitored, so this is less of a problem than that identified on the syringe pumps.

### Panic

Finally the panic alarm has some shortcomings in ergonomic terms and could benefit from redesign. Although there is both auditory and visual alarm outside the ward there is only visual indication within the ward of which patient the alarm refers to. However the procedure of the nurse staying with the patient is designed to overcome the shortfall in the alarm system.

From the shortcomings in the alarm system is is possible to offer a suggestion of how future systems might offer substantial benefits.

### Audible distinction

In principle at least the maintenance of some audible distinction between calls from the panic alarm, ECG monitor, syringe pumps, and blood pressure monitor is not undesirable. It allows for the nursing staff to make some preliminary assessment of what is calling for their attention. However, it is desirable that some effort be put into an integrated approach for presenting this information. A methodology for designing auditory symbolic information has been proposed elsewhere (Stanton, 1993). Such that the auditory channel could contain 'symbolic' reference to the source of reference.

### Integration

The visual alarm system could be integrated also. For instance, each bed could have a panel above each patient similar to that suggested in figure 1. This would not only indicate the type of alarm, but the source also. More detailed information could be provided at the bedside or central monitoring unit if necessary.

| Assistance | ECG | Syringe pump | Blood pressure |
|---|---|---|---|

Figure 1. Proposed visual alarm unit.

### Conclusions

Given that only a small percentage of signals require intervention, it must be difficult for the staff to treat each alarm as if it were an emergency. In some cases the protocol was to telephone the ward and request a nurse to check the patient. In a real emergency this procedure might waste important seconds that could be devoted to recovery techniques. Further it was noted that the spatial representation of patients on other wards was inadequate for them to be identified from the VDU, and that a sticker system had been developed. Thus the nurses had to translate the screen representation to the ward position before telephoning the ward. Clearly there is the need to reconsider screen layout to mimic ward layout to save time in this translation.

# SAFE SURFACE TEMPERATURES OF DOMESTIC PRODUCTS

K C PARSONS

Department of Human Sciences
Loughborough University
Loughborough, Leics
LE11 3TU.

Surface temperatures of 'non functional parts' of domestic products should not achieve levels that cause burns to the skin. To design safe products, designers require information about skin reaction to contact with surface temperatures in terms of relevant factors such as user group, user behaviour, material type and condition and likelihood of contact and its duration. A review of literature is presented and a recently proposed European standard (PrEN 563) described. The standard is concerned with developing limiting values for the surface of machines and if adopted would provide greater protection for industrial workers than for users of domestic products.

## INTRODUCTION

People come into contact with surfaces around them throughout their lives either directly with bare skin or through clothing. The temperature of the surface, along with other factors such as material type and condition, will influence the skin reaction to any contact in terms of thermal sensation and comfort or in extreme heat or cold, in terms of pain or skin damage. Some products are heated as part of their function. For example it is necessary for the hob on a cooker to become very hot to cook food. It is not the function of other parts of products to be hot, for example the front of an oven door or control knobs. These non-functional parts however, can become hot when the product is in operation. To design products with surface temperatures that are 'safe' it is useful to know which surface temperature would produce a burn, for example, when touched by bare skin. It is also useful to know the likelihood and nature of contact and by whom. These are considered below.

BURN THRESHOLDS

A burn threshold is the temperature of skin above which a superficial partial thickness burn would occur and below which it would not. The most notable attempts to determine burn threshold data for human skin were carried out in the 1940s. Leach et al (1943) and Sevitt (1949) exposed anaesthetised guinea pigs and some rats with shaved backs to a heated brass cylinder at a number of temperatures. Henriques and Moritz (1947) conducted a series of experiments with anaethetised pigs and humans using flowing water across the skin. They plotted curves of average skin temperature, to produce a burn, against exposure time and provided the main source of data still used today. Siekmann (1989,1990) used a thermesthesiometer ('artificial finger') to measure contact temperatures of heated discs on a range of materials. Using the data of Henriques and Moritz he then proposed burn thresholds in terms of surface temperatures of typical materials. This work was used in the development of a proposed European Standard PrEN 563 (1992).

The principle of contact temperature is that when two surfaces of different temperature come into contact heat will flow from the hotter to colder surface. Although temperatures will change, there will be a temperature at the interface that will be achieved instantaneously and will not change. This is contact temperature Tc and is given by:

$$Tc = \frac{Tsk.bsk + Th.bh}{bsk + bh}$$

where

       $Tsk$ = temperature of skin
       $Th$   = temperature of the hot body
       $bsk$ = thermal penetration coefficient of skin
       $bh$   = thermal penetration coefficient of hot body

The thermal penetration coefficient is the square root of the product of the density, specific heat and thermal conductivity of a material. The contact temperature demonstrates the effects of different material types. For example if skin touches wood at 100 ºC a low contact temperature would be calculated and it would show less severe reaction than if it touched metal at 100 ºC where a high contact temperature would be calculated. Parsons(1992) considers the use of a modification to the contact temperature equation for practical use. He demonstrates how it could be used to calculate an equivalent contact temperature index (Tceq) and lead to a method for determining a reliable scale for establishing safe surface temperatures. This technique has potential but requires development and is as yet untried. No assessment method has used it to date.

## ESTABLISHMENT OF SURFACE TEMPERATURE LIMITS

PrEN 563 (1992) allows the establishment of temperature limits for machinery in support of the European Machinery Directive (89/392/EEC and amended 91/368/EEC). Although restricted in application to machinery this standard is based upon data concerned with skin reaction and is not specific to a particular context such as machines. It would be just as valid to use it to assess domestic products whether machines or not. Other standards however provide limit values for domestic products. For example Harmonization Document CEN HD 1003 considers the front surfaces of domestic cooking appliances burning gas, that can be touched accidentally.

As an Illustrative example, consider a cooker assessed as if it were a machine used in industry and also as if it were a domestic product.

i) Assessed as a machine: PrEN 563 would apply and the following method used.

1) Identify persons who may touch the surface. Include those who will use the appliance (eg. adults) and those who will not use it but may still come into contact with it (eg. adults and children in the home or cleaners and on-line maintenance workers at work). Perform a task analysis to see who would come into contact with the surface and the likelihood of contact.

2) Identify materials from which the surface is made. (eg. smooth enamel in this example)

3) From the task analysis establish likely and maximum contact periods (eg. 4 seconds in this example).

4) Establish burn thresholds from figures 1 and 2. Figure 1 presents burn thresholds for bare (uncoated) metal. For a contact time of 4 seconds, these range from 58°C below which a burn would not be expected to occur, to 64°C, above which a burn would be expected to occur. Figure 2 provides the increase in burn threshold if they were coated in 160μm enamel. For a contact period of 4 seconds, this increase is 2°C. The burn threshold spread in this example is therefore 60 °C to 66°C.

5. The surface temperature limit value is between 60 and 66°C. A judgement must be made taking into account the context and area of application. For domestic use and risk to children a limit close to 60°C would seem reasonable. For industrial use a limit closer to 66°C may be reasonable.

ii) Assessed as a domestic gas cooker: HD 1003 would apply and the following method used:

The cooker would be operated under standard test conditions in a room at 20°C. To conform to the 'standard' the INCREASE in temperature measured in contact with frontal surfaces of the appliance which can be touched accidentally, must not exceed the following limits:

- Metal and painted metal :  60 K
- Enameled metal          :  65 K
- Glass and ceramic       :  80 K
- Plastic                 : 100 K

That is for a starting temperature of 20 °C, after 60 minutes of operation, the surface temperatures must not exceed 85 °C for enameled metal. According to PrEN 563 this would produce a burn.

SAFE SURFACE TEMPERATURES

A safe product is one that will not cause injury. A safe surface temperature could be defined as one that would not produce damage to the skin (eg. a burn). This does not mean that it may not contribute to an accident. For example reaction to contact with a surface causing pain or simply surprise may lead to an accident. What is safe must therefore be considered in the context of a particular product, its use and its environment. All products will carry with them an associated risk. Risk assessment is a developing subject and has not been used in the context of safe surface temperatures. Thomas(1992) cites the UK Health and Safety Executive in providing the following equation:

Risk = Hazard Severity x Likelihood of occurrence

This can be used as a basis for design decisions. For example if the likelihood of occurrence is low then depending upon consequences, higher surface temperatures may still provide an acceptable risk. The question for the design of domestic products therefore is what is acceptable? It would seem unlikely however that acceptable risk should be greater for domestic products than that for industrial machines.

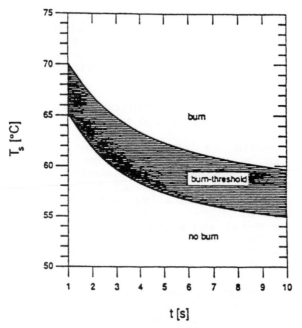

Figure 1:  Burn threshold range for skin contact with
           bare uncoated metal.

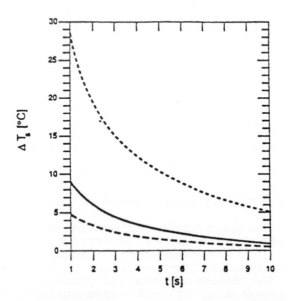

Figure 2:  Increase in burn threshold for metals coated
           with 400μm Rilsan(.....),90μm powder(_____)
           and 160μm porcelain enamel(-----).

The appropriate design of products, involving safe surface temperatures, requires fundamental ergonomics research and knowledge about the reaction of human skin to contact with surfaces. It also requires methods in Applied Ergonomics, both to represent estimates of hazard severity, such as likelihood of burning (eg. Equivalent Contact Temperature - Tceq) and to quantify likelihood and overall consequences of contact leading to risk assessment, acceptable risk and hence design criteria in terms of safe surface temperatures.

REFERENCES

CEN PrHD 1003 (1988) Heating in contact with the front of the domestic cooking appliances burning gas. European Committee for Standardization, CEN, Brussels.

CEN PrEN 563 (1992) Draft for the safety of machinery: Temperatures of touchable surfaces:Ergonomics data to establish temperature limit values for hot surfaces. European Committee for Standardization, CEN, Brussels.

Henriques F C and Moritz A R (1947) Studies of thermal injury I. The conduction of heat to and through the skin and the temperature attained therein. A theoretical and experimental investigation. Am.J.Pathol.23,531-539.

HMSO (1992) The single market : Machinery. Department of Trade and Industry, March 1992, London.

Leach E H, Peters R A and Rossiter R J (1943). Experimental thermal burns, especially the moderate temperature burn. Quarterly journal of experimental physiology, 1943,32,67-89.

Parsons K C (1992) Contact between human skin and hot surfaces: Equivalent Contact Temperature (Tceq). In proceedings of the fifth Int. Conf. on Environmental Ergonomics, W A Lotens and G Havenith (EDs), pp144-145. TNO, Netherlands. ISBN 90-6743-227-X.

Sevitt S (1949) Local blood-flow changes in experimental burns. J. Path. Bact, 61. 427-442.

Siekmann H. (1989) Determination of maximum temperatures that can be tolerated on contact with hot surfaces. Applied Ergonomics, 20,4,313-317.

Siekmann H. (1990) Recommended maximum temperatures of touchable surfaces. Applied Ergonomics,21,1,69-73.

Thomas N (1992) Legal use of personal protective equipment and assessment of risk. In proceedings of the fifth Int. Conf. on Environmental Ergonomics, W A Lotens and G Havenith (EDs), pp48-49. TNO, Netherlands. ISBN 90-6743-227-X.

# AN ERGONOMICS APPRAISAL OF THE PIPER ALPHA DISASTER

W. H. GIBSON          and          E.D. MEGAW

Four Elements Limited                School of Manufacturing and
Greencoat House                     Mechanical Engineering
Francis Street                      The University of Birmingham
London SW1P 1DH                     Edgbaston
                                    Birmingham B15 2TT

The following study is based on a review of the transcripts of the Public
Inquiry into the disaster on the Piper Alpha offshore oil platform. The
analysis is from an ergonomics perspective and provides ergonomics
information about both the platform and the disaster. Four main areas of
concern were identified from the analysis. These were the management
of safety, interface design, training, and human behaviour in response to
fire and evacuation. Inadequacies were found in the ergonomic
consideration given to these areas on Piper Alpha prior to the disaster.
These areas provide examples of the possible impact that poor
ergonomic design can have on the safety of offshore installations.
Finally this paper discusses the validity of using the transcripts from the
inquiry as a method for ergonomic evaluation.

## INTRODUCTION
### Overview of the Piper Alpha disaster event sequence

Because of the death of key personnel and destruction of a major section of the
platform during the disaster, there is uncertainty as to the causes and precise
progression of the accident sequence. The following summary is therefore based
around the most likely scenario presented in the Cullen report (1991).

On the 8 June 1988 an operator attempted to start a pump which was an integral
part of the main production process. A failure in communication meant that the
operator was unaware that a valve directly linked to the pump had been removed
for maintenance and therefore the pump should not have been started. Prior to
this a blind flange was used to cover the open pipework left by the valves
removal but this had probably been inadequately secured by an operator. When
the pump was started highly flammable hydrocarbons leaked past the
inadequately secured blind flange and into a production module .

The hydrocarbons exploded, various engineered safety systems failed and
further explosions and fires ensued. The primary cause of death to personnel was
a large pool of flaming oil and gas fires which gave off large quantities of smoke,

engulfing the platform within fifteen seconds of the initiating event. This made evacuation almost impossible. The majority of personnel died through smoke inhalation. This is the largest offshore disaster to date and lead to the death of 165 platform personnel.

## Literature concerning the Piper Alpha disaster

The Cullen inquiry raised a number of issues which clearly deserved consideration form an ergonomics perspective. In particular chapters of the final report covered the permit to work system and shift handovers, training for emergencies and the management of safety. Also the design of evacuation equipment and evacuation procedures were key issues. Despite this input there was a surprising lack of expert evidence by human factors specialists during the inquiry. Literature directly relating to ergonomics and the disaster is briefly summarized below.

Tombs (1991) discusses the accident in terms of 'distorted communications', in particular, management's failure to act upon relevant warnings available prior to the disaster. Fitzgerald et al (1990) present a human factors approach to the effective design of evacuation systems supported by information from the inquiry transcripts. Munley & Williams (1990) discuss the failure of a senior manager on Piper Alpha to effectively direct the evacuation of personnel as an example of cognitive failure.

It is clear that ergonomics considerations were central to the disaster but, unlike such incidents as Three Mile Island, the extent of published human factors literature is relatively small.

## METHOD

The study was centred around a review of the transcripts of the Piper Alpha inquiry. They are a written account of all that was spoken at the inquiry, numbering roughly 1500 pages. This study was concerned with the evidence given by the 61 survivors and those who were directly involved with managing, and working on, the platform prior to the disaster.

## RESULTS

From the analysis, the areas found to contain sufficient information for further consideration were, the management of safety, training and interface design a further area, human behaviour, fires and evacuation, is treated in detail in Gibson (1992).

## Management of safety

One of the central features of the management of safety is the importance of feedback to management as to the status of the system under their control. Also feedback must be processed in the right manner such that the information can be used effectively in improving systems safety. On Piper Alpha examples of problems with effective feedback and the failure to act on safety information prior to the disaster were found.

A technician was working under the permit to work system on a valve prior to the disaster. As he began to remove the valve, he was hit by a strong blast of flammable gas which blew an operators hat off but luckily did not ignite. This leak occurred because the equipment had not been adequately isolated. On reporting this incident to the maintenance superintendent, he was told that he had been hit by a blast of water and that no action would be undertaken. The information was thus suppressed even though the operators, who were completely dry after the incident, knew the diagnosis to be incorrect (39, 83)[1]. Feedback on an event with clear safety significance was thus suppressed. Possible problems with both the permit to work and isolation procedures could have been brought to the attention of the relevant safety authorities if feedback channels had been working effectively. Significantly both the permit to work system and the isolation procedures played a key role in the disaster.

It is clear that information on the potential safety problems on Piper Alpha were available but no action was taken to eradicate or reduce these problems. Cullen (1991, pp 231 - 234) discusses a previous accident on the platform in which an operator died. This had clear implications for the inadequacy of the permit to work and shift handover systems, again key issues during the disaster. The incident was not adequately considered by management and, therefore, recommendations arising from the incident were not adequately formulated to fully tackle these problems. It should also be noted that adequate feedback was not established to monitor the recommendations which had been proposed.

Similarly, a safety report on the platform which highlighted certain large scale design weaknesses was not considered adequately. One scenario presented in the report mirrored 'very closely what happened on Piper...' during the accident (110, 83) but this knowledge was not adequately incorporated into the safety management system and subsequently acted upon. The Piper Alpha disaster had, in some senses, been predicted and knowledge of this was available but ineffectively used by management.

Interface design

When a fire or gas alarm was identified in the control room the precise location of the source of the alarm could only be found from walking behind a panel to read further indicators (48,33). Thus, an ergonomic design principle was violated and functionally related equipment was not placed close together. This layout has important implications for safety because the precise location of any gas or fire problem needs to be quickly discovered. This is highlighted in the accident sequence where the control room operator did not have sufficient time to go around to the back of the panel to check the more precise location of the initial leak.

The frequency with which spurious fire and gas alarms occurred on the platform meant that they had become viewed as untrustworthy. This meant that, during the accident sequence, an operator was despatched to plant to see if he could identify the leak (48, 33). Using human 'leak detectors' is both

---

[1] Transcripts are referenced as ('Day of trial', 'page number')

unsatisfactory and will severely increase the time taken to identify a leak. The impact that a lack of consideration of ergonomic design may have is clearly highlighted in this case and has direct relevance to the disaster.

During the event sequence, operators were attempting to start a pump. In order to achieve this, controls in three different positions around the pump had to be used. Clearly functionally related equipment which was required to be operated in sequence was not located close to each other. This violates a basic ergonomics design principle. This lead to four important consequences:

1. Considerable time was required for a single operator to restart the pump.
2. If, as during the incident, one operator was at each control area, noise and lack of a line of sight between them produced severe coordination problems.
3. Operators could not check if erroneous actions were being carried out by another operator.
4. Possibly due to (2) and (3) above, an identical pump was unable to be started by operators. If this had been achieved the initial leak would not have occurred.

Due to a lack of evidence as to the precise actions at the pumps, the reasons why the identical pump was not started are unknown. From an ergonomics perspective this poor design has a large potential significance in terms of the disaster. During the trial, lack of human factors input into the inquiry meant that witnesses were not questioned as to the problems involved in restarting the pumps.

Training

Examples from the inquiry illustrated the lack of adequate training given for specific jobs. The control room operator stated that he had received no specific training for his job (49,4). A senior platform manager had not had specific training until three years after taking up his position (110,21). He also stated that he viewed the most essential part of his training as 'experience on the platform' (110,4). It should also be noted that both the control room operator and the manager had no official qualifications (49,4 & 110,4) and their knowledge was based around experience offshore. Personnel in these two positions played a vital role during the disaster.

One of the key features of the initial leak was that a blind flange covering open pipework had not been secured properly. The supervisor for this job had newly been promoted and he failed to check that the flange had been fitted securely. Also it is possible that he did not use the permit to work system effectively. Failure to transmit plant information through the permit to work, as to the status of this valve, was also a key factor in producing the initial leak. Clear inadequacies in the training of this supervisor were presented in the transcripts (50,87). As a new supervisor, with inadequate knowledge of safety procedures, the only option was for him to learn from his mistakes. In the context of the disaster these mistakes were to be an important causal factor.

## DISCUSSION

Possibly one of the most important aspects of this study was the insight gained into the usefulness of public inquiry transcripts to allow a deeper understanding of the causes and background to major disasters. For this reason the final section will discuss the usefulness and validity of a consideration of the transcripts as a form of interview data.

### Interviewers

These are the Representatives of Parties who questioned the witnesses. Inherent within the legal framework is the different emphases that representatives will place on factors considered during questioning. Representatives will act on behalf of parties to support their parties interpretation of events. For example, representatives of Occidental, the owners of Pipe Alpha, supported their clients interests by stressing the lack of corporate responsibility for the disaster. Each witness can be cross-examined by any Representative. This system allows Representatives to select the aspects of witnesses evidence to be viewed from various perspectives. It provides a range of views biased to different parties interests but hopefully provides a more objective overall view because the evidence will have been considered in a variety of ways.

Prior to the Inquiry the Representatives had access to a large body of data on the disaster such as witnesses' statements and the preliminary Petrie Report (1989). Thus many issues were already considered in detail before the Inquiry began. For example the accident sequence described above had already been assessed as the most likely course of events in the Petrie Report. Deeper level causes such as the inadequacy of the permit to work system, training for emergencies and shift changeover had already been identified.

Due to this fact and the nature of the Inquiry format, the questioning was highly directed, that is, the interviewer came with a clear idea of the types of questions to be asked and commonly lead witnesses to give a specific response. This is a useful technique to ensure that interviews are conducted efficiently, but also leads to bias in the data collected because there will be a tendency to reinforce previous findings rather than discover new ones.

In summary the present study was limited by the directed nature of the Inquiry in discussing issues that had already been shown to be important rather than covering a broader range of causal factors. Thus as broad a range of ergonomic issues were not presented as were first anticipated.

### Interviewees

The interviews were conducted in public and within the Inquiry format. These factors produced bias in the evidence given by interviewees. For example, because those involved in the accident may feel that they could be made accountable for the accident they may be unwilling to present information which might implicate that them. One instance of this is that a supervisor, whose actions proved central to the most probable accident sequence, was often evasive in his answers to questions. Secondly many of the witnesses were still involved with the offshore industry. This again may bias the data because they were less likely to publicly criticise their employers. Despite aggressive questioning a senior platform manager resolutely refused to implicate the company's management as a causal factor in the disaster despite the evasive answers this commonly required. It should be noted that those who had left the offshore industry may also have been biased in their attitudes against their former employees.

The evidence should also be viewed in terms of its relationship to the disaster. Survivors were still suffering from the adverse psychological effects of the disaster. They were often confused about events both prior to and during the accident sequence and some had to retire form questioning due to the stress created by recounting these events.

Another important issue is that there was no expert witness who gave evidence on the role of ergonomics in the disaster. It is surprising that despite the central importance of human error in various forms and the key focus on such ergonomic issues as training, safety management and procedures that no witness was called to give specific evidence on the role of human factors in the disaster or for recommendations on this issue in the future. This can be contrasted with the high profile of ergonomics within the Nuclear Industry, found, for example, in the analysis of the Three Mile Island incident.

The Inquiry as an interview form is clearly not an ideal source form which to collect data for objective ergonomic study. Bias is clearly an issue in terms of the interviewer and interviewee, despite the broad objectivity and legally enforced honesty that was required. These methodological problems were outweighed by the detailed information available from the Inquiry and the lack of availability of other sources.

REFERENCES

Cullen, W.D. (1991) The Public Inquiry into the Piper Alpha Disaster  (2nd edition) London: HMSO.

Fitzgerald, B. P., Green, M.D., Pennington, J.and Smith, A.J. (1990) 'A Human Factors Approach to the Effective Design of Evacuation Systems' in Piper Alpha Lessons for Life-cycle Safety Management Institute of Chemical Engineers Symposium Series Number 122, pp. 167-180.

Gibson, W.H.(1992) An ergonomics appraisal of the Piper Alpha disaster MSc Thesis: School of Manufacturing and Mechanical Engineering, the University of Birmingham.

Munley, G.A. and Williams, J.C. (1992) 'Cognitive failures: Experiences and remedies' in Human Factors in Offshore Safety - Seminar Documentation Business Seminars International.

Petrie, J.R. (1989) Piper Alpha Technical Investigation Interim Report London: Department of Energy.

Tombs, S. (1990) 'Piper Alpha - A Case Study in Distorted communication' in Piper Alpha Lessons for Life-cycle Safety Management Institute of Chemical Engineers Symposium Series Number 122, pp. 99-111.

# CARPHONE USE AND MOTORWAY DRIVING

A.M.PARKES, S.H.FAIRCLOUGH AND M.C.ASHBY

HUSAT Research Institute
The Elms, Elms Grove
Loughborough
Leics. LE11 1RG.

This study sought to identify changes in driver behaviour due to handsfree telephone conversations carried out during motorway driving. 18 volunteer subjects either drove in silence or whilst completing verbal tasks on a carphone. No evidence for a change in driving behaviour in terms of speed choice, lane occupancy, accelerator use or overtaking manoeuvres was found. However mental workload did increase. The results are presented in relation to other studies, and safety implications are discussed.

## INTRODUCTION

The aim of this study was to identify changes in driver behaviour due to a handsfree telephone conversation carried out during a motorway journey. Previous research has addressed the problem from a variety of viewpoints, using different independent and dependent variables within a range of test environments. Four studies (Quenault 1968, Wetherell 1981, Mikkonen and Backman 1988, Brookhuis et al 1991) indicate no driving performance change whilst using a carphone, and five (Brown and Poulton 1961, Brown et al 1969, Fairclough et al 1990, Alm and Nilsson 1990 and 1991) indicate change; most typically a reduction in driving speed, but also a decrease in ability to control the lateral path of the vehicle, and an increase in driver reaction time. Of the real road studies, changes in driving behaviour were associated with a driving task with high task complexity i.e. complex manoeuvres or urban traffic conditions.

Previous work at HUSAT has concentrated on handsfree carphones in terms of conversational ability and driving ability on urban roads. Parkes (1991), in a study investigating decision making ability in four communication conditions (face to face, telephone to telephone, driver to passenger and carphone to office), showed consistent drop in structured mental task scores when talking on a carphone. In the range of tests used, it was found that when the driving task interacted with the carphone task it resulted in difficulty in remembering verbal or numerical data, and in making correct interpretations from background information. Fairclough et al (1990) found that drivers not only found speaking (to a passenger or into a carphone) and driving harder than driving alone; but they also made strategic reductions in speed in order to cope with the increased mental workload. The increase in workload also manifested itself in an increase in physiological activity (heart rate) associated with the carphone condition.

Alm and Nilsson (1990) used an advanced moving-base driving simulator to assess drivers' behavioural changes in response to use of a handsfree carphone. It was found that a carphone conversation (in the guise of a Working Memory Span Test) had a negative effect on drivers' reaction times. Speed also decreased with the onset of a call. It had a negative effect on drivers' lane position, worse so when the tracking component of the driving task was hard. Mental workload was rated higher when driving whilst using the carphone.

If a consistent pattern can be gleaned from the studies above, it would seem that priority can be given to the primary task (driving) whilst talking on a carphone, without observable decrements in performance, so long as some threshold point is not reached. The reports of Mikkonen and Backman (1988) indicates that drivers may increase their level of activation to cope with the increased workload of doing two things at once. However the studies of Alm and Nilsson (1990 and 1991) are interesting because they indicate that even in low workload driving conditions in a simulator, changes in safety related performance could be detected.

It is impossible to control all intervening variables in a real road experiment, nor is it possible to look at performance at the edges of normal safety margins. Therefore it was decided to investigate low complexity driving in the relatively constrained environment of a three lane motorway, with moderate traffic flow.

Previous studies would suggest that in such circumstances, gross measures of driver behaviour such as average speed and lane choice would be relatively unaffected by the addition of a carphone conversation. We would also expect there to be an increase in reports of mental workload and mental effort, and a short term dip in speed at the onset of the carphone call.

## METHOD

18 subjects were recruited from research and secretarial staff of the research institute. The 10 female and 8 male subjects ranged in age from 21 to 44.

Two experimental conditions were employed in the study. This allowed for comparison between two driving situations; driving on the motorway in silence (control condition), and driving completing one of two tasks via a handsfree carphone (carphone condition).

Within the carphone condition, two verbal tasks were presented to subjects in the course of the experimental investigation. Both involved components of numerical memory and were quantifiable. The first type of task took the form of a Sternberg memory test. The subject was presented with a list of five numbers and then asked to confirm the presence or absence of an individual digit. This task was called the 'NUMBER LIST' task.

The second type of verbal task involved the subjects listening to a list of three-digit numbers. After each number had been read the subjects had to repeat the number and state if it was higher or lower than the preceding number. This task was called the 'NUMBER JUDGEMENT' task.

The subject received six verbal tasks (three of each) in the course of the experimental trial. Two of these tasks were carried out in the stationary vehicle as a practise trial. An experimenter based at the Institute played the role of 'speaker' to administer the verbal task. Each verbal task lasted approximately two minutes in duration.

The vehicle used in the study was an instrumented Vauxhall Cavalier 2.0 GLi fitted with a Motorola 6800X handsfree cellular telephone. Video recordings were made from inside the vehicle of the external scene. As part of the 'handsfree' package, the carphone uses a small microphone and loudspeaker. The microphone was attached to the sun visor in front of the driver and the loudspeaker fitted into a recessed area above the glove box in front of the passenger. This set-up requires no extra head movement on behalf of the driver in order to hear and be heard properly whilst making a call.

The experimental trials took place on the three lane southbound M1 motorway, through Nottinghamshire and Leicestershire from Junction 26 to Junction 20. Trials took place at off-peak times in the mid-morning and early afternoon. On three occasions during the planning stage, traffic flow measurements were taken from different points along the intended route. Average traffic flows per hour were: Inside lane 515 cars, 352 HGV; Middle lane 940 cars 100 HGV; Outside lane 476 cars 0 HGV.

The experimental measures used during the study may be grouped into three categories.

a)  *Vehicle parameters*: Captured electronically at a rate of 2Hz, these measures were speed before/during each telephone call and accelerator position.

b)  *Observation data*: This data was scored from the video recording taken by the in-vehicle cameras. Frequency counts were taken for the number of vehicles overtaken and number of lane changes. The amount of time spent in each of the three motorway lanes was also scored in seconds.

c)  *Subjective workload measures*: Two subjective questionnaires were used in the course of the investigation. The first was a version of the Modified Cooper-Harper Scale (MCH) (Wierwille et al, 1985) which had been specially adapted for the experiment and a modified version of the NASA-Task Load Index questionnaire (TLX) (Hart and Staveland, 1988).

The experiment took the form of a repeated measures design. The subjects driving behaviour was monitored over a single journey during which they received four telephone calls. Each call lasted for a duration of two minutes and was followed by an interval of three minutes. The intervals between the telephone calls were captured as the experimental control condition.

The order of presentation of the two types of verbal task (JUDGEMENT and LIST) were counterbalanced across subjects. Each task type contained three groups of experimental material, this was also counterbalanced across subjects.  The order of presentation of the two subjective questionnaires (the MCH and the TLX) was also counterbalanced across subjects.

The subjects were familiarised with the dashboard controls and the handsfree carphone whilst seated in the stationary vehicle.  They then received a practise trial during which they performed the two verbal tasks in the stationary vehicle.  The subjects then drove to a motorway junction in order to familiarise themselves with the handling characteristics of the experimental vehicle.  This was followed by a fifteen minute practise drive on the motorway during which they received two shortened versions of the two verbal tasks as an additional practise trial.

The subjects stopped at a service station in order to perform the first half of the NASA-TLX questionnaire before performing the experimental drive.  This journey lasted approximately twenty-five minutes, during which time the subjects received four telephone calls.  The operation of the carphone was completely handsfree except for a single button press needed to 'end' the call after the verbal task had been completed.

On completion of the experimental journey, the subjects stopped at a service station in order to complete the TLX and MCH questionnaires.

## RESULTS

The vehicle parameters captured were analysed for significance ($p < 0.05$) using Wilcoxon signed-rank tests.

### Vehicle Parameters

#### Vehicle speed

No significant change in speed was observed between the control and carphone conditions.  Also, no drop in speed immediately after onset of the call was observed.  There was no difference between the two experimental tasks.

#### Accelerator position

No significant changes in accelerator position resulting from a carphone conversation were apparent.

### Driver Behaviour Measures

Subjects' strategic driving performance on the motorway, recorded on videotape, was analysed post hoc for the two minute periods during and prior to the carphone conversation.  Wilcoxon signed-rank testing was used to test for significance ($p < 0.05$).

#### Number of vehicles overtaken

There was no statistical significance in the difference in the number of vehicles overtaken between the control and carphone conditions.

#### Number of lane changes

There was no statistical significance in the difference in the number of lane changes between the control and carphone conditions.

#### Time spent in each motorway lane

There was no statistical significance in the difference in the time spent in each motorway lane between the control and carphone conditions.

**Modified Cooper-Harper Scale**
Wilcoxon signed-rank testing showed significant increases in both of the speaking conditions over the control condition (p<0.01). Mean ratings were 'Control'= 2.7, 'Judgement'= 4.1, 'List'= 3.8

**Task Load Index (TLX) Questionnaire**
The TLX score represents an index of subjective mental workload for driving and the two verbal tasks.  Significant differences (p<0.05) were found between the control condition and the carphone conditions (Wilcoxon signed-rank tests).  There was no difference between the carphone condition tasks. Mean scores were 'Control'= 36.93, 'Judgement'= 48.01, 'List'= 54.38.

## DISCUSSION
No evidence for a change in driving behaviour was found in this study.  Strategic level choices of speed; tactical level choices of lane occupancy; and operational level activity such as accelerator pedal depression, appear consistent across the experimental conditions.

Analysis of the subjective responses of the drivers revealed an increase in perceived workload when the carphone tasks were introduced.  This was demonstrated by a MCH score of under 3 for the control condition (suggesting acceptable driver effort for adequate safe driving) and around 4 for the carphone conditions (suggesting moderately high effort to attain adequately safe driving).  This semantic categorical difference is not large, but is a clear indication of the additional load imposed by the relatively straightforward verbal tasks involved.  It might be expected that a secondary task of more complex open ended negotiation style dialogues would reveal greater subjective differences.  Analysis revealed a clear difference in RTLX scores between the control and carphone tasks.  The combination of MCH and RTLX results indicate that mental workload is increased by the introduction of a carphone task to levels where workload is judged to be high but not approaching the point where maximal levels of driver effort are required to maintain safe driving.  It must be remembered however, that call duration's in this study were strictly limited to periods of two minutes.  Sustaining the subjectively experienced high mental workload for more protracted periods, may have led to greater demands on resources.

This study did not attempt to replicate previous research (Alm and Nilsson 1991) that had shown a decrease in choice reaction time to events presented undertaking a simulated driving and carphone task; due to the difficulties of performing such tests on real roads.

## CONCLUSION
The presence of a carphone call in moderate traffic motorway conditions did not result in a level of workload that produced behavioural adaptation by the drivers, as measured in terms of speed, lane choice, overtaking strategy or accelerator position. Nor was there any indication of an immediate reduction in speed as an alerting response to the onset of an incoming call.

Measures of subjective mental workload revealed that the combination of low difficulty driving, with simple verbal tasks, raised overall workload to the point where drivers began to feel unsafe whilst driving.  Other studies of driving behaviour in simulators have linked a similar rise in mental workload to increased lateral deviation, and increased choice reaction time (Alm and Nilsson 1991).

The findings from this study support the current concern that the act of using a carphone whilst driving may for many drivers, in many traffic scenarios pose an increase in the risk of an accident due to a reduction in safety related driving performance.

## REFERENCES

Alm, H. and Nilsson, L. 1990. Changes in driver behaviour as a function of handsfree mobile telephones: A simulator study. Report No. 47. DRIVE Project V1017. Swedish Road and Traffic Research Institute, Linkoping, Sweden

Alm, H. and Nilsson, L. 1991. The effects of a mobile telephone conversation on driver behaviour in a car following situation. Report No. 73. DRIVE Project V1017. Swedish Road and Traffic Research Institute, Linkoping, Sweden

Brookhuis, K.A., DE Vries, G. & DE Waard D.  1991. The Effects of Mobile Telephoning on Driving Performance. Accid. Anal. & Prev.. Vol 23, No4,  pp. 309-316.

Brown, I.D. and Poulton, E.C., 1961. Measuring the spare 'mental capacity' of car drivers by a subsidiary task. Ergonomics, 4, 35-40.

Brown, I.D., Tickner, A.H. and Simmonds, D.C.V.,1969. Interference between concurrent tasks of driving and telephone. Journal of Applied Psychology, 53, 419-424.

Drory, A., 1985. Effects of rest versus secondary task on simulated truck driving performance. Human Factors, 27 (2), 201-207.

Fairclough, S.H., Ashby, M.C., Ross, T. and Parkes, A.M. 1990. Effects on driving behaviour of handsfree telephone use. Report No. 48. DRIVE Project V1017. HUSAT Research Institute, Loughborough University, U.K.

Hart, S.G. and Staveland, L.E. 1988. Development of the NASA-TLX (Task Load Index): Results of empirical and theoretical research. In P.A. Hancock and N. Meshkati (Eds.) Human Mental Workload. North Holland, Amsterdam, pp 139-183.

Mikkonen, V. and Backman, M., 1988. Use of car telephone while driving. Technical report No. A39. Department of Psychology, University of Helsinki.

Parkes, A.M. 1991. Drivers business decision making ability whilst using carphones. In Lovesey, E.J. (Ed.) Contemporary Ergonomics 1991. London: Taylor and Francis.

Quenault, S.W., 1968. Task capability whilst driving. TRRL Report No. LR166, Transport Research Laboratory, Crowthorne, Berkshire.

Wetherell, A., 1981. The efficacy of some auditory - vocal subsidiary tasks as measures of the mental load on male and female drivers. Ergonomics, 24 (3), 197-214.

Wierwille, W.W., Rahimi, M. and Casali, J.G., 1985. Evaluation of 16 measures of mental workload using a simulated flight task emphasizing mediational ability. Human Factors, 27 (5), 489-502.

# 1994

| Warwick University | 19th–22nd April |
|---|---|
| **Conference Manager** | Janis Hayward |
| **Chair of Meetings** | Dave O'Neil |
| **Programme Secretary** | Sandy Robertson |
| **Secretariat Included** | Margaret Hanson     Rebecca Lancaster |

| Chapter | Title | Author |
|---|---|---|
| **Occupational Health** | Injury in the orchestra – The ergonomic nightmare | E. Colley |
| **Upper Limb Assessment** | R.U.L.A. – A rapid upper limb assessment tool | L. McAtamney and E.N. Corlett |
| **General Ergonomics** | The teleworking experience | B. Dooley, M.T. Byrne, A.J. Chapman, D. Oborne, S. Heywood, N. Sheehy and S. Collins |
| **Formal Methods** | Validation in ergonomics/ human factors | H. Kanis |

# INJURY IN THE ORCHESTRA -
# THE ERGONOMIC NIGHTMARE

## EVE COLLEY, P.T., M.A.P.A.

*Pacific Physical Therapy Inc.,*
*Honolulu, Hawaii, U.S.A.*

Playing in an orchestra is exacting work and many string, wind and brass players display symptoms of muscular and respiratory stress. This arises from long periods of practice and performance using musical instruments that are of unsatisfactory design from an ergonomic point of view. Methods are described of analysing the movements made in playing instruments and the postures adopted in doing so, and of enabling players to optimise these and thus minimise the stresses of playing. Exercises to counter the effects of stresses resulting from certain instruments are also described.

## Introduction

Until recently, professional players in the orchestras of the world have been largely overlooked in terms of the ergonomics of their chosen careers, and the grave risks they face of repetitive strain injury and myofascial pain syndromes.

Most people, including the musicians themselves, fail to realise that to produce the sound made with an orchestral instrument places unusual stress on the performer's body. Because they continue to play every day, often for several hours at a time, they are prime targets for repetitive strain injuries.

Although musical instruments have been developed over a period of hundreds of years, they are not ergonomically well-designed. They may best be described as "user-unfriendly".

Those musical instruments which place the most severe physical demands on their players are those played asymmetrically - consider the violin, viola, cello, double bass, bassoon, flute, tuba and French horn, for example. Further, they have to be supported while they are being played - and this can cause myofascial tension and strain in the supporting musculature.

Prolonged, sustained postures required to play brass instruments and tympani contribute to myofascial tension in these players.

## Individual factors

The situation is exacerbated by the devotion of the players to their work. If we look at the players during a concert performance, we see the habits and techniques ingrained over the years which create muscular strain and which can lead to neurological damage. Nerve entrapment and thoracic outlet syndrome have been described in several publications.

Almost all professional violinists start to play the instrument when they are less than ten years old. They rapidly accommodate to their instruments - and in this early stage every emphasis is placed on notes rather than on developing the techniques and posture which will ensure the minimum of physical difficulty as the player grows.

Abnormal posture may produce a pattern of muscle tension which alters the alignment of the cervical vertebrae and restricts upper rib movement. If the altered posture is practised daily for prolonged periods of time, it is reasonable to expect altered patterns of growth in the vertebrae - giving rise to bone spurs. While I have not found formal studies of this particular problem, I have found unilateral bone spurs on x-ray in violinists, flautists and a bassoon player.

Part of the problem with young musicians is that they consider themselves indestructible. They may tolerate aches and pains - while they resent being told to stand up straight.

Until the players begin to notice limitations in their personal performance brought on by pain or muscular restriction, they strongly resist suggestions to change their playing technique. When I treat violinists in their early teens who show painful repetitive strain injuries, yet who continue to practise daily for long hours in their school orchestras and small groups, I am struck by how widespread this problem must be.

## Opportunities to help

During my professional career, I have worked with many musicians. Over the last three years, I have had a close association with the members of the Honolulu Symphony Orchestra. In order to help them, the first essential has been to analyse the "job": that is, to understand fully the movements the player is required to make. We must watch the performer play their instrument during practice and during an actual performance, noting each aspect of how they support and handle the instrument. We observe how the muscles and joints are used, how posture is altered as the musician tires, how fatigue or concentration alters angles of attack and rest.

Not only must the physical requirements of playing the instrument be considered, but also the emotional and personal background of the player, since this may well influence posture and attitude. This will have a direct bearing on the ability of the muscles to adapt and work. In short, each musician must be assessed as an individual, both at rest and at work - and not simply while playing scales but demanding music, which requires their concentration and full range of skills.

Detailed studies have been made of 35 players: 18 in violins and violas, 9 with other stringed instruments, 4 woodwinds and 3 brass players, and one tympanist.

## Types of stress

Two types of stress have been found to be prominent: musculotendinous and respiratory. Orchestral players are required to play for long periods under public scrutiny, using muscles ill-suited to repetitive work. Most orchestral players have poor breathing habits and the very actions required to play their instruments may well inhibit proper breathing.

Some of the purest examples of muscular stress are seen in violin playing. For comfort and ease, the instrument needs to be supported on the upper chest and clavicle. If, however, the player holds the violin more horizontally and to the side and places pressure on the first rib, this compresses the brachial plexus and causes myofascial tension in the pectoral muscles, scalenes, upper trapezius and levator scapulae on the left side. Over time, this can produce a thoracic outlet syndrome which severely limits the player.

This danger is reduced by using rests designed to fit between the violin and upper chest in addition to the rest normally located between jaw and violin. Each player needs to select carefully the rest most appropriate to their physical structure and playing technique.

Problems may also arise from bowing action: some schools stress a technique which requires the bowing arm to perform a sharp full-range pronation of the forearm with extension of the elbow on the down-stroke. This can cause the pronator teres muscle to develop myofascial tension and it may compress the median nerve. The resulting symptoms closely resemble carpal tunnel syndrome. The pronator teres muscle may also pull the elbow into an exaggerated carrying angle and the ulnar nerve becomes entrapped in a retinacular tunnel at the medical epicondyle.

This produces a neurapraxia of the ulnar nerve as well. It can be prevented simply by altering the technique in the bowing arm - but treating it and curing the symptoms may be a lengthy and complicated task.

The fact that the violin and viola are supported on the upper chest can lead also to respiratory problems. The player tends to lean towards the supporting side, compressing the lung and limiting upper rib excursion on that side.

Stresses amongst woodwind players arise from the need to provide compressed air, often at high pressure, while supporting the instrument and manipulating keys which require uncommon manual dexterity. Flute players must perform with the upper torso rotated and with both arms and neck at awkward angles.

The brass instruments pose different problems. The tuba player, who may need a very large vital capacity, normally supports the instrument by side bending to the right - thereby reducing the capacity of the right lung. To a lesser extent this also applies to the French horn. Back pain, shoulder pain and neck strain are also common with these instruments. This is a result of the combined effects of bending and breathing. In the case of one tuba player I have examined, rib excursion on the right was extremely difficult and side bending to the left and trying to breathe with the right lung was actually painful.

## Prevention of stress and correcting its effects

I have found that musicians themselves are very capable of finding solutions to their problems once they understand how they are caused and how they affect their playing.

We therefore videotape them, playing their instruments while wearing a swimsuit. This allows them to see themselves from several different angles and to hear variations in sound production which may come about from variations in their grip on the instrument or altering their posture while playing.

It further allows us to assess the problems with the musician and explain them in a third-person context so that the performer can stand off and judge impartially the ergonomic approach to solving them.

We discuss a number of approaches to handling the problem and the musician can then try to put these ideas to work and hear the differences in the music which result. We stress the need for the musician to adapt the instrument and its use to the musician and NOT the musician to the instrument. Failure to do this is the source of most problems which arise with beginners.

The player may need, on examination, to hold the instrument at a different angle or may benefit from using a support for it. They may initially reject using a support because they feel it will alter the quality of tone they produce (violinists) - or their freedom of movement (woodwind players) - but they are often pleasantly surprised to find playing is more comfortable and the quality of music they produce is improved.

Ease of playing is also affected by factors such as the height of the music stand and the player's chair, which in turn affect posture and the ability to see the conductor and the score on the stand. Prescription spectacles should be regularly checked: can the player read the score without undue strain?

We also teach musicians specific exercises to counterbalance the abnormal positions they adopt when they play. For instance, violin and viola players show pronation of the right forearm and supination of the left forearm. They need to be taught stretches which include rotary movements of the forearms and shoulders, combined with breathing techniques as part of their exercise.

If a flautist should play with the head tilted to the right and rotated to the left - a fairly common posture amongst flautists - I teach them exercises to stretch the left pectorals and others to take their head and neck through full range of rotation to both sides, plus side flexion to both sides. These exercises need to be performed both before and after playing the instrument.

French horn and tuba players need exercises to provide side flexion to the left while stretching the right forearm forward and overhead to counteract the habitual side bending to the right with their extended shoulder and flexed elbow posture.

When I work with woodwind and brass players, I also look at speech patterns, lip movement and mouth postures. I give specific exercises for mouth, lip and tongue to break habits which develop from compressing these structures. I also place emphasis on correct breathing techniques so that each musician uses all areas of the lungs.

Where a player has an instrument which rests on the upper ribs and limits their normal excursion, I provide specific exercises for the apical segments of the lungs. I have found that teaching the players breathing and relaxation techniques they can use before going onstage to be particularly beneficial to them.

The obvious time to start such training with players is when they are young - from the moment they first pick up their instrument. Failing that, every effort should be made in their early years to train them in correct posture and playing techniques to prevent their developing the bad habits which will restrict them by the time they mature physically.

## Beyond the detail

Orchestral members are receptive, intelligent and particularly sensitive. They often perform under intense strain in public, before an audience able - and often willing - to hear the slightest wrong note. They have to read music while watching the conductor direct the tempo - which may well vary considerably from that of the last rehearsal. The concentration required and the strain for ensemble and solo players alike is enormous.

In dealing with musicians, we are dealing with individuals who are both workers and artists, yet who make demands on their bodies similar to athletes. I have, in fact, found it useful to address the young members of orchestras as though they were athletes and to stress that they should go through the same routine as an athlete with a warm-up period preceding the performance and cooling-down period afterwards. When practising, they should be acutely aware of any aches and pains.

Treating the professional musician is intensely rewarding; almost alone amongst patients they have the dedication and focus needed to practise repetitively and sensitively the exercises needed to remove muscular restrictions, reduce and eliminate pain and which will allow them to return to a productive life of giving pleasure to others.

## References

Brockman, R., Tubiana, R. et al. 1992. Anatomic and Kinesiologic Considerations of Posture of Instrumental Musicians. *Journal of Hand Therapy*. April-June 1992.

Eaton, G. 1992. Entrapment Syndromes in Musicians. Ibid.

Havlik, R. and Upton, J. 1992. Hand and Upper Limb Problems in the Pediatric Musician. Ibid.

Idler, R., Strickland, et al. 1991. Pronator Syndrome. *Indiana Journal of Medicine* - Feb. 1991.

Lederman, R. Nolan, W. et al. 1993. *Medical Problems of Performing Artists*. Nerve Entrapment Syndromes in Musicians. June 1993 (Hanley and Belfus, Inc.)

Markison, R. 1992. Tendinitis and Related Inflammatory Conditions Seen In Musicians. *Journal of Hand Therapy:* April-June 1992.

Novak, C. 1993. Conservative Management of Thoracic Outlet Syndrome in the Musician. *Medical Problems of Performing Artists* - March 1993 (Hanley and Belfus, Inc. ) 16-21.

Roos, B. 1992. Thoracic Outlet Syndromes in Musicians. *Journal of Hand Therapy:* April-June 1992.

# R.U.L.A. - A RAPID UPPER LIMB ASSESSMENT TOOL

## Lynn McAtamney and Nigel Corlett

*Institute for Occupational Ergonomics*
*University of Nottingham*
*University Park*
*Nottingham NG7 2RD*

RULA (Rapid Upper Limb Assessment) is a survey method developed for use in ergonomic investigations of workplaces where work related upper limb disorders are reported. RULA is a screening tool which assesses biomechanical and postural loading on the whole body with particular attention to the neck, trunk and upper limbs. Reliability studies have been conducted using RULA to assess VDU users and sewing machine operators. A RULA assessment requires little time to complete and the scoring generates an action list which indicates the level of intervention required to reduce the risks of injury due to physical loading on the operator. RULA is intended to be used as part of a broader ergonomic study.

## Introduction

This paper outlines the steps taken when using RULA to assess a working posture. A complete description of the method is presented in McAtamney and Corlett (1993). RULA is intended to be used as part of a wider ergonomic assessment and this is described in McAtamney and Corlett (1992).

## Step 1 - Observing and selecting the posture(s) to assess

A RULA assessment represents a moment in the work cycle and it is important to observe the postures being adopted whilst undertaking the tasks prior to selecting the posture(s) for assessment. Depending upon the type of study, selection may be made of the longest held posture or what appears to be the worst posture(s) adopted. In some instances, for example when the work cycle is long or the postures are varied it may be more appropriate to take an assessment at regular intervals. It will be evident that if assessments are taken at set intervals over the working period the proportion of time spent in the various postures can be evaluated.

## Step 2 - Scoring and recording the posture

Score the posture of each body part in groups A and B (shown in Diagrams 1 and 2) and record them in the appropriate boxes found on the scoring sheet (Diagram 3).

If assessment of the right and left upper limbs is required boxes are provided to record the scores. Follow steps 3 and 4 to evaluate them seperately.

For both groups A and B score the muscle use and forces or load experienced using the definitions in Figure 1 and 2. Record these in the appropriate boxes on the scoring sheet.

## Step 3 - Coding the posture scores

The posture scores are coded using Tables A and B which are shown in Tables 1 and 2. Use the posture scores from Group A and Table A to find Posture score A and add the muscle use and force scores to calculate 'Score C'. Repeat the process for the posture scores from Group B using Table B to find Posture score B and add the muscle use and force scores to calculate 'Score D'.

## Step 4 - Calculating the Grand Score and Action Level

The grand score is calculated by using Score C and Score D in Table 3. This grand score can then be compared to the Action Level List below, however it must be remembered that since the human body is a complex and adaptive system, they provide a guide only. In most cases, to ensure this guide is used as an aid in efficient and effective control of any risks identified, the action leads to a more detailed investigation.

**ACTION LEVEL 1**
A score of <u>one or two</u> indicates that the posture is acceptable if it is not maintained or repeated for long periods.

**ACTION LEVEL 2**
A score of <u>three or four</u> indicates further investigation is needed and changes may be required.

**ACTION LEVEL 3**
A score of <u>five or six</u> indicates investigations and changes are required soon.

**ACTION LEVEL 4**
A score of <u>seven or more</u> indicates investigation and changes are required immediately.

## References

McAtamney, L. and Corlett, E.N. (1992) *Reducing the Risks of Work Related Upper Limb Disorders: A Guide and Methods.* Published by The Institute for Occupational Ergonomics, University of Nottingham.

McAtamney, L. and Corlett, E.N. (1993) RULA: a survey method for the investigation of work-related upper limb disorders, Applied Ergonomics, **24**, 2, 91-99.

# GROUP A

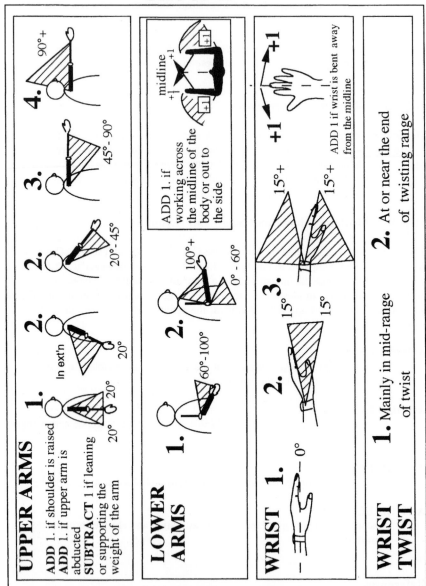

**Diagram 1. RULA Group A body parts**

# GROUP B

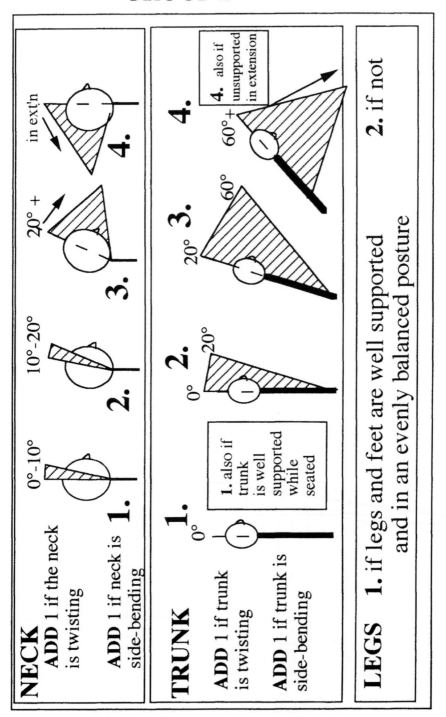

**Diagram 2. RULA Group B bodyparts**

## TABLE A   Upper Limb Posture Score

| UPPER ARM | LOWER ARM | WRIST POSTURE SCORE 1 | | WRIST POSTURE SCORE 2 | | WRIST POSTURE SCORE 3 | | WRIST POSTURE SCORE 4 | |
|---|---|---|---|---|---|---|---|---|---|
| | | TWIST 1 | TWIST 2 | TWIST 1 | TWIST 2 | TWIST 1 | TWIST 2 | TWIST 1 | TWIST 2 |
| 1 | 1 | 1 | 2 | 2 | 2 | 2 | 3 | 3 | 3 |
|   | 2 | 2 | 2 | 2 | 2 | 3 | 3 | 3 | 3 |
|   | 3 | 2 | 3 | 3 | 3 | 3 | 3 | 4 | 4 |
| 2 | 1 | 2 | 3 | 3 | 3 | 3 | 4 | 4 | 4 |
|   | 2 | 3 | 3 | 3 | 3 | 3 | 4 | 4 | 4 |
|   | 3 | 3 | 4 | 4 | 4 | 4 | 4 | 5 | 5 |
| 3 | 1 | 3 | 3 | 4 | 4 | 4 | 4 | 5 | 5 |
|   | 2 | 3 | 4 | 4 | 4 | 4 | 4 | 5 | 5 |
|   | 3 | 4 | 4 | 4 | 4 | 4 | 5 | 5 | 5 |
| 4 | 1 | 4 | 4 | 4 | 4 | 4 | 5 | 5 | 5 |
|   | 2 | 4 | 4 | 4 | 4 | 4 | 5 | 5 | 5 |
|   | 3 | 4 | 4 | 4 | 5 | 5 | 5 | 6 | 6 |
| 5 | 1 | 5 | 5 | 5 | 5 | 5 | 6 | 6 | 7 |
|   | 2 | 5 | 6 | 6 | 6 | 6 | 6 | 7 | 7 |
|   | 3 | 6 | 6 | 6 | 7 | 7 | 7 | 7 | 8 |
| 6 | 1 | 7 | 7 | 7 | 7 | 7 | 8 | 8 | 9 |
|   | 2 | 8 | 8 | 8 | 8 | 8 | 9 | 9 | 9 |
|   | 3 | 9 | 9 | 9 | 9 | 9 | 9 | 9 | 9 |

Table 1.  Table A used to calculate Posture score A

## TABLE B   Neck, Trunk, Legs Posture Score

| NECK POSTURE SCORE | TRUNK POSTURE SCORE 1 | | 2 | | 3 | | 4 | | 5 | | 6 | |
|---|---|---|---|---|---|---|---|---|---|---|---|---|
| | LEGS 1 | LEGS 2 | LEGS 1 | LEGS 2 | LEGS 1 | LEGS 2 | LEGS 1 | LEGS 2 | LEGS 1 | LEGS 2 | LEGS 1 | LEGS 2 |
| 1 | 1 | 3 | 2 | 3 | 3 | 4 | 5 | 5 | 6 | 6 | 7 | 7 |
| 2 | 2 | 3 | 2 | 3 | 4 | 5 | 5 | 5 | 6 | 7 | 7 | 7 |
| 3 | 3 | 3 | 3 | 4 | 4 | 5 | 5 | 6 | 6 | 7 | 7 | 7 |
| 4 | 5 | 5 | 5 | 6 | 6 | 7 | 7 | 7 | 7 | 7 | 8 | 8 |
| 5 | 7 | 7 | 7 | 7 | 7 | 8 | 8 | 8 | 8 | 8 | 8 | 8 |
| 6 | 8 | 8 | 8 | 8 | 8 | 8 | 8 | 9 | 9 | 9 | 9 | 9 |

Table 2. Table B used to calculate Posture score B

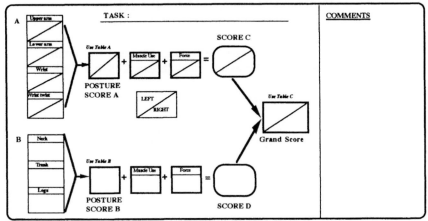

**Diagram 3. RULA Scoring Sheet**

**Figure 1: Muscle Use Scores**

**Give a score of 1** if the posture is;

- mainly static, e.g. held for longer than 1 minute
- repeated more than 4 times/minute

| 0. | 1. | 2. | 3. |
|---|---|---|---|
| • No resistance or less than 2kg intermittent load or force | • 2-10kg intermittent load or force | • 2-10kg static load<br>• 2-10kg repeated load or force | • 10kg or more static load<br>• 10kg or more repeated loads or forces.<br>• Shock or forces with a rapid buildup. |

**Figure 2: Forces or Load Scores**

**Table 3: Table C used to calculate the Grand Score**

**TABLE C**

SCORE D (NECK,TRUNK,LEG )

| SCORE C (UPPER LIMB) | 1 | 2 | 3 | 4 | 5 | 6 | 7+ |
|---|---|---|---|---|---|---|---|
| 1 | 1 | 2 | 3 | 3 | 4 | 5 | 5 |
| 2 | 2 | 2 | 3 | 4 | 4 | 5 | 5 |
| 3 | 3 | 3 | 3 | 4 | 4 | 5 | 6 |
| 4 | 3 | 3 | 3 | 4 | 5 | 6 | 6 |
| 5 | 4 | 4 | 4 | 5 | 6 | 7 | 7 |
| 6 | 4 | 4 | 5 | 6 | 6 | 7 | 7 |
| 7 | 5 | 5 | 6 | 6 | 7 | 7 | 7 |
| 8 | 5 | 5 | 6 | 7 | 7 | 7 | 7 |

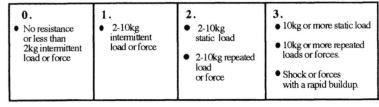

# THE TELEWORKING EXPERIENCE

[1]Barbara Dooley, [1]Marie Therese Byrne, [2]Antony J Chapman,
[3]David Oborne, [2]Sarah Heywood, [4]Noel Sheehy & [2]Sylvie Collins

[1]*University College Dublin*
[2]*University of Leeds*
[3]*University College of Swansea*
[4]*Queen's University of Belfast*

'Teleworking' may be defined as the performance of work
activities at an independent, usually distant, location which
is facilitated by the use of information and communications
technologies. Although in its relative infancy teleworking
offers considerable benefits to both employers and
employees, but its effectiveness is reduced if the
organizational, social and ergonomic features of this new
kind of working relationship are not considered carefully.
This paper outlines some of the more salient results arising
from a comprehensive series of questionnaire and interview
surveys of teleworkers, office-based workers and managers
across Europe, designed specifically to study the social and
psychological experiences of teleworking.

## Introduction

Many commentators have suggested that the key to organizational success
is the ability to adapt and be flexible. Indeed, in contemporary global, economic
climates, the 'survival of the fittest' must be the organizational watchword. In the
pursuit of flexibility many organizations have endeavoured to develop new
working methods and to build the skills needed to cope with the constantly
varying circumstances presented in today's unstable environments. Teleworking
is seen by many to offer the necessary flexibility that will enable organizations to
make the adjustments that are necessary.

Teleworking is certainly on the increase and is soon likely to become a
potent force within future organizational strategies. A recent survey has
estimated that nearly 41 million individuals in the US work from home (reported
in *Teleworker*, November 1993). This amounts to about 32.7% of the US labour
force and is 4% higher than the equivalent 1992 figure. The European Union is
developing an action plan to stimulate European-wide teleworking within Europe.
In its Fourth Framework Programme, for example, it is investigating strategies
for research into the decentralization of large organizations and the development

of smaller networked telecentres and small business opportunities.

Like all new forms of working, however, teleworking requires appropriate organizational, individual, and ergonomic structures to facilitate its effective development. As part of the European Commission's Third Framework Programme (within the ORA -- Opportunities for Rural Areas -- action line) the PATRA (Psychological Aspects of Teleworking in Rural Areas) project was conceived to investigate the human-centred enhancers and inhibitors of current and potential teleworking systems across Europe. By considering the organizational, social and ergonomic factors that contribute to teleworking effectiveness, the approach was to gather material that would ensure not just the survival of the fittest but almost the development of a 'fit test' for the effective survival of teleworking systems. The conference poster is a development of this theme; the purpose of the paper is briefly to provide some background to the work.

## Organizational Factors

If an organization is technologically able to implement teleworking it has still to attend to psychological and organizational factors, including the individual teleworker's perceptions of the organization. In our survey teleworkers' perceptions of their organizations, including relationships with superiors, and the nature of their jobs were assessed using scales relating to the organizational culture, managerial style, job satisfaction, job characteristics, and job composition.

Results suggest that for a teleworking regime to flourish the organizational culture has to incorporate a positive teleworking construct, often effected through a 'teleworking champion'. The organizational infrastructure should embody an open culture, with commitment, effective communication channels, participation and trust. Irrespective of the location of their work teleworkers were found to perceive the climate of their organization positively, with positive feelings of trust, communication, and commitment. The successful communication of an organization's culture normally requires frequent face-to-face interactions with many individuals. However, it was found that employees who reported spending the majority of their work/time away from the office did not differ significantly in their perceptions of the organization compared to others who spent greater amounts of time in the organization. This may indicate that the culture has been strongly communicated prior to telework up-take, and that it is sustained either through regular visits to the organization or via electronic forms of communication such as e-mail, or perhaps through both of these methods. In fact e-mail usage was found to affect perceptions of direct contact and communication with management. This finding emphasizes the importance of perceiving and using information technology as a communication tool rather than allowing it to be perceived as some kind of threat.

Management's ability to prescribe clear objectives and activities is an essential precondition to successful teleworking. The more remote the employee the greater the perceived need for effective tele-management: for example, rural-based teleworkers rated their managers higher on task orientation than did other groups of teleworkers. The majority of the sample reported being satisfied with their work and lives in general.

## Social Factors

One of the most distinctive characteristics of teleworking is that individuals work at an independent location from the centre. This raises a number of important considerations concerning how the teleworker communicates and interacts with work colleagues and how some of his/her interactions with family members may be changed.

Our survey indicated that teleworking affects the pattern of interactions with both work colleagues and family members. In respect of family interactions, teleworkers who spent the least time working in the office and who worked in less populated geographical regions perceived less interference in their family life and their relationships with friends. This illustrates a potential benefit of flexibility in work patterns for teleworkers. The farther away a teleworker is from the central organization the less likely s/he is to interact with work colleagues and the more with family members. However, the reduced level of interaction with work colleagues was not found to hinder communications to the extent anticipated. Plainly telecommunications contribute considerably to the effectiveness of interactions with colleagues.

In particular, greater use of e-mail and modems was linked with beliefs that communications were more accurate and timely. Use of the telephone was particularly important, greater use being associated with having more contact with one's boss, and of being aware of organizational issues and aware of office news. It thus appears that relationships between communication and teleworking are mediated by the use of telematic devices and that on the whole these have a beneficial impact.

No significant effects of location of work on stress data were found. The conclusion here was that work-related stresses and general daily hassles, tend to be a function of the intrinsic nature of working life, rather than the local environment. The important factor is that people should have choices about where they live, where they work, whether they commute, etc.. Teleworking, of course, provides increased choice and opportunity.

Our survey findings, amongst others, suggest a need to build infrastructures that are compatible with these kinds of social concerns. Such infrastructures may be socially based (to overcome social disruptions), and/or they may be technologically based ones (to enhance the possibility of getting better work-based social support through to the teleworker). We need also to note the need for an appropriate training-based infrastructure.

## Ergonomic Factors (including health and safety)

When both teleworkers and non-teleworkers were asked about their general use of technological equipment a number of differences were found. First, an important difference was found between the two groups in terms of the flexibility afforded them to choose their working times. Teleworkers do not appear to work significantly longer than non-teleworkers, just more flexibly. This can have important implications, both positive and negative, for family interactions.

Obvious differences exist between the two groups in terms of their use of information technology. However, differences were also shown to occur between them in terms of how they do their job. Teleworkers spent less time giving instructions and providing information, both when interacting on the telephone and in a face-to-face mode, than did office-based workers.

Teleworkers in our sample had rarely received training for functioning as a teleworker. From responses received training packages should attend to at least the following four factors: adequate time management/job scheduling; the maintenance of social contacts; the need to develop self-discipline/motivation; and the importance of retaining a work/home divide.

Results also suggest that teleworkers use their equipment in a more appropriate (ergonomic) fashion than do office-based counterparts. Both groups were asked to record the heights of their desk top from the floor, their chair seat from the floor, the bottom of the computer screen from the desk, and their own height. No statistically significant differences were found between the two groups in respect of these heights. This suggests that as far as the static aspects of the environment are concerned, teleworkers and office-based workers use the same kinds and designs of office furniture.

In relation to the dynamic aspects of the office environment, respondents were asked whether their normal desk chairs have castors, arm rests, and adjustable backrests. For each of these features significant differences were obtained between the two groups. In each case significantly smaller proportions of teleworkers reported possessing the extra features. One might therefore expect teleworkers to have higher incidences of physical strains than office workers.

This was not shown to be the case, however. On almost all of the problems, the distribution of responses from the teleworking sample suggests that teleworkers suffered from some complaints less frequently than office-based workers: bad back ($p < 0.01$), eye strain ($p < 0.01$), headaches ($p < 0.01$), aching wrists ($p < 0.05$) and neck ache ($p < 0.01$). Of course, the differences in the occurrence of physical complaints may not reflect differences in the design of the office furniture so much as the flexibility in the pattern of working hours and the easier lifestyle reported by teleworkers.

## Conclusion

Asking teleworkers and managers about their experiences within both established and newly-formed teleworking systems has proved invaluable for suggesting aspects of different infrastructures that are important from the viewpoint of this new kind of working behaviour. By examining the social, organizational and technological infrastructures we are now in a position to develop relevant training and information schedules to enable pan-European teleworking to have greater chances of success in the future -- which is where the test of the fit will lie.

# VALIDATION IN ERGONOMICS/HUMAN FACTORS

## H. Kanis

*School of Industrial Design Engineering*
*Delft University of Technology*
*Jaffalaan 9, 2628 BX Delft*
*the Netherlands*

In a selection from Ergonomics/Human Factors literature many papers only pay lip-service to the issue of validation. About one-third of the selected studies actually deal with validation on the basis of empirical research. However, presuppositions and theoretical considerations about both the emergence and the absence of bias are hardly ever questioned. Thus the reporting of much empirical work is reduced to plain registration, in particular of central tendencies, which shows a sharp contrast with the proliferation of jargon. This masquerade echoes an unfortunate tradition in the social sciences. A number of suggestions may help to render (in)validation studies more fruitful.

## Introduction

Terms like repeatability, reproducibility, reliability and validity can be found in numerous research papers. This is logical, since these notions refer to two distinctive phenomena that are basic in empirical research:
- random variation in outcomes of a repeated measurement, and
- systematic deviation between a measurement outcome and what is intended to be measured (cf. Carmines and Zeller, 1979).

As a matter of fact, both issues regularly seem to become confused. Additionally, a jargon has developed within some disciplines which tends to obscure substantial elements of validation, whereas clearness would be highly desirable here. Measuring is always fundamental in empirical research. This includes the evaluation of its outcomes in terms of reproducibility and validity. Ambiguities and mystification can only be detrimental to the possible impact of measurement outcomes, especially in a multidisciplinary area like Ergonomics/Human Factors.

In this paper, the issue of validation is examined on the basis of a study into Ergonomics/Human Factors (E/HF) literature. The study focusses on how systematic deviation in outcomes of measurement plus conclusions derived from those outcomes are dealt with in that literature (the topic of random variation is discussed elsewhere (Kanis, 1993)).

# Method

To begin with, a number of concepts have been developed that should serve as starting points in the study of E/HF publications. For this purpose the relevant literature in the research that can be considered as constituents of E/HF, i.e. the social sciences and the technical sciences, were reviewed. Then three volumes (1989, 1990, 1991) of *Applied Ergonomics*, *Ergonomics*, *Human Factors*, *Contemporary Ergonomics* and *Proceedings of the Human Factors Annual Meetings* were screened. Publications that feature the words 'repeatability', 'reproducibility', 'reliability' or 'validity' in the title or the abstract were selected.

## Definitions and concepts in social and technical sciences

One can only decide upon deviation of a measurement outcome by contrasting this outcome with independent evidence that is external to the method that was applied to produce the outcome in question. As a rule, any conclusion about the (non-)existence of a systematic deviation is expressed in terms of validity. Lack of validity emerges as bias.

This definition of the concept of validity has been developed mainly within the social sciences, particularly in psychology and in educational research. An important issue is always whether, and to what extent, one has measured what was intended to be measured, for instance in an intelligence test or a reading test. Current expressions to indicate (lack of) validity in this respect are 'concept validity', 'construct validity' or 'conceptual validity' (Cronbach, 1971; Nunnally, 1978).

In addition to this, validation may also involve particular conclusions derived from measurement outcomes, e.g. the extrapolation of outcomes to other circumstances, or a prediction of the success in life on the basis of some sort of test. Expressions in use here are 'external validity', 'concurrent validity' and 'predictive validity'.

The technical sciences have nothing substantial to add to the approach in the social sciences, simply because the notion of validity is rarely discussed in this area (cf. Ebel, 1961). In the technical sciences, reproducibility is the topic (ISO, 1986), which deals with the random variation in outcomes when a measurement is repeated.

Thus, three notions have come to the fore that will serve as starting points in the attempt to chart the validation issue in E/HF literature:
- the availability of an independent criterion;
- the type of validation, i.e. a measurement outcome as such or a conclusion associated with a measurement outcome, for example an extrapolation;
- the jargonizing of validation in the social sciences, indicated by the diversity in expressions featuring the word 'validity'.

In the present study a certain reproducibility is conceived as a prerequisite in order to be able to discuss the very issue of validity. This reproducibility means that the variation across outcomes of a repeated measurement should be limited. On the other hand, however, a limited range of random variation provides no clue whatsoever as to the occurrence of bias in measurement outcomes, be it quantitative or qualitative. In other words: a good reproducibility does not generate any credibility in terms of validity. Actually, the more reproducible measurement outcomes are, the narrower the range within which outcomes cannot be identified as being biased. Finally it is noted that validation is a matter of degree instead of a case of 'all or nothing'.

## Validation in E/HF literature

The search in the literature has resulted in a selection of 63 papers which feature the terms 'validity', 'valid(ly)' and/or 'validate/-ion'. The findings can be summarized as follows.

### The availability of an independent criterion

▶ In 27 papers the topic of validity is raised in a general or prospective manner. In a number of cases the required or suggested independent criteria are not (yet) available. Insofar as empirical work is discussed in the papers, outcomes of measurement do not have any bearing on the validity topic as dealt with in these papers.

▶ In 13 papers the validity topic is addressed in an idiosyncratic or intractable way, in a number of cases once again without any reference to empirical work. Some examples from reviewed journals are:

"highlighting validity" which seems to be the percentage of times that a target on a display is highlighted instead of a non-target (Fisker and Tan, 1989);

"the reliance on usual cues is no longer valid", for action in case of human error (Rasmussen", 1990); similarly "the validity of those cues" (Brehmer, 1990) ['validity' and 'valid' may be meant in the sense of 'sound' or 'safe', or any other everyday connotation].

A few illustrations (without references) from proceedings include:

"the validity of automatically generated links in hypertext" [it is possible that the effectiveness of those links is meant];

"statistically valid subpopulations", to interview about a task in a complex interactive system [meaning not clear];

"the validity of the effectiveness of HF-guidelines" [...];

"the validity of a classification scheme", to categorize service reports [a classification may be inappropriate, but cannot be invalidated as such, since the identification of any criterion 'out there' is bound to be illusory];

"... while data reliability refers to the validity of individual data values" [...].

▶ In four papers the issue of validity is wrongly addressed in terms of '(lack of) reproducibility' or '(...) reliability'. This is confusing indeed, since these two terms have been adopted as criteria for random variation in outcomes of a repeated measurement, be it in different research areas: 'reliability' in the social sciences, and 'reproducibility' in the technical sciences, see Kanis (op. cit.) for further discussion.

In the rest of the papers, which amounts to 23 studies, systematic deviation in measurement is actually treated on the basis of a comparison with independent evidence. These papers provide proper insight into the approaches to actual validations in E/HF literature.

### Types of validation

Given the variety in the objects studied and the diversity in the nature of human involvement in E/HF research, the following types of validation can be identified in the 23 papers (see above):

1 validity of measurement outcomes as such;
2 validity of a generalization of measurement outcomes;
3 validity of an inference from measurement outcomes.

The validity of measurement outcomes as such (1)

Validation of measurement outcomes is a basic notion. An example is the estimation by subjects of frequencies on different scales vs. the actual frequency that is known and is used as criterion (Hancock and Klockars, 1991). Validation takes place by contrasting an outcome of measurement with an independent criterion that is of the same dimension - or is at least supposed to be so - but that is established in an essentially different way, e.g. by a different measurement technique. The reverse case, in which measurement outcomes serve as criterion, has been encountered in a few studies, for example the (in)validation of calculations of intradiscal pressure on the basis of a biomechanical model vs. published measurement outcomes (Jäger and Luttman, 1989). In sum, seven papers featured this type of validation.

### Validity of generalizations (2)
This type concerns the extrapolation of measurement outcomes to other conditions, other populations, other times, other environments. Examples are: behavioral compliance with different types of warnings as observed under laboratory conditions vs. observations in a field study (Wogalter and Young, 1991), and drivers' pick-up of traffic sign information in a lab vs. outdoors (Macdonald and Hoffmann, 1991). In the selection of E/HF papers this type of validation was encountered in five papers, while in one of these also type 1 occurred. In each of the five papers, outcomes of a field test served as criterion. Unlike the previous case (type 1), this criterion was established by basically the same measuring method that produced the original data.

### Validity of inferences (3)
This is the most frequently encountered type of validation which was presented in 12 of the 23 publications. Examples are: the use of heart rate to estimate oxygen consumption in different conditions (Maas et al., 1989), and physical signs established by clinical methods for determining musculoskeletal dysfunctions as indicators of inconveniences experienced by employees working under different conditions (Törner et al., 1991). Contrary to the preceding case (type 2), here essentially different phenomena are compared, which is parallelled by significant differences between corresponding measuring methods. Validation of any inference includes basic theoretical considerations, relating the one phenomenon to the other.

## Jargon
Authors generally apply the term 'validity' as if it always operates on measurement outcomes. This is inaccurate when a generalisation (type 2) or an inference (type 3) is validated. In both cases, the validity of outcomes that are generalized or that serve as a basis for an inference, is indeed a prerequisite (cf. Chapanis, 1988). Once that is established, measurement outcomes cannot be 'blamed' for generalisations or inferences which turn out to be biased. In this respect, current terms like 'external validity' and 'predictive validity' miss the point. As a matter of fact, these terms are examples of a wide-spread jargon about validation. In Table 1, thirty-two expressions are compiled that were found in the whole sample of papers (63). The table shows that in one-third of the papers the authors resort to occasional/ haphazard jargon, that for the greater part was encountered in papers from proceedings. This masquerade sharply contrasts with the sparsity of any discussion about factors that may or may not have caused bias, and by no means does it reflect the differences in validation that were outlined above as characteristic from an empirical point of view. Aside from a few papers in which a mechanism is described or a model is specified as a means for simulation, in the great majority of the 23 studies that actually deal with validation, hardly anything is questioned. This concerns for example the amenability of intended behaviour in fictitious circumstances by a questionnaire (Greenwood, 1989), and also why in-lab results (do not) differ from outcomes in the field (Wogalter and Young, op. cit.) or why the one phenomenon can(not) be inferred from an other (Maas et al., op. cit.). This lack of questioning

validations by leaving presuppositions or theoretical considerations implicit, results in plain, isolated registrations. These registrations lose further impact when authors reduce their analyses to central tendencies rather than to explore also possible relationships between observed phenomena and the criterion values, for example by comparing cases individually. In this respect promising possibilities to gain deeper insight into the processes under investigation seem to have been overlooked in several studies (cf. Törner et al., op. cit.).

Table 1. Jargonized validities

| Validity of a measurement outcome (7 papers) | Validity of a generalization (5 papers) | No empirical reference/prospective validation (27 papers) |
|---|---|---|
| concurrent v. | external v. | |
| initial v. | real world v. | conceptual v. |
| model v. | validity* (3x) | construct v. |
| rating v. | | content v. |
| validity* (3x) | | data v. |
| | | differential v. |
| | | ecological v. |
| | | extensional v. |
| Validity of an inference (12 papers) | '[...] validity' used in an idiosyncratic/intractable way (13 papers) | face v. |
| | | human performance v. |
| | | input v. |
| behavioral v. | | intensional v. |
| concurrent v. | classification v. | model v. |
| content v.(2x) | content v. | operational v. |
| external v. | empirical v. | output v. |
| initial external v. | face v. | predictive v.(3x) |
| internal v. | highlighting v.(2x) | simulation v. |
| predictive v.(2x) | predictive v. | simulation model v. |
| strength (test) v. | validity* (9x) | structure v. |
| test v. | | validity* (21x) |
| validity* (7x) | | |

* no adjective or noun added, i.e. just 'validity'

## Summary and conclusions

The issue of validation, which is a particularly empirical notion, is addressed accordingly in about one-third of the selected papers from E/HF literature. To a large extent the remainder of the papers only pays lip-service to validation. The unfortunate tradition in the social sciences to jargonize the issue of validity is echoed in E/HF literature. The proliferation of jargon encountered, occasional and haphazard as it is, sharply contrasts with the lack of questioning the (in)validations about the underpinning of presuppositions and theoretical considerations. This type of plain registration is regularly further reduced by the habit to present only central tendencies.

To anticipate these deficiences the following suggestions may be helpful.

▸ The question whether the issue of validity should be raised indeed seems to be appropriate, particularly when there is no empirical evidence to refer to.

▸ The discussion of validation can be based on the tripartition encountered in the literature: the validity of a measurement outcome, of a generalization or of an inference. All the jargon in the selected papers can be duly eliminated by referring to this distinction, and by just dropping non-committal/unverifiable expressions like 'content validity' and 'face validity'; these expressions often only feature as post hoc qualifications by so-called experts.

▸ In order to gain more in-depth knowledge of the processes that are studied, it seems to be necessary to question and to make explicit presuppositions and theoretical considerations about both the emergence and the absence of bias. Obviously, such an incentive is much more difficult to comply with in the case of mental processes than for more 'tangible' phenomena that can to some extent be modelled.

▸ Besides the identification of central tendencies one should consider to pay ample attention to a detailed exploration of the data in order to generate new hypotheses.

To conclude it is noted that the approach outlined in this paper cannot be seen as definitive. Adjustments may be necessary due to new evidence. An obvious extension comprises the combination of type 2 and type 3 validations, that is: the generalisation of an inference (cf. Cook and Campbell, 1979). In this respect the charting of the issue of validity resembles validation itself: a process that may never be shown to be finished.

## References

Brehmer, B. 1990, Variable errors set a limit to adaptation, *Ergonomics*, **33**, 1231-1239.

Carmines, E.G. and Zeller, R.A. 1979, *Reliability and validity* (SAGE, London).

Chapanis, A. 1988, Some Generalizations about Generalization, *Human Factors*, **30**, 253-267.

Cook, T.D. and Campbell, D.T. 1979, *Quasi-Experimentation. Design & Analysis Issues for Field Settings* (Rand McNally, Chicago).

Cronbach, L.J. 1971, Test Validation. In R.L. Thorndike (ed.), *Educational Measurement*, (American Council on Education, Washington D.C.) 443-507.

Ebel, R.L 1961, Must all tests be valid? *American Psychologist*, **16**, 640-647.

Fisher, D.L. and Tan, K.C. 1989, Visual Displays: The Highlighting Paradox, *Human Factors*, **31**, 17-30.

Greenwood, K. 1991, Psychmetric properties of the Diurnal Type Scale of Torsvall and Åkerstedt (1980), *Ergonomics*, **34**, 435-443.

Hancock, G.R. and Klockars, A.J. 1991, The effect of scale manipulations on validity: targetting frequency rating scales for anticipated performance levels, *Applied Ergonomics*, **22**, 147-154.

ISO 1986, ISO Standard 5725: *Precision of test methods* (International Standard Organisation, Geneva).

Jäger, M. and Luttmann, A. 1989, Biomechanical analysis and assessment of lumbar stress during lifting using a dynamic 19-segment human model, *Ergonomics*, **32**, 93-112.

Kanis, H. 1993, Reliability in Ergonomics/Human Factors. *Contemporary Ergonomics*, (Taylor and Francis, London) 91-96.

Maas, S., Kok, M.L.J., Westra, H.G. and Kemper, H.C.G. 1989, The validity of the use of heart rate in estimating oxygen consumption in static and in combined static/dynamic exercise, *Ergonomics*, **32**, 141-148.

Macdonald, W.A. and Hoffmann, E.R. 1991, Drivers' awareness of traffic sign information. *Ergonomics*, **34**, 585-612.

Nunnally, J.C. 1978, *Psychometric Theory* (McGraw-Hill, New York).

Rasmussen, J. 1990, The role of error in organizing behaviour, *Ergonomics*, **33**, 1185-1199.

Törner, M. Zetterberg, C. Andén, U., Hansson, T. and Lindell, V. 1991, Workload and musculoskeletal problems: a comparison between welders and office clerks (with reference also to fisherman), *Ergonomics*, **34**, 1179-1196.

Wogalter, M.S. and Young, S.L. 1991, Behavioural compliance to voice and print warnings, *Ergonomics*, **34**, 79-89.

# 1995

| University of Kent | 4th–6th April |
|---|---|
| **Conference Manager** | Janis Hayward |
| **Chair of Meetings** | Dave O'Neil |
| **Programme Secretary** | Sandy Robertson |
| | |
| **Secretariat Included** | Margaret Hanson |

| Chapter | Title | Author |
|---|---|---|
| **Anthropometry** | Anthropometry of children 2 to 13 years of age in the Netherlands | L.P.A. Steenbekkers |
| **Drivers and Driving** | Musculoskeletal troubles and driving: A survey of the British public | D.E. Gyi and J.M. Porter |
| **Training** | Teaching older people to use computers: Evolution and evaluation of a course | D. James, F. Gibson, G. McCauley, M. Corby and K. Davidson |
| **Drivers and Driving** | Is risk assessment a necessary decision-making tool for all organisations? | D. Walker and S. Cox |

# ANTHROPOMETRY OF CHILDREN 2 TO 13 YEARS OF AGE IN THE NETHERLANDS

**L P A Steenbekkers**

*Delft University of Technology*
*Faculty of Industrial Design Engineering*
*Jaffalaan 9, 2628 BX Delft*
*The Netherlands*

During the design of products for children it might be useful to have a computer child-model to evaluate the design in an early phase of the design process. Results of a national study on development of children, aged between 2 and 13 years, are used to develop such a child-model. Diversity in characteristics of children, relevant to the development of such a computer model are described.

## Introduction

In order to be able to design products for children designers of daily life products need data on development of children. Much is known concerning growth and development of children, especially in auxology and medical sciences (see for example Tanner, 1990). For designers, however, this information is hardly suitable to be used during the design process. Therefore a study was set up to get normative data on developmental characteristics of children in the Netherlands based on empirical assessments and suitable to be used in the design process. These characteristics were: body dimensions, force exertion, motor performance, physical flexibility, technical comprehension and temperament. A second goal of the study was to investigate whether mutual relationships could be found between these developmental aspects and between each of them and the liability to have accidents. A third goal was to provide suggestions for design and evaluation criteria that will lead to safer daily-life products for children (Steenbekkers, 1993). In this paper results on the anthropometric variables will be discussed.

## Method

In this study 40 dimensions were chosen to be measured, see figure 1. Determination of the method of measurement of the dimensions was based on existing methods in order to be able to compare our results with data from other sources. These sources were: the Dutch Standard NEN 2736 (1987), the German

Standard DIN 33402 (1981), Anthropometric Source Book (1978) and Snyder et al. (1977). When no definition could be found in these sources, we defined our own method of measurement.

**Figure 1.** Dimensions measured in the KIMA project.

A national representative sample of 2245 children aged between 2 and 13 years was measured at health care centres for infants and toddlers (children between 2 and 4 years of age) and at primary schools all over the country. Some 206 children were measured twice in order to get insight into the reproducibility of the measurements. The measurements were taken by trained observers and the data were entered into the PC semi-automatically. For this purpose equipment was developed in our own laboratory.

## General results

Almost all mean values for the dimensions per age group and sex increase with age, which is not unexpected, because a characteristic of growth is enlargement of the body. An exception to this rule are the head dimensions. These dimensions do not increase statistically significant at all ages during the period between 2 and 13 years.

Some analyses were performed in order to describe differences between boys and girls for different dimensions. They show that differences between the mean values per age group for boys and girls do exist but neither at all ages nor for all dimensions. Exceptions are the dimensions of the head and some 'skeletal' measurements, such as knee breadth and thumb breadth. The mean values of these dimensions are larger for boys than for girls at all ages. Only dimensions related to the upper leg are larger for girls than for boys, at least at some ages.

At the ages of 3 and 9 years the differences between boys and girls are most prominent, i.e. the largest number of differing dimensions was found.

## Results important to consider when making a computer child-model

One application of the anthropometric data of this survey is to make ADAPS computer child-models, to be used to evaluate a product design in an early phase (Ruiter, 1995). Some of the results of the present study are of influence on the decisions which have to be made concerning the number of models, given the differences between children between 2 and 13 years of age. These aspects will be discussed below.

### Sex differences

The differences in mean values between boys and girls are hardly relevant in relationship to the development of a computer man-model, because they remain within the possible errors of the model. It is therefore acceptable to develop a 'child-model' instead of models for boys and girls separately.

### Individualization

This refers to the fact that persons become more and more different with increasing age. Diversity between persons becomes larger. This is especially apparent in physical characteristics of persons. An example of these differences is shown in figure 2, in which the results on body weight are presented. In the beginning the differences between children of the same age are small and become larger with increasing age.

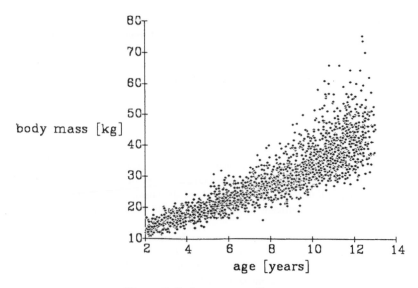

**Figure 2.** Body mass according to age.

*Overlap between age groups*

This is related to the previous point. An increase in diversity implies an overlap between age groups. Children of the same age have very different body dimensions and on the other hand, children with the same, say, stature may differ very much in age. For example: the largest 6-year-old child is as tall as the smallest 12-years-old child in our sample. This is illustrated in figure 3, in which for some age groups the number of children per stature group is presented. In the stature group 130-140 cm. children of 6, 9 and 12 years of age are present.

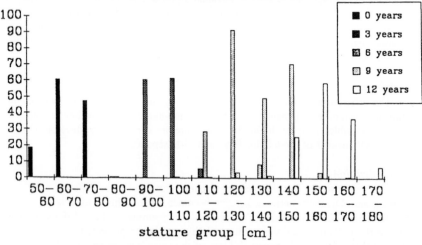

**Figure 3.** Number of children of different ages according to stature group.

*Relative differences between dimensions at different ages*
        The differences as described above are not in all dimensions clearly present. It
is especially demonstrable in the length-related dimensions. The dimensions of the
head, however, hardly increase implying smaller differences. In figure 4 body
dimensions of 12-year-old children are presented relative to the dimensions of the 2-
year-old. The mean values of this latter group are set at 100%.

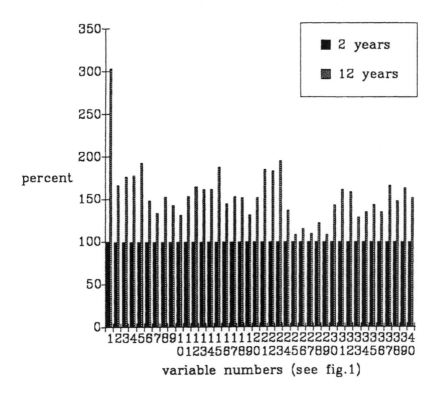

**Figure 4.** Mean values for children 12 years of age, relative to 2-year-olds.

*Differences in proportions*
        The proportions of children differ during the ages between 2 and 13 years. At
2 years the head is relatively large and the legs are relatively small; at twelve years,
however, the legs are comparatively large. These are general rules based on mean
values (see figure 5). When persons are considered individually, again many
differences can be demonstrated because no children can be found that are 'mean'
or '5th percentile' for all dimensions. A child with P5-stature, might have a P30-leg
length in his or her age group. This implies that a computer child model scaled to
P5-stature, having all other dimensions equal to P5 for that age group, will represent
a child that does not exist. This does not, however, imply that such a model is
useless. It can be applied for the purposes it is made (as a design tool), and the user
has to realise that differences between children in the age group concerned are
larger than the model suggests.

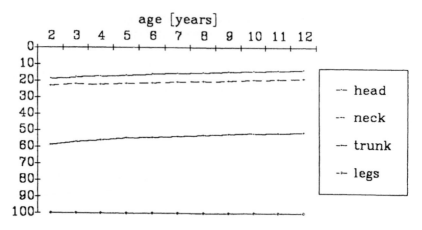

**Figure 5.** Change in the proportions of children.

## Discussion

The results of the anthropometric part of the study on development of children in the Netherlands (KIMA) are currently being used to develop computer child-models. The result will be a number of child-models, each for a different age group.

When using such a model one has to remember the differences between individual children in the age group the model represents. Of course models are a simplification of reality, but nevertheless they are very useful in certain phases of the design process, when data in, for example, tables do not give appropriate information.

A future development of ADAPS child-models might be a feature to adjust body proportions of the child-model used. This would give the user the opportunity to evaluate a design with a model which is more according to reality.

## References

*Anthropometric Source Book* 1978, Volume 2, NASA Reference Publication 1024, (National Technical Information Centre (NTIS), Springfield, Virginia, USA).

*DIN 33402, Körpermasse des Menschen* 1981, (Berlin).

Nederlands Normalisatie Instituut 1987, *Nederlandse vertaling en bewerking van ISO/DIS 7250, document nr. 20* (Translation in Dutch of ISO/DIS 7250), Commissie 30113 Antropometrie en Biomechanica, Delft.

Ruiter, I.A. 1995, *Development of computer man-models for Dutch children*, this volume.

Snyder, R.G. et al. 1977, *Anthropometry of infants, children and youth to age 18 for product safety design*, (Society of Automotive Engineers, Inc. Warrendale).

Steenbekkers, L.P.A. 1993, *Child development, design implications and accident prevention*, (Delft University Press, Delft).

Tanner, J.M. 1990, *Fetus into man, Physical growth from conception to maturity*, Revised and enlarged edition, (Harvard University Press, Cambridge, Massachusetts).

# MUSCULOSKELETAL TROUBLES AND DRIVING: A SURVEY OF THE BRITISH PUBLIC

Diane E Gyi and J Mark Porter

*Vehicle Ergonomics Group,*
*Department of Human Sciences,*
*Loughborough University of Technology,*
*Epinal Way,*
*Loughborough ,*
*Leics LE11 3TU.*

In order to explore the relationship between car driving and musculoskeletal troubles, interview data were collected from 600 members of the general public based on the Nordic Musculoskeletal Questionnaire. The results clearly showed that exposure to car driving was significantly related to reported sickness absence due to low back trouble. There was also a higher frequency of reported discomfort as annual mileage increased. Drivers of cars with a flexible driving package had fewer reported musculoskeletal troubles. It seems from the results that those who drive as part of their job appear to be more at risk from low back trouble than those whose jobs primarily involve sitting (not driving). The results indicate an urgent need for the training of managers in the importance of measures to reduce this problem, for example, selection of the car with respect to postural criteria.

## Forward

This research was carried out as part of the Brite Euram European Initiative (Project 5549). Loughborough University was one of several European based partners in the consortium (which also included car manufacturers and seat designers) whose joint objective was to improve car seat design. The overall purpose being to support national and international standards relating to issues about designing high quality seat systems.

## Introduction

It is nothing new to say that low back discomfort frequently accompanies driving. Porter, Porter and Lee (1992) in a study of 1000 drivers at Motorway Service Stations in England, found that 25% of all drivers and 66% of all business drivers were suffering from some low back discomfort at the time of the interview. Furthermore Pietri, Leclerk, Boitel, Chastang and Morcet and Blondet (1992) in their study of commercial travellers found a significant relationship between car seat comfort (in drivers who drove between 10 and 20 hours a week) and the incidence of low back trouble . Kelsey and Hardy (1975) also carried out a well documented study and found that men who had ever had a job where they spent more than half their time driving were nearly three times as likely to develop an acute herniated lumbar disc. These and other studies indicate that the relationship between musculoskeletal trouble and driving warrants further investigation.

Epidemiological studies examining the relationship between driving and musculoskeletal troubles however are scarce, perhaps indicative of the difficulties of identifying causal factors. As driving is now so much a part of our culture, it is difficult to advise 'giving up' driving and due to costs, to advise 'changing a vehicle' just in order to investigate if driving a particular vehicle was causing the problems. Also the driving task involves prolonged sitting, postural fixity and vibration any of which individually could lead to musculoskeletal trouble. Variables such as gender, lifestyle, work tasks and motivation may also have an effect on reports of symptoms in the lumbar region. It is probable however that symptoms arise from multiple relationships and influences. Pheasant (1992) hypothesised that the pattern of occurrence of musculoskeletal troubles could be described like a pyramid, with a large proportion of people (prevalence 70-90%) at the bottom who suffer task related musculoskeletal trouble but do not complain very much and usually do not develop serious clinical conditions and a few people at the top who suffer severe pathological effects, but between these extremes are a continuum of people with problems many of which could be prevented by redesign of the work task.

An interview survey was carried out in August 1993 of 600 members of the British public in order to investigate the relationship between driving and the prevalence and severity of musculoskeletal troubles including reported sickness absence. The general public were selected at random (roughly within the strata of age and gender) from public places throughout England for example shopping malls, motorway service areas and holiday resorts. They were not told the exact purpose of the survey thus avoiding selection bias. The sample contained individuals who were non-drivers, low mileage drivers, high mileage drivers and people who drive as part of their job. A complimentary study of 200 police officers was also carried out, the results of which are reported elsewhere (Gyi and Porter 1994).

## Methods

The survey was based on the standardised format of the Nordic Musculoskeletal Questionnaire or NMQ (Kuorinka, Jonsson, Kilbom, Viterberg, Beiring-Sorenson, Andersson and Jorgensen 1987). The NMQ consists of a general questionnaire for the analysis of the prevalence and severity of musculoskeletal trouble in different anatomical areas and optional, more detailed, question sheets which concentrate more thoroughly on the common sites of musculoskeletal troubles i.e. neck, shoulders and low back and the severity of the impact of this trouble on work and leisure activities. The questionnaire was not intended to be used for the diagnosis of musculoskeletal disorders and it is accepted that a medical examination is required for this, instead the term 'musculoskeletal troubles' is used to mean aches, pains, discomfort or numbness experienced in the different body areas. Firm diagnosis of low back pain is difficult anyway for example a disc protrusion seen on a CT scan may be asymptomatic (Conte and Banerjee 1993).

The NMQ has been tested for reliability and validity both in Scandinavia (Kuorinka et al 1987) and more recently by the Health and Safety Executive in England (Dickinson, Campion, Foster, Newman, O'Rourke and Thomas 1992). It has also been used in several published studies, for example Beiring-Sorenson and Hilden's (1984) study of low back trouble in the general population; Anderson, Karlehagen and Jonsson's (1987) study of Swedish bus drivers and shunters and Burdoff and Zondervan's (1990) study of low back pain in crane-operators. The NMQ is short, can accommodate different workforces and individuals and has shown itself to be non threatening and accepted.

Questions were added to the NMQ based on work by Hildebrandt (1987) regarding other possible risk factors for low back pain such as age, prolonged sitting, lifting and previous back complaints. A list of occupational task demands was taken from Pheasant (1992). A list of sports felt to be high risk for neck and back ailments was taken from a study by Porter and Porter (1990) of the views of physiotherapists, osteopaths and chiropractors. Kelsey, Githens, O'Conner, Weil and Calogero (1984) have found an association between cigarette smoking and back pain; a question was therefore included regarding cigarette smoking. Scales for measuring factors like job satisfaction and motivation were considered but were felt to be too lengthy and invasive for this type of interview study. A single question about job satisfaction with a 5 point scale was therefore included. Finally a series of questions regarding the age, type and the adjustment features of any vehicles driven regularly were included, with details of their exposure to driving in terms of annual mileage, distance driven as part of work, length of journey to work etc. These questions were asked at the end of the interview to avoid any clues being given as to the reason for the study. The authors are aware of the difficulties of obtaining quality data about many of these factors without the backup of objective measures.

## Results

*Personal details*

**Table 1.** The age distribution of the sample by gender.

| Group | Mean (SD) | Age range |
|-------|-----------|-----------|
| Males (n=303) | 38.48 (13.09) | 17-73 |
| Females (n=297) | 38.47 (13.65) | 17-74 |

The sample of the general public consisted of a wide range of age groups (Table 1), annual mileage, vehicle types, heights, Body Mass Indices etc. All the results reported in this paper refer to driving a car.

*Exposure to driving*
The results clearly indicate that exposure to driving does have a potential effect on reported sickness absence due to low back trouble. Figure 1 shows that the mean number of days ever absent from work with low back trouble was 22.4 (SD 111.3) for high mileage drivers who drove more than 25,000 miles in the last 12 months, compared with 3.3 days (SD 14.7) for low mileage drivers who drove for less than 5,000 miles. Male drivers with longer journeys to work, perhaps representing regular daily exposure, also experienced significantly more low back trouble in the last 12 months (p<0.001). This figure was approaching significance for females.

Considering those whose job involved driving as part of their job, the results again clearly showed that the number of occasions and days ever absent with low back trouble was higher in those with the greatest exposure to driving. Figure 2 shows that the mean number of days ever absent with low back trouble was 51.4 days (SD 192.9) for those who drove more than 20 hours a week as part of their job, compared with 8.1 days (SD 34.2) for those who drove less than 10 hours as part of their job . Also the mean number of days ever absent from work with low back trouble was nearly 3 times higher for those who drove for more than 500 miles a week as part of their job compared with those who drove less than 200 miles .

Discomfort was reported in at least 1 body area by 54% of car drivers. There was an increased frequency of reported discomfort with higher annual mileage with 20% of high mileage drivers (over 25,000 miles a year) 'always' or 'often' having discomfort with their vehicle compared with only 7% of low mileage drivers (under 5,000 miles). The most frequently reported discomfort areas were the low back (26%) and the neck (10%),

which is comparable with the work carried out by Porter et al (1992) where the figures
were low back (25%) and neck (10%).

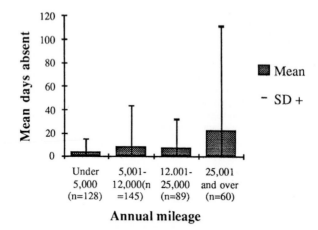

**Figure 1.** Number of days ever absent from work with low back trouble for car drivers
according to annual mileage (n=422)

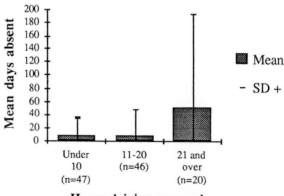

**Figure 2.** Number of days ever absent with low back trouble for car drivers according to
hours driven as part of work (n=113).

*Comparison of driving with other working postures*

Comparisons were made between those whose job involved driving a car for more
than 20 hours a week with 3 other separate groups; those whose work involved sitting
(not driving) for > 4 hours a day, a group whose job involved standing for > 4 hours a
day and a group whose job involved lifting for a large part of the day. Surprisingly it
was found that the driving group experienced more low back trouble in the last 12
months than the sitting and standing groups, however no significant differences were
found with the lifting group. For example 34% of the driving group compared with only

16% of the sitting group and 26% of the standing group experienced low back trouble for more than 8 days in the last 12 months ($0.1 > p > 0.05$).

## Adjustability of the car

The improved postures and freedom of movement permitted by an adjustable lumbar support, adjustable steering wheel, cruise control and automatic gearbox were found to have a beneficial relationship with the measured sickness absence criteria. For example drivers of cars with an adjustable steering wheel or automatic gearbox had less days absent from work with neck and shoulder trouble in the last 12 months than those drivers without these features.

Drivers who reported not enough headroom, poor pedal position, poor steering wheel position and no backrest angle position reported a significantly higher frequency of discomfort with their car, although no differences were found regarding any of the sickness absence measures.

The 3 most common vehicle types were then compared; Supermini, Small Family Car and Large Family Car. Despite drivers of the Large Family Car being of slightly older age group and having a higher mean mileage, the number of days being prevented from carrying out normal activity due to neck or shoulder trouble was higher for both of the smaller vehicle types. This could be hypothesised to be due to the higher mean number of adjustments in the Large Family Car (3.2 adjustments, SD .85, compared with 1.18 adjustments, SD .85).

## Confounding variables

Having shown a clear association between driving and low back trouble, it was important to look more closely at the influence of some of the possible confounding factors.

No significant differences were found in this study between males and females for any of the prevalence or sickness absence measures of low back trouble. However the differences between males and females for reported point prevalence (last 7 days), period prevalence (last 12 months) and severity (last 12 months) of neck, shoulder, upper back and wrist/hand trouble were highly significant, with females reporting more trouble.

Our data did not reveal a statistically significant relationship between age and low back trouble, nor did exposure to driving correlate with increasing age in our sample. It can therefore be assumed that the effect of age on driving and low back trouble is minimal. There was however an increased prevalence of musculoskeletal troubles of the large joints such as the hips, ankles and elbows with increasing age.

For males only, there was a positive correlation between Body Mass and the number of occasions and days ever absent from work with low back trouble. Although, as Body Mass in males did not show a significant correlation with exposure to driving, it is unlikely to be the main cause of low back trouble in high mileage drivers. As may be expected due to its weight bearing capacity, Body Mass was found to be related to the point prevalence, period prevalence and severity of knee trouble.

Significant positive correlations were found between sports activity and low back, neck and shoulder sickness absence criteria. However sports activity did not significantly correlate with exposure to driving and therefore confounding is likely to be minimal.

No significant correlations were found between driving and cigarette smoking or job satisfaction.

## Discussion and Conclusions

The results of this study implicate that exposure to car driving should be taken seriously with regard to reported sickness absence, particularly with individuals who drive as part of their job. Employers need to be made aware of the potential 'hidden costs' incurred if individual postural comfort is not considered. The adoption of a good posture with efficient movement patterns is essential to delay the onset of discomfort and to help avoid more serious health problems. Affordable, highly adjustable driving packages are needed which can be adjusted with minimum effort (even during a journey if necessary) with guidance provided on how to adjust the seat and controls for optimum postural comfort.

People who drive as part of their job appear to be more at risk from sickness absence due to low back trouble than those jobs primarily involved sitting (not driving) and standing tasks. Literature, posters and other media are available warning of the dangers of lifting or sitting for long periods at computers, but nothing exists to inform drivers of the importance of ensuring that the cars they drive allow them to select comfortable and efficient driving postures. Managers with the responsibility for purchasing cars for use by others also need training in the importance of careful selection of the car with respect to postural criteria and the demands of the job, time allowed for exercise regimes and active participation in back care programmes. The potential benefits of such training of managers and their employees then needs to be fully researched and evaluated. It is hoped that as awareness increases, car manufacturers will be under increased pressure to offer suitably adjustable driving packages or risk a fall in their market share.

### References

Andersson, K., Karlehagen, S. and Jonsson, B. 1987, The importance of variation in questionnaire administration, *Applied Ergonomics*, **18**;3, 229-232.

Beiring-Sorenson, F. and Hilden, J. 1984, Reproducibility of the history of low back trouble, *Spine*, **9**;3, 281-286.

Burdoff, A. and Zondervan, H. 1990, Epidemiological study of low back pain in crane operators, *Ergonomics*, **33**;8. 981-987.

Conte, L.E. and Banerjee, T. 1993, The rehabilitation of persons with low back pain, *Journal of Rehabilitation*, **Apr/May/Jun**, 18-22.

Dickinson, C.E., Campion, K., Foster, A.F., Newman, S.J., O'Rourke, A.M.T. and Thomas, P.G. 1992, Questionnaire development: An examination of the Nordic Musculoskeletal Questionnaire, *Applied Ergonomics*, **23**;3, 197-201.

Gyi, D.E. and Porter, J.M. 1994, Musculoskeletal troubles and driving in police officers. In *Proceedings of the IVth Annual Conference on Safety & Well-being at Work 1994: A Human Factors Approach*, Loughborough University 87-94.

Hildebrandt, V.H. 1987, A review of epidemiological research on risk factors of low back pain. In P. Buckle (ed.), *Musculoskeletal Disorders at Work*, (Taylor & Francis, London) 9-16.

Kelsey, J.L. and Hardy, J. 1975, Driving of motor vehicles as a risk factor for acute herniated lumbar invertebral disc, *American Journal of Epidemiology*, **102**;1, 63-73.

Kelsey, L., Githens, P.B., O'Conner, T., Weil, U. and Calogero, J.A. 1984, Acute prolapsed lumbar intervertebral disc. An epidemiological study with special reference to driving automobiles and cigarette smoking, *Spine*, **9**;6, 608-613.

Kuorinka, I., Jonsson, B., Kilbom, A., Binterberg,, H., Biering-Sorensen, F., Andersson, G. and Jorgensen, K. 1987, Standardised Nordic questionnaires for the analysis of musculoskeletal symptoms, *Applied Ergonomics*, **18**;3, 233-237.

Pheasant, S, 1992, *Ergonomics, Work and Health*, (Taylor & Francis, London) 57-76.

Pietri, F., Leclerk, A., Boitel, L., Chastang, J.F., Morcet, J.F. and Blondet, M. 1992, Low back pain in commercial travellers, *Scandinavian Journal of Work and Environmental Health*, **18**, 52-58.

Porter, J.M. and Porter, C.S. 1990, Neck and back ailments: A survey of practitioners views, *Vehicle Ergonomics Group*, Loughborough University (unpublished report).

Porter, J.M., Porter, C.S. and Lee, V.J.A. 1992, A survey of driver discomfort. In E.J. Lovesey (ed.), *Contemporary Ergonomics 1992*, (Taylor & Francis, London) 262-267.

# TEACHING OLDER PEOPLE TO USE COMPUTERS: EVOLUTION AND EVALUATION OF A COURSE

David James[1], Faith Gibson[2], Gerry McCauley[3], Michelle Corby[1]
and
Karen Davidson[1].

[1]Department of Psychology,
[2]Department of Applied Social Studies,
University of Ulster,
Magee College,
LONDONDERRY BT49 OJG.
[3]Causeway Institute of Further & Higher Education,
BALLYMONEY,
County Antrim, N. Ireland.

Older people are partially isolated from society to the extent that they are ignorant of, and intimidated by, developments in information technology. A key to breaching this isolation may be to give older people sufficient reason for wanting to use computers. By offering the recording of personal life history as such a reason, a course and manual were developed through which we (and others) have successfully taught several cohorts of older people to use computers for word processing and other relatively simple functions. The course has primarily been evaluated in terms of a) objective and subjective measures of the skills imparted to participants; b) the effect of participation in the courses on attitudes to computers; d) the effect of course participation on general mental health; and e) critical evaluation of the courses and manuals by participants and tutors.

## Introduction.

In a recent review, Sharit and Czaja (1994) highlighted the importance of issues concerning ageing and computer-based task performance to the development and practice of ergonomics in the work environment. Our own work with older computer users has stemmed from a still broader concern: computer-based technology is now so pervasive that to be computer-illiterate is, to a greater or lesser extent, to be forced to live at the margins of modern society.

An ironic aspect of the marginalisation induced by computer-illiteracy is that computer-based technology could provide uniquely powerful aids to keeping isolated individuals in touch with society. Given the relative ease-of-use of modern interfaces, it would take only a minimal degree of mastery of the technology to operate systems that could radically improve the quality of life for some of those, such as the disabled elderly, who currently feel most excluded from the lives of others. Furthermore, such is the flexibility of the potential applications of IT that there are few interests or talents which could not be enhanced by access to computer technology. Even outside the world of work, therefore, there is good reason to help older people gain familiarity with IT. Why, then, has this sizeable group of potential computer user been relatively neglected?

Temple and Gavillet (1986) indicated a vicious circle of events that has led to the inaccessability of potentially-beneficial IT to many older people. Because the technology has been developed so recently, relatively few older people have had any formal training in IT, nor, in many cases, any direct contact with computers prior to being confronted by an application (such as an automated bank teller or a computerised library catalogue) which might well not have exuded user-friendliness. The confusion and hesitancy caused by encountering an unknown piece of technology can create anxiety and even animosity within the individual and these natural responses are interpreted by others as demonstrating a fundamental inability or unwillingness to use new technology. Thus, older people are to an extent "written off" by IT developers. Indeed, some computer companies that we have approached have emphasised that they wish their products to be associated in people's minds specifically with the younger generations.

Nevertheless, there are examples of IT developments intended to serve older people, but these are of little benefit if the intended users have an antipathy to the technology. We therefore considered that the best way of breaching the vicious circle might be to provide our elderly volunteers with a strong reason for wishing to use computers. We noted complementarities between the beneficial effects of reminiscence among older people, the enthusiasm that it engenders and the suitability of word processors for recording personal life history, where memories can be added to text as and when they are recalled without disrupting what has already been written. We therefore embarked on a pilot study in which we sought to teach a small group of elderly participants to use computers, initially as a means of recording their personal life histories. This pilot course has been developed to the point where it is now a regular part of the programmes offered by Further Education centres.

The development was an iterative process which involved the evaluation of several aspects of the programme. The evaluations were guided by four principal considerations. Firstly, it was reasonable to assume that the project's aim of providing optimal help, encouragement and conditions for older people to acquire basic computing skills could only be accomplished if the particular learning characteristics and requirements of this group were known. We therefore aimed to discover those aspects of the technology and the course which learners found most difficult or discouraging, and those which were easiest or most encouraging. Secondly, we were interested in possible incidental consequences of pursuing the course, such as the general mental health of the participants, the social ramifications of participating in the course, and possible changes in participants' attitudes to modern technology. Thirdly, the efficacy of this specific course as a means of introducing computing to older people had to be assessed. And finally, a major assumption at the start of the project was that older learners would require particularly salient motivation if they were to undertake a course in IT successfully (as Bourdelais, 1986, pointed out). We therefore tried to find out the aspects of the courses that were most important in encouraging participants to enrol and to continue attending. Assessment of the variables listed below, using the methods indicated, was attempted on the assumption that they were pertinent to the four factors outlined above. (The number of participants that we have been able to assess is not yet sufficient to allow a factor analysis of their responses, which might confirm or deny the importance and relative independence of the assumed four factors. Such an analysis is planned for the future when numbers permit.) The variables assessed were:

    1. Learners' views on the ease-of-use of the hardware and software, (questionnaire survey and tutors' observation). A comparison group of younger learners of word processing was also assessed;

    2. the learners' views of the effectiveness and relative importance of the manual, software-based assistance and tutorial assistance as means of acquiring computing skills (questionnaire survey and observation by tutors);

    3. the learners' assessment of the strengths and weaknesses of the course in all its aspects, (questionnaire survey);

    4. the learners' views of the benefits (or otherwise) imparted by attending the course (questionnaire survey and observation by tutors);

    5. the change in learners' attitudes to modern technology over the duration of the course (questionnaire survey);

    6. the change in general mental health of the learners (and, in one instance, tutors) over the duration of the course (questionnaire survey and observation by tutors);

    7. critical evaluation of the manual and software by comparison with accepted human-computer-interaction guidelines;

    8. the learners' subjective estimates of the skill levels they have achieved as a result of taking part in the course (questionnaire survey and observation by tutors);

    9. relatively objective measurement of skill levels achieved by learners at the conclusion of the course (observation of performance on a word-processing exercise).

## Methods.

*Participants*

    Of the six participants in the pilot study, five were residents of an old people's home and one a neighbour or the proprietors. In all subsequent courses, which have been run in six locations throughout Ireland and in Denmark, participants have been volunteers enroled through word of mouth or (mostly) through local publicity about the courses. The vast majority have been totally computer-naïve. Ages have ranged from 56 to 88, with a modal age of approximately 70 years. Numbers have reached over 200 and on most courses women have slightly outnumbered men. A group of 18-30 year old computer-naïve participants, all female, were used as a comparison group in the early stages.

*Hardware, software and manual*

    Initially, the courses used Microsoft Works on Apple Mac computers. Additionally, courses are now given using Microsoft Works for Windows, employing pc.s with 386 cpu.s or better. The pilot study rapidly demonstrated that sharing machines was impracticable. Within the above constraints, the equipment used has simply been whatever has been available at each location. A disk with exercises to be carried out was issued to each participant. A manual was produced as a result of experience teaching the pilot course and was subsequently modified in the light of further evaluations. (A version has now been published; McAuley, McAuley, Gibson, James and Sturdy, 1994.)

*Assessment materials and questionnaire administration*

In the investigation of possible changes in general mental health, the well-established GHQ was employed in either its 12- or 30-question form. All other questionnaires were based on the appropriate literature (e.g. the CAL literature, HCI literature, computer anxiety literature, etc.), were kept as brief as possible and were compiled in accordance with the principles and guidelines outlined by Sinclair (1975). (These questionnaires can be made available to interested parties.) The questionnaires were administered at the start and towards the end of courses, normally during the first and the penultimate sessions.

*Teaching schedules and environments*

Courses have invariably been taught in two-hour sessions, with a break for tea and chat half way through. Sessions have either been once or twice weekly, the latter being the more popular. A full course has been consolidated at twenty sessions. The environment has always been as informal as possible, though the classroom layout of some locations could not be disguised.

*Teaching methods*

Teaching methods have also been as informal as possible, with as much individual attention as the numbers of tutors and participants permitted. Writing personal histories can evoke strong emotions and time had to be available for these to be shared and talked about with tutors and fellow participants: an informal atmosphere was essential for these processes to take place.

## Results and Discussion.

Every participant who has been tested has shown an improvement in general mental health as measured by the GHQ over the duration of the course. Such improvements have sometimes been dramatic, but more often, since the starting point was usually satisfactory, very small. Nonetheless, the probability of all the tested participants ($N = 41$) showing a change in the same direction by chance is less than one in a billion. At an anecdotal level, the beneficial effects of the course was confirmed by the almost evangelical enthusiasm of participants and not-infrequent comments such as "it's transformed my life" and "it's the best thing that's happened".

Formal assessment showed no significant differences between older and younger groups with respect to their opinions of the ease-of-use of the software and hardware, despite some members of the older groups being very frail and needing to acquire individual techniques to overcome physical problems. Most frequent difficulties reported were typographical errors. However, observation did suggest that a substantial proportion of the older groups took longer to acquire techniques: they tended to forget or discount the fact that they had asked the tutors or fellows for assistance (somewhat in line with the findings of Rabbitt and Abson, 1990, 1991).

One hundred percent of respondents considered that the course was "just challenging enough". However, what each regarded as the strengths of the course varied from location to location. Thus, where much social interaction was possible, this was seen as important: where it was not, other features, such as "learning something new" took precedence. While the course manual (as developed at each stage) was reported as useful, tutorial assistance, co-operation with fellow participants or even referring to self-made notes which paraphrased the manual were all preferred to reference to the latter. Later versions of the manual therefore had room for the participants' own notes, a facility which was frequently used.   Some of the perceived benefits of the course have obvious relevance for the "marginalisation" concept:  thus being able to talk to members of the younger generation about computers was an answer that occurred several times to an open question in the questionnaire, and in conversation.   The commonest criticism of the course was that the sessions were too short and too infrequent.   Twenty five percent of participants (mostly male) would have liked more technical information about computers.   Criticisms of the manual, which were taken into account in later versions, included (i) a degree of inconsistency in the use of everyday or technical language;  (ii) occasional assumptions of knowledge which the reader may not possess;  (iii) some difficulty in finding answers to specific questions;  and (iv) the sequence of instructions occasionally seemed inappropriate to the users' activity.

Since the majority of participants elected to take part in the course, it is not surprising that their attitude to modern technology was already reasonably positive. Changes in responses to the computer attitudes questionnaire over the duration of the course were not statistically significant, but were in a positive direction.

Finally, what level of competence has been acquired by participants?   A striking feature has been the over-estimation, by the majority of participants, of their own level of competence.   No-one has under-estimated her/his own ability. Nonetheless, all those surveyed have reached a level at which independent word-processing is possible, while a proportion (about 30%) have, through choice, progressed far beyond this.   In this context, there is no reason to suppose that subjective estimation of one's ability is any less important than objective assessment.

## Conclusions

However ingenious and potentially useful technological developments for older people may be, they are of little purpose if the intended user group shuns them.   Unless we can breach the vicious circle, caused by lack of opportunity for computing experience and intimidation by novelty, in which many of the older generations are trapped, many developments in IT which could do so much to enhance the quality of life for this section of the population are doomed to disuse. However, when they are given good reason to do so, older computer-naïve individuals may not only acquire basic computing skills, but derive great pleasure and unexpected health benefits from doing so.

Encouraging older people to use IT is only one, albeit an important, part of the problem, however.   There is a growing number of applications appropriate for this group, but if their potentials are to be fully realised, much thought also needs to be given to the provision of access to computers for those older people who wish to use them.   While a few of our participants have bought their own machines and local branches of Age Concern have installed computer suites, these solutions are not available to all.   In particular, the provision of networked machines for the growing numbers of older people who live on their own, is an area where the potential benefits are obvious.

A further question raised by this undertaking is what the most effective motivators would be for inducing those 40 - 65 year olds, whose computer illiteracy must diminish their employment prospects, to overcome their anxiety or embarrassment and undertake a course in computing.   These questions are fundamentally ergonomic in nature, recognising that even the most potentially beneficial technological developments are useless unless their intended users are willing to employ them.

## References

Bourdelais, F. 1986. Age is not a barrier to computing. Activities, Adaptation & Aging, **8**, 45-58.

McAuley, G., McAuley, J., Gibson, F., James, D.T.D. & Sturdy, D. 1994. *Essential Word Processing with Microsoft Works for Windows.* (NEC, Cambridge)

Rabbitt, P.M.A. and Abson, V. 1990. Lost and found: Some logical and methodological limitations of self-report questionnaires as tools to study cognitive ageing. British Journal of Psychology, **81**, 1-16.

Rabbitt, P.M.A. and Abson, V. 1991. Do older people know how good they are? British Journal of Psychology, **82**, 137-151.

Sharit, J. and Czaja, S.J. 1994. Ageing, computer-based task performance, and stress: issues and challenges. Ergonomics, **37**, 559-577.

Sinclair, M.A. 1975. Questionnaire design. Applied Ergonomics, **6**, 73-80.

Temple, L. and Gavillet , M. 1986. The development of computer confidence in seniors: an assessment of changes in computer anxiety and computer literacy. Activities, Adaptation & Aging, **8**, 63-76.

## Acknowledgements

We are most grateful for the support and assistance we have received from the Year of Older People and Solidarity Between Generations of the European Community, the David Hobman Trust, Age Resource Awards, the McCrae Trust and the Faculty of Social and Health Sciences of the University of Ulster, North Antrim College of Further Education, Montague Nursing Home, North Eastern Education and Library Board, Northern Ireland Voluntary Trust Community Arts Awards, Workers' Education Association, Age Concern (N.I.), Age Concern (Coleraine) and Ulster Television (Adult Learners' Award).

# IS RISK ASSESSMENT A NECESSARY DECISION-MAKING TOOL FOR ALL ORGANISATIONS?

**Deborah Walker and Sue Cox**

*Centre for Hazard and Risk Management,*
*Loughborough University of Technology,*
*Loughborough, Leicestershire LE11 3TU*

Risk assessment is an essential feature of recent health and safety regulation in the European Community. It is a proactive process which should ensure that the employer's response to risk control is proportional to the magnitude of the risk. Through a range of training courses and research activities, the Centre for Hazard and Risk Management has gathered data on the implementation and management of risk assessment in a wide range of UK organisations. This paper considers the role of risk assessment in health and safety management. It will seek to support, through practical experience, the hypothesis that, although risk assessment is a vital element in the safety management process, its application varies depending on a variety of factors including size of organisation, industrial sector, economic influence, perceived risks and management attitudes.

## Introduction

Quantified risk assessment was originally developed as an engineering technique used in the nuclear and aerospace industries. Its roots can be traced to ideas put forward by Pugsley (1942) and later developed by Freudenthal (1947). The techniques of quantified risk assessment have become increasingly important over the years and are used today in many high risk industries. More recently, the practice of risk assessment has developed to include qualitative methods which are particularly useful in low risk environments. Such methodologies have also been used in major hazard environments to support general hazard and risk assessment.

The introduction of the Management of Health and Safety at Work Regulations (MHSWR) (1992) brought with them the requirement for employers to conduct risk assessments of all their activities. The Centre for Hazard and Risk Management (CHaRM) at Loughborough University of Technology developed a protocol for qualitative risk assessment which enabled organisations to comply with this legislation. This protocol, published in the Risk Assessment Toolkit, Cox (1992), has been presented to over 250 organisations at public access and in-company courses.

Participants were subsequently contacted to provide follow-up information on their use of these generic techniques. The toolkit has also been used as the basis for the development of risk assessment software, which has been produced in collaboration with a software house. The product design has been based on extensive consultation with a number of organisations who have bought into the Toolkit concept and who have developed individual risk assessment systems based on its protocol. Further developments on the Toolkit protocol are planned on the basis of a follow-up survey (ibid).

In parallel with these system design activities, CHaRM is carrying out complementary research on health and safety in small to medium sized enterprises (SMEs). This research includes an investigation of methods of health and safety management, the application of risk assessment techniques and the development of self-audit systems.

This paper will first discuss the role of risk assessment in health and safety management. It will then review the application of the Risk Assessment Toolkit, Cox (1992), in a variety of organisations and industrial enterprises in the United Kingdom. Some preliminary evidence on risk assessment in SMEs also is presented. The authors will argue that the technique of risk assessment is not only a necessary but also an essential health and safety decision-making tool for all organisations.

## The Role of Risk Assessment in Health and Safety Management

The concept of health and safety management is encapsulated in the publication 'Successful Health and Safety Management', HS(G)65, Health and Safety Executive (HSE) (1991). The advice in this document has generally found favour with organisational management and is now being used to support the development of health and safety management systems. HS(G)65 describes six key elements in the process of health and safety management. They include:

a)    **Policy**, which sets out the organisation's general approach, intentions and objectives towards health and safety issues;

b)    **Organising**, which is the process of designing and establishing the structures, responsibilities and relationships which shape the total work environment;

c)    **Planning**, the organisational process which is used to determine the methods by which specific objectives should be set and how resources are allocated;

d)    **Implementing**, which focuses on the practical management actions and the necessary employee behaviours required to achieve success;

e)    **Performance measurement**, which is the process by which information is gathered which reflects progress towards health and safety goals; and

f)    **Audit and performance review**, which includes the review of necessary information and the processes of reflection. These are the final steps in the health and safety management cycle.

Risk assessment may be considered to be the focus of planning. The outputs of a risk assessment provide a description of the health and safety hazards, an assessment of the risks to all those who may be affected by such hazards and the necessary control actions. In practice this information should enable managers to take appropriate actions to eliminate any unnecessary hazards and to reduce those risks which cannot be removed if the risk assessment process is fully integrated into the health and safety management cycle.

The need for organisations to carry out risk assessment and develop health and safety management systems has further been given impetus by the introduction of the new regulations requiring 'management' approaches, for example the MHSWR (1992) and the EC Framework Directive 89/391/EEC, EC (1989). However, more than a year after the deadline set for its implementation (end of 1992), five European Union member states still have not incorporated the 1989 Community framework directive into their natural laws, Vogel (1994). There is also some evidence from the authors' research that, even in the U.K., compliance is patchy.

Current knowledge about the number and types of organisations actually adopting a health and safety management approach and carrying out risk assessments is limited. A number of projects, including the Lead Authority project, Foster (1993), and the European Foundation project, Vogel (1994), are addressing this shortfall. A preliminary analysis of data gathered from activities carried out by CHaRM (both anecdotal and questionnaire) highlight a number of key issues in relation to health and safety management and the risk assessment process.

These data were gained from the following sources:

a)    one day seminars and workshops on the Risk Assessment Toolkit;
b)    in-house risk assessment system development workshops and software; and
c)    questionnaire surveys.

These are described below.

## Risk Assessment Toolkit Workshops

CHaRM has run a series of workshops (in-company and public) at regular intervals since June 1992. These workshops have provided the basis for organisational risk assessment systems to enable compliance with the MHSWR (1992). A breakdown of organisations taking up this training opportunity is shown in Table 1.

**Table 1.** Analysis of organisations participating in risk assessment training

|  |  | Company Size | | |
|---|---|---|---|---|
| \  Business Sector (by SIC code) | | Large | Medium | Small |
| 0 | Agriculture, forestry and fishing | 0 | 0 | 0 |
| 1 | Energy and water supply industries | 7 | 0 | 0 |
| 2 | Extraction of minerals and ores other than fuels; manufacture of metals, mineral products and chemicals. | 21 | 8 | 0 |
| 3 | Metal goods, engineering and vehicle industries. | 16 | 8 | 0 |
| 4 | Other manufacturing industries. | 14 | 20 | 0 |
| 5 | Construction | 3 | 3 | 0 |
| 6 | Distribution, hotels and catering; repairs | 0 | 3 | 0 |
| 7 | Transport and communication | 15 | 0 | 0 |
| 8 | Banking, finance, insurance, business services and leasing | 2 | 2 | 0 |
| 9 | Other services (total) | 33 | 27 | 4 |
|  | - health care | 3 | 6 | 0 |
|  | - education | 0 | 3 | 2 |
|  | - local government | 11 | 6 | 0 |
|  | - central government | 16 | 4 | 0 |
|  | - emergency services | 3 | 0 | 0 |
|  | - miscellaneous | 0 | 8 | 2 |

The organisations have been broken down by business sector and company size. The definitions for company size are based on those used by the European Union, Boyle and McGrath (1994), and the HSE (1994a). Accordingly, small to medium sized enterprises are those who employ less than 500, have a net turnover of less than ECU38m and less than one third of the ownership is ascribed to a parent organisation or financial institution. Small firms are defined as those with 50 or less employees.

These data show that uptake of CHaRM training is spread across most of the industrial and service sectors with the exception of SIC 0. The majority of organisations were large or medium-sized. The absence of small-sized organisations is not surprising as this type of training is relatively expensive. Follow-up phone calls and questionnaires confirmed that many companies attending training have subsequently gone on to implement risk assessment in varying degrees through a variety of methods. This feedback confirmed the importance of risk assessment in the development of health and safety systems. It also highlighted the benefits for future policy and decision making.

## Risk Assessment Software/Systems Development

CHaRM has recently developed RMS, risk management software in collaboration with Warwick IC Systems (1995). The software was developed in consultation with a number of organisations who have used the Risk Assessment Toolkit as the basis for their own risk assessments. In accordance with the trade name (RMS), this software not only facilitates the recording of risk assessments but also assists in managing the control strategies that are the result of a risk assessment. This is illustrated by the main features of the software which are described in Table 2.

**Table 2.**  Features of RMS

| Feature | Description |
|---|---|
| Risk Assessment | This covers identification of hazards followed by estimation and evaluation of risk |
| Actions/Controls | Actions and control strategies can be recorded to help ensure that on-going requirements can be effectively managed. |
| Personnel Records | This incorporates employment history, training and document issue to provide effective control of employee competence and exposure to defined work environments. |
| Accidents | Details of an incident can be recorded to enable effective and structured analysis. |
| Claims/litigations and notices | This allows for the reporting and analysis of claims, litigations and enforcement notices. |

Analysis of user needs has led to the development of several proprietary risk assessment systems including software such as RMS. These developments reflect a recognition that risk assessment is a vital element within the Risk Management Process, Cox and Tait (1991).

This complex process of risk management can be difficult to manage effectively using simple paper systems. RMS has been designed to encompass both risk assessment and risk management elements to accommodate complex problem solving.

**SME Survey**

The SME Research Group within CHaRM has been collecting data via semi-structured interview and questionnaire from a small sample of SMEs (300) in the Charnwood district of Leicestershire; this is an initial pilot study of a larger survey. The study has looked at safety awareness, methods of health and safety management, risk assessment techniques and self-audit systems in SMEs. These data have been supplemented by a 'think tank' workshop, when a group of 'experts' discussed health and safety management in SMEs, Vassie and Cox (1994). The responses highlighted the fact that the majority of SMEs do not manage health and safety and therefore do not carry out risk assessments. Where SMEs do carry out risk assessment there are a variety of factors influencing their activities. Some of these factors are summarised in Table 3.

**Table 3.** Factors influencing the activities of SMEs

| Factor | Comments |
|---|---|
| Trade/Professional Associations | Such organisations may have produced industry specific guidance and methods of risk assessment. |
| Enforcement Visit | Enforcement officers can provide advice or require evidence of risk assessment. |
| Management Awareness | Positive management attitude is often present when those involved have worked with larger organisations who have a good health and safety record or where managers have participated in some health and safety training. |
| Size of Organisation | Size of organisation is important since at some stage a health and safety specialist will be employed. |
| Type of Ownership | Organisations that are subsidiaries of larger companies are often sent information and instruction from a head office function. |
| Insurance Companies | Insurance companies may require a risk assessment before insuring an organisation. This is often linked to use of external consultants. |
| Degree of Risk | Organisations operating in high risk areas are more likely to have carried out risk assessments. |
| Intuitive Risk Assessment | Risk assessments can be carried out 'intuitively' in that the hazards have been identified and the risks controlled, but not documented. |

**Discussion**

The data from the various activities described in this paper highlight a number of trends and influences. First, risk assessment is increasingly regarded as a vital and integral part of health and safety management and is therefore part of the decision making process within an organisation. This is also reflected in current legislative trends, HSE advice and guidance, expressed industry needs and the subsequent development of RMS.

This paper provides some evidence which suggests that not all organisations are currently performing risk assessments. This is reinforced by data from other studies currently being carried out in the European Community, Vogel (1994).

Second, where risk assessment does occur there are a number of factors which influence this process. These include:

a)    size of organisation - larger organisations are more likely to carry out risk assessment;
b)    industrial sector - this may be linked to risks and perceived risks;
c)    management awareness and commitment - this sets the health and safety culture;
d)    economic influences - particularly acute with SMEs; and
e)    HSE and enforcement officer visits.

Application of risk assessment within SMEs is particularly difficult to assess and the positive influences (see Table 3) presented in this paper are associated with limited success. There are many barriers to implementation that must be overcome. One factor that may be developed is the use of partnerships or networks between SMEs. Organisations such as trade associations, professional bodies, Chamber of Commerce and Business Clubs could also play a vital role in facilitating this process. However, the current situation in health and safety management is extremely fluid, HSE (1994b). Much effort is being extended to raise awareness through educative companies and advisory centres throughout Europe. As large and medium sized organisations demonstrate successes and cascade their enthusiasm for risk assessment the authors are confident that its uptake will penetrate all market sectors.

### References

Boyle, J. and McGrath, D., 1994, How to get research funding in Europe, *Scientific Computing 2*, July, 33-37.

Cox, S.J. and Tait, N.R.S., 1991, *Reliability, Safety and Risk Management - An integrated approach* (Butterworth Heinemann)

Cox, S., 1992, The Risk Assessment Toolkit (Loughborough University).

EC, 1989, Council Directive of 12th June on the introduction of measures to encourage improvements in the safety and health of workers at work (89/391/EEC).

Foster, A., 1993, Local authority Enforcement: Lead Authorities, *Health and Safety Bulletin 210*, June, 9-11.

Freudenthal, A.M., 1947, The safety of structures *Trans.Am.Soc.Civil Engrs.*, 125-127.

Health and Safety Executive, 1991, Successful health and safety management HS(G)65.

Health and Safety Executive, 1994a, Review of regulation working paper No. 1 small firms and the self-employed.

Health and Safety Executive, 1994b, Review of health and safety regulation HSC12.

HMSO, 1992, Management of Health and Safety at Work Regulations and Approved Code of Practice.

Pugsley, A.G., 1942, A philosophy of aeroplane strength factors, Reports and Memoranda No. 1906 (Aeronautical Research Council, London).

Vassie, L. and Cox, S., 1994, Progress report on health and safety: voluntary schemes in a European context, CHaRM report to the European Union 1995.

Vogel, L., 1994, Prevention at the workplace: an initial review of how the 1989 Community framework directive is being implemented (European Trade Union Technical Bureau for Health and Safety, Brussels).

Walker, D. and Cox, S., 1994, Feedback on risk assessment toolkit training, CHaRM internal report 1994/01.

Warwick IC Systems and CHaRM, 1995, RMS Risk Management software (Warwick IC Systems, Ripley, Derbyshire, Tel. 01773-512656) launched 25 January 1995.

# 1996

| Leicester University | 10th–12th April |
|---|---|
| **Conference Manager** | Janis Hayward |
| **Chair of Meetings** | Dave O'Neil |
| **Programme Secretary** | Sandy Robertson |
| | |
| **Secretariat Included** | Margaret Hanson |

| Chapter | Title | Author |
|---|---|---|
| **Thermoregulation** | Heat stress in night-clubs | M. McNeill and K.C. Parsons |
| **Cognitive Quality in Advanced Crew System Concepts** | Cognitive quality in advance crew system concepts: The training of the aircrew-machine team | I. MacLeod |
| **Task Analysis** | Recent developments in hierarchical task analysis | J. Annett |
| **Risk and Error** | Railway signals passed at danger – The prevention of human error | J. May, T. Horberry and A.G. Gale |

# HEAT STRESS IN NIGHT-CLUBS

## Marc McNeill and Ken Parsons

*Department of Human Sciences*
*Loughborough University of Technology*
*United Kingdom*

An Internet survey of behaviour, attitudes and opinions of regular club-goers found that night-clubs were considered to be hot or very hot places where many respondents experienced heat related illnesses. The thermal conditions of a night-club were measured (maximum 29°C air temperature, 90% relative humidity) and simulated in a thermal chamber. Four male and four female subjects danced for one hour. The results showed a rise in core temperature (mean=1.8°C, sd=0.26) and skin temperature (mean=1.34°C, sd=0.48) and a sweat rate of almost 1l/h. Subjects generally felt hot and sticky, preferring to be cooler. The physiological responses compared well with predictions from ISO 7933 and the 2-node model of human thermoregulation (Nishi & Gagge, 1977). The predicted effects of continuous dancing for four hours gave a core body temperature increase to 39.1°C, well above the WHO limit of 38°C in occupational settings. Using ISO 7933 appropriate work-rest schedules for dancing and water requirements were suggested.

## Introduction

Every weekend an estimated half a million people in the UK go to raves (all night dance parties) and night-clubs (Jones, 1994). They go to dance, often for long periods of time. The ambient thermal conditions of the night-clubs they dance in are often hot and humid. This can put considerable heat stress on those dancing.

Since 1988 there have been approximately 16 fatalities in UK night-clubs (Henry, 1992; Arlidge 1995). Whilst drugs were often implicated, in most instances heat-stoke was the actual cause of death. It is likely that there are also many less serious heat related problems.

The principle aims of this investigation were to assess the thermal conditions in night-clubs and to quantify the behavioural, physiological and subjective responses of people dancing in them. The study comprised of 4 parts, a survey of behaviour in, and opinions on the thermal conditions in night-clubs; an assessment of the thermal conditions of a night-club, measurement of subjective and physiological responses of dancers in simulated night-club thermal conditions and an evaluation of the accuracy of predictive thermal models in order predict the physiological responses to night-club thermal conditions over a period of time.

## Survey of behaviour, attitudes and subjective responses to thermal conditions in night-clubs

The population was identified to be predominantly young people (Jones 1994), covering a geographically scattered area. A questionnaire was distributed to subscribers on the UK-Dance discussion group on the Internet and to people outside night-clubs. In total 54 subjects responded to the survey, 65% male, 35% female.

Night-clubs were considered to be hot or very hot places where 61% of people would prefer to be cooler. Respondents generally preferred to drink soft drinks rather than alcohol, many (76%) used drugs such as 3,4-methylenedioxymethamphetamine (ecstasy) for their stimulation. High priced bottled water and disconnected water supplies were given as reasons for low consumption of liquids. In such environments with increased metabolic activity from dancing dehydration was likely, indeed 88% of respondents had experienced heat related illnesses.

### Thermal Audit in a Night Club

Night clubs vary in architecture and interior design and each is unique. The auditorium investigated represented a large, high ceiling type, typical of many institutionalised auditoria. Using a Grant Squirrel data logger fixed in the lighting rig, air temperature, radiant temperature and relative humidity were recorded every five minutes over a period of three nights.

A maximum air temperature of $27^\circ$C and 82% relative humidity were recorded in the auditorium during each night. These measurements were made at approximately 2m above head height. Using a hand held Solex humidity/temperature meter the maximum air temperature and relative humidity amongst those dancing were $29^\circ$C and 90% respectively. Air velocity in the empty auditorium was 0.175m/s. With a total of 180 lamps rated between 150-750 watts a significant radiant heat load was expected, however the positioning of the Squirrel data logger prevented the black globe being placed under any direct radiant load, hence the true extent of this thermal load was not seen in the results.

### Investigation into the Physiological Responses in a Night-Club

*Method*
The thermal conditions of the auditorium were simulated in the thermal chamber. A pilot study was conducted to evaluate and improve the experimental methods. Aural, oral and Ramanathan's four point mean skin temperature (Parsons 1993), metabolic rate, heart rate and amount of sweat loss were all measured. Eight subjects were exposed to the

experimental conditions over two sessions. In each session there were four subjects, two males and two females with a mean age of 21.75 years and wearing clothing of an estimated value of 0.7 clo. They were weighed semi-nude then thermistors were securely attached. Subjects danced for 30 minutes with their metabolic rates being taken using the Douglas bag method for 2 minutes after 25 minutes. A five minute break allowed them to rest whilst the music was changed. They then continued to dance for the remaining time, their metabolic rates again being taken after 55 minutes.

Finally they were weighed semi-nude with their clothes being weighed separately in a plastic box. All measurements were repeated to ensure accuracy. After weighing they were given soft drinks, offered a shower and discussed the investigation.

*Results*

The mean aural temperature rose gradually for 15 minutes before flattening. The inadequacies of the measuring techniques for this application were identified. The ear thermistors for three subjects lost contact and were therefore unreliable. When these results were removed, the mean rose to a maximum of $38.2^{\circ}$C, sd=0.25 (Fig 1). The mean 4 point mean skin temperature rose at a similar rate from $36.9^{\circ}$C-$38.2^{\circ}$C sd=0.5. The five minute break was sufficient to elicit a decrease in skin temperature of $0.56^{\circ}$C.

The anticipated difference in metabolic rate between the two different styles of music was not found. The discrepancy between the actual and expected results may have been due to more athletic dancing in response to preferred music being heard.

Heart rate was sustained at a mean of 140 bpm. The heart rates of the females (who were considered to be fitter) were lower than those of the males. The heart rate was correlated with the measured metabolic rates in order to estimate the mean metabolic rate for dancing to be 238 W/m$^2$.

The results suggested that night-clubs present stressful conditions to those dancing. It was not possible to assess the effects of a prolonged exposure in the simulated conditions. The 2-node model of human thermoregulation (Nishi & Gagge, 1977) and ISO 7933 were therefore used to predict how the human thermoregulatory system responds to night club conditions over an extended period of time. The thermal conditions found in the night-club were used in the models, air and radiant temperature being $29^{\circ}$C, air velocity 0.175m/s and relative humidity being either 70% or 90%, the later being the maximum humidity recorded. A metabolic rate of 238W/m$^2$ was used. The models were run on PC's, the predictions generated being compared with the results from the laboratory investigation to evaluate their accuracy.

The 2-node model over estimated the 4 point skin temperature. It was more accurate with core temperature, (Fig. 1). The predicted core temperature did not account for the five minute rest that was observed in the actual core temperature. Towards the end of the experiment the actual temperature exceeded the predicted temperature. This may have been due to the increase in activity by subjects towards the end of the investigation. ISO 7933 also accurately predicted trends in the rise of core body temperature. The effects of a four hour exposure were then predicted, this being the mean time that subjects

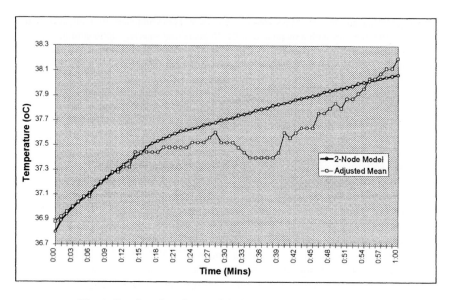

**Fig 1.** *Predicted and actual (adjusted) core temperatures*

| **Table 1.** *Predicted physiological responses to dancing in a night-club from 2-Node model.* | Exposure (Ta=Tr=29°C ) | Final Body Temp (°C) |
|---|---|---|
| | 1 hour (rh=70%) | 38.07 |
| | 1 hour (rh=90%) | 40.71 |

| | Exposure | Alarm Criteria Mins. | Danger Criteria Mins. | SW$_{req}$ (g) |
|---|---|---|---|---|
| | 1 hour (rh=70%) | 43❶ | 52❶ | 997.2 |
| | 1 hour (rh=90%) | 30❶ | 36❶ | 1007.4 |
| **Table 2.** *Predicted maximum exposure times and sweat required to maintain heat balance from ISO 7933 (Ta=Tr=29°C )* | 4 hour exposure, 45 mins. dancing, 15 mins. rest | 124❶ | 149❶ | 756 |
| | 4 hour exposure. 40 mins. dancing, 20 mins. rest | 312❷ | 390❷ | 192.3 |

❶ Rise in core body temp.
❷ Excessive water loss

in the survey danced for. The 2-node model predicted a core body temperature increase to $39.07^0C$ (table 1). This is well above the WHO limit of $38^0C$ in occupational settings. Using ISO 7933 appropriate work-rest for dancing and water requirements were suggested. This can be seen in table 2.

## Conclusions and Recommendations

1. Night-clubs operate at stressful temperatures and humidity. This can predispose those dancing in them to heat strain. Suitable measures such as increased air velocity should be taken to reduce the thermal stress.

2. "Chillout" rooms at lower temperatures to the main dance floor should be provided for rest and cooling of body temperature.

3. Frequent rests should be taken between periods of dancing; after 40 minutes of dancing, a 20 minute rest should be taken in a "chillout" room.

4. Adequate amounts of water should be consumed by those in night-clubs to prevent dehydration. It is suggested that this should be 1 litre/hour for active and prolonged dancing, however over consumption of fluids should also be avoided and advice should be provided.

5. Provision of free, cool, drinking water should be made compulsory.

6. The 2-Node predictive model for human response to thermal environments proved to be a fairly good representation of the actual environment observed. With care this can be used to make further predictions.

7. ISO 7933 accurately predicted the times of exposure before alarm limits were reached and over-estimated times before danger limits were reached.

8. Club goers should be educated as to the risks of heat strain illustrated in this report. This could be made possible with a simple wet bulb globe thermometer (WGBT), displaying the thermal conditions in clubs and the likely effects of dancing in the environment.

## References

Arlidge J (1995) *Ecstasy drug condemned as 'dance with death'* The Independent Thursday 16 February

ISO/DIN 7933 (1989) *Hot Environments- analytical determination and interpretation of thermal stress using calculation of required sweat rate* International Standard Organisation Geneva

Jones D (ed.)(1994) *Equinox November 1994: Rave New World, programme transcript* Channel 4 London

Henry (1992) *Toxicity and deaths from 3,4-methylendioxymethamphetamine ("ecstasy")* The Lancet, Vol. 340 384-387

NIOSH (1986) *Occupational Exposure to Hot Environments* National Institute for Occupational Safety and Health DHSS (NIOSH) Publication No. 86-113 Washington DC

Nishi Y and Gagge A P (1977) *Effective temperature scale useful for hypo- and hyperbaric environments* Aviation Space and Environmental Medicine 48 97-107

Parsons (1993) *Human thermal environments* Taylor and Francis London

WHO (1969) *Health Factors Involved in Working Under conditions of Heat Stress* Technical Report Series No. 412. Geneva

**Acknowledgements**

The authors would like to thank Trevor Cole for technical support during the experimental study. In addition, acknowledgement must be made to all the subjects who gave up their afternoons to dance and to Mark Camp who provided the music.

# COGNITIVE QUALITY IN ADVANCED CREW SYSTEM CONCEPTS:
# THE TRAINING OF THE AIRCREW-MACHINE TEAM

### Iain MacLeod

*Aerosystems International*
*West Hendford, Yeovil*
*Somerset, BA20 2AL, UK*

Training the human in complex skills is expensive and difficult. Training a human team in the complex skills associated with the teamwork needed to operate advanced systems is even more expensive and difficult. To construct a team consisting of human and machine members is currently impossible but is desired. This paper looks at some future problems inherent in the promotion of quality in, and training of, the conceptual human-machine team.

## What is an Aircrew-Machine Team?

A team can be described as a set of two or more people or system components who interact, dynamically, interdependently, and adaptively towards a common agreed goal. In a team specific roles are assigned and the methods of interaction are agreed, trained, practised and modified as necessary through work practice and experience. Therefore, effective teamwork relies on a developed and synergistic goal directed effort by all the team members. In the case of an human-machine team, synergy can only be achieved if::
- the machine elements of the partnership can be made situationally and tactically aware and assist and advise the human in a timely manner conducive to the effective completion of the aircraft mission - a quality Electronic Crew Member (Taylor & Reising. 1995);
- the human elements of the partnership trust and aspire to teamwork with the machine.

The training of the aircrew - machine team is currently taken as the training of the aircrew to operate the aircraft. The aircraft design is seen as fixed and can only be changed through design modifications applied to part or whole of the aircraft. Thus, it can be argued that all existing operational Human-Machine System (HMS) are flawed systems in that they are designed to consist of a human plus a machine with only the human being trainable.

System design should aim for a system performance that is greater than the sum of the performance of each of its component parts. A man-machine team should strive to be a 'Joint Cognitive System' (Hollnagel and Woods, 1983) in that the image of the human to the machine and the machine to the human should support their synergy.

A tenet of this paper is that future advanced HMS will be largely flawed unless they are specifically designed to allow the performance of effective full system training within their role, are contextually aware, and are flexibility in their approach to diverse situations. To reach any maturity, the concept of the aircrew-machine team must aspire to the development and maintenance of a high quality of teamwork. Such teamwork must be understood, specified and trained prior to its improvement through application and practice.

This paper will approach the issues of aircrew-machine teamwork and training from several different perspectives, namely those of:

- team situational awareness;
- tactics and strategies;
- perspectives and influences on HMS team qualities;
- analysis, direction, control and supervision in the HMS team;
- HMS ab initio and continuation training;
- requirements capture and HMS specification

## What is Team Situational Awareness?

Situational awareness (SA) is associated more with complex 'open skills' than with 'closed skills' (Poulton, 1957), in that its sustenance requires a good deal of interaction and appreciation with its operating environment. SA is a necessary and intrinsic property of applied skills. It is also a product of the quality and experience of skill application, but is variable under the influences of certain personal, individual and organisational factors.

Team SA involves a joint management and sharing of information, and the collective projection of that information into a future context. The SA possessed by a team not only relies on the knowledge and skills of the individual team members, and their conception of the perceived working environment, it involves team agreement on the understanding of that environment. This shared agreement implies training, a within team continual awareness of each others' proficiencies and limitations plus a collective appreciation of the standard of current team performance (MacLeod et al, 1995).

Considering future HMS teams, it is suggested that good synergy between human team and machine involves the establishment of extended joint cognitive HMS where physical interfaces are not serious barriers to smooth and proficient system performance.

Moreover, SA supports the maintenance of trust. Trust is essential to sustainable teamwork. Trust is promoted by awareness of the predicaments of both the team as a whole, and of the individual members, as well as awareness of the quality of team performance towards shared goals, through a consensus on the means of their achievement (see Taylor & Reising, op cit).

Team SA relies heavily on the maintenance of common goals. This maintenance can only be achieved by a team through agreed endeavours, a willingness to co-operate, and a consideration of the needs of fellow team members. These agreed endeavours must be based on team training promoting a shared knowledge of tactical and strategic rules, procedures, and guidelines devised to support the team role within their working environment. Tactics and strategies will be discussed next.

## Why Tactics and Strategies?

A tactic is defined as an arrangement or plan formed to achieve some short term mission goal. The goal may be an end in itself or serve as a stage in the progress towards a later mission objective. A strategy governs the use of tactics for the fulfilment of an overall or long term mission plan. Tactics are normally recorded as formal written procedures. When learnt they reside in human memory / cognition. Usually, human work experience with a system leads to modification of any formal system related procedural tactics.

The human perceives the world through the interpretation of information gleaned by the senses. This perception can be achieved through direct observation of the world or by the use of a man machine system's interface with the world. The use of interpreted information is governed by learnt tactics and strategies. These tactics and strategies are tuned through training and experience and are governed by the human roles within a man machine system and the human's interpretation of these roles.

A good quality tactic supports operator task routes to assist skilled performance, is an assistant to the application of all levels of aircrew skill, and allows a flexible and adaptive approach to a changing and potentially hostile environment. In an HMS, the performance of tactics and strategies should be enhanced by system equipments designed to aid the human to select and interpret information contained in the working environment,

Formal Tactics and Strategies were seen as representing an organisations recommended methods of achieving a desired man-machine system performance. They also currently provide a bridge between an operator's performance, assisted by the application of cognitive tactics and strategies, and the control requirements of the engineered part of the man-machine system, as represented by equipment operating procedures.

The different forms of tactical and strategic plans, and equipment operating procedures, should all be as closely associated as possible for the efficient operation of a man-machine system. Knowledge of the equipment operating procedures, and formal tactics and strategies, allows the operator to apply their skills through a formulation and application of cognitive tactics and strategies. Thus cognitive tactics and strategies representing the operator's conception of the means required to achieve man-machine system goals.

Tactics and strategies are continually mediated by both the information that the operator already possesses and information gleaned from the working environment. Currently all HMS performance is under the directed influence of human cognitive tactics and strategies. In the future, cognitive tactics should be devised for work throughout the joint HMS team.

For the conceivable future, only an operator or team can direct an HMS towards its given goals. Too often formal tactics and strategies are devised late in the design of a man-machine system and represent a method of equating problems in the design. However, by explicitly approaching how effectively a system should and might be used by the operators, at as early a design stage as possible, it is suggested that the whole design process and final product must benefit (Macleod & Taylor, 1994).

With reference to an HMS team, there must be perspectives on human teamwork that also can be applied to HMS teamwork. A brief examination of some of these perspectives follows.

## Perspectives and influences on HMS team qualities

It has been proposed that there are seven 'Cs' of good military team performance: Command, Control, and Communication; Co-operation, Co-ordination, and Cohesion; and Cybernation" (Swezey and Salas. 1992) These seven 'Cs' are grouped as follows:

- Command, Control, and Communication ($C^3$);
- Co-operation, Co-ordination, and Cohesion (CO);
- Cybernation.

Thus teamwork is considered through inter-team interactions ($C^3$) and within-team interactions (CO). Cybernation, a term used to refer to highly skilled team performance where that performance far exceeds the sum of the skills of the individual team members, will be argued as a necessary aspiration of the aircrew-machine team.

A manifestation of modern advanced technology is the increasing emphasis in design on efficient HMS communications with many HMS related agents outside the commonly accepted boundaries of the aircraft HMS. A major concern in the $C^3$ area is to provide the support needed to allow necessary team co-ordination and structure to exist between system agents in order that the mission goals can be gained and the results disseminated. Thus the emphasis on $C^3$ is not only on the capabilities of the communication bearer (i.e. the physical form of the means of communication) but on the influence of $C^3$ on team interactions and performance. For example, the poorer the support to control afforded by the design of the HMS, the greater delays and uncertainty in the information presented to the operator and the HMS's operating authority. Poor quality information will adversely affect an operators use of tactics and strategies and their maintenance of appropriate cognitive modes, or necessary conditions for cognitive operations, to effectively support HMS direction and control. [1]

Inter team performance comprises goal directed behaviours by a team during task performance. The team behaviours depend on co-operation, co-ordination and cohesion between the team members to convert the individual behaviours of the team members into the required team behaviour. The required team behaviour will be guided by teamwork as influenced by the formal tactics and strategies applicable to the HMS

Cybernation occurs both within and between teams and represents the high degree of competency that can be achieved when the collective performance of the team is greater than that possible through sum of their individual skilled performances. The means of achieving cybernation is not only through the selection and training of the individuals for the team, it is through effective team training. Cybernation is also strongly influenced by such as organisational mores, inter-team empathy, within team concern for individual members, team spirit, feelings of 'belonging' and friendship to name but a few.

One of the main influences on the degree of teamwork possible is the size of the team with relation to the form and amount of work it is required to perform. Particular team structures and sizes are frequently argued under cost constraints, the argument being primarily supported by statements on the forms, levels, and efficiency of current methods of

---

[1]     Hollnagel describes modes of control as Strategic, Tactical, Opportunistic and Scrambled, For explanation see Hollnagel 1993.

automation of system functions. It appears that automation is being increasingly incorporated into HMS, not to promote a net gain to safe system operation and performance, but to promote a net saving in the monetary life-cycle costs of the system.

Unfortunately, though HMS functions are derived under the 'solid' logic of systems engineering, they fail to fully encapsulate the cognitive logic and functions of the essential human component(s) of the system, the one form of component essential to the performance of system related analyses, direction, control and supervision. The frequent result is an over committed team both with relation to their role within the HMS and their maintenance of good teamwork. It is suggested that a more human centred and multi-disciplinary approach is required to formulate the principles for future HMS design.

Heil (1983), in a particular suggestion to cognitive psychologists that should arguably be applied to most disciplines, stated that practictioners:

" .. might conceivably profit by coming to recognise which of their commitments are due not to a clear perception of the nature of things, but to gratuitous philosophical prejudices that serve no purpose save that of forcing one's thoughts into certain narrow channels and sparing one the intellectual labour involved in coming to see thing s differently."

## Analysis, direction, control and supervision in the HMS team

In military operations, the use of HMS is normally for a particular purpose or mission. In aircraft terms a mission might consist of one or more flights. Each mission and flight has to be planned in detail to meet a strategic aim. This proactive mission planning allows pre-flight rehearsal of the HMS tactics with relation to the expected flight. Once airborne, the proactive planning acts as a reference template to short term reactive planning. Reactive planning is necessary to equate unforeseen perturbations to the proactive mission plan, and to direct and control HMS performance towards the mission objectives.

The need for reactive planning is dependant on human perception and analysis of mission situations. The initial perception of information is influenced by such as HMS design, the environment, the aircraft situation, the vagaries of human attention and the knowledge and efficiency of the adopted tactics. The feedback from situation analyses is used as a basis for the redirection of the aircraft to meet mission goals, or to control the aircraft to improve the HMS performance towards its goals, or to maintain supervision of the aircraft HMS and its situation.

At present, the human is required to perform all the above forms of task without any true assistance from the other components of the HMS. This is because the other components have no situational awareness or applied knowledge of the prescribed tactics and strategies. Current HMS systems act as designed and as used by the operator. HCI principles of display are evoked to support the quality of information presentation to the operator. Further, the diverse type of information produced by such sub systems is a mixture of machine interpreted situation related data, collateral data such as obtained from flight and mission planning, and ephemeris data presenting short term information on flight or sensor performance. The se diverse types of information should not be confused.

The HMS operator is forced to analyse and associate all perceived information. Incorrect interpretations and associations must be avoided (some can be fatal). Interpretations can be dependant on the nature of the environment, on the HMS situation and mode of operation, on the types of information, on correct association of diverse information, and on the idiosyncrasies of equipment design or the human operator.

If a quality HMS team is to be allowed by technology and design, methods must be developed to allow the human and the machine to assist one another as full team members. The difference that must be maintained between man and machine is that the human must always be in a position to lead the team. However, the machine must have the pertinent knowledge and capability to assist the human in a timely and appropriate fashion with regard to the tactical situation. The machine should be 'aware' of the quality of human performance and adapt its performance and advice accordingly; possible examples are the proffering of additional assistance to the normal if an operator is seen to lack skill, attention, or is being distracted from a task or situation by a current performance of an over onerous activity.

As an example, a paper map can be considered as an inflexible but important source of collateral information assisting aircrew planning and anticipation. Moreover, good map design sets appropriate expectations which prime perceptions and direct attention to the critical features and important relationships in the environment.

Understanding of the current situation, the ability to think ahead, and to make effective plans are key to aircrew operational performance. Therefore, what is basically needed is an adopted digital map technology that is an integral part of the HMS team, that helps the aircrew anticipate and manage cockpit and cabin workload, that enhances aircrew and team situational awareness, and supports aircrew planning, decision making, and weapon system control, all in an adaptive, flexible, and intelligent manner (Taylor & MacLeod, 1995).

The HMS team may consist of many components, both human and machine. In the multi operator case, machine team components will have difficulty in proffering advice and adapting to the performance of one human operator. However, there are additional problems if the machine components are required to team with several operators at once. Problems that might occur include questions as to which human operator is currently HMS leader, or which human is currently under performing with relation to task and their interaction with other crew members. Some early suggestions to the solving of such problems by machine have been indicated by Novikov et al (1993).

In association with the above will be a need to keep all team members aware of the performance of the system as a whole, of the quality of their own contributions, and the contribution of others. Such a machine supported feedback is essential to aid the maintenance of the team SA. Such avenues have also been suggested by work on the measurement of cognitive compatibility (Taylor, 1995), where it is indicated that future systems will need to consider dimensions of social interactions (e.g. goals, shared functions, aiding, advice) as well as the traditional human factors considerations of compatibility (i.e. modality, movement, spatial, conceptual compatibility) to order to achieve cognitive quality in joint cognitive systems.

## Advanced HMS ab initio and continuation team training

It has already been discussed that training for a HMS is currently training for the operators only (apologies to maintainers). Usually an engineered system is designed to operate through an interpretation of specifications, often as the designer sees appropriate when there is ambiguity in the specification, and is designed to interface with the human operator(s) in much the same fashion.

In the future ab initio team training for an advanced HMS would consist of :
1) designing the HMS to meet teaming criteria such as outlined in the previous section;
2) training of the human component skills to team with the engineered HMS components.

However, such criteria could not possibly be implemented correctly by design to give the 'right first time' teaming quality required for advanced HMS operations. The design process would need to allow improvement to design through iteration. Moreover, it would only be in the light of actual operating experience, where the tuning of the system is a form of continuation training of both the machine and operator, that the full worth of the teaming concept could ever be realised.

Therefore, to effectively train the machine component of the team would not only rely on a high quality design of the HMS, it would rely on a debrief and tuning of the system after each sortie. This tuning would have to be performed with great care to avoid possible severe degradations in HMS performance. Advice on methods, benefits, and pitfalls inherent in the tuning would have to be included as part of the design process.

## Requirements capture and HMS specification

Requirements capture is a difficult enough process for systems engineers considering current advanced HMS systems. Nevertheless the capture of requirements for any system is essential. It can be said that the quality of a system is the design conformance to the system requirements. Requirements are the foundations for system specification.
Future requirements capture for a teamed HMS will be far more difficult than at present and will require a multi disciplinary approach to design. Further, future requirements capture will require that both machine and cognitive functions are captured.

By cognitive functions we refer to the cognitive representations that are required to transform prescribed operator tasks, dictated by proactive mission planning, work procedures and tactics, to the activities that are actually needed, under the auspices of reactive planning and tactics, to adapt job performance in the light of the appreciated work situation and the mission goals (see Boy. 1995, MacLeod et al, 1994).

Current trends are for the development of knowledge elicitation methods relevant to task (i.e. cognitive task analysis techniques), and also with relation to observable activities (i.e., concurrent verbal protocol analysis). Associated with these current trends are methods of assessing operator mental workload, though it is almost certain that what is measured here is different between task and activity. However, beyond the concepts of workload, there is some evidence that the use of dynamic task analytic modelling techniques may aid elicitation of operator decision processes (Macleod et al, 1993).

Methods of cognitive function capture and analysis are being sought. Without such capture, the cognitive requirements of the machine part of the team cannot be properly considered and specified. It is interesting to consider that cognitive compatibility, currently as assessed by the CC-SART Rating Scale (Taylor, 1995), may very well be indicative of the quality of cognitive functions with relation to their application alongside a particular HMS design - a measure of HMS quality for the future?

## Final Note

The coverage of this short paper is glib considering the complexity and nascence of the subject. However, the coverage is of a subject area where the approach to technological development must be human-centred if it is to succeed. The aim for an Electronic Crewmember exists outside science fiction. The problem is to sensibly maintain the reality of the aim.

## References

Boy, G (1995) Knowledge Elicitation for the Design of Software Agents, in *Handbook of Human Computer Interaction*, 2nd edition, Helander, M. & Landauer, T. (Eds), Elsevier Science Pub., North Holland (In press).

Heil, J (1983), *Perception and Cognition*, University of California Press ltd., London

Hollnagel, E. (1995) *Human Reliability Analysis: Context and Control,* Academic Press Ltd., London.

Hollnagel, E & Woods, D.D. (1983) Cognitive systems engineering: New wine in new bottles, *International journal of Man-Machine Studies,* **18**, pps 583-600.

Novikov, M.A., Bystritskaya, A.F., Eskov, K.N., Vasilyiev, V.K., Vinokhodova, A.G. & Davies, C. (1993) 'HOMEOSTAT -A Bioengineering System' in *Proceedings of 23rd International Conference on Environmental Systems,* Colorado Springs, Colorado, July 12-15, 1993, published by SAE Technical Paper Series, Warrendale, PA.

Poulton, E.C. (1957) ' On the stimulus and response in pursuit tracking', *Journal of Experimental Psychology*, **53**, pps 57-65.

Swezey, R.W. & Salas, E (Eds) (1992), *Teams*, Norwood, NJ: Ablex.

Taylor, R.M. (1995) Experiential Measures: Performance-Based Self Ratings of Situational Awareness, in *Proceedings of International Conference on Experimental Analysis and Measurement of Situational Awareness*, Daytona Beach, 1-3 November. Embry-Riddle Press, Florida (in press).

Taylor, R.M. & MacLeod, I.S. (1995) Maps for Planning, Situation Assessment, and Mission Control, in *Proceedings of The 1995 International Conference of the Royal Institute of Navigation*, London.

Taylor. R. M. & Reising, J. (Eds), The Human-Electronic Crew: Can We Trust the Team? *Proceedings of the 3rd International Workshop on Human-Computer Teamwork*, Cambridge, UK, September 1994, , Report DRA/CHS/HS3/TR95001/01 dated January 1995.

MacLeod, I.S., Taylor, R. M & Davies, C.D. (1995) Perspectives on the Appreciation of Team Situational Awareness, in *Proceedings of the International Conference on Experimental Analysis and Measurement of Situation Awareness*, Embry-Riddle Aeronautical University Press, Florida (In press).

MacLeod. I.S., & Taylor, R.M. (1994) Does Human Cognition Allow Human Factors (HF) Certification of Advanced Aircrew systems? in *Human Factors Certification of Advanced Aviation Technologies,* Embry-Riddle Aeronautical University Press, Florida.

MacLeod, I.S., Biggin, K., Romans, J. & Kirby, K. (1993) Predictive Workload Analysis - RN EH 101 Helicopter, in *Contemporary Ergonomics 1993*, Lovesey, E.J. (Ed), Taylor and Francis, London

# RECENT DEVELOPMENTS IN HIERARCHICAL TASK ANALYSIS

## John Annett

*Department of Psychology,*
*University of Warwick,*
*Coventry CV4 7AL*

Hierarchical Task Analysis (HTA) was developed in the 1960's and grounded in the concepts current cognitive psychology. Used initially for determining training needs of individual operators its scope has been widened to encompass a variety of applications. This paper describes the adaptation of the basic principles to the analysis of team skills with particular reference to Naval Command and Control teams. A model of team skills is outlined which identifies observable team processes and their inferred knowledge bases.The methods used to analyse team skills and to prescribe effective training procedures are briefly outlined.

## Introduction

Hierarchical task analysis (HTA) originated in the early days of cognitive ergonomics and reflected some of the dominant ideas of the time. First, it was an attempt, but by no means the first, to represent significant information processing features of human work in very general terms which did not beg too many questions about specific cognitive mechanisms. Second, HTA was perhaps the first analytical methodology to recognise the importance of feedback in the control of behaviour. Even simple motor acts are not just 'emitted' but depend on the arrival of sensory feedback within 2-300 milliseconds if the action is to continue smoothly and so the description of behaviour had to include not only a characterisation of the action itself and the conditions under which it should occur but also the sensory feedback which assured the operator that the intended action is accomplished. This turned out to be particularly important in process control tasks. Thirdly, the idea that work could be described in terms of a goal hierarchy turned out to be not only a practical necessity but also matched the seminal ideas of Miller, Galanter and Pribram (1960) on the ways in which complex behaviour is organised. Since the publication of the definitive account of HTA in Annett, Duncan, Stammers and Gray (1971) there have been a number of significant developments but in this paper I want to outline a recent adaptation of HTA principles to the analysis of team tasks.


## Team-work

Most published examples of HTA describe the tasks of individual operators yet many of these tasks, including some of those first analysed in the petrochemical industry, are carried out by teams of operators. I became acutely aware of this deficiency when invited to investigate the training of Naval Command and Control ($C^2$) teams for the DRA. I shall not describe the analyses in detail but will show how the basic principle of hierarchical decomposition can be applied team tasks, but first I shall outline a process model of the structure and organisation of $C^2$

### A Process Model

In a powerful theoretical analysis of research on groups and teams Steiner (1972)pointed out that the productivity of the group, and the variables which determine it, are strongly influenced by structure and organisation. In some cases group performance depends heavily on the ability of the weakest member, in others on the ability of the strongest. Adopting Steiner's approach the $C^2$ team can be seen as comprising a hierarchically structured set of sub-teams and individuals. For example in a typical warship there are sub-teams concerned with collecting data on the air surface and underwater environments which contribute to a tactical picture for the Principal Warfare Officer who must decide on the deployment of weapons and other assets. Each of these sub-teams is defined by its *common purpose,* that is the variables it attempts to optimise.$C^2$ tasks are, in Steiner's terms, mostly *divisible* tasks, that is to say the overall objectives can be broken down into different sub-objectives which individuals and sub-teams strive to attain. By contrast a *unitary* task is one in which all members contribute to the overall effort in the same way - a tug-of-war team has a unitary task. The $C^2$ team's task is an *optimising* task, that is the aim is to produce some optimal outcome, an accurate tactical picture (rather than, say, exert maximum force). $C^2$ tasks are to some extent *discretionary*, that is to say although individual roles or functions are largely prescribed by the formal structure there is room for some degree of negotiation. The efficiency of teams of this sort would appear to depend very heavily on (a) the competence of individual operators and (b) the effectiveness of communication between members and with other teams.

A key feature of $C^2$ tasks is that their common purpose relates to the achievement of certain values of the *product* variables by the team processing information concerning the values of other variables representing changing states of the world, the ship and its systems. The work of such teams has been successfully analysed in terms of *control hierarchies*, that is by spelling out the relationships between overall goals and sub-goals of the system. The product of a $C^2$ team, say successful defence against air attack, is achieved by controlling a number of sub-goals. *The analysis of team skills must take into account the processes by which the team achieves its various goals and sub-goals.*

*Team Processes.*

The processes by which teams achieve their goals are *communication* and *coordination*. Figure1 illustrates how these two umbrella terms comprise a number of observable behaviours. Communication includes transmitting and receiving messages whilst coordination includes collaboration, or working together, and synchronization. The direction of the arrows indicates the main routes by which the components of the model exert their influence.

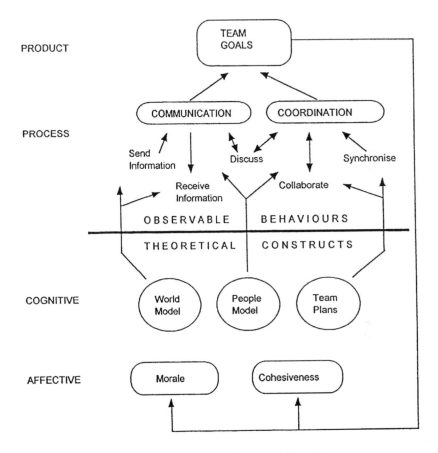

**Figure 1** Model of Command Team Processes.

*Mental Models.*

Figure 1 also recognises a second level of analysis concerned with *cognitive* and *affective processes*. The important cognitive processes are the construction and maintenance of mental models and three types are distinguished in Figure 1. First is a model of the state of part of the world that is relevant to the team task including knowledge of the present disposition of the ship in relation to other friendly and hostile units. Second is knowledge of the other members of the team, what they are doing, or about to do, and even how well they are doing it. This 'people' model will be important in choosing which information to transmit and when to do so and whether or not to offer help. The third represents knowledge of team goals and plans, that is to say what goals the team is aiming to achieve and what strategies and techniques are available and are normally employed. These models are inferred from behaviour. *When any skill is being acquired task-relevant mental models are being learned* and it is therefore important to bear this in mind in the design of training and assessment procedures.

Figure 1 also includes some *affective constructs* namely *group morale* and *cohesiveness.* Arguably his level of description has not, so far, yielded measures powerful enough to justify including in a practical model of team performance. Note the arrow directions suggest that these are influenced by the attainment of team goals. They may well be the *result* of success rather than a contributory cause.

**The Analysis of Team Tasks**

Both individual and team tasks can be analysed by hierarchical decomposition. A task *process* is the means by which the performer (an individual or a team) achieves the task product or goal. For an individual this means that at a given signal, or under a given set of conditions, one or more actions is executed in order to achieve a new set of conditions - the goal. For a team the process by which a goal is achieved is by joint or complementary action, which involves *communication and coordination.*

In individual HTA the basic processes underlying performance are specified as 'input', 'action' and 'feedback' and it has been found useful to add to these fundamental descriptive categories the *plan* which indicates *how sub-operations are linked together in order to constitute a superordinate operation.* Extending this principle to the analysis of team skills some *fundamental categories of interaction* are (i) *communication* - information is passed between team members and (ii) *coordination* - team members carry out joint activities. These general processes can be expanded to identify features of the task which are likely to require special attention in training.

*Communication*

Sent messages may vary in respect of the *accuracy* of the information in the message, the *clarity* with which they are transmitted and their *timeliness.* Timeliness implies that the sender knows enough about the problems of the receiver to be able to judge that the message is required or can be dealt with at that time. Training for this feature is thus bound up with an understanding of the wider picture referred to in the literature as 'situational awareness' and with knowledge of the goals and tasks of other team members. *Receiving* messages depends at least in part on controlled attention to significant sources and this in turn is probably dependent on 'situational awareness', that is an understanding of what is going on in the external world and what other members of the team are doing. Again 'mental models' of the world, of other team members and of team plans could be critical in certain situations. *Discussion* can be related either to the clarification of acts of communication, ensuring that information has been correctly transmitted and received, or it could have as its main purpose arranging some joint activity or changing a plan of action. A properly conducted discussion is likely to result in a modification to one or more of the mental models referred to in Figure 1. Joint decision making is normally based on sending and receiving information but goes beyond simple information exchange to include some *discussion of alternative courses of action.* Team training might well be designed to take this factor into account in assessing team performance. Practice, with appropriate feedback, at reaching genuinely joint decisions might be used to modify the behaviour of team members.

*Coordination*

This is subdivided into *collaboration* and *synchronization.* Collaboration, literally working together, also depends on awareness of the current state of the world, but collaborators must also be aware of the current activities and needs of their fellow team members. The *re-allocation of duties* between team members is likely to arise in special circumstances, such as information overload, or even injury, affecting one member of the team to such an extent that team output will be compromised unless another member takes on part or whole of the duties normally carried out by that individual. One way of developing a model of the collaborators' activities and state of knowledge is to *swap roles* during training so that individuals learn something about the tasks of their colleagues. (b) *Synchronization* means working independently according to a common plan. In the absence of direct communication this depends on the individual's knowledge of *a plan which* may be built up formally by training or may be acquired informally as team members acquire an habitual pattern of actions which may not have been articulated but is known to team members. To the extent that plans are formalised team members need to learn and practise them, but informal plans may only emerge if the team reflects on its own performance in post-exercise debriefing.

*The Stop Rule*

HTA proceeds by 'unpacking' each 'operation' into a set of sub-operations and applying a rule which states that *if the product of the probability of failure and the cost of failure is unacceptable to the system then either propose a solution such that the criterion is met or proceed to a more detailed level of analysis by decomposing the operation into its component sub-operations.* In a training analysis a proposed solution will normally be a training prescription. In the case of *team-work* operations of the kind described in the preceding sections suitable training prescriptions will include learning communication protocols and disciplines and developing situational awareness by exercises designed to develop the mental models referred Figure 1.

*Stress and Environmental Factors.*

The circumstances under which any given operation might be attempted may be unfavourable and may have a significant effect on the probability of failure over and above that attributable to the skill level of the operator. If conditions such as environmental or psychological stress due to workload, danger and so on, are anticipated these must be taken into account in applying the p x c rule and, of course, in the training prescription. For example, in the stress of an engagement there might be a higher probability of communication failures or under workload stress individuals may be less likely to monitor the activities of colleagues with whom they should be collaborating. Conventional Naval task analysis specifies duties and tasks without reference to these factors whilst the HTA procedure *requires* they be considered and suitable training be prescribed to deal with them.

## Analytical Procedures

HTA is a methodology broadly based on principles rather than a rigidly prescribed technique. The analysis of Naval Command Team skills, which is still in progress, is based on interviews with experts, including senior naval officers and experienced ratings, and direct observation of exercises on shore-based simulators and at sea. Although the duties of individual team members are formally specified and well-documented, informal accounts of what happens during a typical exercise give a much clearer impression of the kinds of team processes which are involved and those which are critical. Formal documentation relating to Command and Control teams and the Action Information Organisation specify what should happen, not what does happen nor how things can go wrong. HTA encourages the analyst to pursue just these questions using a variation of Critical Incident Technique and to attempt to identify objective indices of goal success and failure which are currently subjective and impressionistic. Finally, HTA is intended to lead directly to training prescriptions rather than being just a preliminary stage in specifying training techniques and procedures. Since this kind of team training is extremely expensive the identification of critical team skills should lead to the optimal use of training resources. Whether significant savings can be achieved remains to be seen.

## Acknowledgements

This work was supported by a grant from the DRA Centre for Human Sciences, Portsdown. The author wishes to acknowledge the help of David Cunningham and the expertise of the School of Maritime Operations.

## References

Annett, J. Duncan, K.D., Stammers R.B. & Gray, M.J. (1971) *Task Analysis.(HMSO,* London)

Miller, G.A., Galanter, E. & Pribram, K.H. (1960) *Plans and the Structure of Behavior.* (Henry Holt, New York)

ISteiner, I.D. (1972) *Group Processes and Productivity.* (Academic Press, New York)

# RAILWAY SIGNALS PASSED AT DANGER - THE PREVENTION OF HUMAN ERROR

## J. May, T. Horberry, A.G. Gale

Applied Vision Research Unit, University of Derby,
Mickleover, Derby, DE3 5GX, UK. Tel\Fax 44 1332 622287
E-mail: AVRU@derby.ac.uk

The potential for disaster with each Railway Signal Passed at Danger (SPaD) is high. This paper forms a literature review of previous research concerning the occurrence of SPaDs in relation to human error and discusses potential causal factors in relation to situational, individual and perceptual/attentional issues. It is shown that the causes of SPaDs are multifactoral and therefore there is no one solution to SPaDs. The de-skilling of the drivers' task is a serious problem and is not always sufficiently addressed by the introduction of vigilance devices.

## Introduction

A SPaD occurs when a train passes though a red stop signal instead of stopping before the signal as required. While most SPaDs simply result in near miss incidents each one carries with it a high potential for disaster and loss of life and therefore their reduction is of great importance. Some SPaDs are attributed to signal failures or mechanical faults but the more common cause is human error with 85% of SPaD cases being attributable to human causal factors. While vigilance devices and safety systems such as the AWS (Advanced warning system) and ATP (Automatic train protection) have been introduced to monitor the speed of the train and the signals the driver passes through, and warn him accordingly, it may be many years before such systems are implemented in all cabs and on all lines. The problem of SPaDs must therefore still be addressed. Previous literature which has focused on the human error aspects of SPaD causation is revised here and identified causal factors are discussed.

## Situational Factors

### Driver Activity

The activity the driver is performing may make them more susceptible to performing certain types of SPaD. Three distinct driving situations where SPaDs occur have been identified in the literature. These arise from different causal factors and therefore need very different solutions. They are:-
- Driving on a main line - Most SPaDs occur because the driver has seen the red signal but made an assumption that the signal will change to a less restrictive

aspect before he passes through it. These assumptions arise through past experience of driving on that line.

- Starting a train after having stopped completely (e.g. moving away from a station)- When starting away from a location the driver must check a signal which tells them that the line ahead is clear, if they forgets to do this then a SPaD may occur. SPaDs can also arise from poor signal placement with respect to the driver's line of sight and mis-communication with other staff as if the driver cannot see the signal themselves they may ask another member of staff to check it for them.
- Driving in railway yards and sidings - SPaDs often occur because of poor communication with other staff resulting in the driver believing that it is safe to proceed through a red light.

## Day of the Week

No relationship between SPaD occurrence and the day of the week was found in some studies (e.g. Williams, 1977). Driver's who had more frequent shift changes and were working the first day after a change in shift however were noted to be more likely to experience a SPaD. The important factor therefore may not be a particular day of the week, but the timing of when a SPaD occurs in relation to the driver's shift pattern. Van der Flier and Schoonman, (1988) support this by demonstrating that the distribution of SPaD incidents throughout the days of the week corresponded with the change of the drivers' shifts.

## Time of Day

Van der Flier and Schoonman, (1988) found most cases of SPaDs occurred during morning hours (12am to 6am and 8am to 12pm), where as other studies (e.g. Williams, 1977) found no relationship between time of day and SPaD occurrence. It may be possible that this is due to interaction with other factors such as shift rotation and time at work.

## Time into Shift

Several studies (e.g. Williams 1977) have demonstrated that SPaD occurrence is greater after the driver has driven for a certain period of time. Disagreement however arises regarding when precisely in the driver's shift most SPaDs occur and Van der Flier and Schoonman (1988) find no significant effects for time into shift. Again this may be masked by the driver's shift pattern. For example some train drivers may have a tendency to drowse whilst driving (Endo and Kogi, 1975) and the time at which this is most frequent varies depending on the night of the shift rotation (Kogi and Ohta, 1975) with the second or third night being the most common.

## Number of Drivers in the Train Cab.

Recently the number of train drivers present in the cab has often been reduced from two to one. Job analysis techniques revealed that this should pose no additional health and safety problems (Smith and George 1987). SPaDs however occur more frequently in single man cabs (Williams, 1977), and it may be that the second person confirms the signals as the driver passes through the signal.

## Personal Factors

### Personality Factors /Previous SPAD Involvement

Verhaegen and Ryckaert (1986) found positive correlations between error frequencies (delayed reactions to yellow lights and speeding) and the personality dimensions of extroversion, neuroticism and emotional instability. Drivers displaying minor psychiatric, and psychosomatic symptoms were also found more likely to experience a SPaD. Drivers who have had previous SPaD involvement, demonstrate worse performance on multiple reaction tasks, and report less job satisfaction than in a matched control group. They are also more likely to be involved in future SPaD incidents (Van Der Flier and Schoonman, 1988) but it is difficult to explain why this is the case and there is little evidence to identify some drivers as "accident prone".

## Driver Morale and Motivation

SPaD incidents have been attributed to drivers' low morale and poor job satisfaction (Van der Flier and Schoonman, 1988). Many drivers reported feelings of alienation, isolation and redundancy. They also reported anxiety concerning; vandalism, poor pay, working conditions, organisation of work, and relations with management. Drivers blamed SPaDs on low concentration rather than poor equipment design, expressing concern at the de-skilling of the train driving task and the hours worked in a monotonous environment.

## Age/Length of Service/ Track or Rolling Stock Experience

Research regarding age is inconclusive (Williams, 1977). The number of years service which the driver had completed was not important (Van der Flier, 1988) but when driving certain trains for the first time or after a long break the driver was more likely to commit a SPaD. A rise in SPaDs was also associated with introducing new signalling schemes and braking systems, highlighting a need for adequate training.

# Perceptual/Attentional Difficulties

## Signal design/placement

The frequency of missed signals and the drivers' response time have been associated with the signal intensity in relation to; its background, the intensity of the signal light and the frequency of flashing (Mashour and Devine 1977). There is no evidence to determine why any one signal should present a higher risk factor but a high percentage of SPaDs occurred at identified black spots; the most frequently reported hazard was a signal behind a bend (Van der Flier, Schoonman 1988) and more SPaD incidents occurred at signals which have been installed for less than 6 months (which emphasises the importance of the driver's route knowledge).

## Incorrect assessment of track position.

Drivers often use their own route knowledge to anticipate the position of the next signal (Buck 1963). This is important as the signals are often only visible for a short length of time and at varying positions within the visual field. If the driver is unsure of their position then the chances of missing a signal increases. This is particularly important at night or under conditions of poor visibility. In order to determine their position on the track the driver often uses several sensory cues including auditory, visual and vibratory. Newer trains are often insulated against these cues, thus making the drivers' task in this respect, more difficult.

## Selecting the wrong signal.

This often occurs when a number of signals are displayed in one location and the driver selects a signal which is inappropriate for their track. This is determined partly by the signal's position in relation to the track the driver is travelling on and also the drivers' route knowledge (Buck 1963).

*Vigilance/monotony.*

The drivers' task has become more de-skilled and monotonous over recent years because of longer journey times and the growing automation in the cab causing boredom and monotony. The driver's attention may therefore wander away from the task and be diverted for long enough to miss a signal when it is within their potential visual field. The problem of a general lowering of train driver vigilance in a monotonous work environment is frequently mentioned (e.g. Endo and Kogi, 1975). Reason and Mycielska (1982) argue that errors relating to these factors occur because of misplaced competence rather than incompetence. This is often associated with highly skilled or habitual experiences where the task is largely automatic and attentional demand is low. Errors occur when attention is focused on something else or in familiar environments where the expected is the norm and vigilance is therefore reduced. An example of this is when drivers cannot recall the route they have just been driving. The probability of making such errors increases with task proficiency.

Vigilance devices have been introduced to reduce such decrements of attention. These often take the form of buzzers or bells which the driver has to cancel. As a vigilance decrement has been shown to occur long before an auditory decrement the effectiveness of such devices at maintaining the driver's attention to the task is debatable. It is also possible that due to the frequency with which the driver has to cancel the sound they may have learnt to respond automatically without thought. In this case the vigilance device would have failed to bring the driver's attention back to the task and they still may pass through a red light.

Wilde and Stinton (1983) found that certain types of vigilance devices not linked with direct control of the train could in fact, divert the driver's attention away from the task of driving the train to the task of cancelling the warning and thus fail to appropriately focus the driver's attention. They argue that vigilance devices should direct the driver's attention to some specific train driving task such as speed control. This is supported by Buck (1963) who states that certain types of railway vigilance devices are ineffective in relation to attentiveness but are relevant to maintaining wakefulness.

*Driver Assumptions.*

A driver may sometimes have such high expectations of what a signal may be that he may fail to look at it. If this inference is incorrect an erroneous response may be made and he may in fact unconsciously pass through a red light. Even if the driver does look at the signal any false expectations they may have might restrict the perception and assimilation of true information (e.g. a signal may be interpreted as orange when it is in fact red). False expectations usually arise from past experience driving on that line and the fact that normally only approximately 2-3% of all signals through which the driver passes are red (Buck 1963).

*Sensory error/mis-perception.*

Conditions of fog, sun, rain, etc. can distort brightness, contrast and distance perception thus increasing the possibility of error. Colour signals require a precise definition of hue and colour and any possible driver colour blindness needs to be established. Visual acuity is also important and eyesight should be tested regularly.

*Obtaining information from another source*

In some situations, such as in a station, the signal may not be in a position that is easily observed by the driver who may therefore rely on other people to relay the correct information regarding the signal aspect.  If this information is incorrect then it is possible that a SPaD may occur (Buck 1963).

## Severity of SPaDs

Different causal factors of SPaDs may affect the subsequent overrun past the signal. SPaDs with longer overruns have the greater potential for damage or collision with another object.  A SPaD arising from a misjudgement of the braking distance means that the signal has been seen by the driver and they are aware of the need to stop.  A SPaD which occurs because the driver has for some reason disregarded or failed to look for the signal means that the driver is unaware of the need to stop and therefore continues unaware of the potential dangers ahead.  SPaD incidents where the driver is unaware of the need to stop therefore have the greatest potential for disaster and should to be addressed first.  The importance of the driver's actions in determining the outcome of SPaDs is demonstrated in the diagram below.

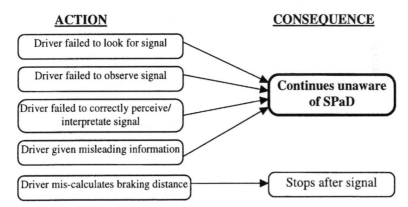

**Diagram 1.**  Driver actions in SPaD causation and the subsequent
awareness of the need to stop.

## Conclusions

Precise data regarding SPaDs are difficult to determine as the driver group is dynamic  and subjected to many external influences (e.g. new traction units, new signalling systems, staff turnover) however several conclusions can be drawn from this review of human factors and SPaD incidents:-
• Many causal and contributory factors  are operating in the occurrence of SPaDs and these may have a cumulative effect.
• Different types of SPaDs occur because of different causal factors and different driver activities. Therefore SPaDs cannot be eradicated by one single solution.
• Different SPaDs have different severity outcomes and those with the highest potential for loss of life and disaster must be addressed first.
• Further research is needed on the role of shift work and organisational factors. Some drivers are more likely to be involved in SPaDs than others, although it is not clear why.  An inability to cope with shift working patterns may be a cause.
• Lack of vigilance and attention are central issues, both appear to be associated with poor motivation and commitment.  It is important that cab design should not de-skill the drivers' task or reduce the motivational levels of the driver who

should be an active and integral part of the driving task.

- Vigilance devices implemented into the cab environment should focus the driver's attention on the task of driving the train and not be distracting. This is important when designing fail-safe equipment such as the ATP system to reduce SPaDs. The sensory cues the driver uses to obtain their track location should not be under estimated and not designed out of the cab completely.

- Proper attention must be given to cab design and internal environmental conditions to ensure compatibility between cab visibility and signal placement and encourage driver alertness and efficiency.

- Drivers should have an active role in the organisational/workplace decision process to encourage feelings of control, commitment and motivation to the job.

- Appropriate training is important regarding the use of new or rarely used equipment. Signals which are passed frequently should be highlighted and addressed even if only by making the driver aware of them during training.

- It is important that good communication systems between the driver and other railway personnel (e.g. the guard, signalman or control room) are maintained. The appropriate personnel should receive training regarding the effective and accurate transmission of information.

The European Commission is seeking to improve the integration and inter-operability of European Transport Networks throughout Europe. Specific projects such as EUROCAB, EUROBALISE and EURORADIO have addressed improvements in the safety and ergonomic aspects of cab interfaces, signalling systems and radio control. Rail signalling is one of the most severe barriers to rail inter-operability and if the project is to be successful the problem of SPaDs will have to be successfully addressed.

## References

Buck, L., (1963) Errors in the perception of railway signals, *Ergonomics* **11** (6).

Endo, T., Kogi, K., (1975), Monotony effects of the work of motormen during high speed train operation. *Journal of Human Ergology* **4**, (2), pp129-140.

Kogi, K., Ohta, T., (1975), Incidence of near accidental drowsing in locomotive driving during a period of rotation. *Journal of Human Ergology,* **4**, (1), pp 65-76.

Mashour, M. and Devine B. (1977) Detection performance and its relationship with human capabilities for information processing. *Reports from the Department of Psychology No. 495.* The University of Stockholm.

Reason J., Mycielska K., (1982) *Stress and fatigue in human performance.* J.Wiley.

Smith M.C., George D., (1987) *Health, stress and locomotive engineers.* O.P.R.A. Ltd. New Zealand.

Van Der Flier H., Schoonman W., (1988), Railway Signals Passed at Danger, Situational and Personal Factors. *Applied Ergonomics* **19** (2), pp1235-1241.

Verhaegen P.K., Ryckaert R.W., (1986). Vigilance of train engineers. *Proceedings of the Human Factors Society 30th Annual Meeting.* Human Factors Society, Santa Monica (California).

Wilde, G.J.S. and Stinson, J.F. (1983) - The monitoring of vigilance in locomotive engineers. *International Journal of Accident Analysis and Prevention* **15** (2) 87-93

Williams J., (1977) Railway Signals Passed at Danger. Paper presented to the Annual Conference of the Ergonomics Society.

# 1997

| Stoke Rochford Hall | 15th–17th April |
|---|---|
| **Conference Manager** | Janis Hayward |
| **Chair of Meetings** | Ted Lovesey |
| **Programme Secretary** | Sandy Robertson |
| | |
| **Secretariat Included** | Margaret Hanson |

| Chapter | Title | Author |
|---|---|---|
| **The Culture of Ergonomics** | The role of ergonomics in development aid programmes | T. Jafry and D.H O'Neill |
| **General Ergonomics** | The ergonomic design of passenger safety information on trains | S. Layton and J. Elder |
| **Health and Safety** | Health and safety problems in computerised offices: The users perspective | R.M. Sharma |
| **The Culture Ergonomics** | The inter-relationship of physiotherapy and ergonomics: Standards and scope of practice | S. Hignett, E. Crumpton and L. McAtamney |

# THE ROLE OF ERGONOMICS IN DEVELOPMENT AID PROGRAMMES

T Jafry and D H O'Neill

*International Development Group,*
*Silsoe Research Institute,*
*Silsoe,*
*BEDFORD, MK45 4HS*

In most development aid projects, economic and social issues are accorded paramount importance. Human-technical issues are often not addressed, or if they are, it is done implicitly rather than explicitly. Incorporating ergonomics into development aid projects can enhance project outcomes by providing greater benefits to poor people. These benefits can be achieved by improving working conditions and designing equipment to be better matched to human characteristics and capabilities. Through a process of creating awareness and providing training to field staff, project managers and other aid professionals, the human-technology gap can be bridged. This paper describes how we tackled the problem of incorporating ergonomics into development aid programmes and associated policy issues.

## Introduction

There is enough evidence to prove that occupational health and ergonomics are two of the major determinants of the state of world health. Forty to fifty percent of the world's population are exposed to work-related health risks (Mikheev 1995), so these work-related health risks need to be properly addressed. However, a lack of awareness of the consequences of; work related diseases results in a total economic loss of up to 10-15% of GNP (Mikheev 1995) and what to do about this remains a problem. We have taken up the challenge of showing how ergonomics can provide benefits by improving the health of people in developing countries, in order to raise the priority of ergonomics activities within the British Overseas Development Administration (ODA) Aid Programme. At present there are few practical examples to illustrate the benefits that can come fromergonomics. In order to determine whether ergonomics should be a matter of policy within ODA, a three stage work plan was developed. The first stage was to establish what ODA Advisers understood by the term ergonomics. Providing practical examples to illustrate how ergonomics can provide benefits to ODA was the second stage. Thirdly, the views of other organisations with an interest in ergonomics, such as the International Labour Organisation, World Health Organisation and the Food and Agricultural Organisation were elicited.

## Creating Awareness

### Informal Interviews with ODA Advisers

Informal discussions were held to assess both the general level of awareness of ergonomics amongst relevant ODA staff and the contribution they perceived ergonomics could make to development aid. This was done using the semi-structured interview technique. The questions put to them followed a standard sequence in order to get a consistent approach with all Advisers. A template of questions was formulated for this but was not actually referred to during the interviews. A total of fifteen advisers was contacted of whom fourteen agreed to be interviewed and one agreed to be interviewed by telephone. A list of the questions asked, together with the responses, are presented in the results section. Prior to the interviews an ergonomics briefing note was circulated which gave a brief explanation of ergonomics and its application.

### ODA Project Screening

In order to illustrate how ergonomics could benefit development aid projects, it was recommended that we work in close collaboration with three existing ODA projects. We therefore had to select three projects with which we could become involved, and, by adding an ergonomics component, provide "case study" information on the relevance and benefits of ergonomics. Three hundred projects from ODA's natural resources project data base were screened. During the initial screening, all projects completely unrelated to ergonomics were eliminated. During the second screening, all projects were scored for their likely ergonomics content using a checklist. Finally, during the third screening, the remaining eight highest scoring projects were scrutinised and discussed with the project managers; from this three projects were selected for ergonomics intervention.

### The Views of Other Organisations with an Interest in Ergonomics

The views of other international organisations, International Labour Office (ILO), World Health Organisation (WHO) and the Food and Agricultural Organisation (FAO), who currently take an active interest in ergonomics for developing countries, were obtained by making short visits to these organisations. Key persons within these organisations were interviewed to determine their views on the role of ergonomics in development aid.

## Results

### Informal Interviews with ODA Advisers

The results of the interviews with the ODA Advisers (UK based) are summarised and given in Tables 1a and 1b.

**Table 1a** Summary of interviews with ODA Advisers (UK Based)

| Question | Response | Specific comments |
|---|---|---|
| Have you heard of ergonomics? | All 15 Advisers had heard of ergonomics. | |
| What do you think ergonomics means? | There was a variety of responses to the question. Some gave more than one response:<br>Very knowledgeable about ergonomics 1<br>Did not know much about it 1<br>Time & motion study 2<br>Workplace design 7<br>Tool & equipment design 3<br>Work on tasks with a repetitive nature 2<br>Physical workload 1<br>Improving efficiency/reducing drudgery 4<br>Manual handling 1 | |
| Have you read the Ergonomics Briefing Note? | Yes 11<br>No 4 | |
| Did you perception of ergonomics change after reading the Briefing Note? (this question put to the 11 that had read the Briefing Note) | Didn't really 2<br>No 5<br>Yes 3<br>No comment 1 | did not realise ergonomics could deal with<br>* tool and equipment design<br>* accidents and injuries |
| Can you see a need for ergonomics in your Programme/Projects? | Yes 11<br>No 3<br>Other: Not in the business of looking out for projects. | |
| What contribution do you think ergonomics can make? | Tool and equipment design 12<br>Women and technology 9<br>Technology transfer 4<br>Work organisation & work Place design 5<br>Others: improve working conditions 1 | |

| Question | Response | Specific comments |
|---|---|---|
| Do you envisage difficulties in trying to implement ergonomics in ODA Programmes/Projects? | No 7<br>Yes 8 | **No problem provided that:**<br>1. The implementation of ergonomics is clear, using thought-out practical applications.<br>2. Ergonomics fits appropriately.<br>3. Ergonomics is integrated well into community problems and financial arrangements.<br>4. There is a need for ergonomics.<br>5. People are involved and the real problems are addressed.<br>6. Ergonomists work closely with the projects.<br>7. No specific comment.<br>**Yes envisaged problems because:**<br>1. Ergonomics cannot so easily be implemented in the fisheries projects (2 Advisers commented on this).<br>2. Ergonomics should be considered for technical co-operation link not for research.<br>3. New ergonomically designed equipment costs money. People couldn't afford it.<br>4. Changing working practices would be quite difficult.<br>5. Ergonomics is not a priority in businesses. Credit more important.<br>6. People trying to survive. People need credit not ergonomics.<br>7. Questioned the importance of ergonomics relative to other subjects.<br>8. No comment. |
| Would you like to know more about ergonomics and its application? | Yes 12<br>No 1<br>No comment 2 | Would like to attend seminar/workshop 11<br><br>Specifically to:<br>* See well-argued case<br>* For clarity<br>* See benefits ergonomics can bring<br>* See good demonstrations.<br><br>Would prefer literature 1 |

**Table 1b** Other comments by Advisers (UK-based)

* Would like to see examples of ergonomics in future ergonomics concept notes.
* People lack awareness of ergonomics. Need to sell it better and increase awareness.
* Need to educate Non Government Organisations on ergonomics.
* ODA addresses ergonomics issues sub-consciously
* Change will become technology-driven. Need to consider the financial limits of change at the level of the individual.
* People would not do a task if it was not productive or efficient for them to do so.
* Would not like to see another checklist.
* Need to define ergonomics better so it stays in people's minds.
* Country specific exploratory study of ergonomics would be useful.
* Ergonomics deals with issues at ground level. ODA has only the capacity for bigger issues e.g economics and empowerment.
* Would welcome any science that reduces drudgery, improves efficiency, releases time to do other things.

Half of the Advisers stated that they did not envisage any difficulties in trying to implement ergonomics in ODA Programmes/Projects. The concerns raised by theother half of Advisers could be alleviated by providing examples on the practical application of ergonomics. It is interesting to note that approximately 75% of Advisers expressed an interest to attend a seminar/workshop to learn more about ergonomics and its application.

*Project Selection*

The three ODA projects selected, where ergonomics intervention could contribute to the project outcomes, were:

* East India Rainfed Farming Project (looking at farming systems in the Rainfed states of Bihar, Orissa and West Bengal).
* Community Participation in Forestry in the Caribbean; ergonomics and charcoal production in St Lucia.
* National Agricultural Research Project in Ghana; ergonomics and the processing of cassava into <u>gari</u> at a women's' cooperative.

For each of the projects, areas were identified where ergonomics could bridge the people technology gap. However, prior to any interventions, the project staff needed to be aware of how ergonomics could benefit their project. Several activities were undertaken to create this awareness.

India
* Ergonomics orientation workshop conducted with Indian ergonomics consultants.
* Training on how to conduct ergonomics survey.
* Recruitment campaign to identify ergonomist to join the project team.

Ghana
* Ergonomics seminar held for agricultural engineers at the University of Science and Technology in Kumasi.
* Discussions held with the women's food processing cooperative to identify problems and priorities.

West Indies
* Field visits to the charcoal producers to discuss problems associated with the production of charcoal.

- Conduct survey and collect work study data.
- Prioritisation of options for improvements.

*Visits to ILO, WHO and FAO*

All the visits had a positive outcome. Each of organisation stated that ergonomics could contribute to development aid. There were some common underlying issues aired by all three organisations as to why ergonomics is not given more recognition. These are:

* There is a lack of **awareness** of the benefits ergonomics can bring.
* There are not enough people trained in ergonomics so ergonomics problems may not be described as such.
* There are not enough data on the incidence of occupational health problems in developing countries.
* There is a need for ergonomics to bridge the human-technology gap but it must be considered as part of an overall strategy within an organisation.

## Discussion

Informal discussions with the ODA Advisers revealed that they have some knowledge of the subject of ergonomics but the full extent of its application was not appreciated. In general, this lack of awareness and the request by Advisers to know more about ergonomics justifies the need to demonstrate how ergonomics can benefit development projects. Comments made by ODA Advisers have highlighted that there is a need and demand for ergonomics to be considered in development projects.

Involvement with the three ODA projects created an opportunity to provide examples of where ergonomics could benefit aid projects. The fact that all three selected project managers requested workshops and training on how to incorporate ergonomics into their projects endorses the view  that *a lack of awareness* is  the reason why ergonomics does not receive more recognition.

The recent experiences gained from this study show that it is important to provide practical examples of the benefits ergonomics can provide. An increase in the number of case studies is still needed to  make a greater impact.

The ILO and the WHO both have  strategies that recognise the importance of ergonomics.  These are *Occupational Health and Safety* and *Occupational Health for All - a Global Strategy* respectively.  Perhaps in the future the ODA will also have a strategy incorporating ergonomics into its development aid programme.  One step towards this is the production and dissemination of *A Guide to Addressing Ergonomics in Development Aid.*  This guide is being developed and tested as a practical tool for aid professionals to use to ensure that ergonomics components in future projects are addressed.

## Acknowledgement

The authors would like to acknowledge the Overseas Development Administration of the Foreign and Commonwealth Office, for funding this work.

## References

Mikheev, M I. 1995, Health at Work- Global Analysis. In E. Juengprasert (ed.), Proceedings of the International Symposium on Occupational Health Research and Practical Approaches in Small Scale Enterprises, 1995 (Division of Occupational Health, Ministry of Public Health, Thailand) 3-10.

# THE ERGONOMIC DESIGN OF PASSENGER SAFETY INFORMATION ON TRAINS

## Simon Layton and Jayne Elder

*Human Engineering Limited*
*Shore House*
*68 Westbury Hill*
*Westbury-On-Trym*
*BRISTOL*
*BS9 3AA*

This paper describes the design and development of on-train safety information for trains with slam doors (or 'central door locking' (CDL)). The levels of passenger awareness of door operation and emergency procedures were established. Solutions for improvement were proposed and validated. Results showed that the new signs and safety information gave significant improvement in passengers' understanding of emergency procedures and door operation. The results support the use of combined text and icons to present information and a phased delivery of safety information.

## Identifying the Problem

The first stage of this programme of work involved conducting surveys and interviews on in-service trains to collect data on passenger knowledge of door operation and emergency procedures and to identify the media, both on-board trains and in the station environment, to which they pay attention. Analysis of the data revealed that there was a lack of awareness and confidence concerning the proper operation of the doors, particularly amongst less frequent travellers. This was also the case for knowledge of emergency procedures. However, on matters of general safety, even *daily* passengers were unaware of some important issues. The main issues to be addressed were identified as:

- Information was required to raise passenger awareness of the operation of the doors and, in particular, whether or not they are locked

- Information was required to raise passenger awareness of general safety issues, but also specific items such as the location and use of emergency equipment

- The signage at the point of need required clarification

- Passengers expressed a preference for safety information to be available when seated

## Design Rationale

New signs, posters and leaflets were designed for trials on an in-service train. The following principles were adopted:

- Clarify messages using pictograms with text

- Add an exterior sign with central door locking information
- Provide general information at seats and carriage ends
- Phased delivery of information

The phased delivery of information aims to minimise the amount of information that the passenger must absorb at the point of need. (The phased delivery should begin before passengers even board the train, i.e. at ticket offices, on platforms, etc.) It is particularly important in cases where the passenger will have to perform a sequence of behaviours or non-intuitive actions. The door opening procedure is the most obvious example. The safety leaflets, read while passengers were still in their seats, were designed to present information which would be recalled at the door, reducing the amount of time the passenger would need to comprehend the sign on the door itself.

A number of the existing signs were re-designed with the emphasis on clarity, use of icons combined with text and appropriate sign location. (Symbols were combined with text as there is evidence that using them in combination is more effective than using either on its own (Edworthy & Adams, 1996).) The situations addressed were:

- Exterior door locking
- Door opening
- Emergency door release
- Danger sign (leaning on doors, etc.)
- Fire extinguishers
- Emergency stop
- Emergency window hammers
- Emergency procedures

### The New Signs

Individual signs were designed for use at the point of need, i.e., at the door, by the fire extinguisher, etc. Due to limitations on space in this paper, only the safety poster/leaflet will be presented here (although the results refer to all new signage). (Detailed reports of the study are contained in unpublished documents; Elder & Gosling.) The poster/leaflet contains general safety information (some of which was also used at the point of need) and was displayed on A3 sized posters and A4 sized leaflets. (Although the results of the initial data collection suggested that passengers would prefer information on labels fixed to the seat backs and window area this was rejected by the train operating companies due to the likely aesthetic impact and cost of maintenance. Therefore, end of carriage notices and leaflets were used instead.) Leaflets were placed in all seat back pockets and posters were located in the vestibule area (between the luggage racks and the external doors.

There are three main areas of information presented on the poster/leaflet as shown in Figure 1.

- Central door locking (CDL)
- Location of emergency equipment
- Emergency procedures

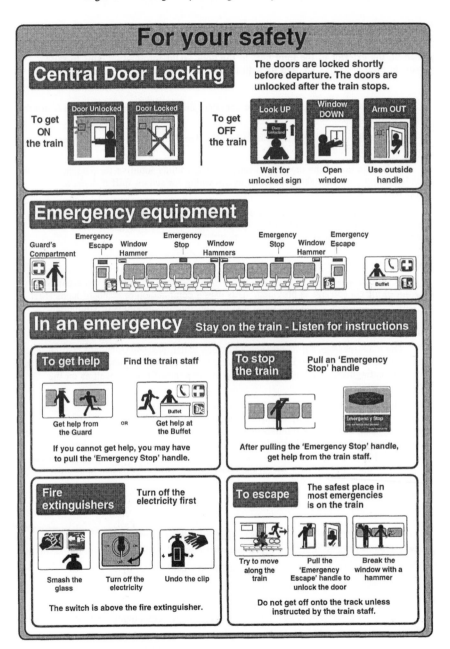

**Figure 1.** Trial safety poster / leaflet

## Central Door Locking

The first two pictograms illustrating getting on the train are also repeated on a sign placed on the exterior of the train next to the door. This information aims to raise passenger awareness of CDL and its indicators on the outside of the train; that is:

- When the orange light is ON, the doors are unlocked and passengers can board the train

- When the orange light is extinguished the doors are locked and passengers cannot enter the train

(N.B. The use of the orange indicator light is counter-intuitive for passengers as the intended purpose of the lights was to act as a warning for station staff if doors are unlocked.)

The second set of pictograms is also repeated on labels on the inside of the door. The original sign contained only text. The new sign includes pictograms and intentionally concise text.

## Location of emergency equipment

To inform passengers of the location of emergency equipment a representative diagram of a train carriage is provided. This uses a combination of standard icons (first aid, fire extinguishers, telephones) and text to illustrate the locations of equipment.

## Emergency procedures

The main message for emergency situations is that passengers should stay on the train. Again the information is presented using a combination of pictograms and concise text. Fire extinguisher information is located at the point of need as well as on the leaflet.

The information on emergency procedures emphasises the fact that passengers should try to get help from train staff. A point to note is that, previously, the Emergency Stop handle was referred to as the 'Alarm' handle. This gave passengers the false impression that it would alert train staff as to where the alarm handle had been pulled and that help would arrive. In fact it simply stops the train and train staff then have to locate the cause of the incident. Therefore, the new sign refers more intuitively to the Emergency Stop handle.

## Validation trial results

### Did passengers notice and read the new information?
- 12% of the passengers surveyed had read a safety poster (located in the vestibule)
- 21.5% of the passengers surveyed had read a leaflet
- 44.9% of the passengers with a leaflet in the seat pocket in front of them read it

For a passenger to read a leaflet they had to have a leaflet in front of them, notice the leaflet and finally choose to read it. The percentages completing each of these steps were:

- 48.5 % of surveyed passengers had a seatback in front of them
- 66% of seatback passengers noticed a leaflet
- 67 % of seatback passengers who noticed the leaflet read it

Although the overall proportions of passengers who read leaflets or posters were small, the numbers should be seen in relation to the accessibility of the information. When information was made available to seated passengers, nearly half of them (45%) spontaneously read it in the short space of time (5-10 minutes) before they were given a questionnaire.

### Did more passengers read the leaflet or the poster?

23% of passengers noticed the poster, whereas 66% of passengers with a leaflet in the seat pocket noticed it.

Of those who noticed leaflets or posters, a slightly greater proportion read the leaflet than read the poster (67% for the leaflet compared to 54% for the poster).

61% of passengers suggested the information should be displayed in locations visible from the seat.

### Did the new signs increase door understanding?

Passengers were asked a number of questions about door operation and central locking indicators while facing the doors, i.e. able to see the signs. Results for passengers viewing the exterior signs showed an improvement in understanding of 30% although this was not statistically significant due to the low numbers of subjects available for this part of the trial.

The results for the new interior door sign show improvements resulting in 80% or 90% correct responses (depending on the specific question asked). Results are statistically significant.

The trial door signs did increase door understanding. The percentage improvements must be considered in relation to the original baselines. The response to the question "Where is the 'Doors Unlocked' sign?" is a good example, with an improvement of 20% from an original score of 72% correct giving a threefold reduction in incorrect answers (from 28% incorrect to just 8% incorrect).

### Did the leaflets and posters increase door understanding?

A large number of passengers (205) completed questionnaires while seated. Passengers with a seat back in front of them read a safety leaflet, while those at tables did not. (All 'seatback passengers' were requested to read the safety leaflet before completing the questionnaire whether they had spontaneously done so or not.) There were statistically significant differences between the two groups when asked about door operation and central locking indicators. Improvements in correct answers were of the order of 15% to 20% resulting in percentages of correct answers between 60 to 70% for poster and leaflet readers.

Some of the passengers who completed questionnaires had read the trial safety posters displayed in the vestibule. These passengers showed percentage improvements in correct answers of 28% for the two main questions about door operation and central locking.

The greater improvement for poster readers compared to leaflet readers may result from the fact that all passengers who read posters did so spontaneously whereas roughly half of the leaflet readers did so only when asked to in the questionnaire.

### Did the new signs increase emergency knowledge?

There were improvements in the percentage of correct answers of the order of 30% in questions relating to the emergency door release when comparing the in-service sign against the trial sign.

Questions relating to the new signs in the carriage (Emergency Stop and window hammer) showed statistically significant improvements of the order of 10 to 14% taking the percentage correct to more than 80% . These improvements are for passengers who did not read the poster or leaflet, so are assumed to be a result of the new 'Emergency Stop' and hammer signs.

For the fire extinguisher, there were no statistically significant improvements. The problem highlighted in the original survey still exists - passengers do not expect to have to turn the electrical supply off before using the fire extinguishers.

### Did the leaflets and posters increase emergency knowledge?

Passengers were asked three general safety questions (safest place in emergencies, fire in carriage, which side to evacuate the train). The percentage improvements in correct answers, when comparing passengers who had read leaflets with those who had not, were from 10% to 22% with 2 of the 3 questions reaching levels over 85% correct for poster and leaflet readers. Results were statistically significant for non-commuters.

Questions relating to the location of emergency equipment showed improvements in percentage correct of the order of 20% for poster and leaflet readers.

### Conclusion

This paper illustrates the successful application of the ergonomic principle that combining images and text is generally more effective than either used on its own. It also highlights the point that safety information cannot be clear unless the procedures themselves are clear and logical. For example, it is unreasonable to ask a passenger to "seek help at the buffet - unless it is closed - in which case find the guard - who may not actually be in the guard's compartment...".

The work also supports the notion that passengers are more likely to read safety information made available at their seats rather than general notices and specific signs at the point of need. If the phased delivery of information is used, signs at the point of need are made more relevant and serve as reminders rather than new instructions.

### References

Edworthy, J. & Adams, A. 1996, *Warning Design: A Research Prospective* (Taylor & Francis, London)

Elder, J. & Gosling, P. 1996, *Central Door Locking Safety Review: Review of Passenger Awareness of Safety and Emergency Regime.* (unpublished)

Sanders, M. S. and McCormick, E. J. 1993, *Human Factors in Engineering and Design*, 7th Edition (McGraw-Hill International)

# HEALTH AND SAFETY PROBLEMS IN COMPUTERISED OFFICES: THE USERS PERSPECTIVE

**Randhir M Sharma**

*Division of Operational Research and Information Systems*
*The University of Leeds*
*Leeds*
*LS2 9JT*

Computers are everywhere, it seems that whatever we do, we cannot escape from them.From the workplace to the supermarket to schools the computer has found a place for itself in all of our lives. Whenever a product becomes an integral part of our lives concerns are raised over the effects on us, our bodies, our environment and those around us. The car is a prime example, nobody thought twice about emissions, lead content or recyclability in the early days of the car, it is only relatively recently that we have seen a sudden increase in concern. However perhaps it is too early to be sure that the blame we attribute to workstations is really justified. The purpose of this work was to obtain an understanding of the level of familiarity amongst typical users of these problems and if problems exist, which factors users feel are to blame. In short this paper summarises problems experienced by users and the components of their working environment that they blame for their problems.

## Introduction

From simple measurable complaints like headaches and stiff necks, to more serious medical disorders such as eye strain and repetitive strain injury through to complex social problems like stress, unemployment and work alienation, people are paying the price of rapid computerisation at home and work(Bawa, 1994). The purpose of this paper and the surveys that it describes is quite simply to determine whether or not we are paying the price. In a similar survey carried out whilst an undergraduate the author was alarmed at the widespread disregard for the subject of health and safety amongst computer users. It was not considered to be serious at all, it was seen as trivial. Users seemed unaware of the numerous cases highlighted in the press. The large library of information on this topic including hundreds of case studies did not concern them. Perhaps it was that these books, journals and reports had no relevance to real computer users. Should we be concerned or do these problems really not affect us? In a famous quote Judge John Prosser once referred to those people who complain of problems after using computers as "Eggshell personalities who need to get a grip on themselves....The RSI epidemic is a form of mass hysteria rewarded and reinforced by the compensation system." (case of Mughal vs Reuters, 1993) The wide range of opinions on this subject and widespread confusion regarding terminology do not aid its progress towards being accepted as a problem which users will either accept or take precautions against. Within the School of Computer Studies at Leeds two members of staff are using voice recognition systems because they can no longer use keyboards comfortably, they are *"normal"* users, if they can be affected, surely it can affect the rest of us.

The purpose of the survey was firstly to establish the scale of the problem, secondly to identify those components held responsible by users for their problems and finally to assess the importance of particular working habits. Two hundred and fifty six replies were obtained, the replies came from a wide range of computer users. One hundred and sixty one replies came from the School of Computer Studies at Leeds. Respondents included undergraduates, postgraduates, academic and non academic staff. Ninety five of the replies were obtained from workers working for a large national newspaper in India. The respondents varied significantly in the tasks they were doing and the number of hours per week they were using workstations. This large variation in the roles and tasks performed by users allows a better simulation of computer users as a whole. The results obtained from this questionnaire are relevant to those using computers in all walks of life, not just programmers and data entry clerks who tend to be the focus of most surveys of this type.

## Questions Asked

One of the most important questions was the number of years that users had been using workstations. This question was asked in order to see whether or not there was a correlation between the number of years a workstation was used and the incidence of health problems. Another question was the number of hours per week that users used their workstations. It has been shown that the greater the amount of VDU use per day, the greater the amount of pain experienced by users. Fahrbach and Chapman (1990) compiled statistics for head, neck, shoulders and back pains and Evans (1985) showed that increased usage affected eyesight.

The distance between a computer user and the screen is a factor which is given a lot of attention. Sitting too close to the screen encourages poor posture and is often held responsible for visual problems arising from the use of VDU's. There is widespread debate concerning the ideal distance between the user and the screen. This in the main stems from a change in viewing angle. Earlier recommendations advocated the monitor being some 20 degrees lower than the line of sight, now the common consensus is that the monitor should be in the line of sight thus removing the risk of neck and shoulder pains as a result of the flexion required to view a monitor which is below the line of sight. The average of all viewing distance recommendations is 20 inches (50 cm) (Godnig and Hacunda,1991)

The time spent working between breaks was also important, regular breaks are recommended by most specialists.Without adequate pauses to rest, the mind loses its clarity and sharpness, and the body suffers wear and tear. You become less productive and more prone to injury(Choon-Nam Ong, 1995). Although there is no specific guideline for the time interval between breaks, most texts do recommend a 10 - 15 minute break every hour. One particular study showed that increasing the number of breaks actually sped up work and improved quality (Stigliani, 1995). The Health and Safety Regulations (Display Screen Equipment) 1992 contained specific guidelines for breaks. Regulation 45c states "Short, frequent breaks are more satisfactory than occasional, longer breaks:eg, a short 5 -10 minute break after 50-60 minutes continuous screen and/or keyboard work is likely to be better than a 15 minute break every two hours".

Stretching is another recommendation that frequently crops up. Stretch before you start work just as athletes do before vigorous exercise, concentrate on your arms, neck and shoulders but stretch your whole body too for general flexibility (Stigliani, 1995). By comparing the incidence of pains in arms and wrists, back pains and neck and shoulder pains amongst users who stretch and those who do not, it should be possible to see whether stretching does reduce the risk.

Another issue that was addressed was that of user familiarity with the topic of health and safety helping to prevent problems. Are those who are aware of the subject less likely to suffer than those who are not? The author was convinced that those who are well acquainted with the possible problems are more likely to take preventative action. The questionnaire asked respondents whether they were familiar with the terms CTS, RSI and WRULD. Although RSI is a very controversial term which we can argue is not really as relevant as WRULD, it is still regarded by computer users as an acronym for muscular problems associated with the use of a keyboard. The definition is vague and questionable but having knowledge of the term is an indication of awareness of the topic of health and safety. The results were then split into groups depending upon the level of knowledge and compared in terms of the incidence of problems.

Finally users were also asked for their opinions on how best to distribute Health and

Safety information so that it would be read. If users are to learn about these problems it is essential that the information is presented to them in a manner that they want and are comfortable with.

## Results

The largest single group of users were undergraduates within the School of Computer Studies. In addition to undergraduates there were also responses from postgraduates and staff. The replies obtained in India came from users in a variety of roles within the newspaper, they included journalists, accountants, graphic designers and secretarial staff. In response to the number of years spent using a workstation, the largest group of respondents (126) had been using a computer for 5 years or less, this figure was influenced by the large number of undergraduates taking part in the survey. 83% of the respondents worked for 10 hours or more per week with a workstation.

Using the criteria that a safe break was one which was taken somewhere between 45 and 60 minutes only 37 (15%) of respondents took breaks within this safe time frame 219 (85%) users exceeded this. The figures for the distance between the user and the screen were also quite surprising, using the distance of 50 cm as a safe distance only 35 (13.7%) of the respondents were sitting a safe distance from the screen, 221 (86.3%) were sitting at a distance which is deemed unsafe.

Only 9% of respondents stretched before working. The small sample size made it impossible to draw any accurate conclusions from the data which provided very similar results for both users who did and did not stretch.

The results in the chart below illustrate the responses to the question asking which problems were suffered by users.

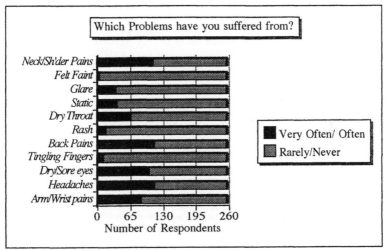

**Figure 1**. Problems suffered by users and the frequency of these problems.

Quite clearly the five key problems identified were neck and shoulder pains, back pains, dry sore eyes, headaches and wrist pains. The results indicate a frequency of between 30 and 40% for these problems. If this chart is modified to combine very often, often and rarely to form a new group of users - those who have suffered at some time this figure rises to between 60 and 75%. This clearly suggests that a large percentage of computer users have at sometime experienced these problems as a result of using a workstation.

The next step was to look at these five most frequent complaints and see if the users who made these complaints had anything in common in terms of their working habits. The habits looked at were the number of years they had worked, the distance they sat from the screen and the frequency of the breaks that they took.

## Headaches

Those respondents who suffered from headaches most frequently sat closer to the screen, and worked the longest amount of time between breaks when compared with their colleagues who did not suffer as frequently. They also worked a longer number of hours. There was no apparent connection with the number of years worked. Next those factors were looked at which were highlighted as being responsible for these problems. The key factor highlighted was heating and ventilation. First impressions were that this result had been influenced by the replies obtained in India. Further interrogation of the spreadsheet revealed that this was not the case. Workers in India take the hot weather for granted, as a result of this all offices are equipped with both air conditioning and fans. The majority of complaints came from the School of Computer Studies. Since this survey was carried out a new air conditioning unit has been installed in the main laboratory used by undergraduates in response to complaints about excessive heat. The other point to raise is that in order to solve the problem we need to be clear as to which factors are responsible for the problems. It is not unreasonable to have expected poor quality monitors to be the factor pinpointed by users as the cause of their problems. The result obtained illustrates that it is not always the most obvious factor which is responsible for these problems, it also illustrates the importance of the working environment. The causes of certain problems may not be workstations but the environment in which we use them.

## Pains in Arms and Wrists

Those respondents who suffered very often from pains in arms and wrists also shared many of the characteristics of those who suffered very frequently from headaches. They sat closer to the screen and took breaks after longer periods of time. They also worked the longest number of hours per week. However in this case there was a correlation between the number of years worked and the frequency of these problems. The greater the number of years worked the more they suffered. One of the noted characteristics of these sort of problems is that they have a cumulative effect. The results obtained from the survey support this claim and also stress the importance of prevention being better than cure. The factors highlighted for these problems were poor chairs, poor furniture and poor keyboards respectively. Poor chairs and poor furniture which do not allow a user to use a workstation comfortably were identified as the most important factors ahead of keyboards which the author had expected to be the most troublesome factor.

## Back Pains

Once again the same three factors surfaced, users who suffered most sat closest to the screen, worked a longer number of hours between breaks and also worked more hours per week when compared with those users who did not suffer as frequently. There was no identifiable relationship between the number of years worked and the frequency of back pain. As expected the key factor identified here was poor chairs, followed very closely by poor furniture in general.

## Neck and Shoulder Pains

The results obtained here were also very similar to those obtained for the previous problems. Users who suffered most sat closest to the screen, worked longest between breaks, and the worked the largest number of hours per week. No correlation was found between the number of years worked and the frequency of these problems. The key factor identified as being responsible for these problems was poor quality chairs. Surprisingly heating and ventilation was in second place followed by poor quality furniture. The surprising prominence of heating and ventilation led the author to the conclusion that fatigue associated with excessive temperatures within offices and laboratories was responsible for encouraging poor posture whilst working.

## Dry Sore Eyes

The results obtained for this problem did not follow the pattern of previous results.

The survey did not replicate results found in other surveys. There was no apparent connection to either of the three factors which were so prominent in the other cases. The conclusion drawn from this was that individual differences in visual quality were more likely to be responsible for these problems than the working habits of users. However one surprising result was found by looking at the factors blamed by users. Heating and ventilation was the factor identified as being most responsible. The statistics showed that there was a huge gap between the blame attached to this factor and any other indicating that this particular factor was considered to be primarily responsible for the visual problems experienced by users.

The issues of whether users who are well informed suffer less or whether responsibility is solely in the hands of service providers who must provide a working environment which is free from potential risk was explored. Respondents were asked about three specific terms, CTS, RSI and WRULD. Only a handful of respondents were aware of all three terms. The results obtained from those respondents who were not aware of any of the terminology were compared with those who were familiar with one of the three terms or a combination. The results indicated that those who had knowledge of the subject worked fewer hours per week and took breaks more frequently. There was however little to choose between the two with regard to distance between user and the screen. The key differences however were found by looking at the frequencies of problems. The table below illustrates the results obtained for the five key problems highlighted earlier in the paper.

**Table 1.** Mean Problem Frequencies

|  | No Knowledge | Some Knowledge |  |
|---|---|---|---|
| **Pains in Arms and Wrists** | 2.39 | 3.22 |  |
| **Headaches** | 2.05 | 2.88 |  |
| **Back Pains** | 2.16 | 2.90 | $p<0.05$ |
| **Dry/ Sore Eyes** | 2.77 | 2.61 |  |
| **Pains in Neck and Shoulders** | 2.21 | 3.04 |  |

Respondents were asked to indicate the frequency of their problems on a scale of 1 to 4 with 1 indicating very often and 4 never. The respondents were grouped into those who knew nothing about the subject and those who knew something. The figures in the table are the means of the two groups. With the exception of dry sore eyes which as mentioned earlier is probably more related to individual differences in vision, those who were familiar with the subject appeared to suffer less. This quite clearly suggests, as with many other problems, that education may the best prevention technique. 80% of respondents felt that they did not know enough about health and safety. Following on from this respondents were asked which techniques they thought should used in order to get the information to them. The current policy employed both within the School of Computer Studies and in India placed most of the emphasis on the user. They had to use their own initiative. Users were asked to suggest mechanisms which they felt would be most effective. The chart on the next page contains the five most preferred techniques. Altogether fifteen different techniques were suggested by respondents. The lack of relevant literature is highlighted by the responses obtained. Relevance to the user is a priority, most users have neither the time or inclination to read books and journals before using a computer. Information should be both concise and clearly presented in a such a manner that it does not become a burden to the user. The fact that users felt that more information and literature is needed suggests that the information currently available is either not easily accessible or is not in a format which will encourage users to read it. The various techniques put forward by users also indicate that a more varied approach needs to be taken. Different users have different needs, and it is important that these are given due consideration.

The results given so far have illustrated that problems do exist and when you consider the variance of the types of use, they also illustrate that it is not only particular types of users who suffer, anyone and everyone can suffer. It is essential therefore that users are educated and are aware of the apparent risks of using a workstation, only by doing this will they be in a position to take preventative measures. In order to educate users, the methods by which information is circulated have to be those which users will read and more importantly understand.

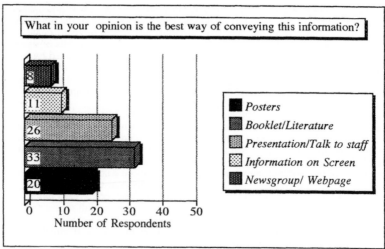

**Figure 2**. Preferred techniques

## Conclusion

The purpose of the work described in this paper was to obtain an understanding of the users perspective. The results obtained indicate that users are suffering and that these problems are not just confined to a handful of individuals. The importance of taking frequent breaks and sitting a safe distance from the screen has been shown. Users have highlighted those factors which they believe are causing them problems and in several cases the key factors are not those which would have been expected. Most importantly the results have shown that those users who have some degree of knowledge regarding the problems are less likely to suffer. Worryingly 80% of respondents felt that they did know enough. Users have indicated which techniques they would like to see employed to educate them. In order to prevent such problems it is important to understand and identify risks associated with the technology, but it is more important to have a full understanding of the user. Understanding and educating users is essential if we are to eliminate these problems. Users need to be be aware not only of the problems but also prevention techniques. The delivery of this information has to consider the needs of individuals to ensure that it stimulates interest. Without careful consideration, we risk endangering the very people we are people we are trying to help...... the USERS.

The next stage of this work is to look at several of the techniques highlighted above in order to find the most effective.

## References:

Bawa J. 1994, *The Computer Users Health Handbook: Problems Prevention and Cure.* (Souvenir Press)
Bentham P. 1991, *VDU Terminal Sickness: Computer Risks and How to Protect Yourself.* (Green Print)
Choon-Nam Ong, 1990, *Ergonomic Intervention for Better Health and Productivity.* in Promoting Health and Productivity in the Computerised Office, edited by S. Sauter. (Taylor & Francis, )
Evans J. 1985,*VDU Operators Display Health Problems.* Health and Safety at Work
Fahrbach , P.A & Chapman L. J., 1990, *VDT Work Duration and Musko-skeletal Discomfort.* AAOHN Journal , 38(1)
Godnig & Hacunda. 1991, *Computers and Visual Stress: Staying Healthy* (Abacus)
HSE 1992. *Display Screen Equipment Work: Guidance on Regulations.*
Huws U. 1987, *The VDU Hazards Handbook: A workers guide to the effects of new technology.* (Calverts Press (TU) Workers' Co-operative)
London Hazards Centre 1993, *VDU Work and Hazards to Health.*
Stigliani J. 1995, *The Computer Uses Survival Guide* ( O'Reilly & Associates, Inc)

# THE INTER-RELATIONSHIP OF PHYSIOTHERAPY AND ERGONOMICS : STANDARDS AND SCOPE OF PRACTICE.

**Sue Hignett**
Ergonomist, Nottingham City Hospital NHS Trust

**Emma Crumpton**
Lecturer, Faculty of Health and Social Care, University of the West of England

**Lynn McAtamney**
Consultant Ergonomist, COPE, Nottingham

This paper explores the relationship between the professions of ergonomics and physiotherapy in their complimentary roles of optimising human performance and well-being. The relationship within the area of physical ergonomics is explored using a model which has 'functional capacity' as the centre point leading through to 'ergonomic design and prevention', and 'rehabilitation and therapeutic care'. Concern is voiced over the scope of practice and the need for continued development of standards and recognition of core competencies to clearly define the boundary between the two professions. In conclusion it is believed that there is a considerable potential benefit from a closer relationship between ergonomics and physiotherapy, but in order for this to be achieved there must be mutual recognition and respect for the specialist skills of each profession.

## Ergonomics

Ergonomics is both a science and a technology using research, knowledge and skills from a range of disciplines (Oborne, 1995). Scientific information about human beings (and scientific methods of acquiring such information) are applied to the problems of design, with the aim of defining the limits of human adaptability and diversity. Ergonomics is a relatively young profession, and as with many professions it is adapting to the changing world by modifying and extending it's scope of practice. Since it's initial inception ergonomics has grown from use in military engineering, through to space applications in the 1950's and onto a wider industrial usage in the following thirty years (pharmaceuticals, computers, cars and other consumer products). In the 1990's ergonomics is much more general with applications in commerce as well as industry. Health and safety legislation has made ergonomic input into most work environments a legal requirement ensuring that this particular role of the ergonomist is a rapidly growing one. This is where physiotherapy has scope for valuable input. One possible way of categorising the scope of practice of ergonomics is:

1.  **Physical ergonomics.** This includes physiology, anatomy, and biomechanics providing information on physical capabilities and limitations, as well as physical dimensions (anthropometry) It also encompasses input from physics and engineering, looking at the design of tools, products, equipment and work environments (lighting, heat and noise).
2.  **Cognitive ergonomics.** This category includes physiological psychology looking at the nervous system and behaviour, as well as experimental psychology providing information about cognitive functions e.g. perception, learning, memory etc.

3. **Organisational ergonomics**. The organisational structure, processes of work (including job design and analysis) and attitude and behaviour of people in the organisation will all effect the effectiveness and efficiency of an organisation.

Ergonomics has a huge scope of practice, and members of the profession come from a variety of background disciplines. In the USA most of the members of 'The Human Factors Society' have a background in psychology (45%) and engineering (19%) with only 3% coming from the medical field (Sanders and McCormick, 1993). This is believed to be a similar proportion to 'The Ergonomics Society' in the UK although the definitive figures are not known. This is in contrast to elsewhere in the world, for example, Norway, where most of the ergonomics practised in industrial settings is by Ergonomists with a background in physiotherapy (Bullock, 1990).

One of the drawbacks of having such a range of professionals, from many different backgrounds and with such a variety of skills and interests, is establishing an agreed level of competency. Bullock, (1994) advocated the definition of core ergonomic competencies (Table 1). She suggested that among professionals practising ergonomics a common basic foundation of knowledge was required to ensure an adequate level of communication which would support the need for Ergonomists to appreciate their own limitations and facilitate assistance being sought when required.

**Table 1.** Core Competencies for Ergonomists

---

| 1. | Demonstrate an understanding of the theoretical basis for assessment of the workplace |
| 2. | Demonstrate an understanding of the systems approach to human integrated design |
| 3. | Demonstrate an understanding of the concepts and principles of computer modelling and simulation |
| 4. | Communicate effectively with clients and professional colleagues |
| 5. | Obtain information relevant to ergonomics from the client |
| 6. | Collect supplementary information relevant to ergonomics relating to the history of the problem and current management |
| 7. | Collect from the work place, in an appropriate manner, quantitative and qualitative data relevant to the perceived problem and ergonomics |
| 8. | Document ergonomic assessment findings |
| 9. | Recognise the scope of ergonomic assessments |

---

The professional backgrounds and specialist skills may reflect in the approach or area of interest of the Ergonomist. A systems engineer may feel that the ability to carry through the design, development and implementation of a specific work station, tool or piece of equipment is of particular importance. Whereas, a physiotherapist may be more concerned with a specific musculoskeletal risk factor and modifications to the work station, task or tool to eliminate or reduce the risk of injury. The specialist skills of the Occupational Health Physiotherapist will be explored with a view to how they can be used to complement ergonomics practice.

## Physiotherapy

Jacobs and Bettencourt (1995) suggest that the fundamental intent in the early days of physiotherapy was to *'assess, prevent and treat movement dysfunction and physical disability, with the overall goal of enhancing human movement and function'*. As with ergonomics the scope of physiotherapy has extended well beyond this initial intent to include, for example, health promotion, fitness training, incontinence treatments etc., and also into the field of occupational health.

The Chartered Society of Physiotherapy (UK) has produced a leaflet (Physiotherapy and Occupational Health) in which they outline the range of skills of a chartered Occupational Health Physiotherapist. (Table 2)

**Table 2.** Skills of an Occupational Health Physiotherapist

---

Expert in human movement
Advice to prevent further injury and helping to speed recovery
Evaluation of human task machine relationships and identification of problem areas which could cause pain
Specialised training in ergonomics and occupational health etc.

---

It is in this area of occupational health that the Physiotherapist has much to offer ergonomics. Physiotherapists are taught to be analytical about injury mechanism in order to be able to apply an appropriate treatment technique to alleviate symptoms. Therefore Physiotherapists have a specialist knowledge and the potential for insight into the scientific study of human work.

A Physiotherapists raison d'être is to restore health and well being. However there is little to be gained by successfully treating someone's back problem or Achilles tendonitis if they are then going back to the same faulty work station and gruelling lifting task, or the same pair of running shoes that led to the problem in the first place. They will continue to have recurrence of their problem until it becomes chronic. The successful resolution of work related musculoskeletal problems requires the elimination of those factors which lead to the continued over use of the bodily structures involved. Ergonomics is relevant to primary prevention and also to those branches of medicine which are concerned with the management of such conditions. Therefore ergonomics and physiotherapy can work together very successfully in order to give a complete approach to injury treatment and prevention resulting in the enhancement of performance.

## Physical Ergonomics

Bullock (1994) proposed a model (outlined in Figure 1) which aimed to optimise human performance by maximising the expertise and input from both the physiotherapy and ergonomics professions in a combination of rehabilitation and therapeutic care together with design and prevention.

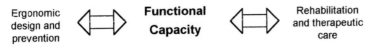

**Figure 1.** Optimisation of human physical performance (modified from Bullock, 1994)

*Functional Capacity*

This is the ability of a client to carry out his / her job. A functional capacity evaluation can be carried out in order to assess and quantify this by relating the limitations of the individual to the demands of the job.

The scope of the Occupational Health Physiotherapist in the U K has tended to be limited to an advisory role with respect to general fitness criteria for return to work after a major sickness absence (Oldham, 1988) In the USA this advisory role has expanded into almost a new profession, with specialist skills, based in the application of work physiology and psychosocial research under the title Industrial Therapy. Industrial therapy is a system which encompasses a wide spectrum of treatment based upon the principles that a person working in industry, as anywhere else, has physical, emotional, vocational, educational, psychological and sociological needs which must be met to gain successful employment or, in the case of injury, rehabilitation and re-employment. This specialisation developed when it was noted that Therapists were moving away from the traditional role of treating injured workers in clinics and starting to provide services including injury prevention and rehabilitation as well as pursuing advanced work in related fields e.g. engineering or ergonomics providing functional capacity evaluations, work conditioning, work hardening within the practice of physical therapy, occupational therapy and certified vocational evaluators (as part of the physical or occupational therapy team).

*Work Conditioning* is the term used to describe a treatment programme which is specifically designed to restore an individual's systemic, neuromuscular and cardiopulmonary functions. It is used for clients who are more than six weeks post-injury and aims to restore physical capacity and function for return-to-work (Hart et al, 1994).

*Work Hardening* is the term used to describe a treatment programme designed to return the person to work. It uses real or simulated work activities and is suitable for clients who are more than twelve weeks post-injury and are physically deconditioned but medically stable (Hart et al, 1994).

*Ergonomic Design and Prevention*

Historically ergonomics has aimed at establishing working capacities of healthy individuals, and enhancing performance by using this data when designing tasks and work places to fit the users or workers. This contributes to functional capacity with a permanent engineering solution to work place problems.

*Rehabilitation and therapeutic care*

The role of physical ergonomics in physiotherapy practice allows a rounded and holistic view of patient care. Physiotherapists with skills and knowledge of ergonomics can not only address the direct therapeutic needs of their patients, but also contribute to the ergonomic design aspects of primary prevention. The following are examples of physiotherapists' involvement in physical ergonomics and illustrate the overlap between physiotherapy and ergonomics.

The first example is the advice that Physiotherapists give to their patients when they are returning to a job or leisure activity following treatment. Unfortunately patients symptoms are all too often addressed with little or no regard to their aetiology, despite the fact that epidemiological data suggests that a very high proportion of musculoskeletal problems are work-related. Often the patients, and therefore the physiotherapists' highest priority is to return to work as quickly as possible. This will frequently involve the alleviation of symptoms only to send the patient back to the very environment and activity that caused the problem in the first place. Primary prevention is obviously the key to breaking this cycle of recurrence, and this is where the physiotherapist may get involved in work place and work station assessments, using ergonomic techniques to recommend improvements to eliminate or minimise the risk of injury. In order to carry out ergonomic assessments there is an onus on the physiotherapist to ensure that they are adequately qualified, both in qualifications and experience, to be able to use and interpret the ergonomic tools. There is a need to recognise personal limitations and ask for expert support when required. Gaining access to a patients work place is in reality unlikely in most out-patient settings, but there is still much that a physiotherapist with the necessary knowledge and skills can do by questioning the patient, assessing the situation and contributory factors. A reasonable appraisal can lead to constructive, practical advice as to which of the factors are under the patients control and can be modified or eliminated.

The second example is the role of the Physiotherapist as an in-house manual handling trainer. There has been a tendency, particularly within the Health Care Industry, to use Physiotherapists as a pool of in-house experts but without always providing the appropriate ergonomics training to enable them to deliver the service required. Frost and McCay (1990) described a cascade system which was used in an NHS Hospital Trust to provide lifting and handling training, which was quoted as including ergonomics, biomechanics, teaching methods and attitudes. They used the hospital Physiotherapists as part of 'The Prevention of Injury Group' and found that one problem was the lack of formal ergonomics training within the group (although some members had limited knowledge). Buckle and Stubbs (1989) have also commented on the tendency within the Health Care Industry to use in-house Physiotherapists, they suggest that the ergonomic input is often left to professional groups that have not received formal training in this subject. In 1995 Crumpton carried out an extensive questionnaire survey of Back Care Advisors in the UK. She found that of the Physiotherapists employed in this role, 57% claimed to have had some training in ergonomics, but for 75% of these the training was limited to 1-3 day courses. She concluded that a greater awareness of the scope of ergonomics practice was needed in order to ensure that personal limitations were recognised.

This trend is of obvious concern to all involved. The vital contribution that ergonomics has to make to this important and growing area could be generally disregarded as ergonomic principles are misused by inexperienced staff and ergonomics is not seen to be doing what it claims to be able to. Also, Physiotherapists are being put into roles which are beyond their capabilities and they are expected to solve complex problems with no extra training by virtue of the fact that they have studied anatomy and biomechanics. This is not good for the physiotherapy profession as the inevitable result is that the problems are not solved efficiently.

The third example is the risk assessment of physiotherapy practice. There have been numerous articles published which have challenged the research basis for many of the techniques used by Physiotherapists. With health and safety legislation in the European Community now focusing on the risk of injury from manual handling activities, the justification for many of the physiotherapy techniques is again being called into question when the safety and well-being of Physiotherapists may be compromised. Ergonomics may have a role to play, using assessment techniques to evaluate the risk to the physiotherapy population (Hignett, 1995) and in the redesign of the work place and working environment (Fenety and Kumar, 1992).

## Conclusion

Foster (1988) suggests that the professions of physiotherapy and ergonomics are divergent and convergent. Divergent in that the scope of ergonomics covers all aspects of peoples' interaction with their environment (physical, cognitive and organisational), whereas physiotherapy is concerned only with the physical well-being of an individual. Convergent in that both aim to optimise human performance and minimise any mis-match in the task requirements and physical ability. However physiotherapy achieves this aim by altering the person (e.g. work conditioning or musculoskeletal treatment), and in contrast ergonomics alters the task to bring it to an appropriate level for the person (including designing for disabled populations). Stubbs (private communication, 1996) concurs with the idea that physiotherapy has much to offer with respect to physical ergonomics, but comments that Physiotherapists tend to focus more on characteristics, capabilities and capacities of the individual rather than of populations. Hence they have what could be considered to be a more narrow perspective. Ergonomics knowledge could serve to broaden the outlook of the Physiotherapist, and ensure a more holistic approach to the patients problem by taking action on the environment as well as the individual

Physiotherapists are experts on the human musculoskeletal system, including work physiology, the effects of loads and forces and the prevention as well as the treatment of musculoskeletal disorders. McAtamney (1991) suggests that they are well qualified to provide advice on aspects of physical ergonomics. McPhee (1984) believes that many physiotherapy skills are wasted if only treatment services are provided, and that worker rehabilitation should form a large part of the treatment. She states that if Physiotherapists are to play a part in occupational health and ergonomics and develop the relevant skills they must first identify the areas in which their expertise can be used most appropriately. This type of role has already been seen in other countries, in particular in Scandinavia where ergonomics is a profession dominated by Physiotherapists (Bullock. 1990). The preventive role includes job analysis, work posture monitoring, task design, personnel selection and placement, education, supervision of work methods, influencing motivation and attitudes, provision of activity breaks and physical fitness programmes. In order to provide this range of occupational health and ergonomic services Physiotherapists need to obtain post graduate qualifications in occupational health and ergonomics and maximise their unique knowledge of the human musculoskeletal system.

There is a huge potential benefit from a closer relationship between Ergonomics and Physiotherapy, but in order for this to be achieved there must be mutual recognition and respect for the specialist skills of each profession.

# References

Buckle, P. and Stubbs, D. 1989, The Contribution of Ergonomics to the Rehabilitation of Back Pain *Journal of the Society of Occupational Medicine* **39** 56-60

Bullock, M. I. 1990, *Ergonomics: The Physiotherapist in the workplace.* (Churchill Livingstone)

Bullock, M. 1994, Research to Optimise Human Performance *Australian Journal of Physiotherapy* 40th Jubilee issue 5-17

Crumpton, E. J. 1995, *An Investigation into the Role of the Back Care Advisor.* Unpublished M.Sc. Dissertation, University of London

Fenety, A.,and Kumar, S. 1992, An Ergonomic Survey of a Hospital Physical Therapy Department *International Journal of Industrial Ergonomics* **9** 161-170

Foster, M. 1988, Ergonomics and the Physiotherapist *Physiotherapy* **74** 9 484-489

Frost, H, and McCay, G. 1990, Prevention of Injury Group Initiative *Physiotherapy* **76** 12 796-98

Hart, D., Berlin, S., Brager, P., Caruso, M., Hejduk, J., Hoular, J., Snyder, K., Susi, J., Wah, M. D. 1994, Development of Clinical Standards in Industrial Rehabilitation *JOSPT* **19** 5 232-241

Hignett, S. 1995, Fitting the Work to the Physiotherapist *Physiotherapy* **81** 9 549-552

Jacobs, K, and Bettencourt, C. M. (eds) 1995, *Ergonomics for Therapists* (Butterworth-Heinemann)

McAtamney, L. 1991, Physiotherapy in Industry in Lovesey E J (ed) *Contemporary Ergonomics* (Taylor and Francis)

McPhee, B. 1984, Training and possible future trends in Occupational Physiotherapy *New Zealand Journal of Physiotherapy* Dec 12-14

Oborne, D. J. 1995, *Ergonomics at Work* 3rd Ed (John Wiley & Sons)

Oldham, G. 1988, The Occupational Health Physiotherapist's role in assessing fitness for work *Physiotherapy* **74** 9 422-425

Sanders, M. S. and McCormick, E.J. 1993, *Human Factors in Engineering and Industry* 7th Ed. (McGraw-Hill Inc.)

# 1998

| **Royal Agricultural College Cirencester** | 1st–3rd April |
| --- | --- |
| **Conference Manager** | Janis Hayward |
| **Chair of Meetings** | Ted Lovesey |
| **Programme Secretary** | Margaret Hanson |

| Chapter | Title | Author |
| --- | --- | --- |
| **Stephen Pheasant Memorial Session** | The combined effects of physical and psychosocial work factors | J. Devereux |
| **Work Stress** | A risk assessment and control cycle approach to managing workplace stress | R.J. Lancaster |
| **Design and Usability** | Pleasure and product semantics | P.W. Jordan and A.S. Macdonald |
| **Drivers and Driving** | The use of automatic speech recognition in cars: A human factors review | R. Graham |

# THE COMBINED EFFECTS OF PHYSICAL AND PSYCHOSOCIAL WORK FACTORS

## Jason Devereux

*Research Fellow*
*Robens Centre for Health Ergonomics*
*EIHMS, University of Surrey*
*Guildford, Surrey, GU2 5XH*

Physical and psychosocial work factors have been implicated in the complex aetiology of musculoskeletal disorders. Psychosocial work factors differ from individual psychological attributes in that they are individual subjective perceptions of the organisation of work. An ergonomic epidemiological study was undertaken to determine the impact of different combinations of physical and psychosocial work risk factors upon the risk of musculoskeletal disorders. Physical work factors are more important determinants of recurrent back and hand/wrist problems than psychosocial work factors. The greatest risk of musculoskeletal problems occurs when exposed to both physical and psychosocial work risk factors. Ergonomic strategies should, therefore, aim to reduce physical and psychosocial risk factors in the workplace.

## Introduction

" We are fiercely competitive in our consumption of goods and services; and our sense of self-worth is tied up in our use of status symbols. This lies at the root of our stress levels."

(Pheasant, 1991)

Stephen Pheasant realised that humanity has created a 'milieu' of self-imposed stress by increasing the demand for goods and for services. Satisfying the demand has been formalised into work organisation goals, culture and beliefs but at what cost to the individual worker ironically from which the demand has originated. The work organisation imposes physical and psychosocial stressors upon the individual. The physical stressors originate from environmental, manual handling and other physical demands and the psychosocial stressors originate from the perceptions of the organisation and the way the work is organised. Models have been proposed that describe the probable pathways by which these work factors can impose a threat such that symptoms, signs and diagnosable pathologies of musculoskeletal disorders can ensue (Bongers *et al*, 1993; Sauter and Swanson, 1996; Devereux, 1997a). The model by Devereux (1997a) proposed that the individual perceptions of the work organisation e.g. the social support, the control afforded by the work and the demands imposed may be influenced by the capacity to cope with such psychosocial stressors. Individual capacity may be affected by a number of factors including previous injury, cumulative exposure to work risk factors, age, recovery, and beliefs and attitudes towards pain. Beliefs, attitudes and coping skills have collectively been referred to as psychosocial factors by some (Burton, 1997), but to minimise confusion, they are referred to as individual psychological attributes in this text.

The relationship between risk factors and musculoskeletal problems may be dependent on the definition of the latter (Leboeuf-Yde *et al*, 1997) and also on the interrelationship between risk factors (Evans *et al*, 1994). For example, many studies have simply considered the effects of either physical or psychosocial work factors upon the risk of musculoskeletal disorders. Some studies have considered both sets of factors but have assumed the relationships between them to be independent when in reality such factors mutually exist. The effect that the mutual existence of these risk factors has upon the risk of musculoskeletal disorders has not been adequately investigated (Devereux, 1997a, Devereux, 1997b). An ergonomic epidemiological investigation was conducted to examine the effects upon the musculoskeletal system of physical and psychosocial work factors acting in different combinations in a work organisation which employed workers engaged in manual handling, driving and sedentary office work. The ethical permission for the cross-sectional study was obtained from the University of Surrey Committee on Ethics.

## Methods

Company work sites from around the U.K were randomly selected to participate in the study. The mixed gender study population (N=1514) was given a self-report questionnaire that included information on personal data and demographics, physical and psychosocial work factors and musculoskeletal symptoms. Questions on physical and psychosocial work factors had been validated elsewhere (Wiktorin *et al*, 1993; Hurrell and McLaney, 1988). Most of the physical scales had a kappa coefficient greater than or equal to 0.4 (except for bent-over posture and trunk rotation) and all the psychosocial scales had acceptable alpha coefficients (0.65-0.95). The musculoskeletal symptom questionnaire included the head/neck, trunk and upper and lower limbs. Items for the lower back had been validated against a physical examination using a symptom classification scheme proposed by Nachemson and Andersson (1982). The kappa values for the 7-day and 12-month prevalence were 0.65 and 0.69 respectively (Hildebrandt *et al*, 1998).

Physical and psychosocial work factors that had been shown in previous epidemiological studies to increase the risk of back disorders by a factor of 2 or greater were selected to classify individual workers into one of four physical/psychosocial exposure groups. For physical exposure the criteria consisted of heavy frequent lifting or relatively lighter but frequent lifting performed as well as driving. The physical exposures were quantified with respect to a level or amplitude, a frequency or duration. For psychosocial work factors, mental demands, job control and supervisor and co-worker social support were used to classify workers into low and high exposure groups. Subjects not satisfying the low/high physical and psychosocial criteria were excluded from the analysis. Recurrent back disorder cases were defined as having experienced problems more than 3 times or longer than one week in the previous year and were also present within the last 7 days at the time of the survey. These back problems were not experienced before starting the present job. Univariate Mantel-Haenzel chi-squared statistics were used to test the hypothesis of no association between the exposure criteria variables and recurrent back disorders. Crude and logistical regression analyses provided an estimate of the risks associated with exposure to different physical/psychosocial work factor combinations. The potential confounding/modifier effects of age, gender and cumulative exposure (defined as the number of years spent in the present job) were controlled for in the logistical regression.

## Results

There were 869 valid responses from the survey (57%) of the total study population (N=1514). Non-respondents did not differ with respect to gender, age or cumulative exposure. Recurrent back disorders were prevalent in 22% of the valid number of survey responses. Of the 869 valid questionnaire responses, 638 workers were classified into low/high physical and psychosocial exposure groups. The gender, age and cumulative exposure for the exposure stratified population and the valid questionnaire population did not differ. The univariate analysis for recurrent back problems showed that heavy frequent lifting > 16 kg ≥ 1-10 times per hour (p<0.001) and relatively lighter but frequent lifting > 6-15 kg ≥ 1-10 times per hour performed as well as driving ≥ half the working day (p<0.001) were associated with recurrent back disorders. Forward bent-over postures > 60 degrees for greater than a quarter of the working day (p<0.05) was also found to be significantly associated with recurrent back problems. Trunk rotation of 45 degrees greater than a quarter of the working day increased the risk of experiencing recurrent back disorders but was not statistically significant at the 5% level. A perceived high workload was associated with recurrent back problems (p<0.05). Mental demands, job control, supervisor and co-worker social support factors were not found to be statistically associated with recurrent back disorders when considered independently.

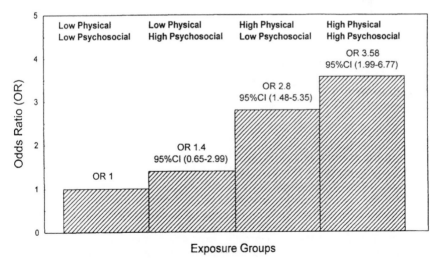

Figure 1. The risk of recurrent back problems for different exposure groups

Figure 1 shows the combined risk effects of physical and psychosocial work risk factors associated with recurrent back disorders. High exposure to physical work risk factors had a greater impact upon the risk of recurrent back problems compared to psychosocial work risk factors. For workers highly exposed to physical work risk factors and with relatively low exposure to psychosocial work risk factors, the approximate risk of experiencing recurrent back problems was approximately 3 times greater than for workers exposed by a lesser extent to both physical and psychosocial work risk factors. The risk increased approximately 3.5 times for workers highly exposed to both physical and psychosocial work risk factors compared to those exposed by a lesser extent to both sets of factors. A similar exposure-risk relationship was observed for self-reported symptoms in the hands/wrists experienced both within the last 7 days and the last 12 months at the time of the survey using the same exposure criteria. The risk to the hands/wrists due to high exposure to physical and psychosocial work risk factors was approximately seven times greater than being exposed to these risk factors by a lesser extent (OR 6.94 95%CI 3.79-12.82). An exposure-risk relationship was not observed for the same definition in the neck region. After controlling for the effects of age, gender and cumulative exposure, a similar exposure-risk relationship was observed for each exposure group and recurrent back disorders except for the low physical-high psychosocial exposure group. The risk associated with this group was equal to unity.

## Discussion

Exposure to a combination of psychosocial work risk factors seems to have a greater impact on the level of risk compared to considering individual psychosocial work factors (Bongers and Houtman, 1995). The greatest risks of experiencing musculoskeletal disorders was derived from high exposure to both physical and psychosocial work risk factors but physical work risk factors were more important determinants than psychosocial work risk factors for recurrent back and hand/wrist disorders. A Swedish epidemiological study also showed that a combination of heavy lifting and a poor psychosocial work environment increased the risk of back pain and neck pain compared to being exposed to neither work factors (Linton, 1990). However, the study design did not permit the analysis of the other possible exposure combinations. In this study, associations with neck problems were not observed. Workers in the low exposed groups performed tasks that were associated with neck and shoulder disorders so they were not truly unexposed with respect to this anatomical region. As a result, it could not be determined whether the combination of high exposure to physical and psychosocial work risk factors increased the risk of neck disorders.

A cross-sectional study does not allow exposures to be measured before the onset of musculoskeletal problems and so reporting biases may have been present for exposures and self-reported symptoms. The influence of these biases was controlled by assessing current exposures and assessing recently experienced symptoms. Self-reported exposures were also tested using observation, instrumentation and interview methods and it was found that workers could provide accurate reports for the exposure criteria (Devereux, 1997a). The cross-sectional study was also limited at examining exposure-disease causation, but temporal data on the outcome measure provided strong evidence that the exposures currently experienced were associated with the development of back problems. The combination risk effects were not limited to the back and have also been shown to be present for the hands/wrists.

# Conclusions

The relationship between physical and psychosocial work risk factors is complex and is not fully understood but reduction in exposure to both sets of factors is needed in risk prevention strategies for musculoskeletal disorders. Ergonomic interventions should be targeted at the organisation of work and the individual worker to reduce the psychosocial work stressors and also the physical stressors. The consumption of goods and services remains unabated and will be driven to higher levels in the years to come. Work organisations should strive to achieve a balance between satisfying the demand and maintaining a healthier workforce.

# References

Bongers, P. M., de Winter, C. R., Kompier, M. A. J., & Hildebrandt, V. H. 1993, Psychosocial factors at work and musculoskeletal disease, *Scandinavian Journal of Work Environment and Health*, **19**, 297-312

Bongers, P.M. and Houtman, I.L.D. 1995, Psychosocial aspects of musculoskeletal disorders. Book of Abstracts, Proceedings of the Prevention of Musculoskeletal Disorders Conference (PREMUS 95), 24-28 September, Montreal, Canada, (IRSST, Canada), 25-29

Burton, A. K. 1997, Spine update - Back injury and work loss: Biomechanical and psychosocial influences, *Spine*, **22**, 2575-2580

Devereux J. J. 1997a, A study of interactions between work risk factors and work related musculoskeletal disorders, Ph.D. Thesis. University of Surrey

Devereux, J. J. 1997b, Back disorders and manual handling work - The meaning of causation, *The Column*, **9**, 14-15

Evans, G. W., Johansson, G., & Carrere, S. 1994, Psychosocial factors and the physical environment: Inter-relations in the workplace. In C.L. Cooper & I.T. Robertson (eds.), *International review of industrial and organizational psychology*, (John Wiley, Chichester), 1-30

Hildebrandt, V. H., Bongers, P. M., Dul, J., Van Dijk, F. J. H., & Kemper, H. C. G. 1998, Validity of self-reported musculoskeletal symptoms, *Occupational and Environmental Medicine*, **In press**

Hurrell, J. & McLaney, M. 1988, Exposure to job stress-a new psychometric instrument, *Scandinavian Journal of Work Environment and Health*, **14 Supplement 1**, 27-28

Leboeuf-Yde, C., Lauritsen, J. M., & Lauritzen, T. 1997, Why has the search for causes of low back pain largely been nonconclusive?, *Spine*, **22**, 877-881

Linton, S. J. 1990, Risk factors for neck and back pain in a working population in Sweden, *Work and Stress*, **4**, 41-49

Nachemson, A. L. & Andersson, G. B. J. 1982, Classification of low-back pain, *Scandinavian Journal of Work Environment and Health*, **8**, 134-136

Pheasant, S. 1991, *Ergonomics, Work and Health* (Macmillan Press, London)

Sauter, S. L. & Swanson, N. G. (1996). An ecological model of musculoskeletal disorders in office work. In S.D. Moon & S.L. Sauter (eds.), *Beyond Biomechanics: Psychosocial Aspects of Musculoskeletal Disorders in Office Work* (Taylor and Francis, London), 1-22

Wiktorin, C., Karlqvist, L., & Winkel, J. 1993, Validity of self-reported exposures to work postures and manual materials handling, *Scandinavian Journal of Work Environment and Health*, **19**, 208-214

# A RISK ASSESSMENT AND CONTROL CYCLE APPROACH TO MANAGING WORKPLACE STRESS

Rebecca J Lancaster

*Institute of Occupational Medicine*
*8 Roxburgh Place*
*Edinburgh, EH8 9SU*

A Health and Safety Executive (HSE) publication proposed that the assessment and control cycle approach, already applied to physical health and safety risks, be adopted to manage stress at work. The Institute of Occupational Medicine (IOM) has developed an Organisational Stress Health Audit (OSHA) using this approach. The feasibility of this was tested in a study commissioned by the Health Education Board for Scotland (HEBS). The OSHA is a three tiered approach. Stage One involves the identification of sources of stress. Stage Two investigates areas of major concern and generates recommendations for risk reduction. Stage Three evaluates the effectiveness of the recommendations made in reducing risk. This paper presents the background to this organisational approach and a discussion of its feasibility in managing workplace stress in an NHS Trust.

## Introduction

Stress is a real problem in the workplace, often resulting in high sickness absence and staff turnover coupled with low morale and performance. Various intervention strategies have been suggested to combat the detrimental effects of workplace stress. Murphy (1988) emphasised the following three levels of intervention which have since been widely accepted: (1) Primary or organisational stressor reduction, (2) Secondary or stress management training, and (3) Tertiary, encompassing counselling and employee assistance programmes (EAPs). Whilst there is considerable activity at the secondary and tertiary levels, primary/organisational reduction strategies are comparatively rare (Murphy, 1988 and DeFrank & Cooper, 1987). An HSE publication (HSE, 1995), providing guidelines for employers on how to manage workplace stress, has recognised identifying and controlling causes of stress *at source* as the most appropriate. The HSE recommend the assessment and control cycle approach for managing physical hazards in the workplace e.g. Control of Substances Hazardous to Health (COSHH) and suggest that this same approach be adopted in controlling psychological stressors. The Institute of Occupational Medicine (IOM) has developed an Organisational Stress Health Audit (OSHA) for the identification and control of work-related stress. The feasibility of this was tested in a research study commissioned by the Health Education Board for Scotland. (HEBS). This study applied the approach in three organisations: heavy industry; telecommunications; and an NHS Trust. This paper describes its application in one of these, namely the NHS Trust.

The OSHA is a three tiered approach, covering hazard identification, risk assessment, review of existing control measures, recommendations for improved control and evaluation of control. Stage One provides an organisational overview by identifying the presence or absence of work related stressors and opportunities for risk reduction, many of which can be implemented by the organisation without further external input. Stage Two focuses on investigating, in more detail, areas of particular concern identified in Stage One. Stage Three involves assessing the extent to which recommendations in Stages One and Two have been implemented and their effectiveness in reducing organisational stress.

A database of known causes of work-related stress was compiled from the scientific literature and this formed the background to the OSHA. Over recent years, numerous researchers have carried out extensive studies designed to validate or indeed refute the existence of work characteristics which impact upon employees' mental health (eg Cooper & Marshall 1976, Karasek & Theorell 1990, Warr 1992, Cox 1993, and Nilsson 1993). In general there is a high level of consensus concerning those psychosocial hazards of work which are considered to be stressful or potentially harmful (Cox, 1993). However, although aspects relating to non-work issues, such as home/work interface, are acknowledged to some degree, there is a general lack of information relating to other factors which have been clearly shown to have a significant impact on mental health, eg physical hazards, industry specific pressures and company policies. The IOM researchers developed their approach based on all four components ie Environmental, Physical, Mental and Social, rather than just the work content/context divide. By addressing these four components the total work sphere i.e. all possible work-related stressors are investigated. In addition, by placing work-related stress within this acknowledged health and safety framework, there is more likelihood of stress being accepted and treated in conjunction with the other types of work-related hazards.

Development of the stressor database involved ascertaining traceability of the various 'stressors' to be included in the IOM approach. In terms of both Environmental and Physical components the majority of health related issues are very definitely enveloped within legislation such as the Control of Substances Hazardous to Health Regulations (1988) and the Management of Health and Safety at Work Regulations (1992). A checklist was collated using the current legislative documents on all work-related physical and environmental hazards. Knowledge concerning the latter two areas, ie Mental and Social, however is not detailed in such a way and, as such, a review of relevant research material was carried out to conclude the full range of work-related factors which should be addressed in a comprehensive organisational stress audit.

The OSHA is centred around semi-structured interviews tailored to the specific needs of the organisation under investigation. Representatives of all levels and functions within the organisation are interviewed. The line of questioning follows those known causes of work-related stress in the database. The interviews are constructed from a database of questions relating to the following areas: Health & Safety, Organisational Structure, Communication, Management/supervisory skills, Training, Staff Support Facilities, Policy, Sickness Absence, Contracts/Terms of Employment, Changes/Incidents, Work Characteristics and 'General'. The interviews themselves are undertaken by Occupational Psychologists due both to the nature of the study and the need to interpret and analyse the interviews as the sessions progress.

## Method

### Preliminary information

In applying the OSHA within the NHS Trust, a certain amount of background information was requested, including: issues pertaining to the economic and competitive climate surrounding the organisation; organisational structure; and data on trends in possible indicators of stress-related problems (sickness absence, staff turnover). This information was then used to determine a profile of the organisation and presented to the IOM internal stress team, which comprises business related staff including those experienced in finance and personnel management as well as appropriate scientific staff such as psychologists and occupational physicians. Issues were raised through this presentation which helped in constructing the semi-structured interviews.

### Stage One

A Steering Group was formed by the Trust to identify the representatives for interview and to discuss how the work would be communicated throughout the Trust. All directorates were to be included and guidance was provided on the roles and functions required for interview. Subsequently, directorate representatives on the Steering Group identified appropriate persons for interview. All selected interviewees were contacted by IOM auditors regarding possible participation and provided with background material on the study. When the final list of participants was collected, the structured interviews were tailored according to the interviewee's function. For example, it would be inappropriate to ask a Managing Director about specific work tasks and an Employee Representative about strategic decisions.

Results of the interviews were then disseminated to the IOM internal team, presenting the sources of stress identified and possible opportunities for risk reduction. All issues were considered with regard to the potential impact both on employee health and on the organisation. A report was then presented to the organisation from which some recommendations have been implemented without further IOM involvement. The report also contained recommendations for Stage Two, proposing detailed investigations of the major concerns.

### Stage Two

A number of recommendations were made for further investigation and, from these, it was agreed that the role of the Charge Nurse should be looked at, in particular their conflicting roles as ward manager and provider of patient care. The investigation involved interviews with focus groups and a sample of charge nurses were asked to complete a number of published scales and a tailored questionnaire. These Stage Two investigations identified training, information and support needs of charge nurses aimed at reducing causes of stress, promoting health and well-being, and optimising performance.

### Stage Three

Due to the time constraints of the project, evaluation of the process within the Trust was limited to a review of Stage One via feedback questionnaires administered to interviewees and the Steering Group Leader. The interviewee questionnaire asked participants whether or not the issues that were important to them had been tackled and how they felt about the interview process. The questionnaire administered to the Steering Group Leader asked more about the organisation's perspective of the process and the outcomes of the audit.

# Results

## Stage One

Stage One was successful in identifying a number of sources of stress, for example: pace of change; uncertainty about the future; and work overload, which is perhaps not surprising given recent changes in the NHS at large. The OSHA also successfully identified a number of sources of stress at a local level including; poor communication among certain groups of staff and poor relations between different professions.

## Stage Two

The Stage Two investigations confirmed many of the findings of Stage One, despite the fact that different individuals were interviewed. The following stressors were identified in both stages: poor communication; lack of feedback and formal appraisal system; lack of clarity of Business Manager role; and lack of support from Occupational Health. In addition the following stressors were identified in this detailed investigation: lack of allocated time to manage; duplication of effort in administrative tasks; poor management of change; and lack of accountability. Recommendation were made to reduce the risk associated with the stressors identified, an example of which is illustrated below:

> *Stressor:* lack of allocated time to manage
> *Recommendation:* There is a need for experienced staff at ward management level, who understand operational issues and yet have sufficient influence within the hierarchy to influence strategic development. Consideration should be given to a supernumerary Charge Nurse role having responsibility for a number of wards, and clinical management devolved to Charge Nurses.

## Stage Three

An average of 94% of interviewees reported that their interview addressed relevant causes of stress in the organisation.

# Discussion

The application of the OSHA in the Trust is part of an ongoing programme of work to tackle workplace stress. There are constant changes in the NHS and it may be argued that the individual Trust is limited in terms of what it can do. This study has demonstrated that there are a number of possibilities for reducing workplace stress at this local level.

The approach adopts the risk-assessment, control-cycle approach in terms of hazard identification; risk assessment; review of existing control measures; recommendations for improving control; and evaluation of controls. The approach has proved to be effective in meeting these steps. Although the Trust showed a willingness to implement the recommendations, due to the limited timescale of the study, evaluation of their impact on reducing stress at source has not been carried out. It is hoped, as part of this ongoing program of work, that the impact of the changes will be evaluated. This evaluation is intended to include a review of the impact of changes on; sickness absence, job satisfaction, and staff turnover throughout the Trust, as well as reviewing the impact on the role of Charge Nurses specifically by re-administering the standard questionnaires.

It was possible to ensure an organisational, cross-functional approach whereby all levels and functions within the Trust were represented in interview. There is the possibility of selection bias as the directorate representatives identified people for interview. However, in many instances, the selection was determined by job title rather than specific individual. It is recommended, in future applications of the approach, that an organisational chart be supplied to the external auditors (IOM Team) complete with job titles and names of post holders, so that they can select the participants in order to eliminate any selection bias.

The success of the approach is due, in part, to its flexibility in meeting the specific needs of the organisation. This is achieved through the development of the company profile to allow tailoring of the semi-structured interview, coupled with in-depth information from representatives of the organisation. These are the main advantages that the approach has over existing audit tools which administer a 'standard' questionnaire to all employees.

Organisations commented on the minimal disruption caused during administration of the approach. The commitment and enthusiasm of the company contact is crucial to the success of the approach and this person should be selected with great care.

## Conclusions

The study reported here has demonstrated the feasibility of addressing stress in the same manner as physical hazards in the workplace and adopting a risk assessment-hazard control approach to reducing stress at source. Using appropriately skilled staff, backed by others to advise on the interpretation and evaluation of findings, it has been possible to identify sources of occupational stress and to indicate avenues for risk reduction. These recommendations have been recognised as practicable by the Trust and some have already been acted upon. The timescales of the work precluded the inclusion of an evaluative phase to determine the success of the outcome in terms of reduced stress at work. However, a full evaluation is envisaged as part of this ongoing programme to tackle stress within the Trust.

## References

HSC 1988, *Control of Substances Hazardous to Health Regulations* (HMSO, London)

Cooper, C.L., Marshall, J. 1976, Occupational sources of stress: a review of the literature relating to coronary heart disease and mental ill health, *Journal of Occupational Psychology*, 49, 11-28

Cox, T. 1993, Stress research and stress management: Putting theory to work. *Health & Safety Executive contract research report No.61/1993*, (HMSO, London)

DeFrank, R.S., Cooper, C.L. 1990, Worksite stress management interventions: Their effectiveness and conceptualisation, *Journal of Managerial Psychology*, 2, 4-10

Karasek, R.A., Theorell, T. *1990, Healthy Work: Stress, Productivity and the Reconstruction of Working Life*, (Basic Books, New York)

HSC 1992, *Management of Health and Safety at Work Regulations*, (HMSO, London)

Murphy, L.R. 1988, Workplace interventions for stress reduction and prevention. In C.L. Cooper, R. Payne (Eds) *Causes, Coping and Consequences of Stress at Work 1988*, (Wiley, Chichester)

HSE 1995, *Stress at Work: A guide for employers*, (HSE Books, Sudbury)

Nilsson, C. 1993, New strategies for the prevention of stress at work, *European Conference on Stress at Work - A call for action: Brussels Nov 1993, Proceedings*. (European Foundation for the Improvement of Living and Working Conditions, Dublin)

Warr, P.B. 1993, Job features and excessive stress. In R. Jenkins, N. Coney. (Eds) *Prevention of Mental Ill Health at Work*, (HMSO, London)

# PLEASURE AND PRODUCT SEMANTICS

## Patrick W. Jordan

*Senior Human Factors Specialist, Philips Design, Building W, Damsterdiep 267, P.O. Box 225, 9700 AE Groningen, The Netherlands*

## Alastair S. Macdonald

*Course Leader, Product Design Engineering, Glasgow School of Art, 167 Renfrew Street, Glasgow G3 6RQ, Scotland*

Human factors has tended to focus on pain. As a profession, it has been very successful in contributing to the creation of products that are safe and usable and which, thus, spare the user physical, cognitive and emotional discomfort. However, little attention seems to have been paid to the positive emotional and hedonic benefits — pleasures — that products can bring to their users.

This paper examines the relationship between product semantics and pleasure in use, within the structure of the 'Four Pleasure Framework'. Studies such as this represent human factors' first steps towards establishing links between product properties and the types of emotional, hedonic and practical benefits that products can bring to their users.

## Introduction

*"I can sympathise with other people's pains, but not with their pleasures. There is something curiously boring about someone else's happiness."* This quote comes from Aldous Huxley's 1920 Novel, 'Limbo' however, it could almost be a motto for ergonomics. Ergonomics journals, conference proceedings, and textbooks seethe with studies of pain: back pain, upper limb pain, neck pain, pain from using keyboards, pain from using industrial machinery, pain from hot surfaces — these are just a few examples from studies that have been presented at recent Ergonomics Society Conferences.

As a discipline, ergonomics has been focused on eliminating pain, whether it be in the form of physical pain, as in the examples above, or the cognitive/emotional discomfort that can come from interacting with products that are difficult to use. Meanwhile, the idea that products could actually bring positive benefits — pleasures — to users, seems to have been largely ignored. So, whilst ergonomics has had a great deal to offer in terms of assuring product usability and safety, it seems to have had very little to contribute in terms of creating products that are positively pleasurable.

The case for ergonomists to take the lead in addressing the issue of pleasure with products has been made elsewhere (Jordan 1997a) and a framework for approaching the issue — the four pleasures — has been proposed (Jordan 1997b). In this paper, this framework will be summarised and illustrated with examples that show the relationship between pleasure and product semantics.

## Pleasure with Products

Pleasure with products is defined as: "... the emotional, hedonic and practical benefits associated with products." (Jordan 1997a)

## The Four Pleasures

The four pleasure framework was originally espoused by Canadian anthropologist Lionel Tiger (Tiger 1992) and subsequently adapted for use in design (Jordan 1997b). The framework models four conceptually distinct types of pleasure — physio, socio, psycho and ideo. Summary descriptions of each are given below with examples to demonstrate how each of these components might be relevant in the context of products.

### *Physio-Pleasure*

This is to do with the body — pleasures derived from the sensory organs. They include pleasures connected with touch, taste and smell as well as feelings of sexual and sensual pleasure. In the context of products physio-p would cover, for example, tactile and olfactory properties. Tactile pleasures concern holding and touching a product during interaction. This might be relevant to, for example, the feel of a TV remote control in the hand, or the feel of an electric shaver against the skin. Olfactory pleasures concern the smell of the new product. For example, the smell inside a new car may be a factor that effects how pleasurable it is for the owner.

### *Socio-Pleasure*

This is the enjoyment derived from the company of others. For example, having a conversation or being part of a crowd at a public event. Products can facilitate social interaction in a number of ways. For example, a coffee maker provides a service which can act as a focal point for a little social gathering — a 'coffee morning'. Part of the pleasure of hosting a coffee morning may come from the efficient provision of well brewed coffee to the guests. Other products may facilitate social interaction by being talking points in themselves. For example a special piece of jewellery may attract comment, as may an interesting household product, such as an unusually styled TV set. Association with other types of products may indicate belonging in a social group — Porsches for 'Yuppies', Dr. Marten's boots for skinheads. Here, the person's relationship with the product forms part of their social identity.

### *Psycho-Pleasure*

Tiger defines this type of pleasure as that which is gained from accomplishing a task. It is the type of pleasure that traditional usability approaches are perhaps best suited to addressing. In the context of products, psycho-p relates to the extent to which a product can help in accomplishing a task and make the accomplishment of that task a satisfying and pleasurable experience. For example, it might be expected that a word processor which facilitated quick and easy accomplishment of, say, formatting tasks would provide a higher level of psycho-pleasure than one with which the user was likely to make many errors.

### *Ideo-Pleasure*

Ideo-pleasure refers to the pleasures derived from 'theoretical' entities such as books, music and art. In the context of products it would relate to, for example, the aesthetics of a product and the values that a product embodies. For example, a product made from bio-degradable materials might be seen as embodying the value of environmental responsibility. This, then, would be a potential source of ideo-pleasure to those who are particularly concerned about environmental issues. Ideo-pleasure would also cover the idea of products as art forms. For example, the video cassette player that someone has in the home, is not only a functional item, but something that the owner and others will see every time that they enter the room. The level of pleasure given by the VCR may, then, be highly dependent on how it affects its environment aesthetically.

## Product Semantics

Product semantics refers to the 'language' of products and the messages that they communicate (Macdonald 1997). Product language can employ metaphor, allusion, and historical and cultural references, whilst visual cues can help to explain the proper use or function of a product. What follows are a number of examples, demonstrating the link between product semantics and pleasure with products.

### Karrimor's Condor Rucsac Buckle

This side release buckle closes with a very positive 'click'. Visual, audio, and tactile feedback combine to ensure that the rucsac looks, feels and sounds good. These physio-pleasures are part of projecting the benefit of a reliable and reassuring fastening.

### Global Knives

Global knives are a new concept in knives, designed and made in Japan. The blades are made from a molybdenum/vanadium stainless steel and are ice tempered to give a razor sharp edge. The integral, hollow handles are weighted to give perfect cutting balance with minimum pressure required. The comfort of the knife in the hand and the aesthetic sensation of the finely balanced weight are both physio-pleasures. Because of their smooth contours and seamless construction, the knives allow no contours for food and germs to collect and thus are exceptionally hygienic. This provides the user with a feeling of reassurance — a psycho-pleasure.

### NovoPen™

Traditionally, those suffering from diabetes had to use clinical looking syringes and needles. The NovoPen™ is a device for the self-administration of precise amounts of insulin. Its appearance is rather like that of a pen — this provides a more positive signal than that of the hypodermic syringe, which is coloured through medical and drug abuse associations. This offers the user both ideo- and socio-pleasure, by playing down any stigma that the user and others may associate with syringes and/or the medical condition. The NovoPen™ also incorporates tactile and colour codes which refer to the different types of insulin dosage that may be required. These provide sensory back up and contribute to the product's aesthetic profile. The technicalities of administering precise dosages have been translated into easy human steps and a discrete but positive click occurs when the dose is prepared for delivery. This provides the psycho-pleasure of reassurance to the user in what might otherwise be a rather daunting task. Finally, the NovoPen™ also provides physio-pleasure through its tactile properties — the pen is shaped to fit the hand comfortably and the surface texture, achieved through spark erosion, is pleasant to the touch.

### Samsonite Epsilon Suitcase

Journeys through air terminals can be fraught with stress — both physical and psychological. The three handle options on the Samsonite Epsilon Suitcase provide a number of comfortable options for lifting, tilting or trailing. The handle material is a non-slip rubberised coating which does not become sweaty or slippy in use. These features, providing physio-p, reduce the stress associated with the situation. The design of the suitcase's coasters allow a controllable and responsive movement in the 'trailing' mode, the suitcase 'obeying' the needs of the user, providing a degree of psycho-p over other suitcases who do not obey their owners' will.

*Mazda Car Exhaust*

'Kansei Engineering' is a term coined by Nagamachi (1995) for turning emotions into product design, and has been extensively employed in automotive design. The Mazda team has engineered the sound emitted by the MX5 Miata to evoke association with classic (British?) sports cars, satisfying ideo-p (macho, youthful associations), and socio-p (I have arrived, have I not?!).

Table 1 gives a summary of the benefits associated with the products within the context of the four pleasure framework.

**Table 1. Four pleasure analysis of the benefits associated with the example products.**

| PRODUCT | PHYSIO-P | SOCIO-P | PSYCHO-P | IDEO-P |
|---|---|---|---|---|
| Karrimor's Rucsac Buckle | • positive click on closing | | | |
| Global Knives | • comfortable to hold<br>• weight balance | | • reassuringly hygienic | |
| NovoPen™ | • pleasant to hold | • plays down negative associations (from others) | • easy and reassuring to use | • plays down negative associations (from user) |
| Samsonite Epsilon Suitcase | • comfortable, lifting, tilting and trailing | | • responsive movement — suitcase 'obeys' user | |
| Mazda Car Exhaust | | • status symbol | | • positive, youthful self-image |

## Designing Pleasurable Products

The examples given have demonstrated that, through their semantics, products can provide different types of pleasure to their users. Even from this little selection, it is clear that products can bring practical, emotional and hedonic benefits to users which go beyond those associated with concepts such as ergonomic design and usability.

The four pleasure framework gives a useful structure within which to approach the issue of pleasure with products. In particular, it has proved useful at the beginning of the product creation process, as a vehicle for discussion and agreement between human factors, design, product management, marketing, engineering and market research, as to what the main benefits delivered by a product should be. These agreements lead to the unity of purpose, that is so important to creating products that deliver clear benefits and tell a clear 'story'. For example, Jordan (in preparation) describes a set of benefits that might be agreed for a new photo camera, with young professional women as the target group. They are summarised in table 2.

Having agreed that these are the benefits to be delivered, the entire product development team can then concentrate on these. Having these common aims in mind when developing the technology, design and marketing material for a product, ensures that all disciplines are working to a common goal.

## Conclusions

A number of examples have been given, illustrating links between product semantics and pleasure in use. This was based on a qualitative analysis within the context of the Four Pleasure Framework. Because of its simplicity and accessibility, this framework is a useful tool, suited to the multi-disciplinary nature of product development. It supports constructive, focused and progressive co-operation to move the design along. Such approaches enable human factors to move beyond usability to support the creation of products that are a positive pleasure to use — products that will delight the customer.

**Table 2. Four Pleasure Analysis of product requirements for a camera aimed at young women of high socio-economic status (from Jordan, in preparation).**

| PLEASURE | USER REQUIREMENTS FOR CAMERA |
|---|---|
| Physio-P | fits hand well |
| | material pleasant to touch |
| | moving parts give optimal level of resistance |
| | fits nicely into handbag or pocket |
| | doesn't jab into body when carried |
| | doesn't jab into face when taking photos |
| | 'fingernail friendly' catches |
| Socio-P | reflects users' high socio-economic status |
| | demonstrates users' good taste and success |
| | many automated functions |
| | quiet motor drives |
| Psycho-P | good quality (but not professional quality) photos |
| | 'point and shoot' operation |
| | easy to use/ guessable |
| | ergonomically designed |
| Ideo-P | post-modern organic design language |
| | feminine styling |
| | non-patronising design |
| | environmentally responsible |

## References

Jordan, P.W., 1997a, Putting the pleasure into products, *IEE Review*, **November 1997**, 249-252

Jordan, P.W., 1997b, A Vision for the future of human factors. In K. Brookhuis et al. (eds.) *Proceedings of the HFES Europe Chapter Annual Meeting 1996,* (University of Groningen Centre for Environmental and Traffic Psychology), 179-194

Jordan, P.W., in preparation, The four pleasures — human factors for body, mind and soul, submitted to *Behaviour and Information Technology*

Macdonald, A.S., 1997, Developing a qualitative sense. In N. Stanton (ed) *Human Factors in Consumer Products*, (Taylor and Francis, London), 175-191

Nagamachi, M., 1995. *The Story of Kansei Engineering,* (Kaibundo Publishing, Tokyo)

Tiger, L., 1992, *The Pursuit of Pleasure,* (Little, Brown and Company, Boston)

# THE USE OF AUTOMATIC SPEECH RECOGNITION IN CARS: A HUMAN FACTORS REVIEW

**Robert Graham**

*HUSAT Research Institute,*
*The Elms, Elms Grove,*
*Loughborough, Leics. LE11 1RG*
*tel: +44 1509 611088*
*email: r.graham@Lboro.ac.uk*

Automatic speech recognition (ASR) has been successfully incorporated into a variety of domains, but little attention has been given to in-car applications. Advantages of speech input include a transfer of loading away from the over-burdened visual-manual modality. However, the use of ASR in cars faces the barriers of high levels of noise and driver mental workload. This paper reviews some of the likely in-car applications of ASR, concluding that its widespread adoption will be driven by the requirement for hands-free operation of mobile phone and navigation functions. It then discusses some of the human factors issues which are pertinent to the use of ASR in the in-car environment, including dialogue and feedback design, and the effects of the adverse environment on the speaker and speech recogniser.

## Introduction

Automatic speech recognition (ASR) technology has been successfully incorporated into a variety of application areas, from telephony to manufacturing, from office to aerospace. However, so far, little attention has been given to in-car applications. This is perhaps surprising given that one of the major advantages of speech input over manual input is that the eyes and hands remain free. The task of safe driving could clearly benefit from a transfer of loading from the over-burdened visual-manual modality to the auditory modality. Indeed, numerous studies have confirmed the potential adverse safety impacts of operating a visual-manual system (e.g. a mobile phone or car radio) while on the move. This situation is likely to be exacerbated by the rapid growth of Intelligent Transportation Systems (ITS) such as navigation or traffic information systems, which require complex interactions while driving. As well as improving driving safety, ASR could increase the accessibility and acceptability of in-car systems by simplifying the dialogues between the user and system, and the processes of learning how to use the system.

The main difficulty facing the incorporation of speech into in-car systems comes from the hostile environment. Noise (from the vehicle engine, road friction, passengers, car radio, etc.) can adversely affect speaker and speech recognition performance. The car is also characterised by a variety of concurrent tasks to be carried out while using speech (particularly the primary task of safe driving), and varying levels of driver mental workload. As well as these and other human factors issues, Van Compernolle (1997) suggests that the automotive industry is a slow acceptor of new technologies in general, and that there has been confusion in the past about which speech applications to implement.

Despite these barriers, ASR may be useful for a number of different in-car applications, each with particular requirements (in terms of vocabulary, dialogue, etc.) These applications are outlined in the sections below. There then follows a general discussion of human factors issues relevant to the use of ASR in cars.

## Applications of ASR in Cars

Likely in-car applications for ASR can be put into 3 groups - standard vehicle functions (including the stereo), phones and navigation/ information systems. Van Compernolle (1997) rates the importance of incorporating ASR into these functions as low, high and essential respectively.

### Standard Vehicle Functions

Any non-safety-critical vehicle control may benefit from the incorporation of speech recognition, particularly those whose interfaces have multiple control options. A prime candidate is the car stereo (radio/ tape/ CD). For example, Haeb-Umbach and Gamm (1995) discuss a system in which speaker-independent continuous-speech is employed to access various functions (e.g. "CD four, track five"), and speaker-dependent recognition allows users to define their own names for radio stations (e.g. "change to BBC now"). Using speech for the car stereo benefits from compatibility in input and output modalities; that is, the auditory input of a speech command results in the auditory feedback of the change in radio or CD output.

Other applications include the car's climate control system, and the mirrors, windscreen wipers, seats, etc. Although speech input to such basic car systems may not have significant advantages over manual input for most users, it could allow drivers with physical disabilities to use their arms and/or legs solely for the most important tasks of safe driving.

### Phones

In recent months, a number of high-profile legal cases have argued the dangers of operating a mobile phone while driving. These have led to adjustments in the Highway Code in the UK, and the belief that legislation preventing the manual operation of phones on the move is inevitable (Vanhoecke, 1997). Consequently, much effort is being invested towards the development of voice-operated, hands-free kits, initially for keyword dialling and eventually for all phone functions. The former application requires speaker-dependent recognition, allowing the user to dial commonly-used numbers through keywords (e.g. "mum", "office"). The latter needs speaker-independent, continuous-word recognition for inputting numbers or commands. The ASR capability may be either incorporated into the phone, accessed over the mobile network, or pre-installed by the car manufacturer.

### Navigation and Travel/Traffic Information

Technology such as navigation systems (which aid drivers in planning and finding their destinations) or travel/traffic information systems (which inform drivers of local 'events' such as accidents, poor weather, services, etc.) have the potential to greatly increase the complexity of driver-system interactions. Current systems require the driver to input information, such as a journey destination, while on the move, and often use an array of buttons or rotary switches to accomplish this. The near future is also likely to see a rise in the prevalence of driver-requested services (for example, the ability to investigate the availability of parking spaces in a town, or the location of the next petrol station), for which ASR could be even more useful.

Fully voice-operated navigation requires that a user can input thousands of possible geographical names to specify a destination. Apart from the obvious technical difficulties of large-vocabulary, phoneme-based recognition, there are added problems that the system must cope with multi-national names, and poor pronunciation by users unfamiliar with the place name they are speaking.    One solution is to incorporate standard word recognition for commonly-used names, with a fall-back to a spelling mode for less frequent inputs (Van Compernolle, 1997).    Of course, spelling itself is a complex process for a speech recogniser due to the highly confusable 'e- set' (b, c, d, e, g, etc. all sound similar).

## Human Factors Issues of ASR in Cars

Much has been written in the past about the human factors of ASR (see, for example, Hapeshi and Jones, 1988; Baber, 1996). The following sections discuss those issues which are particularly pertinent to the incorporation of ASR in cars.

### User Population

It is generally accepted that a variety of user variables (age, gender, motivation, experience, etc.) may affect the success of the speaker in operating an ASR system. The avionics industry is probably the closest to the automotive industry in terms of the environmental demands on speech recognition (concurrent tasks, noise, etc.); however, whereas aircraft pilots tend to be highly-motivated, well-trained, younger and male, car drivers make up a heterogeneous sample of the general public.    Indeed, it should be noted that there are very few successful public applications of ASR, probably due to the wide variation in speaking style, vocabulary, etc.

Perhaps the most important factor is experience. Both the user's experience with the specific recognition system, and their experience with technology in general may affect performance. Users who are computer literate may well adapt to speech systems more readily than naive users (Hapeshi and Jones, 1988), but they may also have over-inflated expectations of the technology. The implication for the design of in-car ASR is that both pre-use and on-line training must be provided. For example, the system designed by Pouteau *et al* (1997) allows the user to ask the system for assistance at any stage of the dialogue, to which the system responds with its current state and allowable operations.    This system also provides automatic help if the user falls silent in the middle of a dialogue.    As well as help for naive users, the interface should adapt for expert users; for example, shortening the dialogues as the user becomes familiar with them to avoid frustration.

### Dialogue Initiation

Leiser (1993 p.277) notes that "an unusual feature of user interfaces to in-car devices is that there will be a combination of user-initiated and system-initiated interaction. For example, interaction with a car stereo will be largely user-initiated. A car telephone will demand a roughly equal mixture...an engine monitoring system will be largely system-initiated". Both types of dialogue initiation must be carefully designed.

Because of the prevalence of sounds in the car which are not intended as inputs to the ASR device (e.g. the radio or speech from/ with passengers), user-initiated dialogues must involve some active manipulation. One possibility is a 'press-to-talk' button mounted on or near the steering wheel. However, this may result in one of the major potential advantages of voice control over manual control (hands-free operation) being lost. An alternative is some keyword to bring the system out of standby mode (e.g. "wake up!", "attention!", "system on!")

In system-initiated dialogues, care must be taken not to disrupt the user's primary task of safe driving.    Unless an intelligent dialogue management system which estimates the driver's spare attentional capacity is incorporated (see Michon, 1993), the system may request information when the driver is unable to easily give it. Therefore, system prompts should be designed to reassure the driver that a dialogue can be suspended and successfully taken up again later (Leiser, 1993).

## Feedback

Feedback is any information provided by an ASR system to allow the user to determine whether an utterance has been recognised correctly and/or whether the required action will be carried out by the system (Hapeshi and Jones, 1988). As a general human factors principle, some sort of feedback should always be provided, and it has been shown that this increases system acceptance (Pouteau *et al*, 1997).

For many in-car applications, feedback will be implicit ('primary feedback'); that is, the action of the system (e.g. change of radio station, activation of windscreen wipers, phone ringing tone) will directly inform the user what has been recognised. In these cases, additional feedback from the speech system may not be required.

If explicit ('secondary') feedback is necessary for cases when system operation is not obvious, or when the consequences of a misrecognition are particularly annoying, there are a number of possibilities. A simple system of tones has been found to be efficient and well-liked for certain ASR applications, but in the car environment there are likely to be a variety of easily-confusable abstract tones present. Spoken feedback is transient and makes demands on short-term memory (Hapeshi and Jones, 1988). It is also impossible to ignore, and may be irritating for the user. Visual feedback via a simple text display has the advantage that it can be scanned as and when required, but requires the eyes to be taken off the road. A combination of spoken and visual modes may be preferable (Pouteau *et al*, 1997).

## Effects of Noise

The failure of ASR devices to cope with the noisy car environment is probably the main reason why in-car applications of speech input have been unsuccessful in the past. Noise can adversely affect both speaker and speech recognition performance at a number of levels. First, noise can impact directly on the recognition process by corrupting the input signal to the recogniser. Second, speakers tend to sub-consciously adapt their vocal effort to the intensity of the noise environment (the 'Lombard Effect'), which then adversely affects recognition accuracy. Third, noise can cause stress or fatigue in the speaker, which affects the speech produced, which in turn affects the recognition accuracy. And so on. Noise can also impact on cognitive processes outside speech production, which may affect the ability of the user to carry out the required tasks concurrently.

ASR in the noisy car environment may be less problematic than other environments such as offices or industrial settings, as it is more predictable. Although in-car noise comes from a variety of sources (engine, tyres, wind, radio, passenger speech, etc.), the speed of the vehicle can give reference points for reasonably effective noise reduction (Pouteau *et al*, 1997). Technological solutions for coping with noise include selection of appropriate microphone arrays, acoustic cancellation (especially from known sources such as the radio), and active noise suppression through masking or spectral subtraction (Van Compernolle, 1997).

Poor accuracy associated with the Lombard Effect can be reduced by training the speech recognition templates in a variety of representative noisy environments. However, because in-car recording can be expensive, some success can be found by artificially degrading speech with environmental noise (Van Compernolle, 1997). Also, for a given individual, the effects of noise on speech production are relatively stable; therefore, speaker-dependent training of particular template sets for 'noisy speech' may be an effective solution (Hapeshi and Jones, 1988).

## Effects of Workload and Stress

Speech has also been shown to be vulnerable to the effects of speaker workload or stress (e.g. Graham and Baber, 1993). Sources of driver mental workload include (a) the driving task itself (e.g. lane-keeping, speed choice, keeping a safe headway and distance from other vehicles), (b) the driving environment (e.g. traffic density, poor weather, road geometry, etc.) and (c) the use of in-vehicle systems (e.g. the presentation, amount and pacing of information to be assimilated and remembered). For in-car applications of ASR, this implies that the speech recogniser may fail just when it is needed most; in high workload conditions where the driver cannot attend to visual-manual controls and displays.

Similar to the strategy adopted to overcome noise, enrolment of speech templates under a variety of representative task settings may reduce the effects of stress or workload on ASR performance. As users tend to revert to speaking more-easily-recalled words under stress, system vocabularies should be designed to be 'habitable'. The size and complexity of the vocabulary might also be reduced in stressful situations (Baber, 1996).

## Conclusions

Despite the barriers, it seems very likely that ASR will be widely incorporated into cars in the near future. Legislation relating to the use of mobile phones on the move and the rapid growth of the ITS market will drive its adoption. However, little research has been directed towards the use of ASR in the car environment. Further work is required into ASR dialogue design for in-car applications, particularly with respect to the mode and timing of feedback while the user is engaged in the concurrent driving task. Work is also required into the effects of the particular sources of noise and mental workload found in the in-car environment on the speaker and speech recogniser.

## Acknowledgements

This work was carried out as part of the SPEECH IDEAS project, jointly funded by the ESRC and the DETR under the UK Government's LINK Inland Surface Transport programme. For further details of the project, please contact the author.

## References

Baber, C. 1996, Automatic speech recognition in adverse environments, *Human Factors,* **38**(1), 142-155

Graham, R. and Baber, C. 1993, User stress in automatic speech recognition. In E. J. Lovesey (ed.) *Contemporary Ergonomics 1993*, (Taylor & Francis, London), 463-468

Haeb-Umbach, R. and Gamm, S. 1995, Human factors of a voice-controlled car stereo. In *Proceedings of EuroSpeech '95: 4th European Conference on Speech Communication and Technology*, 1453-1456

Hapeshi, K. and Jones, D.M. 1988, The ergonomics of automatic speech recognition interfaces. In D. J. Oborne (ed.) *International Reviews of Ergonomics*, (Taylor & Francis, London), 251-290

Leiser, R. 1993, Driver-vehicle interface: dialogue design for voice input. In A. M. Parkes and S. Franzen (eds.) *Driving Future Vehicles*, (Taylor & Francis, London), 275-293

Pouteau, X., Krahmer, E. and Landsbergen, J. 1997, Robust spoken dialogue management for driver information systems. In *Proceedings of EuroSpeech '97: 5th European Conference on Speech Communication and Technology, vol. 4*, 2207-2210

Van Compernolle, D. 1997, Speech recognition in the car: from phone dialing to car navigation. In *Proceedings of EuroSpeech '97: 5th European Conference on Speech Communication and Technology, vol. 5*, 2431-2434

Vanhoecke, E. 1997, Hands on the wheel: use of voice control for non-driving tasks in the car, *Traffic Technology International*, April/May '97, 85-87

# 1999

| Leicester University | 7th–9th April |
|---|---|
| **Conference Manager** | Janis Hayward |
| **Chair of Meetings** | Ted Lovesey |
| **Programme Secretary** | Margaret Hanson |

| **Secretariat** | Paul McCabe | Paula Rogers | Phil Bust |
|---|---|---|---|
| | Aris Basiakos | Taran Barrington | Kate Taylor |
| | Damien Forrest | Timo Bruns | Dorothy Chan |
| | Mark Young | | |

## Social Entertainment

Janis Hayward's last conference as manager and her favourite dance record, YMCA, was played more than a few times at the post annual dinner disco. An unofficial 'Late Night Party' developed with some assistance from the person who became the sponsor of the official event some six years later.

| Chapter | Title | Author |
|---|---|---|
| **Air Traffic Control** | The future role of the air traffic controller: Design principles for human-centred automation | M. Cox and B. Kirwan |
| **The Future of Ergonomics** | The future of ergonomics | B. Green and P.W. Jordan |
| **General Ergonomics** | How I broke the Mackworth clock test (and what I learned) | B. Shackel |
| **Health and Safety** | Ageing, health and work: A framework for intervention | L.A. Morris |

# THE FUTURE ROLE OF THE AIR TRAFFIC CONTROLLER: DESIGN PRINCIPLES FOR HUMAN-CENTRED AUTOMATION

## MARTIN COX[1] AND BARRY KIRWAN[2]

[1]*EASAMS Defence Consultancy, Marconi Electronic Systems Ltd, Lyon Way, Frimley, Camberley, Surrey. GU16 5EX*
[2]*Air Traffic Management Development Centre, National Air Traffic Services Bournemouth Airport, Christchurch, Dorset. BH23 6DF*

There is concern that progressive automation within air traffic control may compromise the central role currently played by Air Traffic Controllers in the management of air traffic. Human Centred Automation (HCA) is a concept that has evolved in the field of aviation. It relates to the difficulties of designing semi-automated systems within which humans can still operate successfully. This paper discusses how and why human centred automation has evolved, and its importance for future Air Traffic Management (ATM) systems. Principles of HCA for ATM are developed, and evaluated against a sample of prototype future system designs.

## Introduction

The advent of so-called 'glass cockpits' (computerised flight decks) saw a series of incidents and accidents, many of which related to a fundamental mismatch between the computer system (the automation) and the mental models of the pilots, often resulting in tragedy. Human-Centred Automation (HCA) is a set of principles through which such mismatches, and their consequences, can be avoided, or at the least minimised.

HCA is relevant to Air Traffic Management (ATM) system development because of the move to more computerised support for air traffic controllers, in order to achieve required capacity levels safely. Quite simply, ATM agencies will not wish to go through the same learning curve that early computerised aviation experienced (and to some extent is still experiencing). There is a need to ensure that controllers can not only avoid misunderstandings associated with the computerised support they will receive over the next decade, but also that they can maximise the operational effectiveness of the system. The experience from aviation suggests that this will only be achieved through developing the computerised tools around the controllers and their mental models. Research at National Air Traffic Services (NATS) has explored how this can be done and has lead to the conclusion that HCA principles can in fact be incorporated into

NATS' design strategy, and that these principles would benefit NATS and help to protect ATM in the UK from 'automation-assisted failures'.

## Key Automation-Related Problem Areas

Based on a review of relevant literature and associated incidents (e.g. Garland and Wise, 1993; ICAO, 1994; Riley et al, 1994; Billings, 1997), the key automation impacts on human performance were identified. These are outlined in Table 1. However, to make them more understandable, they have been cast in a framework of human performance in ATM. This framework represents the ATM system at three levels: the mental functioning of the controller; the interface (computer and workstation); and the ATM organisation. This structure is in a sense hierarchical, with the first level 'inside' the human, relating to the controller's mental functioning, situation awareness, picture, and decision-making etc. The next level concerns the human's interaction with the machine, and relates to the controller's interface, the computer software and functionality. The third level concerns the broader picture, for example how the automation affects teamwork, but also other aspects such as legalistic implications of automation.

**Table 1.  Automation impacts on human performance**

| Framework Level | Impact of Automation - Problem Areas |
| --- | --- |
| Level 1 - Mental functions | Visual channel overload, Lack of situation awareness, Lack of mode awareness, Lack of trust, Complacency, False positives & misses, Insufficient knowledge of automation limitations, Skill degradation & immature skills, Increased training requirements, Workload problems, Automation bias, Less active (involved) output. |
| Level 2 - Interface | Lack of understanding of what the computer has done and why, Impact on decision-making, Interface complexity and usability (ergonomics), Indirect information, Intolerance towards, and magnification of, skill-based errors (slips) |
| Level 3 - Work Situation & Organisation | Responsibility and legal considerations (i.e. *'Who's in charge ?'*), Job design & the human role in the system, Secondary tasks and time management, Transition, Task-paced versus self-paced work, Crew Resource Management (CRM) issues, Recovery from system failure |

Figure 1 shows the three-level framework. Level 1 (Mental Functions: see Wickens, 1992) comprises the following basic human functions: perception (principally visual and auditory modes); thinking (comprising executive control, where planning and assessment and judgement takes place, which operates within the working memory - this is where situation awareness or 'the picture' resides); long term memory (including knowledge of facts, opinions, past judgements based on experience, and skills); attentional resources (our mental capacity constraints, which are susceptible to overload or underload, and vigilance constraints); response (our output modes, usually tactile (e.g. activating something using a mouse or keyboard) or vocal).

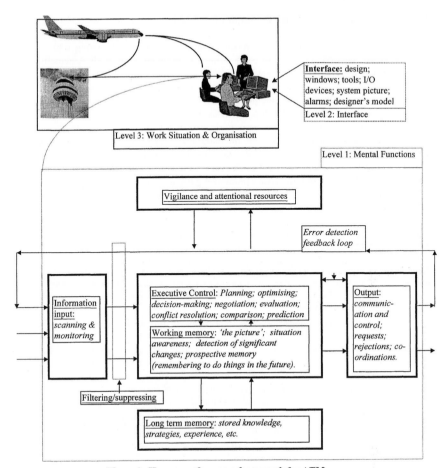

**Figure 1.  Human performance framework for ATM**

Level 2 represents the interface and software design aspects that can lead to problems - they affect (via level 1) the previous functions, but it is useful to focus on the system design aspects themselves.  Level 3 represents the controller's job and work situation, including how controllers work together and interface with the pilots/aircraft etc.

## Human-Centred Automation - Principles for ATM

Having briefly explored automation in aviation, and the pitfalls of automating without adopting an HCA-based approach, research defined in more detail what HCA actually means in practice, by developing a set of HCA design principles for ATM. These have been derived from the available literature and from a consideration of the key problem areas as stated in Table 1. The principles themselves are organised into six areas: the role of automation (the relationship between human and machine); automation (functional design - what the system aims to do and how it does it); interface design (usability and error tolerant aspects of the interface); implementation and training support aspects (achieving and maintaining good implementation and acceptance); team and organisational issues (fitting computerised support into the ATM team infrastructure); risk and reliability issues (minimising negative safety impacts).

Table 2 highlights a number of the 46 relatively high level principles for HCA in ATM that were identified.

**Table 2.  Examples of HCA Principles for ATM**

| Area | Human-centred automation principles for ATM |
|---|---|
| The Role of Automation | Automation should support the controller: with housekeeping and nuisance tasks; by enabling more rapid and accurate task execution; enabling more complex tasks to be achieved safely; by enhancing flexibility of operations; supporting with prediction and interpolation tasks |
| | Automation should enhance situation awareness.  It should not draw the controller into thinking more about the automation than the aircraft. |
| Automation Functional Design | In order to help the controller maintain the picture, displays must still present the basic information currently used (e.g. location; identity; vector; temporal characteristics, etc.). |
| | Automation should not be in conflict with controller mental models, goal hierarchies, or strategies, and preferably should build upon them. |
| Interface Design | There should be feedback when automation is operating, so that the user always knows 'who is in control', and when the automation has finished a task. |
| | Usage of different 'modes' for automation support should be minimised.  If different modes are used, it must be compellingly clear which mode the system is in. |
| Implementation and Training Support | Trust in tools needs to be addressed when designing, developing and introducing them.  This will require early involvement of prospective users, and diverse and changing user groups throughout the system development life cycle. |
| | The controller's training should encompass how the automation works (not merely how to work the automation), and its limitations (i.e. when it won't work), what system failures will look like, and how to compensate for them. |
| Team and Organisation | Consideration needs to be given as to what back-up skills new controllers will need (i.e. for controllers who will only know the automated environment). |
| | The role of automation and the controller needs clear demarcation. |
| Risk and Reliability | The system should inform the controller if it is compensating for some system aberration or failure, so as to avoid loss of such compensation if the controller takes control of the task. |

## Examples of Uptake of HCA Principles

Making the derived HCA principles operational, i.e. making them 'happen' in real design projects, will require adaptation by the project design teams and operational customers. In order to ascertain the practicality of using the HCA design principles posited in this research, prototype future systems with which NATS has involvement were briefly examined.

There is evidence that system developers and designers are already beginning to embrace a number of the HCA principles. For example, the Programme for Harmonised Air Traffic Management Research in EUROCONTROL (better known as PHARE) has been developed upon the premise that the controller is to remain at the centre of the ATM "world" - retaining ultimate authority and decision making function. Its tools generally aim to support the controller by allowing him to plan with more certainty and dispense with some mundane monitoring tasks (amongst other things). Thus, the skills of the controller are both supported and indeed enhanced. The progressive application of the HCA principles highlighted by this research should further ensure that automation complements the controller, rather than competes with him.

## Conclusion

Automation or computerised support for the controller will become a reality within the next decade, in order to realise predicted capacity demands. The aviation experience has shown that there are downsides to such automation, and that the solution is to adopt a Human Centred Automation (HCA) approach. This research has tried to extract what can be most usefully learned from the experience of aviation, ATM's closest relation in industry, and one which has learned painful lessons over the past two decades.

A set of principles of HCA for ATM has been distilled. A good number of these principles are already being adopted within NATS' future systems design and development programmes. It is therefore recommended that such practices continue, and that the principles and methods highlighted by this research are considered and developed for all future automation or computerised-support-tool projects.

## References

Billings, C.E. 1997, *Aviation Automation: The Search for a Human-Centred Approach.* (Lawrence Erlbaum, Mahwah, New Jersey)

Garland, D.J. and Wise, J.A. (Eds.) 1993, *Human Factors and Advanced Aviation Technologies.* (Embrey-Riddle Aeronautical Univ. Press, Daytona Beach, FL.)

ICAO 1994, Human Factors In CNS/ATM Systems: The Development of Human-Centred Automation and Advanced Technology In Future Aviation Systems. *Human Factors Digest No.11: Circular 249-AN/149.* (International Civil Aviation Organization, Montreal)

Riley, V., Lyall, E. and Wiener, E.L 1994, Analytic methods for flight-deck automation design and evaluation, phase 2 report: Pilot use of automation. *Contract report DTFA01-91-C-00039.* (FAA Washington, DC)

Wickens, C.D. 1992, *Engineering Psychology and Human Performance. 2nd edition.* (Harper Collins, New York)

# THE FUTURE OF ERGONOMICS.

## Bill Green[1] and Patrick W. Jordan[2]

*1. Delft University of Technology, The Netherlands.*
*2. Philips Corporate Design, Groningen, The Netherlands.*

This paper discusses the some of the issues which were raised during a symposium at the International Ergonomics Association conference in Tampere in 1997. It reviews the progress of the discipline and suggests some of the ways in which fruitful development may take place using domestic products as an example.

## Introduction.

The historical background of ergonomics is well known. Deriving from the ancient notion of 'fitness for purpose', developing through the stages of technological and sociological progress under a number of different banners, and finding a semantic home in the United States as 'Human Factors' and in England and Europe as 'Ergonomics' in the late forties, it is now a well established discipline with university level courses all over the world. It seems to be enjoying an unprecedented 'vogue', as exemplified by its incorporation into safety legislation and industrial practice. In the development cycle of human concepts there seems to be a natural progression of an almost gaussian nature; the rise followed by the peak followed by the decline. If we are correct in this assessment, then ergonomics is approaching its peak. The next phase is decline; a condition with which no one, least of all the ergonomics profession, can be comfortable.

## Ergonomics and its 'raisons d'etre'

It follows from the above that the discipline of ergonomics/human factors is at an important point in its development. It is a discipline which has always been linked with design, whether of work practices, products or systems, but developments in the recent past have placed a new emphasis on the nature of those links. Domestic and other products have always to some extent been the subject of anthropometric and biomechanical scrutiny, at however crude a level, and other aspects of human characteristics have been routinely employed as limiters in the design of complex systems, usually for expert or trainable users. The electronic revolution has precipitated an unprecedented invasion of hitherto esoteric problems into the common domain, and together with the greying of the population of that domain, has created new demands on ergonomists and the design profession. So-called 'intelligent products' have sometimes been seen to be remarkably stupid, and the strategies of traditional ergonomics revealed as inadequate to deal with the problem.   The statement that 'allowing the possibility of use says nothing about how a product will be used' (Green, 1999) may be seen as a truism, but it helps to focus on the changes which are taking place in ergonomics and design research and practice. Concentration on the user as a set of quantifiable characteristics is giving way to acceptance of the user as an unpredictable part of a complex interaction, and in turn the emphasis is being placed on researching what has been termed the ecology of that interaction. This ecology is never more complex than when domestic products and users are the central issue. It is clear that ergonomics has a powerful base in, for example, military systems design and evaluation, and also in similar, if less well controlled situations in industrial practice, which deal with traditional and important concerns such as task analysis, man/machine task distribution, and the application of human quantifications. The view of ergonomics taken in this paper is a broad one, and steps over the boundaries of the classic work efficiency definitions to deal with the complete user/product interaction. Domestic products are unashamedly used as the vehicle, if for no other reason than the extremely limited potential to control their purchase, functionality and use. An attempt is made to identify a number of issues that the profession will be called upon to address in the years to come. Whilst the discipline is currently thriving to an unprecedented degree, its continued success is likely to be dependent on the effectiveness with which it meets these new challenges.

**New Challenges.**

There are many challenges remaining for the proper establishment of a body of ergonomics knowledge; most of these are not new. The development and maintenance of anthropometric data bases, for example, is an ongoing concern for researchers and practitioners alike as population shift and grow and coalesce. Human characteristics data is never complete, and must constantly be updated and added to. These have been central issues for ergonomics for many years and should never be neglected. However, in recent times the shift in emphasis towards the particular problems caused by the way people use products has gathered momentum. Usability research is now well established, but is it, in its current form, enough?

*1. That was then, this is now!*
From the point of view of customers, usability has moved from being a 'satisfier' to a 'dissatisfier'. A few years ago, users would probably have accepted low levels of usability, provided that a product performed well, contained useful functions, was durable and so on. If a product had good levels of usability, users would have seen this as a plus point (in marketing terms, as a 'satisfier'). The increasing emphasis in the popular media on what started out being called 'user friendliness' - usually applied to computers only,  but now a point of critical judgement of almost anything -

has led to the buying public tending to take it for granted that a product will be usable and 'user friendly', and being disappointed if it is not (in marketing terms a 'dissatisfied'). The concept of 'friendliness' is flexible, and Jordan (1997) has argued that if human factors is to help create positive benefits that users will really notice, it must now move beyond usability, to more holistic 'pleasure-based' approaches. In recent articles, Bonapace (1999) and Taylor (1999) have given examples of such approaches and their practical application, and the related Japanese technique of Kansei Engineering has received some coverage in the recent ergonomics literature. Of course, such concerns are not new. They have been one of the central issues of product design for decades, but have been pursued in the mystical way of the artist, with the 'knowledge' invested by some common consent in a small group of individuals whose approach owed virtually nothing to conventional science and, far from attempting to achieve repeatability, actively eschewed it. The post-modernism of Sottsass and his followers, and what might be termed the post-post-modernism of Starke and others, owes nothing to ergonomics in its traditional definition, but has nonetheless had an enormous influence on the form of products and hence on their usability. Design and ergonomics are converging, and the power of ergonomics is surely its ability to invest approaches to product design with a firm base in human centredness. To do this, ergonomics must move towards an understanding of the real imperatives of design; to embrace the idea that there is no clear dividing line between our physical and emotional responses to products, and the ways in which we use them, and to find ways of researching these issues. Perhaps then our design icons will actually work.

## 2. Getting older.

A second major issue is the ageing of the population. The elderly are becoming an increasing large and financially powerful group, and 35% of UK consumer spending is currently controlled by those over 50 years of age. It is clear from the work reported by Etchell (1999), that many manufacturers are still putting products onto the market with little or no thought for the requirements of their elderly customers. A number of strategies have been outlined for approaching design for an ageing population and Freudenthal (1999) has proposed some basic guidelines for designing for the elderly. In spite of some reservations, and recognising clearly the difficulties of a broad spectrum approach, it is still possible to assert that design for the elderly can be part of an inclusive approach to design, which supports the creation of products to be used by all, regardless (in as far as possible) of the user's age or (dis)abilities. We would go further, and assert that inclusive approaches to design are both a financial and a moral imperative for the profession. There is no doubt that the issues are complex, and that there is no glib and comprehensive formula for dealing with them. It is equally clear that use of the methodologies of user trialling together with an enlightened and holistic view of users' needs, can assist designers to make decisions which are sustainable and supportable from a moral, a sociological, and a commercial point of view.

### 3. Using the right tools in the right place.

The third major issue is the pragmatic role of human factors in the product creation process. Human factors techniques must be relevant and practical in a commercial manufacturing context. Green (1997) described some of the conditions for the successful integration of usability issues in product development, warning in particular against over reliance on exhaustive 'experimental' trialling — an approach which can prove extremely costly. There is a wide range of different evaluation techniques, with different strengths and weaknesses and different situations under which each is best used. It is surely important that those carrying out usability evaluations have a portfolio of techniques to address each project in the most appropriate and cost effective manner. Currently, demonstration of appropriate techniques is perhaps best focused on case studies, for the simple reason that theory development is lagging behind the empirical process. This theoretical shortfall has the effect of slowing down progress to the pace at which a new user trial becomes absorbed into the usability lexicon and contributes a new insight. Green and Kanis (1998) have explored this issue in another forum with the conclusion that a comprehensive theory of product interaction is unlikely to appear 'top down', and that attempting to add to the 'body of findings' is the best contribution anyone can make; keep the possibility of theory development in mind, and meanwhile maintain the empirical exploration of the usability process.

### 4. And another thing...............

An issue which follows the academic search for a sound theoretical base is the tension between academe and industry, and this emerged very clearly during the symposium discussion on usability at IEA 1997 in Tampere. Such tensions are not new, and are part of the fabric of most disciplines. However, domestic product development seems to be more than usually vulnerable to them as a consequence of the actual closeness of the fields. The connections between academic research and industrial practice in usability are so evident that when they don't work well, it is plain to all involved. In the symposium at IEA Tampere, the divergence between the rigour of academe and the pragmatism of industry  was debated by a group of remarkably homogenous people: ergonomists whose primary interest was design, and designers with a strong background in ergonomics. At one point it was described as a 'war': this reflected much of the discussion, and the war was no less bloody for being a civil one! The positions may be summed up as follows: academics regard industrial approaches as sloppy and lacking in rigour and validity, whilst industrialists regard academic practice as over complex and impractical. They want 'off the shelf' methods which are instantly applicable and have some acceptable level of 'validity' (a term which should be used with caution, given its various conotations in both scientific and common parlance - see Kanis 1997) A common complaint was that academics have been slow to respond to these requirements, and that the tools and techniques are 'sledgehammers to crack nuts'. This is a difficult assertion to counter, since many of the academics involved are all too conscious of the problems of applying their preferred techniques in a high pressure commercial environment.

*5. Working together.*

Finally, it is worth re-stating the obvious. The relationship between designer and human factors specialist is central to the creation of products which fulfil usability criteria in the broadest sense of products which are safe, usable, and provide the elusive 'joy in use'. In a recent study, Thomas and van Leeuwen (1999) demonstrated how close co-operation between interaction designers and human factors specialists can lead to the creation of a product widely recognised as having achieved the very highest standards of usability. This relationship operates best at an individual level when a designer, design engineer or product development team has a strong background in ergonomics and experience of trialling techniques (it seems rare for a human factors specialist to have a strong design background or to be employed as a professional designer) and when the design and development efforts are supported by professional ergonomics expertise and are conducted in a mutually supportive and non-hierarchical environment, with a management who recognises the important competitive advantage of truly excellent design for people. There are many examples in the design/craft tradition, where tools developed over many years achieve a level of usability which transcends mere use and becomes a deeply satisfying experience. The challenge for modern ergonomics and design is to achieve this in radically compressed time scales and in the face of an exponentially expanding technical potential.

**References.**

Buti, L.B., Bonapace, L., and Tarzia, A. 1997 *Sensorial Quality Assessment* in From Experience to Innovation IEA '97. Proceedings of the 13th Triennial Congress of the IEA.

Etchell, L. 1999 *Designing Domestic Appliances for Everyone.* In 'Human Factors in Product Design', Green and Jordan (eds) Taylor and Francis in press.

Freudenthal, A. 1997 *Testing new design guidelines for all ages, especially menu design on home equipment.* In 'From Innovation to Experience' IEA 97.

Green, W.S. 1997 *Essential Conditions for the Acceptance of User Trialling as a Design Tool.* In 'From Experience to Innovation IEA 97. Proc. of the 13th Triennial Congress of the IEA, Tampere, Finland.

Green, W.S. and Jordan, P.W.1999 *Ergonomics, Usability and Product Development* in Human Factors in Product Design. Green and Jordan (eds) Taylor and Francis in press.

Green, W.S. and Kanis, H. 1998. *Product Interaction Theory: A Designer's Primer.* In Global Ergonomics 1998. Scott, Bridger and Charteris (eds) Elsevier.

Jordan, P.W. 1997. *Products as Personalities.* In Contemporary Ergonomics 1997 S.A. Robertson (ed.) Taylor and Francis.

Kanis, H. 1997 *Validity as Panacea?* in 'From Innovation to Experience' Proceedings of the 13th Triennial Congress of the IEA, Tampere, Finland 1997.

Taylor, A.J.. 1999. *Understanding Person-Product relationships – A Design Perspective.* In Human Factors in Product Design Green and Jordan (eds) Taylor and Francis in press.

# HOW I BROKE THE MACKWORTH CLOCK TEST
## (and what I learned)

**Brian Shackel**

*Department of Human Sciences & HUSAT Research Institute,
Loughborough University, LE11 3TU*

Vigilance is still an aspect of human performance worthy of study, despite the many studies from the beginnings of Ergonomics. Classic studies during and after WW2 were concerned with performance decrement in radar scanning and similar vigilance tasks as a function of time on watch and also sleep loss. For laboratory studies simple simulations were needed and the so-called Mackworth Clock Test appeared valid and was extensively used. While a 'guinea pig' in a pilot study of three nights without sleep, leading to a television programme in 1952, the author chanced upon a way of achieving no performance decrement at all in the Clock Test – a unique result yielding an unjustified reputation but for which no reasons were sought. This result has not hitherto been reported in print but provides several salutary lessons, which still have validity for the design and conduct of experiments today.

## Introduction

Let me teleport us back 47 years to November 1952, to a Cambridge with no parking meters yet; Watson and Crick were unravelling the double helix - but most of us did not know it at the time. The Medical Research Council's Applied Psychology Research Unit (APRU) was in the Department of Psychology. This was the last year of Sir Frederic Bartlett (FB) as Head of Department. Some of us were in the Department building, and some of us were in an Annex in Carlyle Road (across the river off Chesterton Road). FB was Director and Mackworth (or Mack) was the Deputy Director, located in the Department. Many whose names you may recognise were in Carlyle Road - Broadbent, Conrad, Fraser, Gibbs, Poulton - and I was very new and in an Annex to the Annex because Carlyle Road was full.

There we all were, fairly young but most of us had been in the Services; we were 'bright eyed and bushy tailed', beavering away at various studies. One focus of the work was sleep deprivation and especially the problems of how to measure any effects of sleep loss. The effects were quite elusive and we did not yet understand so much about it. Various measures were being tried. Some were perceptual tests, mainly of the vigilance type, and of course the Mackworth Clock Test was one of those used. A brief view of the work on vigilance and fatigue at that time at the APRU is given by Broadbent (1953).

Also at that time there was black and white television, developing but not yet widespread. Many academics rather looked down on it and did not have TV sets of their own. Nevertheless, it was just becoming respectable for scientists and researchers to appear on TV - and in October, I think, out of the blue an invitation arrived from the BBC to provide material for a programme on sleep in about a month's time. A full 45 minute programme was envisaged - a live programme remember, no pre-recording at that time. The final programme plan was to focus upon the sleep loss work at the APRU and also upon some of the physiological work on sleep at Oxford with Roger Bannister involved.

At the APRU it was treated as a pilot study, and during the planning discussions someone (probably from the BBC) had the idea of getting a volunteer to stay awake for three nights and be tested at the Unit for any effects and then appear on the programme. Being the newest recruit to the APRU - guess who was volunteered! So the programme was scheduled, I think for only a little over a month away in late November. Various tests were planned; fortunately not only vigilance perceptual tests and the Clock Test were chosen but also some performance tests particularly the 5-choice dotter.

## Methodology

Sleep loss regime - for two nights I was trusted to go without sleep at home but punching a portable time clock every 15 minutes. I occupied the time usefully by putting up shelves in the sitting room and the kitchen in the cottage we had 10 miles north of Cambridge up the old Roman road to Ely in the village of Stretham. For the third night without sleep I was in the Unit doing the various tests and accompanied throughout the night - except during the Clock Test because one of the rules of the Clock Test was that the subject had to be without his watch and entirely alone.

Now the testing regime - on day one, fresh and with no sleep loss of course, there were tests to gather data of 'normal' performance. On days 2 and 3 and on the morning of day 4 there were test regimes after one, two and three nights without sleep; the test regime comprised a series of the tests in the same order lasting about three hours in all. There was the Clock Test for one hour and the 5-choice dotter for half-an-hour and Poulton's multi-channel intercom intelligibility/confusion test, and several others which I do not remember now. Each day I went through the test sequence each morning and afternoon; I was accompanied all the time and was allowed to read in between times. For the nights I was on my honour to stay awake and was occupied and monitored during the first two, as I said, and accompanied throughout and tested intermittently during the third night.

So that was the procedure for the whole pilot study; at the end during the final morning came the decisions on what to report and demonstrate in the TV programme itself. At lunch time we drove down to London for rehearsals in the afternoon and then the programme live, I think it was 7.30, that evening. Afterwards Ron Lewis drove me back to Cambridge, asleep all the way; I don't remember the few days afterwards nor the sleep recovery pattern. But I do remember how people were rather nice to me afterwards, although there was a certain distance with Mack for a time - which I'll come to shortly.

## The Clock Test

Now to the Clock Test (cf. Mackworth, 1950). Hands up anybody who knows what the Clock Test is and how it works – very few. I must explain how it works or you will not understand the full point of the story. Well, the essence of the Clock Test was an attempt to simulate a really dull vigilance-type task where you have to monitor the display all the time in order to detect a low discrimination signal. The stimulus was a large black pointer on a large circular white background, with the pointer moving round in small jumps (like a clock's seconds hand) roughly but not precisely every second. Every so often, very infrequently of course, the pointer would move a double jump (two jumps in one) and that was the signal to be reported – see Figure 1 below.

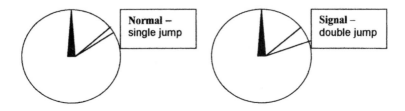

**Normal –**
single jump

**Signal –**
double jump

**Figure 1. Sketch of the Clock Test basic features**

Typically the sailor subjects sat there for two hours (similar to a naval watch period). I was only given one hour sessions, presumably because the task was thought to be impossible for me after three nights without sleep. Plenty of data already showed that typically after half-an-hour there was significant deterioration in performance. The double jumps to be detected were somewhere between 3 and 5 in each half-hour. You were seated in a medium sized room with a push-button on the arm of the chair, and you had to push the button when you thought you had seen the stimulus. If the button was pressed within 15 seconds (I think it was) of the appearance of the stimulus that was accepted as a correct response - and of course if it was pressed after 15 seconds or at any other time that was treated as a false positive. That was the testing regime. So there was I sitting in this room, without a watch, peering at this pesky pointer, during the normative trial on the first day and during the test runs on the morning and afternoon of day 2 and 3 and on the morning of day 4.

So, what happened? Well, I was not myself working on any of the sleep deprivation studies nor with any of the related performance measures; and of course I was going to show that, especially as an ex-naval officer, I could perform just as well, even after three nights without sleep, as anyone else could. To some surprise I did indeed show very little deterioration in performance. This must have been especially puzzling to Mack because the data were clear that everyone after the first half-hour showed, even when fresh, a definite deterioration in performance. I began to realise on the last morning that there were discussions going on, because I was showing no major deterioration even after 72 hours without sleep. At the very least I was highly abnormal statistically.

So there was real concern about what we could truthfully report and demonstrate on the television programme; and that became clear at lunchtime on the 4th day, because I had then finished the testing and was allowed to join the discussions. After all, I was going to be there on the programme and supposed to demonstrate something of what the scientists had been able discover about performance deterioration after sleep loss. But I had not shown any deterioration - at least not on the Clock Test.

Well, as I said earlier, quite a number of tests were used and fortunately the 5-choice dotter was one of them. Who knows what the 5-choice dotter is? Noone. So I must describe it. Basically it consists of an upright flat plate (like a screen) and a horizontal flat plate attached to it (like a small table in front of the screen). On the upright are five light bulbs located in the shape of a pentagon and on the flat plate are five fairly large brass discs also shaped as a pentagon and corresponding exactly to the lights.

The subject has a metal stylus and when a light is on should tap the corresponding disc; as soon as you tap a disc, whether you are correct or not, the light goes out and another comes on, and you go on tapping as quickly as you can. The errors are recorded on counters, but no knowledge of results is given; you are not required to do other than go on tapping as accurately and as quickly as you can for half-an-hour typically.

So that was the task, and the typical scores from the 5-choice dotter were total number of taps in the half-hour and average tap rate for each successive 5-minute period. But there were no digital computers readily available in 1952; it was all done with telegraphic electro-mechanical relays and counters, so the best one could do was to count taps for five minutes on one counter and then switch to another counter. However, I believe that this was the first occasion that Alfred Leonard, who was using the 5-choice dotter at that time, connected to it Norman Welford's SETAR (Successive Event Timer And Recorder). This was a specialist electronic device which enabled the capture and subsequent slow printout of the time interval between successive events, at better than $1/10^{th}$ second accuracy. As a result not only were the total and 5-minute rates recorded but for the first time the exact intervals between each tap and the next. This 'saved the day' as it were. The total taps and the 5-minute tapping rate did not vary much between my sleep-deprived and my fresh performance; but the SETAR showed up what had been suspected but not measured exactly before, that there were many occasions of high variability when there would be no tap for two or three seconds and then a rapid increase in tapping. Thus the tap rate even over five minutes did not show much change but the variability over short periods was large, and this was then linked with the earlier work on mental 'blocks' by Bills (1931). So now something had come out of this study and we had results to present on the TV programme, which went very well (as I learned later after I had recovered).

## Solution

But what about the Clock Test, I hear you ask? So let me tell you the solution – for the first time in print and for only the third time ever. Mackworth never knew; he never asked me to explain. If he had asked I would have felt obliged to tell all. I believe I was regarded as a minor phenomenon. Well here's the answer. On my very first trial I quickly found the task to be utterly boring and after only five minutes or so I had started to walk around, while still keeping an eye on the white board and pointer; I do not know why, but I started counting in time with the pointer movements and the faint click of the mechanism; and I noticed that the count appeared to come to 25 when the pointer had gone through a quadrant. So I tried again - 25 and it's at three o'clock, 25 and it's at six o'clock and so on.

I went on counting for something to do, and then - hey that's the double jump but still only one click; I saw that one and carried on counting and it's only got to 24 counts! This was when I was fresh, as I said, only about 10 minutes into this first dreadfully boring hour. So now I had a hypothesis and went on counting in time with the clicks, and every time I got to 25 I looked at the pointer; and by about the first half-hour I had broken the Clock Test. Whenever I did see a double jump the count only reached 24 at the quadrant position on the clock face; so whenever I reached the count of 25 and the pointer was one space beyond the quadrant position then I hit the button.

So, after each night's sleep loss what was I doing? I was walking around the room doing little exercises and even literally banging my head to stay awake, counting aloud with the clicks and at 25 if the pointer was one step past the quadrant then I hit the button.

On average this was inside the 15 second time limit and I believe I did not have any false positives, so my performance was as good as it had been when fresh. That is how I broke the Clock Test.

## Learning Conclusions

From this experience I learned a number of valuable lessons, several of which had not been embodied in the Cambridge experimental psychology course training at that time. All were concerned with ways of ensuring that the measured results do represent the realities of the test situation, and of understanding clearly what the results actually mean.

- Searchlight effect – all subjects think they are being personally evaluated so will find any way, however unexpected or unintended, to give their best performance – eg. how does the subject know that the experimenter's expectation is for him/her to fail?
- That and other motivation factors (eg my prior experience as a naval officer) can sustain a subject through long, stressful tasks (because I can tell you that after that amount of sleep loss the hour with the Clock Test certainly was stressful).
- Therefore, to ensure good understanding and interpretation it is essential to –
    Check results subject by subject for any unexpected data
    Always interview subjects afterwards for any needed explanations
    Watch subjects, covertly if necessary, for a time especially if data are unusual
    Run pilot trials of tests by intelligent colleagues for them to find faults
    Double check equipment to avoid unintended cues (cf. the Clock Test clicks).

The other conclusion was the successful exact measurement of variability with the 5-choice dotter, and that was what added to the thinking at the APRU about Bills' blocks and helped some of the sleep loss work back to them (later termed 'microsleeps'), so this pilot study was scientifically useful.

Another aspect of the experience which interests me is the parallel I later recognised with Christopher Poulton's findings on problem solving; how problems which are easy when under normal atmospheric pressure can be difficult or even impossible to solve when first experienced under as little as only one-and-a-half atmospheres. I think it likely that I chanced upon a 'solution' for the Clock Test because I was fresh, and probably would not have done so when fatigued or after even one night's loss of sleep.

Finally, I learned much more about how to run tests better, and above all the need to include proper consideration of the whole human subject in one's planning and not merely the phenomena under investigation.

## References

Bills, A.G. 1931, Blocking: a new principle in mental fatigue. *American Journal of Psychology,* **43,** 230-245

Broadbent, D.E. 1953, Neglect of the surroundings in relation to fatigue decrements in output. In W.F. Floyd & A.T. Welford (eds.) *Symposium on Fatigue,* (H.K.Lewis, London), 173-178

Mackworth, N.H. 1950, *Researches in the Measurement of Human Performance.* Medical Research Council Special Report N. 268 (HMSO, London)

# AGEING, HEALTH AND WORK:
# A FRAMEWORK FOR INTERVENTION

**L A Morris**

*Health and Safety Executive,*
*Magdalen House,*
*Bootle, L20 3QZ, UK*

Demographic ageing together with increased participation of older workers (>45 years) will result in an increase in the average age of the labour force over the next decade. It is generally accepted that task demands and work environments should be matched to the capabilities of older workers in order to prevent occupational ill health and to maintain safe working conditions. Ergonomics has a role in achieving an optimal fit, mitigating the functional decline associated with physiological ageing and age-related disease by adapting workplaces and tasks to meet the needs of the older worker. The traditional approach has been to consider capabilities within age bands but research has shown that variability in performance increases with age and that an approach focused on individual capability is needed.

This paper assesses the implications of an ageing work population for occupational health and safety and proposes a framework for maintaining the health and well-being of older workers. This framework is based on a model of functional capability and health which considers changes both within and over an individual's life span.

## Introduction

Ageing has been a central theme in the development of ergonomics as a discipline and there is likely to be renewed interest in this topic due to a predicted increase in the average age of the population in many countries of the world. The changes in functional capabilities associated with ageing are relatively well known and result from the combined effects of biological ageing, age-related disease and lifestyle factors. Ageing is also a social process and both society's and individuals' assessments of physical and psychological capabilities are often shaped by cultural factors.

## Ageing of the UK population

Over this century, both the total population and the labour force in the UK have aged and this trend is likely to continue (Fitzpatrick, 1998). Population projections for the UK (Ellison *et al*, 1997) indicate that the number of people aged 45-59, the ages of peak earning capacity, will increase by 13% by 2006. This reflects the ageing of the large cohorts born in the 1950s and 1960s. Labour force projections also indicate increases of 1.5 million in the 35-54 age group and 0.8 million in the over-54 age group with a corresponding fall of 1.1 million in people aged under-35. The decline in the numbers of young people entering the workforce reflects an increased uptake of further education by the 16-24 age group (Fitzpatrick, 1998).

Significant numbers work beyond state retirement age (770,000 in 1996), particularly the self-employed and female workers, eligible for retirement at 60 (Ellison *et al*, 1997, Fitzpatrick, 1998). This trend is likely to continue as people supplement retirement income through work opportunities in the service sector.

Clusters of older workers may also develop in particular sectors as a result of recent employment practices and future skill shortages. Restrictions in recruitment in the offshore industry, for example, have resulted in a increase in the average age of installation workers (Parkes, 1998). Recent changes in patterns of recruitment and a predicted decline in young people entering training may have a similar effect in the nursing profession (Buchan, 1998).

## Implications for Health and Safety

The ageing of the UK population and the labour force has implications for occupational health and safety. Research has shown that physical work capacity declines from peak levels usually attained around 20-30 years of age (Shephard, 1997). Similar but slower patterns of decline in sensory and cognitive capabilities have also been reported (Rabbitt, 1991). Loss of muscular strength may increase the risk of developing musculoskeletal injuries, in tasks where workloads approach the individual's maximum capacity (Haslegrave and Haigh, 1995). There is also some evidence, to suggest that older workers may be more vulnerable to the effects of psychosocial stressors, although other moderating factors are involved (Griffiths, 1998).

There is general agreement (World Health Organisation, 1993) that task demands and work environments should be matched to the capabilities of older workers in order to prevent occupational ill health and maintain safe working conditions. This presents ergonomists with a challenge in assessing the effects of the ageing process on work performance capabilities and making appropriate adjustments to workplace and job design.

Human factors data is available, mainly from cross sectional studies, mapping average changes in various physiological and performance indices against age but experience has shown that chronological age is a poor guide to individual work capabilities (McFarland, 1973). While the ageing process is traditionally viewed as a gradual process of functional decline, there is considerable variation between individuals in the rate of change. The process can be considered as a trajectory across an individual's life span (Rabbitt, 1991, Yates, 1993) with lifestyle and other environmental influences interacting with the physiological and psychological effects of biological ageing to determine performance capabilities. Age-related disease can also have an impact on work performance eg., arthritic conditions of the hands reduce dexterity (Sharit and Czaja, 1994) and individuals vary in their susceptibility to this. There have been various attempts to develop a measure of functional or biological age which can be used to assess the effectiveness of interventions but there is as yet no agreed gold standard (Shephard, 1997).

While there is likely to be some debate about the most relevant factors to include in such indices, a common feature which emerges is the need for a multi-faceted approach to workplace intervention.

## A Framework for Intervention

It is clear that problems are likely to arise when job demands exceed the capabilities of older workers. It is known for example that older workers face difficulties from working under time pressure, heavy physical work, night-time shiftwork and poor environmental conditions (Haslegrave and Haigh, 1995). Workers faced with an irreconcilable mismatch are likely to migrate towards less demanding tasks or may seek early retirement. However, on a more positive note, there is some evidence to suggest that, where sufficient flexibility exists in the design of the task, older workers can intuitively achieve an optimal fit between demands and capabilities and maintain acceptable levels of performance (Davies *et al* , 1992).

The ergonomics framework proposed in Figure 1 relates health and performance outcomes to the relative balance between task demands and individual capabilities. It acknowledges that this is not a passive process; individuals can use their accumulated experience to modify task demands through work technique or the pacing of work. Training, often considered less of a priority for older workers (Sharit and Czaja, 1994), can help to reinforce effective strategies and is particularly important when new technologies are introduced.

Ergonomists have traditionally designed workplaces to accommodate groups of individuals and there has been less focus on individual capabilities. The framework suggests, however, that there is much to be gained by working at an individual level. Technological developments may facilitate the customisation of interfaces to suit individuals without excessive cost. Mechanical assistance devices have also been developed to extend diminished operator capabilities in production line work. The theoretical underpinnings of this approach are well developed within the field of rehabilitation.

A more realistic alternative in the short term, may be a more inclusive approach to design, since this will benefit all workers. This approach is likely to be more acceptable to older workers who may view targeted interventions as discriminatory. There may be conflicts between requirements for older and younger workers in terms of viewing distances, working heights etc., but these can be resolved by building in adjustability (Haslegrave and Haigh, 1995).

The framework also recognises that physical and psychological capabilities vary both across and within the life span. A model which envisages a gradual decline in capabilities belies the plasticity which exists in reality. Physical work capacity, work performance and general health can be improved through regular physical exercise and other lifestyle changes. While this type of intervention relies heavily on individual motivation, there is growing evidence to support its effectiveness in delaying the loss of functional capabilities (Shephard, 1997). Complex skills can be learned and retained throughout life even in the presence of physical impairment. Ergonomists should therefore consider the possibility of enhancing individual capabilities in order to maintain performance or in anticipation of changing work demands.

   In conclusion, a management approach is needed, working towards an optimal fit between task demands and individual capabilities and experience throughout working life. The proposed framework suggests a combination of interventions; an inclusive approach to the design of the physical work environment, job design which allows workers some control over job demands, health promotion and training to enhance individual capabilities and monitoring of health and performance outcomes. Ergonomists should recognise the positive contribution that older workers can make to work activities and should aim to accommodate a wide range of capabilities in the workplace.

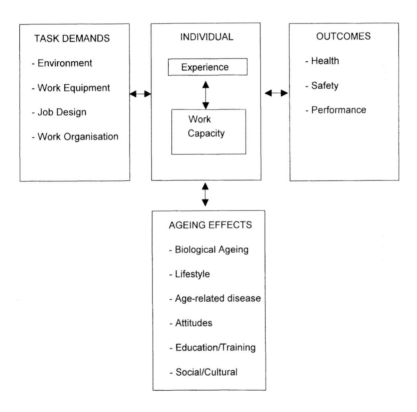

**Figure 1. Interactions between task demands, individual capabilities, ageing effects and health and performance outcomes**

# 2000

| **Stoke Rochford Hall** | 4th–6th April |
|---|---|
| **Conference Manager** | Carol Greenwood |
| **Chair of Meetings** | Ted Lovesey |
| **Programme Secretary** | Margaret Hanson |

| **Secretariat** | Paul McCabe | Phil Bust | Guy Walker |
|---|---|---|---|
| | Tim Dubé | Martin Hicks | Sam Murphy |
| | Damien Forrest | Rachel Harrison | Brendan Ryan |
| | Rachel Jones | Sue Mackenzie | David Gledhill |

## Social Entertainment

The vaulted brick cellars of this grade 1 listed building provided a wonderful setting for a very lively disco.
Delegates were asked to dream up a use for the stone obelisk standing in the 28 acres of gardens surrounding the hall.

| Chapter | Title | Author |
|---|---|---|
| **General Ergonomics** | Long days and short weeks – The benefits and disadvantages | K.J.N.C. Rich |
| **HCI & IT Systems** | Consumer acceptance of internet services | M. Maguire |
| **Legislation** | Public transport and the Disability Discrimination Act 1995 | F. Bellerby |
| **Product and Workplace Design** | Design issues and visual impairment | K.M. Stabler and S. van den Heuvel |

# LONG DAYS AND SHORT WEEKS — THE BENEFITS AND DISADVANTAGES

## Karl J.N.C. Rich

*Human Engineering Ltd, Shore House, 68 Westbury Hill,*
*Westbury-On-Trym, Bristol BS9 3AA, UK*

The desire for flexibility from employers and employees has led to a wide range of industries experimenting with Compressed Working Weeks. The main advantage for employees is increased leisure time and for employers, a move away from an overtime culture and reduced costs. Research in this area reveals equivocal results, with some early successes leading to later failure. Fatigue is a major problem for some workers and moonlighting is prevalent, raising concerns regarding exposure to physical hazards and toxins. Work schedules must take account of legislation such as the EU Working Time Directive and Health and Safety Law.

## Introduction

The 24 hour society, globalisation and industrial deregulation has led to immense changes in working practices. Shiftworking, part-time activity, flexible working hours and self-employment, are perhaps the most obvious. The inherent conflict within all of these practices is the drive for increased productivity and efficiency versus the workers' need for domestic and professional fulfilment.

Harmonisation of these criteria is often difficult and trade-offs are inevitable. One such trade-off is the compressed working week (CWW), where a set number of contracted hours are worked but they are delivered in a non-standard week. CWW has been defined as a trade off between the number of days worked per week and the number of hours worked per day (Ronen, 1984). There are many variants in use across the world (Ronen, 1984, Pierce *et al*, 1989) and some common examples are described in Table One. Most of the schedules employed fall into two categories: those with constant hours and longer working days and those with decreased hours and fewer working days. CWWs present a number of apparent benefits for both workers and management but may also lay traps for the unwary.

**Table 1. Examples of CWW Schedules from Pierce et al where Type A = hours constant, longer days and Type B = shorter hours, fewer days**

| Schedule | Description |
|---|---|
| 4-40 (Type A) | Four ten hour days followed by three days off |
| 5-45 - 4-36 (Type A) | Nine days in alternating five day and four day weeks |
| 4-36 (Type B) | Four nine hour days |

## Uptake

These work systems have been utilised principally in North America (Maric, 1977) with 3% of the total workforce of Canada being involved (BoLaC, 1994). Despite early optimism, CWWs have not broken through into larger proportions of the workforce for reasons which will be discussed below.

## Flexibility

The desire to change to a less traditional and more flexible working week has come from both employers and their workforces. Employers may need to replace expensive arrangements such as overtime and shift pay with cheaper alternatives such as employing part timers or annualising working hours. Extending operating hours may also reduce unit costs. Employees desire for change are twofold: demographic (more women workers, students financing their studies, more diverse life styles) and a rejection of night and weekend work by many workers (OECD, 1996). Most (American) Trade Unions were against the idea of CWWs as they had historically opposed long working days. Some however saw it as an opportunity to reduce working time. The Canadian Labour Congress and UK unions have also sought a shorter working week *and* less hours (Ronen, 1984, Maric, 1977).

## User profile

In the 1970s the pioneers of CWW were usually small (mostly non-union, non-urban) manufacturing firms and service retail companies. At that time the trend appeared to be towards urban centred organisations: hospitals; insurance companies; police departments. Since then industries including manufacturing, telecommunications, banking and printing have utilised CWWs (Ivancevich, 1974).

## Examples

Two major Canadian companies introduced innovative work systems with varying degrees of success (BoLaC, 1994). The Bank of Montreal introduced a range of work options for all staff, including a 3 day compressed working week (3-36), job sharing, part time, teleworking and special arrangements for workers having difficulty balancing the needs of their families with their jobs. Interestingly the Bank reported an uptake lower than anticipated.

Bell CEP negotiated an agreement to reduce working hours and bring in a four day week rather than make staff redundant. Staff worked a nine hour, four day week. This proved popular but the company were concerned that most days off were being taken on Fridays and Mondays. Hardly surprising!

An electrical engineering company adopted 3-36 with a voluntary night shift during the 1973-74 miners strike. This schedule proved popular with the workers but was discontinued. The company was taken over and the new owners brought in a 4-40 system with a night shift. This resulted in reports of increased production, recruitment of more staff (paid for by reduced overtime payment), reduced overheads (lighting, heating etc.), easier access for maintenance, reduced start-up/shut-down cost and improved recruitment potential. The employees incurred less travelling expenses and commuting time (Lapping, 1983).

## Benefits & Disadvantages

Analysis of CWW programmes produce equivocal results. Early studies reported reduced absenteeism, higher levels of production, increased job satisfaction and decreased start-up and shutdown costs. However, in the longer term there was evidence of a return to higher rates of absenteeism and tardiness, increased work related accidents, reports of fatigue and higher employee turnover (Ivancevich & Lyon, 1977, and Tepas, 1985). Despite the mostly contradictory evidence, two major issues stand out unequivocally: leisure and fatigue; CWWs provide workers with larger blocks of leisure time but expose them to longer working days. There are no inherent reasons why CWWs should enhance performance but there are very good reasons why they should cause performance decrements - fatigue and circadian effects.

Reduced absenteeism may not be a result of greater satisfaction with CWWs. The economic consequences of a lost day under for example a 4-40 system is 25% greater than under a traditional schedule. Alternatively, the additional day off allows workers to attend to their personal and domestic needs in their own and not the firm's time (Pierce *et al*, 1989). Whilst job satisfaction is generally higher in CWWs one cannot assume a direct causal link. There may be indirect causes such as unanticipated changes in the job (enrichment through increased learning, greater autonomy or responsibility) or activities outside the workplace.

Generally speaking, CWWs are popular with workers and because of this it may be extremely difficult to go back once a company has implemented a scheme. They should therefore only be implemented after careful diagnosis of the feasibility and the consequences of change (Pierce *et al*, 1989). Tepas (1985) cites many potential disadvantages and advantages of CWWs. Some issues are double edged, demonstrating the ambiguity or conflicting views on these schedules. Commuting for example may favour those with their own transport since urban trips during busy periods are reduced. Workers relying on public transport however may have problems with off-peak travel. Compressed working may suit some activities and not others. Vigilance or inspection tasks may be made more difficult, as would physiologically demanding manual work. Changes in working schedules may require a concomitant change in the tasks themselves i.e. job re-design.

### Fatigue & Circadian effects

Sleep deprivation and fatigue may be a problem if workers are exposed to nightwork. But compressed hours may eliminate or reduce need for night hours. However, even without nightwork, Physical and mental fatigue reduces physiological and psychological performance, resulting in increased error rates, accidents etc. As the working day lengthens, the probability of including circadian troughs rises, with concomitant decrements in performance.

## Moonlighting

Some American studies demonstrate that employees working a four day week are twice as likely to hold a second job (Ronen, 1984). Regulatory standards of exposure to toxic materials or physical hazards (such as noise) may be based on time weighted averages or dose levels, assuming that the normal day is eight hours (ISO, 1990). There is a risk of over-exposure and moonlighting in this context presents a very serious threat to the employee.

## To implement or not

To succeed, CWW scheduling must undergo careful analysis in terms of the nature of the production process or service, the pattern of customer requirements and worker involvement. A major hurdle to wider uptake is the number of hours worked (Blyton, 1985). CWW scheduling is most popular amongst those engaged in light work and who are not fatigued by heavy domestic responsibilities. A substantial reduction in hours worked may make these schedules more popular (Owen, 1979). CWWs will affect each company differently and there can be no one implementation method that will suit all. However, there are a number of methods available and these are discussed elsewhere (Ronen, 1984, Pierce et al, 1989, Wedderburn, 1989, Rich, 1998).

## The Eu Working Time Directive & Uk Health And Safety Legislation

The Working Time Directive sets limits on the number of hours that can be worked per week (48) within a given reference period and entitles workers to breaks during their working day, daily and weekly rest periods and paid annual leave (DTI, 1998). This could interfere with some CWW schedules. Additionally, the onus on employers to deal with moonlighting is explicit and could also cause problems.

The UK Health and Safety at Work Act places a duty upon employers to ensure, so far as is reasonably practicable, the health and safety at work of employees and other persons who may be affected by the work activity. The UK Management of Health and Safety at Work Regulations state that every employer should carry out an assessment of risk of injury or ill health arising from the working activity. Any significant change in work schedule which may affect the level of risk must be assessed prior to implementation. A risk based approach will greatly assist managers when considering such factors as exposure to hazards, toxins etc.

## Conclusions

CWW schedules are popular with employees but their benefits to the business may be unsustainable in the long term. Some of the findings in the literature are equivocal and there are many advantages and disadvantages, some of which appear contradictory. Worker fatigue and moonlighting are major problems, as is the risk of uncontrolled and excessive exposure to toxins or physical hazards. Benefits to some businesses include reductions in overtime, greater worker flexibility, reduced start up and shut down costs and decreased overheads. CWWs must not be viewed in isolation and the global effects on the business should be considered. Careful analysis should be undertaken prior to the implementation of any new schedule and changes in work schedules must consider the Working Time Directive, Health and Safety and other legislation.

# REFERENCES

Blyton, P. 1985. *Changes in working time: an international review.* (Croom Helm, London and Sydney)

Bureau of Labour Information and Communications (Programs) of Human Resources Development Canada. 1994. *Report of the Advisory Group on Working Time and the Distribution of Work.*

Department of Trade and Industry 1998. *Guidelines on The Working Time Directive,* DTI Website

ISO 1999. 1990. *Acoustics - Determination of Occupational Noise Exposure and Estimation of Noise Induced Hearing Loss.* International Standards Organisation, Geneva.

Ivancevich, J.M. 1974. Effects of the shorter working week on selected satisfaction and performance measures. *Journal of Applied Psychology.* **59**, No 6, 717-721

Ivancevich, J.M., and Lyon, H.L. 1977. The shortened workweek: A field experiment. *Journal of Applied Psychology.* **62** No 1, 34-37.

Lapping, A. 1983. *Working time in Britain and West Germany.* Anglo-German Foundation for the Study of Industrial Society. (Belmont Press, Northampton)

Maric D. 1977. *Adapting working hours to modern needs.* International Labour Organisation, Geneva

OECD 1996. *Working Hours,* Working Paper No 82, Vol. 4

Owen, J.D. 1979. *Working hours.* (D.C. Heath and Company, U.S.A.)

Pierce, J.L., Newstrom, J.W., Dunham, R.B. and Barber, A.F. 1989. *Alternative work schedules.* (Allyn and Bacon Inc., Boston London, Sydney and Toronto)

Rich, K.J. 1998. The benefits and disadvantages of the compressed working week. *Shiftworking and Rostering Conference, IIR Ltd, The Euston Plaza, London WC1, 29 - 30 June 1998*

Ronen, S. 1984. *Alternative work schedules: Selecting Implementing and Evaluating.* Dow Jones - Irwin, Illinois.

Tepas, D. 1985. Flexitime, Compressed Workweeks, Other Work Schedules. Ch 13 In *Hours of Work: Temporal Factors in Work-Scheduling* (Eds. Folkard, S., and Monk, T.H.). (J. Wiley and Sons, Chichester, New York, Brisbane, Toronto, Singapore)

Wedderburn, A. 1989. *Negotiating shorter working hours in the European Community.* Bulletin of European Shiftwork Topics, EF/89/29/EN. European Foundation for the Improvement of Living and Working Conditions.

# CONSUMER ACCEPTANCE OF INTERNET SERVICES

## Martin Maguire

*HUSAT Research Institute, The Elms, Elms Grove, Loughborough,*
*Leics LE11 1RG, UK. Email: m.c.maguire@lboro.ac.uk*

This paper addresses peoples' acceptance of the many new electronic and Internet services now available. Consumers can now benefit from a myriad of information resources, buying and banking opportunities, and public information. Yet there are still many people who are reluctant to take up such services. This paper discusses problems of acceptance of Internet services and discusses approaches for encouraging their use by a wider range of society.

## Introduction

There is now a vast range of Internet services available to consumers in the areas of ecommerce (e.g. Internet shopping and banking), broadcasting (e.g. digital TV and radio), social contact (e.g. email and online chat) and government services (e.g. tax filing and welfare benefit information). Consumers can receive such services at anytime and in their chosen setting. Suppliers are now moving rapidly to offer services over the Internet because of the potential savings that may be achieved from their retail outlets and the potential world-wide market that is available to them. Typical trends are as follows:

- Government and public service agencies will, in future, offer more information and services on-line (Hoare, 1999, Rumbelow, 1999).
- Supermarket shopping services are now available across the whole country.
- Items such as holidays, consumer goods, books, cars, will be sold in greater quantities and more cheaply on-line.
- Banks are encouraging customers to take up on-line banking which will be cheaper and more flexible.
- Email is now a rapid, convenient and popular form of communication.

While the future is exciting for those consumers most interested and motivated to use these new electronic services, less technically confident consumers may be slower to take them up, thinking that they lack the technical know-how to understand them, and seeing them as remote and impersonal. Older or disabled people, in particular, may feel that they are falling too far behind these technological developments to benefit from them.

## Trends in Internet use

There are signs that sections of the public may need persuading to take-up the concept of electronic services. It is reported that a £28m project in the UK, piloting government services over the Internet via public kiosks, may be halted because of public indifference (Phillips, 1999). This could be a setback to the Government's plan (presented in the white paper *Modernising Government*) to deliver 100% of public services electronically by 2008. The reasons behind this lack of take-up is reported to be due to poor marketing and technical problems which has undermined public confidence in the system. These difficulties can be avoided but the debate continues about public attitudes to electronic government. A nine-month study of 4,000 citizens and business owners published by the Cabinet Office in Britain in November 1998 found that 20% were "antagonistic" toward electronic public services and a "significant hardcore" rejected the idea of smartcards.

While many articles in IT journals and newspapers predict that the vast majority of the population will be using the Internet and similar electronic services in the near future, it appears that there will remain a group of people unlikely to use even well established services. For example, it is reported that after 30 years, one in four bank customers still do not use bank machines (Derbyshire, 1999). Reasons for non use are: distrust of computers, anxiety about becoming targets for muggers, and forgetting their PINs or secret access numbers. It has also been said that electronic shopping will be the normal method of buying goods in the future. Yet even by 2010, electronic shopping is only predicted to take account of 7% of all retail sales (Weathers, 2000).

Market research companies find it helpful to categorise users and non-users of technology in order to determine which groups companies should focus on. The Henley Centre, categorises the UK adult population into five types of Internet users from advanced users to those who have not yet used it (Ward, 1999). They are:

- The @home group (7% of the population) active users of the Internet at home and work.
- The @ll group (14%) use the Internet at work but are very likely to become home users.
- The Ne@rly group (30%) yet to use the Internet at all but may do so soon.
- The M@ybe group (19%) have less confidence using interactive media and little experience. Probably won't use Internet at home but may do so at work in next 5 years.
- The Not @lls: 30% of the population who are unlikely to use the Net at all.

The Henley Centre predicts that the Ne@rly group is going to be of the greatest significance to companies in the future, as they represent a huge section of the population and have already shown their willingness to use computers at work and at home. Although the Ne@rly group actively use computers, almost half say they don't understand new technology. They prefer to wait and see what other people are doing and then follow their lead. It is of interest to note that 49% of the sample (the M@ybe group and the Not @ll group) will take much longer to use the Internet, if at all.

At the time of writing, one in four adults in the UK have access to the Internet, a proportion that will undoubtedly increase. In every age group from 15 to 54, there are more Internet users than would be expected compared with the population as a whole (Schofield, 1999). But this is not the case with the over 65s who make up 20% of the population but only 4% of domestic Internet users. It is now realised unless older people are able to enjoy the benefits of the connected future, with information available electronically on tap and convenient and inexpensive shopping on-line, this will be a lost opportunity both for suppliers as well as for older people themselves. According to Oftel (the UK telephone watchdog organisation) and disability campaign groups, "A growing grey market containing millions of potential customers is being ignored in the telecoms boom" (Dawe, 1998).

## Issues of acceptance of Internet services

The reluctance of certain groups of users to access new Internet services indicates that the issue of acceptance needs to be considered alongside the design of services which are functional and easy to use. A number of stages of user acceptance can be considered.

Initially people will have *pre-conceptions* about new technologies when they hear about them. A good example is that of biometric techniques for personal identification. When this topic was studied at HUSAT by holding a number of discussion groups with the public, less than half thought that the techniques of voice recognition, automatic signature recognition, retina scanning, and hand shape recognition, would work well as identification techniques. Even iris recognition which has been shown in practice to be an effective method of identification for bank machines (after testing by the Nationwide Building Society) was only thought to be effective by just over 50% of the focus group sample.

Security of the Internet for transactions is another area of concern to the public and is still a major barrier to online shoppers. This is shown by a survey carried out by NFO Interactive (Usability Lab, 2000) based on a sample of online customers. It was found that 31% of the sample had concerns about security, compared to other barriers: prefer to go to shops (21%), not got round to it (14%), like to see things before buying (11%), concerns about data protection (7%). An edition of the BBC's Money Programme in November 1999 also found that 93% of British consumers do not feel secure when submitting credit card details over the Net. This level of concern is perhaps due to the fact that while many argue that Net shopping is generally safe (e.g. Baguley, 1999) examples of security failure still appear such as when one building society was forced to pull the plug on its Internet-based share-dealing service after a technical glitch allowed customers to into other people's accounts (Hinde, 1999).

Reports from America (Vernon, 1999) also show a concern over lack of security and privacy which is deterring people from submitting their income tax returns via their PCs. As with online shopping, America leads the world in online tax preparation and filing as part of the strategy, (common in both the US and Europe) to develop electronic government. Yet research conducted by Jupiter Communications (an analysis company in the US), shows that the number of American Internet users who plan to prepare their tax returns over the Net is less than 2% which is less than 0.5% of all expected tax returns. The main reasons for consumer reluctance to file their taxes over the Web are security, fear of error and privacy. While consumers are becoming more comfortable with electronic filing in general (e.g. completing on-line forms), they are hesitant to rely on the Web for tax returns which is regarded as a far more important transaction. The use of firewalls will improve security but unfortunately will degrade the performance of the system. To encourage online filing of tax returns, the Inland Revenue in the UK are offering discounts on tax bill if they are returned over the Internet (Hopegood, 1999).

*Seeing* a system or service for the first time is the next stage in the process of user acceptance. A first look at a system that is being used by a salesperson or friend is of course a good way to show the user what it does, how it might be useful to them, that they can trust it. Yet there can be negative aspects to this process. A demonstration of a system may reveal too much complexity which deters potential users. For example, if a buyer sees an expert user operating the system too quickly, this may make them feel unable to use it themselves.

When *using* a system or service for the first time, consumers will often compare it with existing processes they perform e.g. comparing an on-screen TV programme guide with newspaper or magazine listings. If by comparison the new system seems problematic, this will discourage them from continuing to use it. These barriers may not be apparent when the system is first demonstrated to them. First use of Internet shopping can also be disappointing. Arends (1999) reports that "despite the hype, so little Christmas shopping is being done electronically. Retail experts Verdict reckon it will be less than 1% of total sales this year". Arends then asks, "What no one measures is how many people tried to do their Christmas shopping over the net but gave up". He then reports on the use of shopping sites provided by leading UK stores such as:

- Having to update a browser to access the site.
- Passing through too many administration screens to provide personal details, credit card check, and selecting a delivery date, even before deciding to buy anything.
- Being given complex access codes that are easy to forget.
- Forcing the user to register for a loyalty card.
- Not offering enough goods to buy online, and too little information about them.
- Not allowing goods to be searched for quickly by name.
- The results of a search not being presented in an appropriate order.
- The site not being available at night!

Arends compares these problems with the slick and effective sites of online start-up companies such as amazon.co.uk. Users may see an Internet service as a challenge and will gain satisfaction in using it successfully. However if, after a period of *continued use,* they find that it does not match their needs, they will use it less. In terms of home shopping, problems such as not updating the site frequently enough, not allowing customisation (e.g. to set up a regular order for goods) and not making information easily accessible will discourage continued use.

## Encouraging user acceptance of Internet services

In order to address the problems of acceptance listed above, this section presents some initial ideas to encourage wider use of electronic services.

### Pre-conceptions

As previously stated, users will have preconceptions about services that may or may not be accurate. It is important that such pre-conceptions are based on correct information by providing more details about the service and possibly demonstrations of it. For example, in the HUSAT study of biometric techniques for user identification subjects had several misconceptions about iris scanning. Firstly they confused this technique, which involves photographing the iris, with retina scanning which involves the use of a low power laser to read the back of a person's eye (a process which seemed more dangerous). Similarly they expected to have to bend down and look into a tube. This was thought would require and awkward posture and would be unhygienic. In fact they only had to stand infront of a camera that took a photograph of their eyes. It is important then to overcome inaccurate pre-conceptions by presenting clear explanatory information about a new service when it is being promoted.

### First look

People's acceptance of a service will also be influenced by their first viewing of it. It is believed that when non users see others successfully using the Internet, they soon realise they can use it too (Ward, 1999). Demonstrations of the system or service should perhaps be a simple overview, followed by a step-by-step description of the main functions. This will help users understand the concept of the system without overwhelming them with detail. After a demonstration, on the BBC's Tomorrow's World programme, of a bank machine which identified customers using iris scanning, members of the public found it quite acceptable. While a salesperson can be briefed about how to demonstrate a service effectively, this process is hard to control. However the use of introductory videos or a demonstration mode within the user-interface can provide a standard means of introduction to a system or service allowing the user judge its value to them objectively.

### First use

Users should be guided to explore a service so they gain an understanding of it before using it. For example, to overcome the problem of home shopping registration, it would be better to allow users to browse through goods or a selection of them before requiring registration. Users should not be forced to focus in too quickly on a particular product to buy. As in a traditional shop, they should be allowed to compare products side-by-side and to browse, before committing to buy. Value added features should be used when developing an existing service. For example in many home shopping sites, use may be made of features such as: personalisation (e.g. clothes size, colour preferences, likes and dislikes), providing extra information such as interviews with book authors, samples of music, etc. Feedback is also important so that receiving an email to confirm that an order has been accepted is a good way to boost confidence. Showing the shop's refund and return policy up front on the Internet site is also important to building trust. Other aspects of first experience are to show the postage and packing costs and allowing the user to take items out of a trolley as well as putting them in.

### Continued use

The service should match the users' needs in the longer term if they are to continue to use it. For example, for an Internet based service, it is important to keep the site up to date, allow customer personalisation and make information easily accessible to encourage continued use. As users become familiar with a service, they will also expect to be able use short cuts to reach the information they require quickly and easily. Efficient fast paths should be provided for such users as demonstrated by mature search engine sites.

## Summary and conclusion

This paper has tried to show the importance of considering acceptance issues for new Internet services. Table 1 lists acceptance barriers and ideas for addressing them.

**Table 1. Overcoming barriers to user acceptance of services**

| Stage of use | Possible barriers | Principles for addressing barriers |
|---|---|---|
| Pre-conceptions | • Not understanding what service is about.<br>• Not appreciating service benefits.<br>• Doubting service security or privacy. | • Be explicit about service content in promotional materials and on site.<br>• Take opportunities to demo service e.g. in public locations and community centres.<br>• Describe how security handled on site. |
| First look/ demonstration | • Service looks complex to use.<br>• Service contains too much information. | • Provide step-by-step demo mode, or online support using 'chat' or video.<br>• Provide personalisation filters to make service more relevant to end users. |
| First use | • Difficulty accessing service (e.g. requiring browser upgrade).<br>• Service is not what it appears or contains hidden pitfalls. | • Base access on typical task scenarios to support ease of use.<br>• Support first access with simple hand-book or local access to browser upgrades.<br>• Describe all costs involved and returns policy upfront.<br>• Describe what the service doesn't provide. |
| Continued use | • Service appears too rigid.<br>• Service too long winded.<br>• Service too general so it makes finding relevant items difficult. | • Allow browsing and comparisons between items before requiring a purchase.<br>• Provide efficient fast paths and personalisation options. |

The key message of this paper is to ensure that the whole community can benefit from Internet services in the future. To quote Ruth Lea, head of policy unit, Institute of Directors (Tozer and Palmer, 2000): "Ecommerce will revolutionise our lives, but I do hope it benefits everybody. If those who are 'on the net' pull ahead in leaps and bounds from those who aren't, that would exacerbate social inequality, which I would very much regret".

## References

Arends, B., 1999, Christmas shoppers shun the World Wide Wait, *Daily Mail*, 20 Dec, p51

Baguley, R., 1999, Go shopping now!, *Internet Magazine*, April, 37-40

Dawe, T. 1998, Big market is there for the disabled, *Times 'Interface' - Telecoms extra*, 7 October, pT4

Derbyshire, D. 1999, Cash machine phobia, *Daily Mail*, 14 July, p31

Hinde, S., 1999, We're afraid the 'e' stands for 'easily cheated', Sunday Express, 28 Nov, p2

Hoare, S. 1998, City hall opens 24-hour hotline, *Times Telecoms*, 29 Sept, p9

Hopegood, 1999, Taxman to offer online discount, *Daily Mail*, 1 Dec

Phillips, S. 1999, Public snubs e-government, *Computer Weekly*, 6 May, p1

Rumbelow, H. 1999, Online NHS heralds interactive healthcare, *Times*, 8 Dec p12

Schofield, J. 1999, Older hands weave the web, *Guardian Online*, 10 June, 2-3

Tozer, J. and Palmer, A., 2000, What I want for the future, *Daily Mail*, 3 Jan, p11

Usability Lab, 2000, Convenience stores, PC Magazine, Ziff-Davis, Jan, **9**, 1, 162-177

Vernon, M. 1999, Form-filling on the web - Filing tax returns over the Internet, *Financial Times, FT-IT Review* , 2 June, p6

Ward, C. 1999, Hoi polloi are next on the net, *Times Interface Supplement*, 12 May, p2

Weathers, H. 2000, Shopping? So different but we will still have an eye for a bargain, *Daily Mail, Millennium Mail Supplement*, 1 Jan, pVI

**Acknowledgement:** This paper has been developed with the support of the EC TEN-Telecom TUSAM project and Esprit EMMUS project. The author also thanks his HUSAT colleagues Kathy Phillips and Colette Nicolle, for providing useful material for the paper.

# PUBLIC TRANSPORT AND THE DISABILITY DISCRIMINATION ACT 1995

## Fiona Bellerby

*(formerly of Loughborough University), Ergonomics Consultant, Davis Associates Ltd, Wyllyotts Place, Potters Bar, Herts EN6 2JD, UK*

The Disability Discrimination Act 1995 (DDA) has received various criticisms since entering Parliament as a Bill in January 1995 for doing little to end discrimination against disabled persons. This project aimed, therefore, to consider the advantages of the Transport section and Goods, Facilities and Services section of the Act and highlight the shortfalls of air, rail, bus and coach travel for those with special needs. Five focus group discussions and three individual interviews were conducted with disabled and elderly people to determine which of their special needs are not being met. To balance the argument interviews were conducted with service providers to determine their disability policies and the actions that are being taken, or proposed for the future, to increase access to their services, as well as the constraints under which they must operate.

## Introduction

The DDA came into force in November 1995. Since then doubts have been raised as to whether it will be fully implemented, and the Act has been criticised for its lack of depth and completeness. This project's aim is to study the Transport sections of the DDA and discuss its relevant components with service providers and users. The outcome of the research is future recommendations to be included in proposed changes to transport systems. This can limit later costs to service providers, improve their public image and enable more people to travel on public transport vehicles.

## Methodology

Twelve representatives, including director of operations, of 10 Leicestershire operating companies including an airport, 5 rail operators, 3 urban bus operators and 1 inter urban coach operator were interviewed individually. In addition, 49 elderly and disabled people were interviewed in groups or individually. They ranged from 16 to over 75 years old; 35 were physically disabled, 7 were visually impaired and 1 person hearing impaired. The remaining 7 users were elderly people who did not like to term their ailments as a disability.

Interviews with service providers and users took place at their premises or meeting place, and with their permission the interviews were recorded on audio tape. The service provider interviews covered the DDA with regard to transport and the user interviews followed a typical journey using public transport to enable group members to visualise their actions. The user group members also completed a questionnaire to record their personal details.

The audio tapes were transcribed and the information provided by the users' was categorised and summarised into tables with the relevant reply from the service providers. Finally, recommendations were produced either from the literature, or by devising a compromise solution between the needs of the users and the restrictions imposed upon the service providers.

## Literature Review

In Britain in 1991 there were 2.1 million people over the age of 80. This will increase to 3.2 million by 2021 (Stewart-Davis, 1996). In 1997 there were almost 1.1 million people who were economically active and who have long-term illness (Great Britain, 1997). Before transport systems become unacceptable to users we need to address the requirements of an ageing population as well as meeting the needs of fringe members of society (Geehan & Suen, 1993). If people with special needs are to be fully integrated members of society, they need access to public transport (ECMT, 1989; Geehan et al, 1992). Without it they cannot access education and business, which in the UK are the greatest reasons for using public transport (Great Britain, 1997).

To travel people need information for pre-planning, in the terminal and in the vehicle. However there are no consistent methods by which information is communicated by operators to users. The amount of information is often insufficient, in-vehicle information is confusing and signage too small, inappropriately placed, of little contrast and poorly maintained. Information is provided visually or auditorily, rarely in both and usually excludes tactile information (Arnold & Wallersteiner, 1994). Information systems need to harness users' capabilities and be compatible with the users perceptual, cognitive and behavioural characteristics.

Regarding physical access, two of the most common complaints about buses are the high steps and the distance between the vehicle and the kerb (ECMT, 1989; Petzall, 1993). The Petzall study determined that uniform steps were needed, as they require fewer and smaller trunk movements. It is also preferable for the bottom step to be in-line with the pavement, as it reduces leg movement, and the pavement edge marks the starting line of steps for visually impaired people.

It is important for service providers to communicate with users to determine their actual needs and identify their perceptions of public transport. When solutions have been proposed it is important to carry out user testing of transportation technologies to avoid designing with one group in mind and presenting barriers to others (Abdel-Aty and Jovanis, 1997).

## Disability Discrimination Act 1995

The Act defines a disability as physical or mental impairment which has a substantial and long-term adverse effect on [a person's] ability to carry out normal day-today activities (1(1)) and therefore a disabled person is a person who has a disability (1(2)).

### Part III Goods, Facilities and Services

This part of the Act came into force in October 1999 and covers access to, and use of, communication and information services. It is unlawful for a service provider to discriminate against a disabled person by "refusing to provide, or by not providing, any service which he provides, or is prepared to provide to members of the public". It must not be impossible or unreasonably difficult for the disabled person to make use of such service. The standard of service, the manner in which, and terms on which it is provided must not be discriminatory. It therefore follows in Section 21 that if it is impossible or unreasonably difficult for disabled persons to make use of a service it is the service provider's duty to take all reasonable steps in order to change the practice, procedure or policy so that it no longer has that effect.

### Part V Public Transport – Public Service Vehicles (PSV's)

Section 40(1), and sections 46 and 47b which refer only to rail vehicles, state that regulations will be produced to ensure that it is possible for disabled persons –

(a) to get on and off regulated public service vehicles in safety and without unreasonable difficulty (and in the case of disabled persons in wheelchairs to do so while remaining in their wheelchairs); and
(b) to be carried in such vehicles in safety and in reasonable comfort.

## Mapping user needs to service provision

### Access to check-in and information desks

There are few check-in and service desks in terminals at the height of a seated wheelchair. Airports are unable to provide these as the desks are designed to protect staff and equipment. Railway stations will provide lower desks as and when stations are refurbished, but in the meantime service providers are willing to serve passengers in the waiting area.

### Access to vehicles

Bus steps are too high for many passengers, and low floor buses will only be introduced as older buses need replacing. One alternative is "castle kerbs" which raise the level of the pavement to that of the bus. However, low floor buses and "castle kerbs" become redundant if the bus cannot park flush with the kerb, and a good-run to the parking bay is necessary to achieve this. Therefore saw-tooth layouts need to be installed to provide a greater run-in.

### Access to stations and platforms

Physically disabled users stated that they were unable to access stations. Information provided by a station leaseholder showed that of their stations 70% are unstaffed, 70% of stations are wheelchair accessible from the main entrance however only 60% allow access to all platforms and 15% do not afford access to any part of the station. These problems should be overcome by Railtrack's £300 million plan to upgrade and refit stations between 1999 and 2020.

### Facilities - Toilets

Disabled toilets are often provided but they generally only accommodate people in wheelchairs no larger than the government's standard wheelchair. The designers of trains find the positioning of disabled toilets difficult because they take up a lot of room and receive a great deal of criticism from groups representing disabled people

### Personal Assistance

The availability of personal assistance can mean the difference between a person being able to travel or not. However many people are not aware of the services that are available, the costs, if they will be subjected to degrading manual handling practices, or forgotten. All operators have staff training schemes in operation, but the operators state that it is not possible to ensure that individual staff members put this into practice.

### Space and Seating

The physically disabled users agreed that there was not enough leg room and too few spaces for wheelchair users. However more spaces and increased leg room will reduce seating capacity causing problems for ambulant disabled persons travelling at peak times, and cause economic problems for the service providers.

### Colour coding and tactile cues

Three of the visually impaired users compared that bright yellow is used for colour coding which causes them eye strain and headaches. As an alternative red would be preferable. The colour choice coincides with the Rail Vehicle Accessibility Regulations. Tactile cues are often provided for visually impaired persons, however none of those interviewed were aware of the existence of many of the cues – because nobody has told them! Users need more appropriate information provided to inform of the presence of such cues.

### In-terminal information

Users with a visual impairment found that too much audible information was presented at once announcements are indistinguishable from background noise. One way around this can be found in new information sources such as telephone-information kiosks which provide information e.g. the next service from that platform. However the unreliability of services make in-terminal information redundant. Bus and train operators are aware of, and are experimenting with, Real-Time scheduling, which provides accurate information transmitted via satellite from the vehicle to the terminal.

### In-vehicle information

Travellers with visual impairments or those in an unfamiliar place are unsure of their location, especially on buses and trains, when information is not always provided and often inaudible and illegible. On buses and trains users require visual and audible information that is clear, timely and accurate.

## Conclusion

There are many aspects of urban, inter-urban and international travel that makes it difficult or impossible for disabled and elderly people to travel independently and there are many feasible solutions to these problems. However the solutions cannot be implemented overnight and many operators will not provide the solutions without guidance from the Government. Therefore this report acknowledges that whilst the needs of disabled and elderly travellers are not being met, the service providers need time and guidance to implement the relevant improvements. The DDA covers many of the areas highlighted by the users for improvement however some important areas, such as personal assistance, are not included in present regulations. The inclusion of the recommendations in this paper which are not already covered by the DDA will increase the usability and accessibility of transport systems for those with special needs, making the British Transport system truly public.

## Recommendations

- The provision of lower service desk for wheelchair users,
- Improved access to vehicles with castle kerbs, low floor buses and sawtooth access to bus stops.
- Toilet facilities need to be more accessible to those in larger wheelchairs,
- More flexible seating needs to be introduced to allow spaces for wheelchairs to be created with ease.
- Audio-visual information that is distinguishable and readily available is needed for all journey stages from pre-planning to the destination for people of all abilities.

## References

Abdel-Aty, M.A. and Jovanis, P.P, 1997, Using new technologies to meet the transportation demand of the special needs travellers – a framework, *Proceedings of the 30th international symposium on automotive technology and automation*

Arnold, A.K. and Wallersteiner, U., et al, 1994, Human factors evaluation of information systems on board public transportation vehicles: implications for travellers with sensory and cognitive disabilities, *Proceedings of the 12th triennial congress of the International Ergonomics Association*, 15th -19th August.

ECMT, 1989, Transport for People with Mobility Handicaps, *European Conference of Ministers of Transport Seminar, Dunkirk*, 19th November.

Geehan, T., Arnold, A.K., Suen, L., 1992, An ergonomic assessment of assistive listening devices for travellers with hearing impairments, Proceedings of the 6th international conference.

Geehan, T., Suen, L., 1993, User acceptance of advanced traveller information systems for elderly and disabled travellers in Canada, *Proceedings of the IEEE-IEE vehicle navigation and information system conference, Ottowa, Canada*, 12 – 15 October.

Great Britain, 1997, Transport: Social Trends 27, *Office for National Statistics*.

Great Britain, 1998, *The Rail Vehicle Accessibility Regulations*, London: HMSO,

Petzall, J., 1993, Traversing step obstacles with manual wheelchairs, *Applied Ergonomics*, 24(5), 313-326.

Stewart-David, D., 1996, Planning transport for octogenarians, *Global Transport*, 66-68.

# DESIGN ISSUES AND VISUAL IMPAIRMENT

## Katie M. Stabler & Sabine van den Heuvel

*Royal National Institute for the Blind, Product Development Department, Bakewell Road, Orton Southgate, Peterborough PE2 6XU, UK*

There are about 1.7 million people in the UK with a serious sight problem. 90% of these are over 60 years of age, and this is the only age group which is growing. Some estimate that by 2001 over 20 million people in the UK will be aged 50 or over, and this segment will hold around 75% of the nation's wealth. In purely commercial terms it therefore makes good sense for designers and manufacturers to produce products that are suitable for older people, yet today little account has been taken of the characteristics of this group in mainstream product design. This paper aims to highlight the key design issues which must be addressed in order to make products more accessible to visually impaired people; namely, the nature and quality of visual, tactile and auditory product information.

## Introduction

There are many myths about blindness and partial sight. Here are some of them:

| Myth | Fact |
|---|---|
| 'Blind people see nothing' | Only about 18% of visually impaired people are classed as totally blind, and most of these can distinguish between light and dark. |
| 'Blind people have special gifts' | Many blind and partially sighted people have a poorer sense of hearing or touch than sighted people, especially if they are older. However, many blind people have learnt to listen more carefully, or to make more use of their sense of touch. There is no such thing as a 'sixth sense'. |
| 'All blind people read Braille' | The number of people who can read Braille is quite small – about 13,000. To read Braille, it is necessary to develop a good sense of touch and this is often difficult for older people. Many visually impaired people use large print and tape recordings instead of Braille. |
| 'Most blind people own guide dogs' | There are roughly 4,000 guide dog users in the UK, representing a tiny fraction of the visually impaired population. Many prefer to use alternative mobility aids, such as a white cane. |

## What is Visual Impairment?

Many people find it hard to see even after having an eye test and wearing the right spectacles or contact lenses. There are 1.7 million people in the UK who have a serious sight problem which significantly affects the way they live; from reading the daily newspaper and cooking at home, to getting out shopping and socialising. Generally, 'blindness' is regarded as a substantial and permanent lack of sight. 'Partial sight' is a less severe loss of vision. A person can register as partially sighted if they can only see the top letter of the eye chart at a distance of six metres or less, wearing corrective spectacles. Sight loss is one of the commonest causes of disability in the UK. Blind and partially sighted people come from all sorts of backgrounds. They go to school, university, get jobs, bring up families, watch TV, enjoy holidays, friends and hobbies etc - but they may need help to do some or all of these things.

Only 8% of blind and partially sighted people are born with impaired vision. Most visually impaired people have gradually lost their sight in later life as a result of the ageing process. In fact, four out of every five people with impaired vision are over retirement age (see Figure 1), many of whom may also have other disabilities or illnesses such as hearing loss or arthritis.

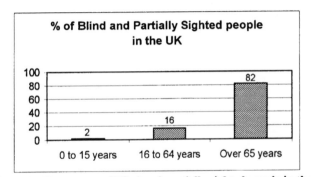

Figure 1 – Percentage of blind and partially sighted people in the UK

The different eye conditions may have a variety of effects on individuals. Very few people see nothing at all. The four most common eye diseases causing low vision in the UK are: macular degeneration, which results in a loss of central vision; diabetic retinopathy, which can result in 'patchy' vision; glaucoma, which can cause loss of peripheral/side vision; and cataracts, which often cause 'misty' vision. So some people with impaired vision can see enough to read this article, although they might have difficulty crossing the road.

## Design issues and visual impairment

When designing a product it is important to focus on all product-related issues, and not just on the product itself. For example, even a well designed product can be badly let down if the user is unable to open the packaging, read, or understand the instructions etc.

The following is a list of aspects to take into consideration when designing a product. We trust that you understand the list is not complete, but it should get you thinking about other aspects of product accessibility for visually impaired people.

- **Information/advertising** - How do people know about the product in the first place? Is the advertising appropriate? (e.g. is a TV advert purely visual, or does it have a verbal commentary?).
- **Packaging** - How easy is it to find where and how to open the box, where are the top and bottom, etc.? (consider people with, for example, arthritis or dexterity problems).
- **Instructions** - Are the instructions easy to read (e.g. large enough print size, Braille or audio tape)? Are they understandable? (lots of complicated diagrams and schematic drawings do not help). Is there a customer help line?
- **Assembly** - Does the product require assembly by the customer? (e.g. does a table lamp need to be attached to its base / have a bulb inserted etc before it can be used).
- **Guarantee/return information** - Is it clear what to do if the product does not work? Where to send it to, who to contact etc.?
- **Cleaning** - Is the product easy to clean? Bear in mind that visually impaired people might find it more difficult than a sighted person to know whether a product is properly cleaned (this is particularly important with equipment which will come into contact with food, or for young children's toys).
- **Safety** - Even if a product complies with all the British and European safety regulations this does not ensure that it is necessarily safe for visually impaired people to use (consider warning lights, sharp edges, finger traps and moving parts). As an extreme example, consider a fully compliant motorcar – it would most likely be unsafe for a blind or partially sighted person to drive on the public road.
- **Replacement parts** - How easy is it to get replacement parts like batteries etc.? Are they readily available, and how expensive are they?
- **Stigma** - It is important that a product does not scream out 'for use by visually impaired people only'. Some people, especially in the younger age groups do not like being labelled by their products as blind or partially sighted. They have just as much right to attractive 'sexy' products as their sighted peers.

### How do you make a product accessible?

In this paper, we address how to make products more accessible for people with a visual impairment only, but in real life it is important to take other age-related disabilities into account for this group of people (e.g. hearing loss, arthritis, dexterity problems, etc.). The following three areas should be taken into consideration when designing for visually impaired people.

### 1.  *Visual information*

As most visually impaired people have some useful residual vision (72% are able to read large print – 14 point Arial). Attention should be paid to the visual aspects of the product's design. The following points are important to make a product 'easy to see'.

- Effective colours: Yellow seems to be a particularly easy colour to see and is often the last to be 'lost'. It stands out in many situations, especially against black.
  Red stands out in good lighting conditions, though soon disappears as light fades.

- Contrast in tone and colour are essential. A difference in tone is usually more effective than a colour difference. For example, bright red and blue contrast greatly in terms of colour, but very little in tone. Very light grey and very dark grey have no colour contrast but very good tonal contrast.

Red / Blue   Dark / Light

- As a general rule for lettering, use dark characters against a light background; although the reverse can work well on larger signs, e.g. yellow letters on a matt black background.
- Glare should be avoided at all times. Shiny surfaces reflect light in a way that can be confusing or uncomfortable to the eye.
- Single upper case letters are easier to read than single lower case. **H h**
- Words and sentences in mixed case lettering are easier to read than WORDS AND SENTENCES WHERE THE WORDS ARE ALL UPPER CASE.
- Use a plain font with open letters, such as Arial *(Don't use fancy fonts)*.
- Don't leave too large spaces between the words. And don't use justified text, as both these make documents difficult to read for some people who have a limited field of vision and/or who may be using low vision aids (e.g. a high power magnifier, which might focus just a few letters at a time).
- ## A minimum of 14 point font size is recommended.
- Touch screens and membrane key pads are very difficult for visually impaired people to use as it is not always possible to see or feel exactly where to press, and there is sometimes no tactile feedback. It is, therefore, important that the size and contrast of the characters on the screen/keypad are maximised to make them more suitable for visually impaired people.

### 2. Auditory information

- It is useful for a click to be heard/felt as confirmation of pressing a button.
- Use varied volume, pitch and duration in auditory signals to distinguish between the product's various functions (e.g. on = 1 bleep, off = 2 bleeps).
- For hearing impaired people it is important to use signal frequencies that can be heard. Consider that as a result of the ageing process, people tend to lose the ability to hear higher frequencies first, therefore, a male voice, is generally preferred.
- When using speech, it is important to consider accents and languages in terms of understanding and personal preference (e.g. some English people find it difficult to understand a synthetic voice which has an American or Oriental accent).
- A volume control and/or headphones are recommended for privacy (consider talking bathroom scales, or using a talking watch in a meeting).

### 3. Tactile information

The number of Braille readers in Europe is less than 0.02% of the population. So although useful for some blind users, Braille is not a total solution for visually impaired users. (John Gill, 1998). In the UK only around 13,000 visually impaired people read Braille. Aside from achieving the standard Braille profile, it is important to get tactile features right, particularly when considering the needs of elderly visually impaired people. With age, tactile sensitivity may be reduced due to loss of feeling in the fingertips (common age-related causes of which, are diabetes, strokes and circulation problems etc).

- Tactile markings need to be much bigger than their printed equivalent.
- Orientation cues are crucial (to ensure you have the product the right way up/round).
- If two markings have to be distinguished from one other, it is important to make them feel as different as possible. Shape, size, height and texture can be used to differentiate between markings.
- It is important that tactile information does not cover up visual information, as people may use both.
- Vibratory output can be used for deaf-blind people.

Each of the above areas is key to ensuring that a product is accessible for as many people as possible. A good rule of thumb, when designing for visually impaired people, is to use a combination of different types of feedback. Nevertheless, feedback given by all sources must be identical (e.g. the analogue hands on a clock should read exactly the same time as its speech output).

## Conclusion

This paper has outlined some simple and cost effective design solutions to make products more accessible for visually impaired people.

The RNIB Product Development team has had a decade of experience in designing and user testing specialist products to meet the needs of visually impaired and elderly people. We appreciate that there will always be a need for specialist products, but our long-term goal is to reduce the need for RNIB in-house design, by advising mainstream designers and manufacturers of the needs of visually impaired and older people. By actively promoting a 'design for all' approach, we hope to improve the lives of older, disabled and visually impaired people by making mainstream products affordable and accessible. Indeed, many mainstream products can easily be made more accessible without necessitating major design changes or costs. Take for example, the buttons on a phone : simply by adding some colour contrast with the background, upping the print size, and using a mix of upper and lower case lettering; buttons are immediately easier to see for visually impaired people. If mainstream products are more accessible for visually impaired people, it follows that they will also be more accessible for everyone. This is especially true for those of us, who, having lost our reading glasses, might find it a strain to find a button on a remote control, or read a small printed label.

For more detailed design guidelines, and specific information on product design and evaluation, please do not hesitate to contact us at the RNIB Product Development Department in Peterborough. Tel: 01733-375168/5155
For general information on visual impairment see our web page:   www.rnib.org.uk

## References

Gill J. 1998, *Access prohibited*, (Royal National Institute for the Blind, on behalf of Include)
RNIB, 1990, *General needs survey*, (Royal National Institute for the Blind)
RNIB website, 1999, www.rnib.org.uk

# 2001

| **Royal Agricultural College Cirencester** | 10th–12th April |
| **Conference Manager** | Carol Greenwood |
| **Chair of Meetings** | Sandy Robertson |
| **Programme Secretary** | Margaret Hanson |

| **Secretariat** | Paul McCabe | Phil Bust | Guy Walker |
| | Tim Dubé | Catherine Ejiogu | Jenny Boughton |
| | Adam Parkes | Thea Vigli-Papdaki | Michael Hartley |
| | Dan Makhan | Nikki Bristol | |

## Social Entertainment

The Victorian Gothic style hall, of the first agricultural college in the English speaking world, was the setting for both the Taylor and Francis drinks reception and the post annual dinner disco.

| Chapter | Title | Author |
| --- | --- | --- |
| **Health and Safety** | Actions of older people affect their risk of falling on stairs | R.A. Haslam, L.D. Hill, P.A. Howarth, K. Brooke-Wavell and J.E. Sloane |
| **Education** | The teaching of ergonomics in schools | A. Woodcock and H. Denton |
| **Selling and Communicating Ergonomics** | The costs and benefits of office ergonomics | S. Mackenzie and R. Benedyk |
| **Additional Papers** | A profile of professional ergonomists | R.B. Stammers and E.J. Tomkinson |

# ACTIONS OF OLDER PEOPLE AFFECT THEIR RISK OF FALLING ON STAIRS

**R A Haslam, L D Hill, P A Howarth, K Brooke-Wavell and J E Sloane**

*Health & Safety Ergonomics Unit*
*Department of Human Sciences, Loughborough University*
*Loughborough, Leicestershire, LE11 3TU*
*R.A.Haslam@lboro.ac.uk*

An interview and environment survey was undertaken to investigate how older people keep and use their stairs, with regard to stair safety. Visits were made to 157 older people, aged between 65-96 years, in their own homes. Information was collected under three headings (1) behaviour involved in direct use of stairs, (2) decisions and actions which change the stair environment and (3) behaviour affecting individual capability. The findings indicate that behaviour-based risk factors for falling on stairs are widespread among older people, while only 13% of those interviewed were able to remember ever receiving advice about stair safety. Simple, inexpensive measures exist which individuals can take to reduce their risk of falling on stairs. Primary care providers have an important contribution to make in encouraging implementation of these.

## Introduction

Older people falling on stairs in the home is a serious problem, resulting in up to 1000 deaths and 57,000 hospital A&E attendances each year (DTI, 2000). In 22,000 of these incidents, casualties suffer a fracture, concussion, or otherwise require admission to hospital for more than a day. The cost to the health services and wider community of caring for these patients is substantial. Also, falls have serious psychological and social consequences for the individual, affecting mobility, confidence and quality of life. Although the personal and environmental factors involved in falls on stairs are well known, the influence of behaviour has received much less attention from researchers (Connell and Wolf, 1997). The aim of this investigation was to improve understanding of how older people keep and use their stairs, considering the implications for stair safety.

## Behaviour and falls

A series of preliminary focus groups, conducted at the commencement of the project, generated an initial model demonstrating how the behaviour of older people might contribute to risk of falling on stairs (Hill *et al*, 2000). The model suggests that increased risk of falling on stairs arises from the way individuals interact with stairs directly, as a

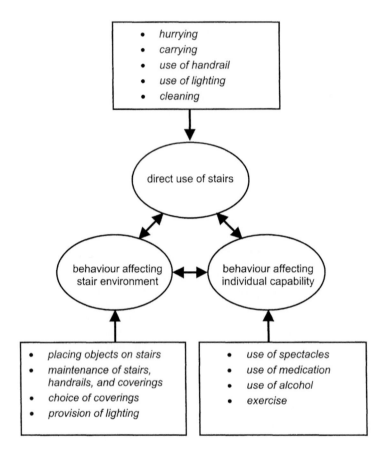

**Figure 1. Behaviour-based risk factors for older
people falling on stairs**

consequence of actions which modify the stair environment, or through behaviour affecting individual capability, figure 1. This framework underpinned the main element of the research, a home interview and stair environment survey.

## Interview and stair environment survey

An interview and stair environment survey was undertaken to validate the findings from the focus groups and to quantify different risks. Semi-structured interviews were conducted with 157 community dwelling individuals, aged 65-96 years, in their own homes. The interviews collected qualitative and quantitative information on the behaviour of respondents on and around their stairs, awareness of safety factors, and history of falling. During each home visit, lasting approximately 2 hours, details were recorded and photographs taken as to the design and repair of stair coverings, number and condition of handrails, objects on and around the stairs, and provision of lighting. Measures were also taken of health status and vision.

Participants were recruited on a quota basis, according to age and gender, using estimated UK population figures from the Office for National Statistics. Likewise, dwellings were selected based on national estimates of housing stock, with respect to age and type of dwelling.

## Results and discussion

### Direct use of stairs
A majority of participants (63%) reported hurrying on stairs on occasions, despite 89% acknowledging this as increasing the likelihood of falling. The most common reasons given for hurrying were to respond to someone at the door, answer the telephone, retrieve items left upstairs or to use an upstairs toilet.

Although 92% of interviewees recognised that carrying items up and down stairs could be hazardous, 29% said they would still attempt to carry something that might cause them difficulty. Among the items interviewees reported carrying on stairs were laundry, vacuum cleaners and walking sticks.

Adequate lighting is important for safety on stairs. Low day time illuminance (<50 lux) was found in 61% of the survey households. Almost a quarter (23%) of interviewees living in these homes said that they do not switch on stair lights during the day. Interestingly, nearly one fifth of participants (18%) indicated that they do not switch on lights when using the stairs during the night. Reasons given for this included that sufficient light was already present, eg from street lighting outside, or not wanting to disturb a sleeping partner, or to make it easier for the person to get back to sleep.

Cleaning stairs appears to be a struggle for many older people, with 58% of interviewees identifying difficulties with this activity. It was reported that problems arise due to a combination of difficult access (eg landing windows) or the need to use heavy equipment, such as a vacuum cleaner.

### Behaviour affecting stair environment
Although many participants (64%) recognised that objects on stairs may be a hazard, items were found on the stairs in 29% of households, see figure 2, for example. Objects on stairs may either be permanent, for example furniture, or placed there on a temporary basis. Permanent items were more common in smaller dwellings (p<0.001), presumably because space is at a premium. Discussions with interviewees revealed that temporary items are often left on stairs for taking up or down later.

While most participants (86%) reported their stair covering to be in reasonable condition, in 29% of cases the stair covering was judged by the researcher to be in need of replacement or repair. It seems likely that either the interviewees had a lower threshold for what constitutes reasonable condition, or else they had not noticed wear and tear that may have happened gradually over many years.

Although not quantified, many examples were seen during the home visits where the carpet colour and pattern on stairs makes it difficult to distinguish edges of steps, figure 3. Generally, coverings light in colour and non-patterned make steps more visible, while those that are heavily patterned may cause problems. There was low awareness that this might be an issue among survey participants.

**Figure 2.  Objects placed on stairs**            **Figure 3.  Carpet design can camouflage step edges**

*Behaviour affecting individual capability*

Among study participants, 82% were taking at least one prescribed medication daily, with 16% reporting that their medication makes them feel drowsy, dizzy or affects their vision. Over a third of participants (38%) reported drinking alcohol when taking prescribed medications. Almost all of the sample (99%) had spectacles, with 16% indicating these cause visual difficulties on stairs, as a result of distortion from bifocals, for example. Although not prompted during the interview, a number of participants (10%) specifically mentioned using the stairs as a form of daily exercise. Although exercise is considered beneficial in reducing falls among older people (Health Education Authority, 1999), advice encouraging use of stairs for exercise needs to consider the increased risk this might create for some individuals of falling.

## Conclusions

Falls on stairs typically involve an interaction of events (Templer, 1992), illustrated by the following accounts from interviewees:

> *"I was carrying the washing basket downstairs and feeling unwell. I had a lot on my mind and slipped and fell all the way down to the bottom."*

*"I got into bed, I'd had a drink and realised that I hadn't got any meat out for Sunday, rushed downstairs and slipped. The shopping trolley was at the bottom, I wouldn't have hurt myself if it wasn't for the trolley ... I cracked my ribs. But that was my own fault, I'd had a drink, and I had nothing on me feet."*

Prevention depends on eliminating or reducing as many different risks as possible. Although it may be difficult to improve fixed features of the environment, such as steep or awkward stairs, there is scope to address behaviour-based risk factors. Encouragement comes from randomised trials, which have shown interventions combining medical assessment, home safety advice and exercise programmes to reduce falls (Close *et al*, 1999; Steinberg *et al*, 2000).

The results of this study indicate that occurrence of behaviour-based risk factors for falling on stairs is widespread. While many participants were able to recognise and appreciate risk factors once prompted, many had not previously given much thought to stair safety. A notable finding is that only 13% of the sample recalled ever having received advice about stair safety. The UK government 'Avoiding Slips, Trips and Broken Hips' campaign is a welcome initiative in this respect.

## Acknowledgements

This research was funded by the Department of Trade and Industry (DTI). The 'Avoiding Slips, Trips and Broken Hips' falls prevention (for older people) campaign is run by DTI in association with Health Promotion England. An information pack 'Step up to Safety' is available from DTI Publications Order Line on 0870 1502 500.

## References

Close J., Ellis M., Hooper R., Glucksman E., Jackson S. and Swift C. 1999, Prevention of falls in the elderly trial (PROFET): a randomised controlled trial, *The Lancet*, **353**, 93-97

Connell B.R. and Wolf S.L. 1997, Environmental and behavioral circumstances associated with falls at home among healthy elderly individuals, *Archives of Physical Medicine and Rehabilitation*, **78**, 179-186

Department of Trade and Industry (DTI), 2000, *Home accident surveillance system including leisure activities, 22nd annual report 1998 data* (Department of Trade and Industry, London)

Health Education Authority (HEA), 1999, *Physical activity and the prevention and management of falls and accidents among older people* (Health Education Authority, London)

Hill L.D., Haslam R.A., Howarth P.A., Brooke-Wavell K. and Sloane J.E., 2000, *Safety of older people on stairs: behavioural factors* (Department of Trade and Industry, London), DTI reference 00/788

Steinberg M., Cartwright C., Peel N. and Williams G. 2000, A sustainable programme to prevent falls and near falls in community dwelling older people: results of a randomised trial, *Journal of Epidemiology & Community Health*, **54**, 227-232

Templer J. 1992, *The staircase: studies of hazards, falls and safer design* (MIT Press, Massachusetts)

# THE TEACHING OF ERGONOMICS IN SCHOOLS

## Andree Woodcock[1] and Howard Denton[2]

[1]*VIDe Research Centre, School of Art and Design, Coventry University ,Gosford Street ,
Coventry, A.Woodcock@coventry.ac.uk*
[2]*Department of Design and Technology, Loughborough University, Loughborough,
H.G.Denton@lboro.ac.uk*

The relationship between ergonomics and the disciplines it informs has always
been tenuous. Woodcock and Galer Flyte (1997) hypothesized that teaching
ergonomics in schools would lead to a greater acceptance and willingness to learn
and use ergonomics techniques during tertiary education and once in professional
practice. This paper discusses the relationship between ergonomics and design,
and considers the teaching of ergonomics in secondary schools. Preliminary
results of surveys conducted with first year undergraduates to investigate the
teaching of ergonomics they received at both 'GCSE' and 'A' level are presented
which indicate that most ergonomics education occurred, as expected, in design
and technology courses, but was also present, though patchy, in other disciplines.

## Introduction

Gaining acceptance and use of ergonomics by designers and other professions has been
difficult. This seems somewhat strange, firstly because humans have what appears to be an
almost innate capacity for understanding and applying the principles of ergonomics. For
example, Australopithicus Prometheus selected pebble tools and made scoops from antelope
bones to make tasks easier to perform; in their use of computers for distance co-operation,
designers will select those features which enable them to complete the task as effectively and
efficiently as possible (Scrivener, Chen and Woodcock, 2000); additionally it is believed that
children, even at pre-school level perform user centred design activities in their play and make
activities (Woodcock and Galer Flyte, 1998). Secondly, ergonomics has proved to be
beneficial when used to inform product and work place design. Thirdly, industry is seeking
graduates who are able to design for niche markets.

There have been attempts to improve communication between ergonomics and the
disciplines it seeks to inform. Woodcock and Galer Flyte (1998) reviewed these approaches
which have included in-house training schemes (Shapiro, 1995), production of designer-
friendly literature (Chapanis, 1990), development of techniques which could directly benefit

many stages of the design process (Simpson and Mason, 1983) and the integration of ergonomics into other undergraduate courses (Woodcock and Galer Flyte, 1997). These are gradually changing the design and engineering climate. However, these approaches focus on practicing designers or students in tertiary education.

Informal observation has revealed some antagonism towards ergonomics amongst undergraduate engineering and design students. If unchallenged, this attitude might predetermine the use of ergonomics in later careers. For example, Meister (1982) showed that the attitude of senior managers affected the tone of the department, and that one of the ways this could be changed was through the greater integration of ergonomics and the continuing education of managers. If undergraduate engineers and designers are unappreciative of the benefits of ergonomics then we need to consider the reasons for this. This may include multidisciplinary course structure; assessments emphasising the mastery of skills and creativity at the expense of user evaluation. It is hypothesized that the failure to appreciate the value of ergonomics may be partly based on pre-university experiences of ergonomics.

Previous research suggests that pre-school children are cognisant of the need to consider others in their play and make activities and that this empathy continues in primary schools, but is downplayed in secondary school curricular. Secondary schools are largely structured around the examination of individuals. Long tradition of solo endeavour may result in undergraduates who (for the most part) are antipathetic in their views towards user issues. It is argued that this is a lost opportunity and that the discipline of ergonomics has much to offer teachers, pupils and curriculum developers not only as a discipline in its own right, but also as a means of integrating diverse areas of the curriculum and enhancing learning experiences. The rest of the paper discusses investigations undertaken to assess the way in which ergonomics is taught in secondary schools..

## Secondary education

A computer based search of the National Curriculum requirements for all subjects at Key Stages 3 (age 11-14) and 4 (age 14–16) showed no specific mention of 'ergonomics', 'anthropometrics' or 'human factors' (National Curriculum, 1999). Currently the Qualifications and Curriculum Authority (QCA) is conducting a major review of National Curriculum requirements and qualifications such as 'A' level. Their guidelines for 'A' level subjects were searched for the terms 'ergonomics', 'human factors' and 'anthropometrics' in the subjects: design and technology, physics, maths, biology, physical education and art. The only hit scored was on 'ergonomics' in 'A' level design and technology guidelines. Here the requirement is for all new 'A' level Design and Technology syllabi to include a section on 'planning and evaluating' and within this is: section c. 'use ICT appropriately for planning and data handling, for example, the use of data base, drawing and publishing and design software. Interpret design data such as properties of materials, ergonomics and nutritional information.' (Midland Examining Group, 1999). 'A' level grade descriptions for Design and Technology is the statement (grade A) that: 'when developing and communicating ideas, take into account functionality, aesthetics, ergonomics, maintainability, quality and user preferences.....'

The overall position, however is clear: a student with 'A' levels in maths or a science may have looked at ergonomics, but it is more likely that they will not. A student with 'A' level 'Design and Technology' should have learned something about ergonomics and should have applied this knowledge to project work. The following investigations were carried out to verify this in terms of undergraduate knowledge and attitudes towards ergonomics.

## Investigation 1

The aim of this study was to investigate further the current teaching of ergonomics in schools at GCSE and 'A' level to ascertain the subjects in which ergonomics is being taught, and the methods used to teach it. Such information is necessary for curriculum developers who might wish to co-ordinate activity across subjects, educational product designers who might wish to develop teacher support material and university staff who wish to better understand the prior learning of students. The questionnaire consisted of five main sections:

1. Personal details relating to the age and gender of the respondent, academic course, and willingness to participate in follow up interviews.
2. School details - to enable a follow up survey of selected schools, to discover how ergonomics is taught in these schools.
3. Subjects taken at 'A' level. This also attempted to establish the amount of ergonomics in these subjects (coded into a four point scale from nil to high) and the manner in which that ergonomics was taught (video, reading, lesson, applied in a project etc
4. Subjects taken at GCSE, using the same format at (3) above.
5. Ergonomics questions - to gain an impression of the respondents depth of understanding of ergonomics and perception of the value of ergonomics in their future careers.

Most of the sample were enrolled on engineering courses-Mechanical Engineering, Mechanical Engineering and Manufacturing (MEM), Civil Engineering, Engineering, Science and Technology (EST), and Aeronautical and Automotive Engineering (AA) with a high proportion from Industrial Design/Industrial Design and Education courses, and Product Design / Product Design and Manufacturing (PDM). A third, smaller group consisted of ergonomists, psychologists and human biologists. Over 80% of the respondents were male, of age 18 or 19, with very few mature entrants. This is a fair reflection of the trend within Loughborough University. The following results are based on over 350 responses.

At 'A' level it is only in Design and Technology that ergonomics is taught to any great extent, as is to be expected. However, it is interesting to note that ergonomics was being taught in some schools in other disciplines, most notably IT, sport and geography. Only in a few cases were single methods (e.g. lecture, handouts) used to teach ergonomics. It was shown, not surprisingly, that students who had been exposed to a greater number of teaching methods had a higher level of understanding of ergonomics. These results would seem to indicate that multiple teaching methods appear to be more effective than the single use of any method including project work.

Other trends in the data revealed that younger respondents had slightly more knowledge of ergonomics than older ones, but the sample size for older participants was too small for this to achieve any levels of significance. With regard to the extent to which ergonomics was considered to be important in a later career, 299 thought it was important and 32 that it was not. Of this latter group, 20 of were civil engineers and 9, mechanical engineers. The results also indicated a gender difference, with females having a higher regard of the value of ergonomics to their future careers (mean = 2.11), than male respondents (mean = 1.86).

Only those students who have completed a full GCSE in Design and Technology and particularly those who have taken an 'A' level in Design and Technology can be assumed to have studied ergonomics at school. Individually, students may have covered aspects of ergonomics in independent study or where ergonomic data is used as an illustration of data handling. Older teachers of Design and Technology will not have learned ergonomics as part of their initial training. Students are dependent on these staff having read up on the subject when it entered these syllabi in the 1970's; not all can be relied upon to have done this.

At GCSE level less than half the sample had some remembrance of ergonomics being taught, for example, in technology, art, PE, dance and drama, and the humanities (most notably geography), as well as in the more design related courses. However it formed a very minor element unless developed by the teacher out of personal interest.

*Investigation 2*

In 2000 a second questionnaire survey was conducted with 66 consumer, product and transport design students to gain further insight into the teaching methods and their level of understanding of ergonomics when they entered the first year of their university studies. The responses from the mainly male (86%) respondents may be summarised as follows.

83% would have liked more ergonomics tuition at school either because it would have made their present course easier (30%) or because they believed it to be an integral part of the design process (50%). Those 17% who did not want ergonomics tuition at school perceived it to be boring and uninteresting, as restricting the creative part of the design a premium was placed on appearance at 'A' level, the use of ergonomics never really changed the end design that had been planned, and it detracted from the more important aspects of the course.

In terms of the teaching of ergonomics, 38% felt that they had been taught well, but a third thought that the teaching had been poor. The remaining third had had no experience of using or being taught ergonomics. As in the previous investigation, teaching was again through multiple teaching methods (11%, 1 method; 15%, 2 methods; 9%, 3 methods and 20% having received ergonomics teaching through multiple methods (4-7 methods)). These included handouts (17%), projects and discussions (both 14%) and reading. Other methods included lectures, video and television, tutorials, past exam papers, practicals and design journals. Topics which were covered included anthropometry (30%), human machine interface (16%) and others including vision, comfort, aesthetics, texture, psychology and social aspects.

When asked to consider the most appropriate method to teach ergonomics, just under a third of the students felt that it should be a more hands on experience, with examples and interactive lectures. 9% felt that if ergonomics was to be taught then it should be considered an important, integral part of design and that this should be reflected in assessment. Under 5% felt that there was scope for ergonomics to be considered earlier than at secondary school.

In terms of the subjects where ergonomics was taught the results confirmed the earlier study. Ergonomics was never taught in languages, classics, maths, humanities or economics. There was considerable amount of variation (possible due to the interest and knowledge of the teacher) in the amount taught in computing, art, IT, 3d design and physics. Most ergonomics was taught in design/technology, art and design, design, design and graphic communication.

In the design related subjects ergonomics was mainly taught through large projects such as transport design (e.g. car, lorry, scooter), chairs (orthopaedic, bench, chair, wheelchair), product (lectern, shrine, sundial, multi-gym). In small projects, design was more varied and product related. In terms of when ergonomics should be used or had been experienced, 34% answered during planning, 32% during design and 20% in evaluation. This implies a naiveté in the sample as to where ergonomics could/should be used during the design cycle.

In the first investigation it had not been possible to clearly ascertain the level of ergonomics knowledge of the undergraduates. Here, when directly asked 17% admitted to having no knowledge at all, 26% claimed little knowledge, 39% some and 17% thought they had a good deal. At the end of the questionnaire students were asked to consider ergonomics aspects of the interior of vehicles in terms of broad categories (e.g. seating, dashboard design), and then to break these down into subcategories (e.g. visibility, adjustability, comfort, reachability). The extent of ergonomics knowledge displayed by some of the respondents was surprisingly high and diverse. This might have been an artefact of the cohort who hoped to specialise in automotive design.

## Discussion and future work

The results presented show that ergonomics is being taught in schools in design related disciplines. Clearly some teaching occurs in other subjects where appropriate. A diverse range of teaching methods are employed. The results indicate that where students are exposed to a diversity of educational material, teaching and learning methods they may gain a higher understanding of the discipline.

Some students in the second investigation expressed an antipathy towards ergonomics, and felt it was not part of the design process. There is a slight indication that this may be due to the way in which it is taught in secondary schools. The results show a need to develop student centred, experiential material and to consider the manner in which ergonomics is assessed if it is taught so it is not seen as detracting from the main focus of design.

Future work should address the development of student centred resource packs which will enable students to grasp the main principles of ergonomics in an enjoyable manner and what works well in current practice. From the research we know which schools were rated highly by students in their teaching of ergonomics and these can be targeted for further study.

No evidence emerged in these investigations for the use of ergonomics in cross curricular activities. This might have been due to the manner employed in the data collection, an inability of the students to identify those areas in the curriculum where they had been taught ergonomics by a different name, or it may simply not be happening. For example, with children and schools developing their own web sites, there are opportunities to consider hci issues; to develop cross cultural studies and experimentation through the use of shared resources; in the development of more complex areas such as citizenship. For example a safe driving campaign could consider ways to make passengers, pedestrians and drivers safer, consider vehicle dynamics and the human factors issues surrounding accident causation and what happens during accidents.

As the National Curriculum is flexible in terms of the precise nature of what is taught and how it is taught teachers could use ergonomics as a vehicle for learning in several subjects, for example, within maths, ergonomics could be a subject for statistical data management and interpretation; physical education and art might also raise the subject in various ways. Such educational resources could not be developed by the teachers themselves but possibly could in partnership with ergonomists.In conclusion it would appear from our work that little attention has been given to the teaching of ergonomics to the under 18s and that this area is ripe for the development of innovative, challenging and multidisciplinary teaching methods and resources.

# References

Chapanis, A. 1990, To communicate the human factors message you have to know what the message is and how to communicate it, *Communiqué, Human Factors Association of Canada,* 21/2, 1-4

Denton, H.G. and Woodcock, A. 1999, Investigation of the teaching of ergonomics in secondary schools - preliminary results, In Juster N.P. (ed) *The Continuum of Design Education,* (London: Professional Engineering Publishing), 129-138

Meister, D. 1982, Human factors problems and solutions, *Applied Ergonomics,* **13**, 3, 219-223.

Midland Examining Group Design and Technology: Resistant Materials 1999. Page 10.

The National Curriculum 1999, Department for Education and Employment, (http://www.qca.org.uk/) April

Scrivener, S.A.R, Chen, C.D. and Woodcock, A. 2000, Using multimedia mechanism shifts to uncover design communication needs, in S.A.R. Scrivener, L. Ball and A. Woodcock (eds) *Collaborative Design, CoDesigning 2000,* September 13th-15th, Coventry, 349-359

Shapiro, R.G. 1995, How can human factors education meet industry needs, *Ergonomics in Design,* 32.

Simpson, G. C. and Mason, S. 1983, Design aids for designers: An effective role for ergonomics, *Applied Ergonomics,* **14**, 3, 177-83.

Woodcock, A and Galer Flyte, M. 1997, Development of computer based tools to support the use of ergonomics in design education and practice. *Digital Creativity,* **8**, 3 & 4, 113-120

Woodcock, A. and Galer Flyte, M. 1998, Ergonomics: it's never too soon to start, *Product Design Education Conf*erence, University of Glamorgan, 6th-7th July

Woodcock, A. and Galer Flyte, M. 1998, Supporting the integration of ergonomics in an engineering design environment, *Tools and Methods for Concurrent Engineering '98,* 21-23rd April, Manchester, England, 152-168

# THE COSTS AND BENEFITS OF OFFICE ERGONOMICS

## Sue Mackenzie[1] and Rachel Benedyk[2]

[1] *Nestlé UK Ltd., Hayes, Middlesex, UB3 4RF*
[2] *University College London, 26 Bedford Way, London, WC1H OAP*

The aim of this project was to identify the costs and benefits of office ergonomics. The project centres on a computerised office where operators input newspaper advertisements. Since 1992 the company concerned has realised the potential of this work to put staff at risk of developing musculoskeletal disorders. Aware that the ergonomic intervention within the office has been costly though beneficial, the Health and Safety staff were keen to compare the costs and benefits.

The analysis showed that the resulting benefits would equal the cost of the intervention in approximately two years following the implementation. The study also revealed a number of additional relevant benefits, which could not readily be costed or included in a financial analysis.

In addition the study concluded that there are a number of aspects regarding the use of cost benefit analysis methods for ergonomic interventions which require further investigation.

## Introduction

Loot is a leading free advertisement paper in the UK. The Free Ads department employs approximately 170 members of staff – the majority of whom are involved in copytaking - inputting adverts received via post, fax, mail, live telephone, recorded telephone messages and e-mail (these staff are referred to as ad takers). This work is repetitive in nature and often involves working to deadlines. Workstations are shared, rather than owned by ad takers, as space in the department is limited. Loot early on recognized the potential for their employees to develop WRULDs (work related upper limb disorders) as a result of this work.

Since 1992 Loot have made efforts to improve the working environment of their offices with prevention of WRULDs in mind. A considerable amount of time, money and effort has gone into these improvements. A consultant Ergonomist assessed the work environment and suggested a number of changes to the work environment. The changes ranged from alterations in the routine of ad takers, changes in work organisation, through to recommendations for lighting and adjustable furniture to suit all workers.

## Cost and benefit comparison

*Selecting a tool*
Simpson and Mason 1995 suggest the identification of core costs arising from health and safety issues to illustrate the benefit of ergonomics. This involves calculating the sickness and disruption costs avoided by the use of ergonomics. The pay-back method (Oxenburgh 1991) goes one stage further than this, calculating the effect of sickness and disruption on the productivity of staff and producing a period in which the cost of the ergonomic changes is repaid. This has the advantage of being a method and term with which company managers will be familiar. Both of these methods concentrates on cost avoided, and do not attempt to cost any benefits gained which are not avoided costs.

*Pay-back period*
This analysis is based upon data gathered concerning the costs of the work force. These are usually calculated for the current or pre-intervention state and then reworked for the predicted or post-intervention state. The costs can then be compared and cost of the changes included in the calculation. Any improvements related to prevention of injuries will be shown as a reduction in absence time, decreased staff turnover and related indirect costs.

*Data selection and collection*
The main aim of the Health and Safety staff was to assess the costs associated with the development of WRULDs and compare it with the cost of the changes. After studying the literature, including the core costs proposed by Simpson and Mason (1995) a comprehensive list was developed of possible costs to the company of a staff member needing to leave their job due to WRULDs. Further items were added following staff interviews and some items were discarded as irrelevant in the situation at Loot.

It was estimated that an injured worker would be absent from work on full pay for two weeks. If WRULDs is diagnosed, then Loot would then pay for extended sick leave - six weeks in each case. It is also likely that prior to taking any sick leave the employee would be less productive at work - in the calculation this period is three months. If the medical diagnosis confirms that the injured worker has WRULDs, Loot will offer the injured employee physiotherapy, counseling and retraining.

As few pre-intervention data were available, two scenarios were developed in order to estimate the pay-back period that may apply to Loot:
1) The current situation - using current data on staff turnover, absence and number of staff employed as a base line.
2) The current situation data is reworked to include upper limb injury rates prior to the ergonomic intervention. In 1991-1992 the number of staff leaving their job in Free Ads due to WRULDs was between 2 and 5 per year. The ergonomic changes started after the consultation of an Ergonomist at the end of 1992.

The entire cost calculation comprises of four elements or groups.
Group 1 : the average number of hours for which each employee is productive per year.

Group 2 : the average cost of the wage for each employee per productive hour.
Group 3 : costs associated with losing staff and recruiting new staff.
Group 4 : calculated costs associated with covering or hiring temporary staff to cover for
sickness and reduced productivity related to poor equipment or injury caused by the work
place.

These are first calculated for scenario 1 and then reworked for scenario 2 so that the
difference between the two scenarios can be identified. If the costs are greater in scenario 2,
then the difference will indicate the benefit achieved through the reduced injury rate in
scenario 1.

### Table 1. Pay-back period calculations

|  | Current situation Scenario 1 | Including injury rate from 91-92 Scenario 2 | Benefit value of Scenario 1 over Scenario 2 (Difference) |
|---|---|---|---|
| Group 1 Average productive hours per employee per year | 910 | 907 | 3 hours/year |
| Group 2 Average wage cost per employee per productive hour | 8.11 | 8.14 | 0.03 £/hour Total, £4,859 / year |
| Group 3 Cost of recruitment | 121,817 | 131,187 | 9,370 £/year |
| Group 4 Cost of productivity losses due to injured workers with WRULDs | 0 | · 5,374 | 5,374 £/year |
| Net benefit. This is the total benefit from groups 2,3, and 4. |  |  | 19,603 £/year |

Table 1 shows that a saving of £19,603 per year is achieved in scenario 1 (i.e. with the
ergonomic intervention). This is calculated by adding the differences or benefits from group
2,3 and 4 calculations (group 1 calculations are necessary only to feed into the calculations
of group 2,3 and 4). This saving would occur for every year that the injuries in scenario 2 are
avoided.

When the differences are examined on a per employee basis they seem to be small, e.g.
the difference in average wage cost per employee is only an extra three pence per productive
hour in scenario 2. When these are worked up for the total workforce of ad takers the
differences are more substantial, nearly five thousand pounds per year all together.

The pay-back period is calculated by dividing the cost of the improvements by the net
benefit:

Pay-back period = $\dfrac{\text{Cost of improvements}}{\text{Net benefit}}$

The cost of the ergonomic intervention was £43,270. This includes the cost of the Consultant Ergonomist, all the furniture and equipment purchased and any necessary maintenance.

$$\text{Pay-back} = \frac{\text{Cost of improvements}}{\text{Net benefit}} \qquad = \frac{43,270}{19,603} \qquad = 2.2 \text{ years}$$

The result in this case is 2.2 years, i.e. the cost of the ergonomic measures is recouped in two years and 3 months. Beyond this period, actual savings are realised.

During discussions with staff a list was developed of costs to the company of a typical WRULD sufferer. As some of these were not easily identifiable in the pay-back calculations, they were listed and approximate costs were then calculated. This is presented in table 2, note that the costs are shown in the equivalent number of hours of the injured employee. The wage cost for the employee is taken at £7.10 per hour.

**Table 2. The cost of one worker developing WRULDs**

| Cost item | Cost in Employee hours equivalent | Explanation |
|---|---|---|
| Normal sick pay | 40 | 2 weeks (part time staff working 20 hours/week) |
| Extended sick pay | 120 | 6 weeks |
| Cover for sick leave | 160 | 8 weeks |
| Decreased productivity, due to injury | 60 | Estimated at 25 % reduced productivity for three months prior to sick leave |
| Visits to Doctor/therapist | 6 | Estimated at 2hrs away from work x 3 |
| Doctors report fees | 12 | 2 letters @ £45 each |
| Cost of physiotherapy | 25 | 4 sessions @ £45 each |
| Cost of recruiting new staff member | 125 | Advertisements, interviews, administration and 25 % lowered productivity for first 3 months due to inexperience |
| Cost of training new staff | 85 | Induction and on the job training for 3 months |
| Discussions with Health and Safety staff or Personnel | 8 | Estimated @ 1 hr of employees time and 2 other staff (paid 3.5 times injured staff salary) |
| Counseling | 16 | 4 sessions @ £30 each |
| Retraining programme for injured staff member | 300 | Estimated wages of another staff member training plus injured staff member on full pay |
| Loss of skill and knowledge | | Unable to estimate cost |
| Impact on other staff, morale, motivation | | Unable to estimate cost |
| Compensation possible plus cost of resulting negative publicity for the company | | Unable to estimate cost |
| Total | 957 hours | Employee hours equivalent |

**Table 3: The cost of ergonomic changes**

|  | Employee hours equivalent | Explanation |
|---|---|---|
| Cost of Ergonomic changes | 6090 hours | For the same cost as the ergonomic changes (£43,270) a worker could have been employed for 6090 hours (at £7.10) |

The cost of the changes can be seen to be equivalent to 6090 hours of the injured employees time (table 3). When this is compared to the total cost of the worker developing WRULDs and leaving their current job (957 hours), it can be seen that the cost of the changes are approximately 6 times the cost of a single worker becoming injured and leaving.

### Factors missed in a purely financial analysis

It has not been possible to estimate certain items such as the impact of the injury on other members of staff which may result in decreased productivity and poor morale. This was a point raised by one of the interviewees, who remembered a time in 1992 when five employees were reporting symptoms of WRULDs. The cost to the company of lowered morale at that time was thought by the interviewee to be large, in terms of reduced performance and productivity of the staff in the Free Ads department.

Injuries sustained by individual workers carry costs which must be borne by the individual (e.g. reduced quality of life) and by the state (e.g. medical treatment, sickness benefit) (Cherniack and Warren 1999, Levenstein 1999). Often the extent of suffering and personal cost may be hidden due to a reluctance of workers to report injury. None of the methods reviewed include specific guidelines for including these individual or society costs.

Some staff believed that the ergonomics intervention may have contributed to the image of the company as a caring employer. The attention that the company pays to looking after its staff attracts potential employees to apply for jobs with Loot and current staff are encouraged to stay at Loot as employees due to this culture. A recent survey of staff in similar working environments gave the top five factors for retaining staff as: a caring company culture; team spirit; competitive salary; supportive and effective team leaders (Hills 2000). At Loot the use of ergonomics has been part of a wider health and safety department campaign to ensure the well being of the staff and this health and safety policy has now been integrated into the overall functioning of the business. This has created a good image for the company externally and internally.

There were few negative aspects highlighted by the interviews. Some staff commented on the way in which staff have raised expectations of their employer and come to take the efforts toward preventing WRULDs for granted. One commented that the attitude of 'I would rather have the money' (than have better equipment) existed, and that many were not very appreciative. On this point, Oxenburgh (1991) believes that if workers did not expect much from the employer, if they considered that injuries were part of the job, then the solutions implemented would be superficial and in the long term less viable.

## Concluding remarks

It is concluded that there have been many benefits to Loot as a result of the use of ergonomics within the Free Ads department. The monetary benefits were demonstrated using the calculation of the Pay-back period, which showed that the cost of the ergonomic changes would be recouped in approximately 2.2 years, by avoiding the costs associated with WRULDs. Beyond this period actual financial benefits would be realised. In the example of Loot, the use of office ergonomics has proved beneficial, it is suggested that it has played a major role in decreasing injury and that has resulted in large cost avoidance. There were also some benefits revealed, during interviews, which were not costs avoided, such as an increase in morale or improved company image. These benefits are less tangible and difficult to apportion to ergonomic intervention.

Cost benefit analysis for ergonomics is not straightforward. There are few validated methods for carrying out a cost benefit analysis for use by ergonomists and none that give guidance or attempt to include more subjective factors. The analysis also requires much data to be gathered which may be difficult to obtain either because figures are not kept or because of the sensitive nature of some company records. The lack of suitable data means estimates are made which may bring into question the full validity of the analysis.

## References

Cherniack, M and Warren, N. 1999, Ambiguities in office related injury: the poverty of present approaches.
*Occupational Medicine: State of the Art Reviews*, Vol. 14, No. 1, 1 – 15

Hills, F. 2000, Human resources, east meets west.
*Connect*, June 2000, 56

Levenstein, C. 1999, Economic losses from repetitive strain injuries.
*Occupational Medicine: State of the Art Review*, Vol. 14, No. 1, 149 – 161

Oxenburgh, M. 1991, *Increasing productivity and profit through health and safety.*
(CCH International, NSW Australia)

Simpson, G. and Mason, S. 1995, Economic analysis in Ergonomics. In Wilson J. and Corlett E.N. (ed.) *Evaluation of Human Work* Second Edition (Taylor and Francis, London), 1017 – 1037

# A PROFILE OF PROFESSIONAL ERGONOMISTS

## RB Stammers & EJ Tomkinson

*Centre for Applied Psychology, University of Leicester*
*Leicester, LE1 7RH, UK*

The results of a survey on UK ergonomists is reported. The sample was derived from the 1997/8 Ergonomics Society Membership Directory and a return rate of 56% from a postal questionnaire was obtained. There is a wide variety of professional specialisms within the respondent sample, with the practitioner appearing as the most numerous sub-group. Researchers make up the next largest group. A variety of qualification routes to membership are indicated, with the MSc being the most common one. Information on years of experience and on other topics is presented.

## Introduction

There is a need to know more about the practice of ergonomics. This paper will focus on the UK, but the issue is an international one. Many individuals and groups in the UK, from time to time, make statements on the practice of the subject, but all too often the information on which this is based is sketchy. As part of a European project, a study has been carried out to assess the current situation with regard to university courses in ergonomics (Stammers & Tomkinson, 2001).

A postal survey was used to gather information from qualified ergonomists working in the UK. Whilst the main role of the questionnaire was to collect information on courses, the opportunity was also taken to collect information to help in building up a picture of contemporary professional ergonomists.

The Ergonomics Society (ES), as a collaborator in the research, could have given access to up-to-date names and addresses of members. It was thought important, however, to contact, (a) people some time after graduation, and, (b) people who might have left the ES and/or the profession. The 1997/1998 ES Membership Directory was therefore chosen as the source of potential participants. This was published in October

1997 and at the time of the survey was over 2 years old. Since its publication, a number of people would have left the Society and a number would have joined (these numbers would be roughly equal). Of the latter group, most would have recently finished their degrees and were of less interest for the survey. Of those who had left the Society, it was possible that they were still practising as ergonomists and their responses were therefore of interest for the survey.

The final version of the questionnaires was sent, via the post, to people who had been randomly chosen from the 1997/8 Ergonomics Society Membership Directory. Members were rejected if they lived outside the UK or were not graduate, registered or fellow members. Eighty-two questionnaires were returned out of the 156 sent out. However, 9 were also returned due to wrong addresses. The response rate (82/147) was therefore calculated as 56%.

The results were categorised using the respondent's professional area. In the first section of the questionnaire, respondents were asked what their major area of activity was. They could choose from, Management, Research, Consulting/Advising or Teaching. There was also the option of specifying their major area of activity if the other options were not suitable.

Forty-one respondents chose Consulting/Advising, 19 chose Research. Only 7 chose Management and only 3 chose Teaching. Also, 12 people fell into a General group, these were those that had chosen more than one category and those that did not fit into the other categories e.g. unemployed. Due to the small numbers for Management, Teaching and the General group, these were all placed into one group labelled 'General'. Three groups of respondents were thereby created, Consulting/Advising with 41 people, Research with 19 and General with 22.

These groups were of most use in relation to the course survey, this is described in an accompanying paper and in the full report. The focus in this paper will be on information that was collected that will help build up a picture of ergonomists at work in the UK in the year 2000.


## Findings

Before discussing the findings, some comments on the sample and on possible bias is necessary. The numbers chosen to send the questionnaire to were approximately a 25% sample of the graduate, registered and fellow members. However, members not living in the UK were rejected from the random sample and replaced. A higher percentage of the UK-resident members was therefore sampled, estimated at over 30%.

It is possible to speculate on possible sources of bias. Ex-members are more unlikely to respond, especially if they are no longer working in ergonomics. Against this background a return rate of 56% was felt to be a healthy rate, even if it was possibly biased in its makeup towards working ergonomists.

The first finding has already been covered above. This was the division of the sample depending on professional specialism. Fifty percent of the sample saw themselves as consultants/advisors, 23% described themselves as researchers and 27% of the respondents could be described as generalists.

Thus practitioners, as opposed to those based in ergonomics in other ways, make up a sizeable majority of professionals in the field.

One of the questions asked was whether the respondent's work directly involved the use of Ergonomics knowledge.   Thirty-six out of 41 Consultants (87.80%) indicated that it did.  Eighteen out of the 19 Researchers (94.74%) and 19 out of 22 Generalists (86.36%) also indicated that it did.  Thus 11% of the overall sample were not directly using ergonomics knowledge in their work.

The respondents were also asked if they are currently members of the Ergonomics Society.  The following breakdown was obtained.

| Consultants | Researchers | Generalists | Overall |
|:-----------:|:-----------:|:-----------:|:-------:|
| 90% | 100% | 86% | 91% |

The obvious question is what is the overlap between ex-members and those no longer using ergonomics in their work?  Of the 9 respondents who indicated non-use, 4 were no longer in the Society.  This leaves 5 who have kept their membership despite no longer using ergonomics in their work.

Although a concern to the ES, it is hard to determine how widespread non-membership of the ES by practising ergonomists is.   The small number of 3 professionals who have left the Society, but still use ergonomics knowledge is too small to draw any conclusions.

There was also a question on membership grade.  The numbers did not differ greatly across group categories and are given below as overall percentages, but are contrasted with the latest membership figures taken from the ES Annual Report 2000. These latter percentages refer to these membership categories only, as these were the groups sampled from.  The odd 3% of 'others' in the sample row refers to two people who had changed their membership grade.

Given that the 2000 Report figures will include overseas members, then the sample might be considered to be fairly representative, although graduate member numbers appear somewhat low.

| | Graduate members | Registered members | Fellow | Others |
|:---|:---:|:---:|:---:|:---:|
| **Sample** | 15% | 59% | 23% | 3% |
| **2000 Report** | 37% | 48% | 15% | - |

Another question asked fellows and registered members how many years it was since they qualified for a 'professional' grade of membership.  This did not apply to graduate members, who would not normally have reached this stage of experience yet. Again, there were no substantial differences between the groups, so the overall findings are presented below.

| **Years** | 0-5 | 5-10 | 10-15 | 15-20 | 20+ |
|:---|:---:|:---:|:---:|:---:|:---:|
| **Percentage** | 33% | 26% | 13% | 13% | 15% |

Another question asked whether respondents were qualified as "European Ergonomists".   The overall figure of 21% indicates a sizeable uptake of this qualification by UK ergonomists.  An notable point here was there was a higher percentage of the Researchers (26%) than the Consultants/Advisors (17%) holding this qualification. The Generalists come between the other two samples (23%).

Data on membership of other professional societies was also collected. The following percentages of respondents were members of the societies indicated. Some of the respondents were not members of the ES but they did not indicate membership of any other society. These figures do not seem very high and indicate a fair degree of single society membership status for the ES members.

| Society | BPS | HFES | IOSH | BOHS |
|---|---|---|---|---|
| Percentage | 20% | 10% | 10% | 4% |

Another area of questioning concerned the route to qualification. Over the years a number of routes to membership have been in place. The majority of the sample qualified with an MSc in the subject (54%), those qualifying with a BSc numbered 18% and those coming though the experience and/or research degree route totalled 20%. Of those qualifying with an MSc, about half had some work experience before taking the degree.

A question that focused just on those that had completed a taught degree course in ergonomics, enabled an overview on number of years since completing the course. This reinforced the findings given above on years since qualification (by any route) and showed a wide spread of experience in both the sample, and presumably in the overall membership.

| Years since grad. | 4-5 | 6-10 | 11-15 | 16-20 | 21-25 | 26-30 | N/A |
|---|---|---|---|---|---|---|---|
| Percentage | 8 | 18 | 16 | 16 | 21 | 15 | 6 |

The remainder of the questionnaire focused on course related issues which are being reported on separately.

## Discussion

Given that there is no existing clear profile of the ergonomics-qualified people in the UK, then the results presented have no real points of contrast. Some usable suggestions emerge, however. These give reasons for both satisfaction and concern for the ES.

Practitioners are largest sub-group within the Society, with researchers being the only other sub-group of substantial size. Whilst it must be a concern that there are some ex-members who are still involved in ergonomics, there are also existing members who are not now using their ergonomics professionally. The Society remains an "interest" and a professional society, but not one that can claim exclusivity for its activities.

The sample exhibits what could be called a healthy spread across age and experience categories. This a broad-based membership in terms of its experience bodes well for the Society in the future.

Most members do not have "divided loyalties" with other societies. The MSc is the most common route to qualification, although it is notable that there have been a variety of ways of achieving membership.

## Reference

Stammers, R.B. & Tomkinson, E.J. 2001, **A study of ergonomics training in the UK.** (Leicester: Centre for Applied Psychology, University of Leicester).

# 2002

| Homerton College Cambridge | | 3rd–5th April |
|---|---|---|
| **Conference Manager** | | Carol Greenwood |
| **Chair of Meetings** | | Sandy Robertson |
| **Programme Secretary** | | Paul McCabe |

| **Secretariat** | Phil Bust | Elizabeth Hofven-schiold | Guy Walker |
|---|---|---|---|
| | Tim Dubé | Martin Robb | Jenny Boughton |
| | Amanda Roach | Suzanne Fowler | Minna Laurell |
| | Ozhan Oztug | | |

## Social Entertainment

Delegates were challenged with something more that trivia. Tina Worthy and Ann Brooks took centre stage on the first night of the conference to host a 'Quiz Night' in the bar.

| Chapter | Title | Author |
|---|---|---|
| **Hospital Ergonomics** | Hospital ergonomics: Organisational and cultural factors | S. Hignett and J.R. Wilson |
| **Motorcycle Ergonomics** | Motorcycling and congestion: Quantification of behaviours | S. Robertson |
| **Work Design** | Workplace bullying: An ergonomics issue? | N. Heaton and V. Malyon |
| **Warnings** | Orienting response reinstatement in text and pictorial warnings | P. Thorley E. Hellier, J. Edworthy and D. Stephenson |

# HOSPITAL ERGONOMICS:
# ORGANISATIONAL AND CULTURAL FACTORS

## Sue Hignett[1] and John R. Wilson[2]

[1]*Ergonomics and Back Care Advisory Department*
*Nottingham City Hospital NHS Trust, Nottingham NG5 1PB*
[2]*School of 4M, Management and Human Factors Group*
*University of Nottingham, Nottingham NG7 2RD*

This paper describes an exploration of the organisational and cultural factors in the practice of hospital ergonomics in a qualitative interview-based study with twenty-one ergonomists (academics and practitioners). The analysis identified three themes: organisational; staff; and patient issues.

It is suggested that hospitals present a particularly complex setting in which to practice ergonomics. This is partly due to the organisational structure (with multiple professional and managerial lines), but also to the core business of health care.

The area of female workers in hospital ergonomics was found to be under-researched both with respect to the type of work and to social and cultural issues about gender stereotyping.

## Introduction

Ergonomics has been described as a socially-situated practice (Hignett, 2001a & b). In order to achieve the goals of 'design' and/or 'change' the ergonomics practitioner has to have an understanding of the culture of the industry or organisation in which they are working. Hospital ergonomics is a relatively new area of practice, but has an enormous potential scope of practice. The National Health Service (NHS) is the biggest civilian employer in Europe, employing more than 1.1 million people, 5% of the UK working population. It is the largest employer of women, with approximately 75% female workers, with nurses accounting for 50% of all staff.

The following questions were explored:

- What are the characteristics of the health care industry with respect to the organisational and cultural factors?
- How do these characteristics impact on the practice of ergonomics in hospitals?

The paper is intentionally written in the first person to emphasise the use of a qualitative (or interpretative) approach throughout the study. Establishing one's position by writing in the first person is supported by a tradition in the social sciences and education (Wolcott, 1990; Webb, 1992). The literature review is embedded throughout the findings and discussion to give a more interactive analysis (Wolcott, 1992) and to facilitate the testing of the data against the literature (inductive analysis).

## Methodology

Qualitative methodology was chosen as a suitable approach to explore the two aims. This approach enabled the questions to be explored by giving:

1. Access to information through interactive interviews, with the flexibility to develop the questionnaire both during an individual interview and throughout the study.
2. An inclusive position to reflect on the diversity of the perspectives held by academics and practitioners involved in ergonomics.

The choice of qualitative methodology was supported with a middle ground philosophical stance, giving an ontological position of subtle or transcendental realism. This allows that there is a physical structure beyond our minds, '..things exist and act independently of our descriptions ... objects belong to the world of nature' (Bhaskar, 1975), but also that different people will have different perceptions of them, the idea of non-competing multiple realities (Murphy et al, 1998). This accepts the view of Hammersley and Atkinson (1995) that '.. there is no way in which the researcher can escape the social world in order to study it'. So two people may interact with the same situation or product and have very different experiences and perceptions of it, and both can be equally valid.

## Methods

Twenty-one semi-structured interviews were carried out with academics and practitioners using a questionnaire proforma which developed iteratively over the 18 months of the project. The interviews were audio-taped and transcribed verbatim. The transcripts were returned to the interviewee for an accuracy and confidentiality check before analysis.

Contact data sheets were completed after each interview to capture my immediate thoughts and summarise the main points from the interview. These were used as the first stage of data reduction. As the study progressed the new/target questions started to develop into questions for the data analysis rather than the interviewees.

## Sampling strategies

A progressive four stage sampling strategy was used starting with purposive sampling to spread the net. Suggested contacts were then followed up (snowball sampling), before the third stage of intensity sampling to focus on subjects with specific experience in hospital ergonomics. A final strategy of analysis sampling sought extreme and deviant cases to test the analysis and interpretation.

All the interviewees agreed to participate but unfortunately for three I was unable to arrange a convenient time. Other interviews were booked to try and achieve saturation from different discipline areas although input from psychology remained limited.

## Analysis

The analysis used the three steps of data reduction, data display and conclusions drawing/verification (Miles and Huberman, 1994). The interview transcripts were imported into a qualitative data management tool, NUD*IST N₄ (Gahan & Hannibal, 1998). The data were summarised, coded and broken down into categories using qualitative classification (Miles and Huberman, 1994; Sanderson and Fisher, 1997).

## Findings

The analysis resulted in three categories: organisational, staff and patient issues.

**Hospital Themes**

**Organisational issues**
*Size*
*Complexity*

**Workers**
**(Staff issues)**
*Multiplicity of professions*
*Gender*

**Caring for People**
**(Patient issues)**
*Dirty and emotional work*
*Patient expectations*
*Life, death and mistakes*

**Figure 1.   Hospital Themes**

*Organisational Issues*
The organisational issues included both the size and complexity of the National Health Service. For example, three hierarchical lines were identified in the management structure: an administrative line, a professional line and a patient-focused clinical management line.

> '..there are three sources of power from the management which is, of course, connected to the health authorities, and you have the doctors, and then you have the nursing staff which is also a source of power and you cannot do anything if you can't make agreement with all these three...'

The three-way hierarchy adds to the complexity with respect to accountability, authority and power. Quantitative measures of performance are applied at the level of units of provision, or 'cost centres' (directorates). This means that meeting the clinical targets may not be the responsibility of individual nurses. NHS managers may not directly control the work of nurses through performance. But they probably control the supply of other things, for example: (1) the context in which the nurses carry out their work and exercise their professional autonomy; and (2) the number of patients and therefore the amount of time (staff-patient staffing ratio).

The recurring themes of the complexity and interface with the patient seem to be fundamental in how health care differs from other industries.

> 'you start to try and draw your person, equipment interaction and always there's another person in the picture as well, so it's actually a people-people interactions are, are quite a big focus '

Van Cott (1994) called this difference 'people-centred and people-driven' in contrast to other industries which are technology-centred where the human role is to monitor the equipment or supervise small numbers of other staff. This relates to the core business of the hospital providing the public service of health care (to include both public and private sector organisations).

> 'the product that the hospital has, as a business, is caring for patients, and caring for patients isn't seen as a 'product'. So whereas, whereas in industry or in commerce you're producing something which you're selling or a service that you're providing..'

Implementing change is often a key part of ergonomics projects and it was suggested that 80% of the effort when working in hospital ergonomics was needed to progress the project and with only 20% on understanding the problem. The reverse was perhaps the more usual model for ergonomics projects, with 80% of the time spent on understanding or solving the problem and only 20% on progressing the project.

## Staff Issues

Two main areas were explored. The first related to the multiplicity of professions found in health care, and the second to the high proportion of female workers. The literature points to evidence for gender stereotyping for care tasks, but the case study generated very little data in support.

> '..there may be cultures that are specific to predominantly female professions and semi-professions which may be about sacrifice, and all of that stuff, that actually may not be true of, I don't know, car workers in the Midlands...'

Paid care work has been considered to be a low status occupation and almost an extension of housework (Giddens, 1993; Miers, 1999). This has led to dubious assumptions, for example that 'women are equipped to deal with bodily substances and that they enjoy this work as an extension of their 'natural' role and engage in it by choice' (Lee-Treweek, 1997).

There was a general feeling that there was a lack of data or information on women within ergonomics.

> 'if you look at any of the standard texts there's really, there has never been, in my view, sufficient general data gathered on either females, or anything more than the fit population which is invariably youngish ..'

## Patient Issues

The patient issues incorporated three dimensions associated with the caring role: the type of work; expectations; and possible outcomes. The work tends to be dirty and emotional, with a professional subculture to allow the handling of other peoples' bodies.

Other aspects of handling people include the emotional impact of other people's nakedness. Lawler (1998) looked at the nurses' first experience of suffering, disfigurement and death, and suggested that speed was often used as a method to manage difficult or potentially embarrassing situations. Technical vocabulary or jargon is also used to cope with the full significance of handling bodies.

> 'in many, many situations you have to deal with an interaction between people which both parties have to really have very high belief in, where there are, can be, very strong emotional influences at a level which is just about as sharp as you can get I think in terms of interactions between people'

This subculture was linked to a 'coping' attitude where staff put the patients' needs and well-being before their own, and has resulted in staff taking risks. The change in patient expectations (from being apologetic through to demanding their rights) is mirrored in the cultural move from paternalism to partnership (Boseley, 2000). This change might also fit the two models of care described by Miller and Gwynne (1972) with respect to risk-taking. A minimum risk environment was called the 'warehousing model of care', whereas a more stimulating, riskier environment was described as the 'horticultural model of care'. In order to provide both care and cure there are different type of service provision required.

There is a growing field of application of human factors in medicine, especially in the area of human error. This growth was discussed by Caldwell (1996) and a parallel was drawn between medical practice and 'other technologically dynamic, error-critical systems', e.g. aviation, nuclear and petrochemical industries. At the moment approximately one in ten patients are known to suffer adverse consequences as a direct result of their admission to hospital (Department of Health, 2000) and there are initiatives to change this through audit, further research and education.

## Discussion and Conclusion

It is suggested that hospitals are different to all other industrial organisations. Health care is a service industry like banking, but additionally it is also a public service (like the railways). The difference for ergonomics practice may lie in the definition of the 'user' in the context of a user-centred design or task analysis. Every member of the United Kingdom population is a potential user of the NHS so the definition of the user group is difficult for many areas. As a service industry the clients (patients) are not paying at the point of contact (unlike banking or transport services) and they do not have to be there (unlike education or the prison service). For banking and transport services the 'users' are all either paid employees or paying customers. A closer comparison might be education, but here the 'users' are paid employees (teachers and support staff) or children, who are legally required to attend the school. The prison service again has a complex user definition with the inmates giving an additional interface to the employees, but again the prisoners have not chosen to be there, that choice is made for them. This makes the definition of 'user' very difficult and creates complex interface. My conclusion is to suggest that the complexity of the organisation and culture needs to be taken into account for ergonomics practice in the health care industry.

## References

Bhaskar R. 1975, *A realist theory of science* (Leeds, Leeds Books)

Boseley S. 2000, *Doctors and nurses to be trained together to relax the elitist divide,* The Guardian, Monday 15 May 2000, 7

Caldwell B.S. 1996, Organisational bridges from research to practice: cases in medical practice, Proceedings of the Human Factors and Ergonomics Society 40th Annual Meeting, (Philadelphia), 530-532

Department of Health. 2000, *An organisation with a memory*, (Norwich, The Stationary Office)

Gahan, C. and Hannibal, M. 1998, *Doing Qualitative Research using QSR NUD\*IST*, (London, Sage Publications)

Giddens, A. 1993, *Sociology*, Second Edition, (Cambridge: Polity Press)

Hammersley, M. and Atkinson, P. 1995, *Ethnography: Principles in Practice*, (London and New York, Routledge)

Hignett, S. 2001a, Embedding ergonomics in hospital culture: top-down and bottom-up strategies, *Applied Ergonomics*, **32**, 61-69

Hignett, S. 2001b, *Using Qualitative Methodology in Ergonomics: theoretical background and practical examples,* Ph.D. thesis, University of Nottingham

Lawler, J. 1998, Body Care and Learning To Do for Others. In M. Allott and M. Robb (eds.) *Understanding Health and Social Care. An Introductory Reader,* (London, Sage Publications and The Open University), 236-245

Lee-Treweek, G. 1997, Women, resistance and care: An ethnographic study of nursing auxiliary work, *Work, Employment and Society*, **11**, 1, 47-63

Miers, M. 1999, Nurses in the labour market: exploring and explaining nurses' work. In G. Wilkinson and M. Miers, (eds.), *Power and Nursing Practice*, (Basingstoke: Macmillan), 83-96

Miles, M.B. and Huberman, A.M. 1994, *Qualitative Data Analysis: An Expanded Source Book,* Second Edition. (Thousand Oaks, CA: Sage Publications)

Miller, E.J. and Gwynne G.V. 1972, *A life apart: A pilot study of residential institutions of physically handicapped and the young chronic sick*, (London, Tavistock).

Murphy, E., Dingwall, R., Greatbatch, D., Parker, S. and Watson, P. 1998, *Qualitative research methods in health technology assessment: a review of the literature*, Health Technol Assessment, 2, 16

Sanderson, P.M. and Fisher, C. 1997, Exploratory Sequential Data Analysis: Qualitative and Quantitative Handling of Continuous Observational Data, Chapter 44. In G. Salvendy (ed.), *Handbook of Human Factors and Ergonomics*, Second Edition, (NY, John Wiley and Sons), 1471-1513

Van Cott, H. 1994, Chapter 4. Human Errors: Their Causes and Reduction. In M.S. Bogner, (ed.), *Human Error in Medicine*, (New Jersey, Lawrence Erlbaum Associates)

Webb, C. 1992, The Use of the First Person in Academic Writing: Objectives, Language and Gate keeping, *Journal of Advanced Nursing*, **17**, 747-752

Wolcott, H.F. 1992, Posturing in Qualitative Inquiry. In M.D. LeCompte, W.L. Milroy and J. Preissie (eds.), *The Handbook of Qualitative Research In Education*, (New York, Academic Press), 3-52

Wolcott, H.F. 1990, *Writing up qualitative research. Qualitative Research Methods Series 20.* A Sage University Paper, (Thousands Oaks, CA, Sage Publications Inc.)

# MOTORCYCLING AND CONGESTION: QUANTIFICATION OF BEHAVIOURS

**Sandy Robertson**

*Centre for Transport Studies*
*University College London, Gower St. London WC1E 6BT*

It has been suggested that increased use of motorcycles could result in lower congestion, reduced fuel consumption and thus lower emission of pollutants. DTLR wished to explore this proposition and commissioned a study. This paper reports on part of that study in which a set of previously defined behaviours of TWMVs (two wheel motor vehicles) are quantified in congested conditions. This work was funded by DTLR as part of a larger study into TWMVs and congestion. TWMV users appear to make use of strategies to reduce delay which appear not to impede other traffic. Up to 90% of TWMVs were observed to filter in traffic. The views expressed in this paper are those of the author alone.

## Introduction

The Government encourages alternative modes of travel that could help to reduce overall levels of congestion and pollution. Various groups have suggested that the increased use of TWMVs (two wheel motor vehicles) could result in lower congestion, reduced fuel consumption and thus lower emission of pollutants. The Department of Transport Local Government and the Regions wished to explore this proposition. The project was awarded to Halcrow Group with UCL providing input for the behavioural study. The background to this study is described in Robertson (2002) which paper describes the behaviours used in this study together with the process by which they were identified. The aims of this part of the study were to quantify the behaviours that might contribute to reducing delay or congestion used by TWMV users.

## Method

Behaviours were recorded by video cameras at five London sites. Four sites were signal controlled junctions and one was a roundabout. These are shown in Table 1. A journey time survey and traffic counts were undertaken at the same time as the video recordings were made as part of the DTLR project (see Martin, Phull, and Robertson 2001). Video recordings were undertaken in the morning peak period (7am to 10am). At each site, a single video camera was used, usually pointing upstream so as to include the junction. The frequency of behaviours as described in Robertson (2002) were recorded for the 3 hour period for every TWMV observed. The TWMVs were also categorised as being a motorcycle or a scooter. This level of disagregation was the finest that could be undertaken given the constraints of video location and quality

### Table1. The sample

| Site | N (TWMV) | % TWMV | Weather | Junction type | Lanes |
|------|----------|--------|---------|---------------|-------|
| Farringdon St | 280 | 7.8 | rain | Signal | 2 two way |
| Commercial St | 308 | 2.1 | dry | Signal | 1 two way |
| Lwr. Thames St | 952 | 15.5 | dry | Signal | 2 dual |
| Upr. Thames St | 498 | 9.7 | rain | Signal | 2 dual |
| Monmouth St | 107 | 6.7 | rain | R'dabout | 1 one way |

## Results

The results are shown in Tables 2 and 3. Table 3 indicates differences between behaviours observed for motorcycles and scooters and statistical tests on these are shown in Table 4. The results for each behaviour are discussed in turn.

### Move to head of queue.

The proportion of TWMV users who exhibited this behaviour varied from none at Monmouth St to over half the riders at Commercial St. The data suggests that at signal controlled junctions in London, at least one in five TWMVs reach the head of the queue. There were no differences in the proportion of behaviours exhibited by different TWMV types. The implications of this are that, at typical signal controlled junctions in London, the users of between one quarter and one half of TWMVs will not contribute any delay at traffic signals (based on estimates quoted by Powell 2000). This may, however be a questionable assumption. This reduction to delay in the road network might not have much impact where there are only a small percentage of TWMVs in the traffic. Where higher proportions of TWMVs use the road system, there is potential for reducing congestion.

### Table 2. Proportion of TWMVs users exhibiting behaviours.

| Site | Behaviours | | | | | | | |
|------|------------|-----------|----------|----------|----------|---------|-----------|------------|
| | Move to head of queue | Filtering: stationary | Filtering: moving | Lane changing | Inaction | Balking | Wriggling | No observed behaviour |
| Farringdon St | 0.19 | 0.19 | 0.20 | 0.56 | 0.05 | 0.23 | 0.01 | 0.26 |
| Commercial St | 0.56 | 0.69 | 0.36 | 0.50 | 0.08 | 0.35 | 0.08 | 0.06 |
| Lower Thames St | 0.20 | 0.46 | 0.92 | 0.23 | 0.02 | 0.56 | 0.00 | 0.04 |
| Upper Thames St | 0.18 | 0.32 | 0.57 | 0.27 | 0.08 | 0.22 | 0.02 | 0.24 |
| Monmouth St. | 0.00 | 0.14 | 0.22 | 0.30 | 0.10 | 0.25 | 0.03 | 0.50 |

**Table 3. Proportion of TWMVs of each vehicle type, Motorcycle (M) or Scooter (S), exhibiting behaviours.**

| Site | Behaviours | | | | | | | | Type |
| | Move to head of queue | Filtering: stationary | Filtering: moving | Lane changing | Inaction | Balking | Wriggling | No congestion | |
|---|---|---|---|---|---|---|---|---|---|
| Farringdon St | 0.19 | 0.17 | 0.23 | 0.51 | 0.05 | 0.23 | 0.02 | 0.29 | M |
| | 0.20 | 0.24 | 0.15 | 0.68 | 0.04 | 0.23 | 0.01 | 0.21 | S |
| Commercial St | 0.57 | 0.70 | 0.37 | 0.50 | 0.06 | 0.35 | 0.07 | 0.08 | M |
| | 0.52 | 0.66 | 0.31 | 0.50 | 0.17 | 0.36 | 0.09 | 0.02 | S |
| Lower Thames St | 0.19 | 0.45 | 0.92 | 0.233 | 0.02 | 0.54 | 0.00 | 0.04 | M |
| | 0.23 | 0.49 | 0.92 | 0.226 | 0.03 | 0.61 | 0.01 | 0.03 | S |
| Upper Thames St | 0.17 | 0.32 | 0.55 | 0.30 | 0.08 | 0.21 | 0.02 | 0.26 | M |
| | 0.21 | 0.34 | 0.61 | 0.21 | 0.09 | 0.25 | 0.01 | 0.21 | S |
| Monmouth St. | 0.00 | 0.16 | 0.24 | 0.37 | 0.14 | 0.26 | 0.04 | 0.42 | M |
| | 0.00 | 0.10 | 0.19 | 0.13 | 0.00 | 0.23 | 0.00 | 0.71 | S |

**Table 4. Chi squared test on the proportion of motorcycles and scooters exhibiting behaviours, df 2. (only values significant at p>0.05 are shown.)**

| Site | Behaviours | | | | | | | |
| | Move to head of queue | Filtering: stationary | Filtering: moving | Lane changing | Inaction | Balking | Wriggling | No congestion |
|---|---|---|---|---|---|---|---|---|
| Farringdon St | | | | 7.51 | | | | |
| Commercial St | | | | | 8.91 | | | |
| Lower Thames St | | 4.214 | 5.7 | 6.3 | | 24.3 | | 29 |
| Upper Thames St | | | | 4.65 | | | | |
| Monmouth St. | | | | 6.02 | 5 | | | 7.3 |

## Filtering in stationary traffic

There were a range of proportions of TWMVs observed to filter in stationary traffic ranging from 14 to 69 per cent. This behaviour appeared to be associated with the length of queue on the link (see Figure 1). Possible reasons for this are: a) This behaviour is associated with the opportunity to do so, and longer queues mean that there is more opportunity and b) Riders may be more likely to undertake such behaviours where there are long queues leading to perceived delay for them. These findings indicate that a high proportion (4 out of 5) TWMV riders are filtering at some sites and that not less than one in six riders were filtering at the sites in this study. Filtering is a behaviour that appears to be a classic 'congestion busting' behaviour which is associated with the presence of congestion as indicated by the presence of queues of traffic. Filtering by TWMV may be seen as a method of making use of the small size and manoeuvreability of the TWMV to make effective use of parts of the roadway that could not be used by larger vehicles.   In terms of time saving and queue length, there is some relationship between queue length and time saving estimated as 38 seconds per 100m of queue (See Martin, Phull and Robertson 2001).

*Filtering in moving traffic*
This was more prevalent than filtering in stationary traffic. The site with the highest proportion of TWMVs filtering was also the site with the highest flows.

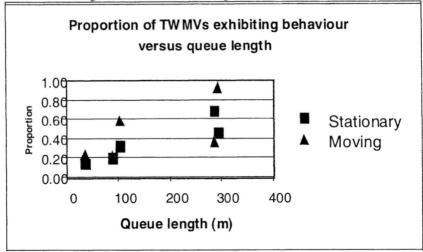

**Figure 1 Proportion of TWMVs filtering in traffic vs. queue length.**

A plot of average queue length versus the proportion of TWMVs filtering (Figure 1) shows a similar broad pattern to that observed. While traffic is moving it cannot, by definition, be queued, but there may be traffic moving in one lane but not the other. The small vehicle width of a TWMV means that it is possible to filter in streams of traffic that are moving.

One may describe filtering behaviour in terms of 'virtual lanes' or additional streams of traffic which may allow additional capacity to be obtained from the existing network. Observations indicated that the TWMVs would sometimes position themselves such that they were in a position between the conventional lanes, even if they were not actively making progress. Such positioning of the TWMVs may provide additional viewing distance and possibly braking distance. The latter would of course only be if the existing streams of traffic did not change their lateral position on the road under braking.

*Lane changing*
There were a range of proportions of TWMVs exhibiting this behaviour. For the purposes of this study, the lanes were defined as streams of traffic. There did not appear to be a relationship between the queue length and the proportion of TWMVs that were observed to change lanes. Lane changing behaviour was, therefore, observed equally within and outside queues.It is possible that this behaviour allows TWMVs to select the fastest moving traffic stream at any

given time or get access to a 'virtual lane' as a precursor to filtering behaviours (though this has not yet been quantified). A higher proportion of motorcycles (compared to scooters) appeared to change lanes at 3 of the 4 sites that had statistically significant differences. These differences were, however, small in magnitude.

### Inaction

Few TWMVs were observed not to act when an opportunity presented itself. It may be concluded that TWMV users generally take up an opportunity to reduce delay where it was presented. These observations also support the suggestion the TWMV users take an active approach to their driving and make use of opportunities to make progress where possible.

### Balking

The proportion of TWMV users being balked was between 0.22 and 0.56. Balking would appear to occur frequently, but as it was often a transitory phenomenon, it is not yet clear to what extent it might reduce any time savings made by TWMV users.

### Wriggling

This behaviour was rarely seen at most sites but Commercial St was an exception. It appears that wriggling is a site specific activity relating to the need of the TWMVs to change traffic streams. Wriggling may have occurred at Commercial St due to a slight pinch in the road some distance back from the signals such that filtering became impossible in the stream originally chosen and that the TWMVs had to move to a different stream to make progress.

## Discussion

As this study only covers 5 sites caution should be exercised in generalising these findings. Some behaviours appear to be more prevalent in different road types. For example, filtering in moving traffic was more prevalent at the Upper and Lower Thames St. sites. These are dual carriageway sites which were conducive to filtering in the middle. At the other sites, which were not grade separated, there was a smaller proportion of TWMVs that filtered. This pattern was not, however, repeated for filtering in stationary traffic.

TWMV users appear to make efficient use of the road space when queuing. Observations indicate that a great number of TWMVs may get close to the stopline by a) forming a line across the road at the stopline and b) by queuing back between cars waiting. While the numbers have yet to be quantified the initial observations indicate that 3-4 TWMVs may queue across a single lane and that 2 or more TWMVs may queue in the space between 2 cars. It must be emphasized that these are rough estimates and that a count will need to be undertaken to confirm these figures.

## References

Powell. M (2000) A model to represent motorcycle behaviour at a signalised intersections incorporating an amended first order macroscopic approach Transportation Research 34A 497-514, 2000

Martin B, Phull S. and Robertson S (2001) Motorcycles and Congestion. Proceedings of the European Transport Conference  10-13 September 2001  Homerton College Cambridge. Proceedings published as a CD, PTRC, London

Robertson S.A. (2002) Motorcycling and Congestion: Behavioural Identification. In P.T. McCabe (ed.) *Contemporary Ergonomics 2002,* (Taylor and Francis, London).

# WORKPLACE BULLYING: AN ERGONOMICS ISSUE?

## Nigel Heaton and Vicky Malyon

*Human Applications, 139 Ashby Rd, Loughborough*
*Leicestershire, LE11 3AD*
*enquiries@humanapps.demon.co.uk*

Failing to take into account the potential for bullying can have a major impact on the health, safety and welfare of employees and subsequently impact their performance at work. Ergonomists are often involved in the redesign of the workplace and the underlying work system and therefore have an opportunity to give advice to prevent or manage the problem.

This paper looks at the problem of bullying in the workplace, methods of identification and how to deal with complaints. More importantly it outlines a proactive approach. It looks at how Ergonomists need to take into account bullying issues when designing new systems of work. It suggests that the Ergonomist, working as part of a team, must create appropriate policies, procedures and an audit trail that demonstrates how workplace bullying is managed.

## Introduction

Ergonomics is primarily concerned with fitting the task or activity to the person. Ergonomists believe in 'user centred' design, that is designing with people in mind, often via consultation and interaction. One of the problems with this approach is that the resulting system of work often focuses on the physical and even when psychological aspects are considered they tend to be at an individual level rather than an interpersonal level.

There are many aspects of an Ergonomist's job where the interpersonal could be considered. Throughout this paper the methodology behind identifying and managing workplace bullying will be explained.

There have been many attempts at defining 'bullying.' The behaviours that categorise the term are open to individual interpretation and therefore it is difficult to devise an all-encompassing definition. By examining a number of definitions it is possible to identify the elements that are shared. Bullying is usually defined in terms of the effect it has on the recipient rather than the intention of the bully. This is usually compounded by a description of the negative effects on the recipient and the fact that the behaviour is persistent. Bullying behaviour is not characterised by one-off incidents it tends to continue over a period of time, sometimes years.

One of the factors on which definitions focus is that it is a 'systematic abuse of power' (Smith and Sharp, 1994). When we think of a bullying scenario we usually imagine a subordinate being bullied by a superior. In the majority of incidents this is the case but it must not be ignored that it can happen the other way round where subordinates bully their managers!

When dealing with interpersonal or personnel issues bullying and harassment tend to be grouped together and dealt with in the same way. However, the two behaviours are inherently different in their nature. Bullying is not age, gender, race or physical ability related it is usually based on competence. It is psychological as it is rarely obvious and tends to happen behind closed doors. As a result it can take the recipient weeks to realise that they have become a victim. What the victim fails to realise is that they are actually a threat to the bully and the behaviour that they are subjected to is the bully's only way of making themselves feel better about their position, either within society, the family or the organisation.

There has also been some debate about the importance of defining the difference between bullying and strong management. Legitimate and constructive criticism and the odd raised voice can be seen to be acceptable but it is not acceptable to condone bullying under the guise of strong management.

## Legislation
The difficulty with recognising bullying as a manageable problem is that there is no specific legislation under which to prosecute. However, there are a number of Acts and Regulations under which bullying may be covered.

The Health and Safety at Work Act 1974 made it a legal responsibility for employers to ensure that the health, safety and welfare of employees at work is protected. More specifically, section two requires employers to provide a safe place and safe system of work to ensure the physical and psychological well being of employees. Additionally, section seven requires employees themselves to take the responsibility of ensuring that they take reasonable care of their own health and safety and that of others.

More recently, the Management of Health and Safety at Work Regulations 1992 (revised 1999) places a legal duty on employers to carry out an assessment of the risks workers face whilst at work (this would include bullying) and to take any necessary preventative measures. There is also the requirement to provide training, information and supervision.

There is also non-discriminatory legislation. However, these Acts are aimed at harassment rather than bullying. The main difference is that harassment can be a "one-off" and is based on discrimination or abuse. The three Acts that are most important in this context are:
*   The Sex Discrimination Act 1975
*   The Race Relations Act 1976
*   The Disability Discrimination Act 1995

It is sometimes possible to use these Acts to support a bullying claim but it is far more likely that the case will be one of harassment.

The introduction of the Human Rights Act, which states that 'no one shall be subjected to torture or to inhuman or degrading treatment or punishment', may have significance in bullying cases. Although, this would not strictly be used in an employment case its introduction highlights the increasing importance placed on such interpersonal behaviours.

## Scale of the Problem

A study by UMIST (Cooper and Hoel, 2000) found that 1 in 4 people have been bullied at work and 1 in 10 have been bullied in the last six months. The TUC launched a hotline for people to call to report "bad bosses" and problems at work (including pay and conditions). The TUC were surprised to find that the number one complaint was not pay but bullying.

The costs to the organisation from incidents of bullying are huge and must not be underestimated. The main effects are an increase in absenteeism and staff turnover, demoralisation and lack of motivation and a decline in productivity and profit.

It is difficult for organisations of any size to identify the specific cause of absenteeism, (especially with the surge in musculo-skeletal problems). However, it is even harder for them to identify the cause as interpersonal problems such as bullying, violence and aggression and as a result stress related absence. It is important to consider the relationship between bullying and work-related illness. In one case, an employee successfully sued a company for "RSI" and was awarded a six-figure sum. Part of her evidence included a statement that "the company not only ruined my health but tried to intimidate me out of bringing a claim". The judge noted that the employee had been required to come in early, work during her break-times and go home late. Interestingly, after the case was lost, the manager responsible was sacked for bullying!

The negative effects of bullying on the recipient obviously play a major part in the costs to the organisation. They usually begin with a loss of confidence both in themselves and their work, which can lead to demotivation and as a result poor work quality and reduced output. Victims of bullying often resign from work with stress related ill health soon after their plight has been unearthed. Cases are often left unidentified until the recipient reaches breaking point.

In recent years there has been an increase in personal injury claims where the injury is psychological in nature. The increase in the "blame" culture and our litigious society indicates that further increases in such claims can be expected.

A personal injury claim based negligence will succeed if an individual can prove they have suffered loss, and that they were owed a duty of care, and there was a breach of the duty of care and that the injury was a result of the breach.

The awards made in recent stress claims could be seen to set a precedent for the penalty paid if interpersonal relationships are not managed within organisations. In

March 2000, the Ministry of Defence was made to pay out £745,000 to a soldier who was pushed out of a window in an extreme bullying incident 11 years before. A teacher who was humiliated, excluded, embarrassed and prevented from doing his job by the Head Teacher was awarded £101,028 in an out of court settlement after retiring in medical grounds.

## Negating Ergonomics Interventions

As Ergonomists we therefore need to consider the impact of ergonomic interventions on interpersonal relationships. If we are involved in management structures and systems of work it is imperative that we consider the possibility of the abuse of power and the opportunity for bullying. Management systems are hierarchical in nature. Individuals report upwards. Much (but not all) bullying occurs between a worker and their immediate line manager or supervisor. When we design systems to make the workplace safer, more productive and more comfortable, we often assume that this relationship is a "productive" one i.e. that both parties wish to achieve a safe, productive and comfortable workplace. However, if a manager or supervisor believes that this can only be achieved by bullying the worker, either implicitly or explicitly, then our interventions fail.

We must include fail-safe mechanisms, for example bullying policies, reporting procedures and alternative "routes" for those who feel bullied to report problems and have those problems dealt with in a constructive and sympathetic manner.

## Methods of Identification

There are a number of methods that can be utilised in order to gain information about the situation within an organisation. There is not the scope within this paper to go into great detail but a brief description will be included.

Employee surveys are probably one of the most common methods of attempting to reap qualitative data but sometimes their validity is dubious due to the sensitivity of the subject involved. Although, often confidential there is great stigma associated with either admitting that you have a problem or reporting that you are aware that someone else has a problem and therefore the responses may not be a true reflection of the situation.

Exit interviews and return to work interviews may also be used in order to gain qualitative data. If managed correctly they can be used to identify problems that may be causing the employee to leave.

Another popular method is to provide a hotline where employees can contact a third party in confidence and share their experiences. However, the calibre of callers must be considered as only those who feel confident will call. Nevertheless, details of incidents, however subjective can provide a valuable insight into the causes and incidence of the problem.

A lot can be said for 'gut feel' and the 'grapevine,' things that many of us rely on most of the time. If an employer has noticed that there has been a decrease in productivity or morale in certain individuals or groups they may decide to look more closely at the interpersonal relationships between those concerned.

## A Risk Management Framework

Once we understand the extent of the problem, systems need to be designed that focus on prevention. Controls need to consider the potential for bullying and ensure that organisations go "as far as is reasonably practicable" to reduce the risks associated with bullying. In short, a proactive risk assessment must be undertaken.

We understand the nature of the hazard and the extent to which it can cause harm. We need to be clear about how likely it is that the harm will be realised and then to apply simple controls to manage and reduce the problem.

In practice organisations must conduct an "a priori" study of bullying, the extent and nature of the problem, design controls and then regularly conduct post hoc assessments (and amend controls as required).

In the event of a civil claim, claimants' legal experts are well aware of the need for an organisation to demonstrate that it is managing risks appropriately. Organisations will be expected to:

- Demonstrate how they have eliminated bullying throughout their organisation. For example, how have bullying cases been dealt with, what has happened to both "victims" and "offenders".
- Show how the effects of bullying have been reduced. For example by the introduction of early reporting procedures (e.g. through the use hotlines, personnel councillors, etc.). Early reporting is an important way of reducing the effects of bullying.
- Design systems of work to take account explicitly of the potential for bullying – this will include the production of a bullying policy, the procedures to manage bullying incidents and the provision of appropriate training and information.
- Provide appropriate training and information to all levels of staff. Note that this occurs at the bottom of the hierarchy and is not to be seen in isolation from the controls detailed above.
- Demonstrate a management strategy for dealing with people after they have been bullied (e.g. counselling, re-introduction strategy, etc.).
- Show how the above is audited regularly and how it changes.

If an organisation is not able to demonstrate this proactive approach and a problem goes unrecognised until an individual becomes severely harmed, then the scale of the compensation may reflect the courts view of how far the organisation has failed.

## Policy and Procedure

A survey by the Andrea Adams Trust and Personnel Today in 1999 found that whilst 93% of personnel managers acknowledged that bullying occurred within their organisations, only 43% had policies to deal with it. The basis of a proactive approach is the development of a policy and procedures to ensure that all employees are aware of the organisations stance with regard to bullying and the procedures that will be followed if such behaviour is to occur.

The policy should broadly define what is regarded as bullying, although this definition should not be too prescriptive. It should outline what it is intended to achieve and the underlying philosophy that bullying behaviour will not be accepted and the commitment of the organisation to enforce that philosophy. The policy must also assign responsibilities, which should include the requirements of individuals and the organisation's responsibilities.

The policy may refer out to a procedure to be followed in the event of bullying behaviour or this could be included within. It should give recipients the confidence to report incidents and therefore should include the arrangements for the support of 'victims.'

Lastly, no policy should be without details of performance measures and arrangements for the review and monitoring of the policy in order to create an audit trail.

Inevitably, once bullying behaviour has been identified action must be taken. Firstly, it is important to identify the difference between deliberate and unconscious bullying. Due to the variation in individual interpretation and sensitivity, what the recipient may consider to be bullying the bully may see as harmless fun. In most cases, however, the bullying behaviour is rarely misinterpreted and an official route must be taken. Once reported to a third party the third party may confront the 'accused' and inform them of the complaint, this may occur without mentioning names.

It is then imperative that an investigation takes place. The alleged bully should be confronted with the facts and their response should be recorded, regardless of the disciplinary action taken some form of monitoring must also be agreed as the 'complainant' may be at greater risk once the investigation has been completed.

## Conclusion

Regardless of the absence of specific legislation and the fact that it is still a 'taboo' subject bullying in the workplace is evidently a big problem that we should ensure is taken into consideration.

Systems need to be in place to ensure that
- policy and procedures are developed, measured and monitored
- all staff are trained and aware of the content and implications of the policy and procedures, and,
- if an incident does occur the correct level of support is available

As Ergonomists we need to ensure that each time we are involved in a project where there is an organisational problem we consider the interpersonal effects and the relationships within the working environment.

## REFERENCES

The Andrea Adams Trust, www.andreaadamstrust.org
Cooper, C. and Hoel, H. (2000) Destructive Inter-Personal Conflict at work.
Smith and Sharp, (1994) School bullying: Insight and Perspectives. London: Routledge.

# ORIENTING RESPONSE REINSTATEMENT IN TEXT AND PICTORIAL WARNINGS

Paula Thorley[1], Elizabeth Hellier, Judy Edworthy and Dave Stephenson

*Department of Psychology, University of Plymouth*
*Drake Circus, Plymouth, Devon, PL4 8AA*
[1] *now at Serco Assurance, Thomson House, Warrington WA3 6AT*

An experiment is reported which demonstrates both habituation and an orienting response to visual warning signs. Using skin conductance as a response measure, subjects were repeatedly exposed to a single warning sign over twelve trials. Subjects were presented with warnings that were either pictorially- or text-based. In the experimental trials, the warning changed to a slightly different format on the tenth trial and reverted to its original form on the eleventh and twelfth. Control conditions were also used where the warning format did not change at all throughout the twelve trials. The results showed a consistent decrease in response from trial one to trial nine, followed by an increase in trial ten for the experimental subjects only. These results suggest that habituation to warnings is a measurable phenomenon.

## Introduction

There is some evidence demonstrating that people can and do habituate to warnings (e.g. Thorley et al, 2001). The study reported here further investigates the effects of repeated exposure to visual warnings. Skin conductance response was measured as a determinant of the orienting response, serving as an indicator of both habituation and dishabituation.

The design of the study was similar to that of Ben-Shakhar *et al.* (2000), where the stimuli were primarily either text- or picture-based and included standard stimuli and test stimuli. It was hypothesised that there would be habituation effects following repeated exposure to a standard stimulus for both text- and picture- based visual warnings. Moreover, it was hypothesised that the orienting response would be reinstated following the introduction of a novel test stimulus subsequent to the habituation trials. Following the test stimulus, it was expected that dishabituation effects toward the standard stimulus would be demonstrated by an increased skin conductance response.

## Method

### Participants
An opportunistic sample of sixty (28 females, 32 males) participated in the experiment for payment. The sample was selected partly from a student and partly from a non-student population. All participants were advised of their right to decline or withdraw, although none did. No other details were recorded.

*Materials*
A constant voltage system and two silver-silver chloride electrodes (9mm Diameter) measured skin conductance. In the absence of availability of an electrode gel mixture to the recipe provided by Fowles *et al* (1981, as cited in Ben-Shakhar *et al.* 2000) a substitute saline gel was used; 'Johnson & Johnson KY Jelly'.

The experiment was conducted in an air-conditioned, soundproof laboratory at the University of Plymouth and was monitored via an adjacent laboratory housing the recording equipment. A PC was used to control the stimulus presentation and another computer recorded skin conductance changes. The stimuli were displayed on a Hewlett Packard HP71 colour monitor, approximately 70cms from the participant's eyes.

*Stimuli*
All stimuli contained 3 components based on the recommendations of Rogers *et al.* (2000). The text-based stimuli contained a signal word, instructions of how to avoid the hazard, and the potential risk. The picture-based stimuli contained a generic warning sign, instructions on how the hazard should be avoided and the potential risk. The experimental stimuli differed from the control only in either the signal word or sign. The instructions and potential risk remained the same for both groups.

*Design and Procedure*
The stimulus sequences used in this experiment were comprised of text-based (signal word) and pictorially based (icon) visual warnings. A test stimulus (TS), created by substituting either the signal word or icon on the standard stimulus (SS), was introduced after nine repetitions of SS, followed by two additional repetitions of SS. There were thus 12 exposures to the stimuli in all.

The between-subjects factor was the components of SS that were substituted to create the TS (0, where no change was made, or 2, where both the signal word/symbol and colour were altered in order to create a warning with higher perceived urgency). The dependent variables were the skin conductance responses (SCR) elicited by TS (OR reinstatement) and by the SS immediately following TS (dishabituation).

Participants were allocated randomly to the 4 conditions, 15 in each group (text control, text experiment, picture control, picture experiment). All participants were fully briefed and debriefed. Each participant was presented with only one of the stimulus sequences. Two electrodes were attached to the volar side of the index and middle finger on the participant's non-preferred hand using surgical tape applied with a comfortable pressure. An earthing strap was also attached to the inside lower arm of the non-preferred hand in order to maintain accuracy of measurement. Participants were requested to sit at ease and told that the computer program, to which they should pay attention, would begin in a few minutes' time, and that they would not be told when it was about to begin.

All participants were given a three-minute rest period before the program began. For all groups a flash of light was presented on the screen for 500ms, this was the range correction stimulus to eliminate individual differences. The control groups were presented with 12 visual warnings that differed in neither appearance nor intensity from one another. The warnings were displayed for 5 seconds each with a fixed inter-stimulus interval of 15 seconds. The experimental groups saw the same as the controls except in Trial 10, where the warning differed in both appearance and intensity to the previous nine. At the end of the experiment, a thank you screen was displayed, participants were debriefed as the equipment was removed from their person and they were then paid for their participation.

## Results

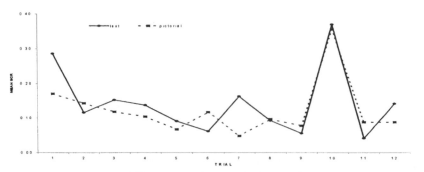

**Figure1. Comparison of text and picture based visual warnings**

Figure 1 compares the mean skin conductance responses between the text and picture based visual warnings. Habituation effects are evident across both modalities as is demonstrated by the downward trend in responses between Trial 1 and Trial 9. The increase in mean SCRs at Trial 10 indicates an increase in the orienting response for both modalities; however, dishabituation effects are not obvious at Trials 11 and 12.

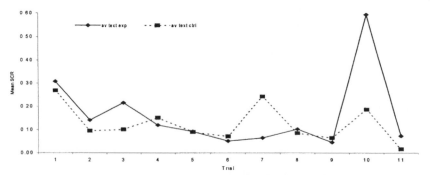

**Figure 2. Comparison of text control and text experimental groups**

Figure 2 compares the mean skin conductance responses between both control and experimental conditions for the text based visual warnings. Only Trials 1 to 11 are represented here, as they are the main trials of interest. Again, habituation effects between Trials 1 and 9 are evident across both conditions and disparities in responses have been recorded at Trial 10. Trial 11 does not indicate dishabituation effects.

Figure 3 compares the mean skin conductance responses between both control and experimental conditions for the picture based visual warnings. Habituation effects between Trials 1 and 9 are evident across both conditions and disparities in responses have been recorded at Trial 10. Dishabituation effects do not appear to have been demonstrated at Trial 11.

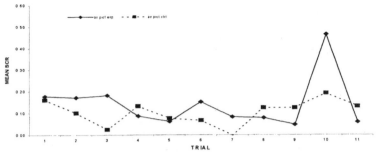

**Figure 3. Comparison of pictorial control and pictorial experimental groups**

In order to determine habituation effects, Trials 1 and 9 were subjected to a repeated measures mixed analysis of variance (ANOVA) (trial as the within subjects factor and modality and condition between subjects factors). Descriptive statistics are summarised in Table 1.

**Table 1. Descriptive statistics for habituation**

| | TEXT BASED WARNINGS | | | | PICTURE BASED WARNINGS | | | |
| | Control | | Experimental | | Control | | Experimental | |
| Trial | Mean | SD | Mean | SD | Mean | SD | Mean | SD |
|---|---|---|---|---|---|---|---|---|
| 1 | .794 | .455 | 1.13 | 1.31 | .660 | .723 | .627 | .477 |
| 9 | .475 | .955 | .163 | .488 | .523 | 1.08 | .441 | 1.08 |

The ANOVA revealed a significant main effect of trial, $F(1, 36) = 4.796$, $p < .05$. There was neither a trial by modality nor a trial by condition interaction.

In order to determine orienting response reinstatement Trials 9 and 10 were subjected to a repeated measures mixed analysis of variance (ANOVA) (trial as the within subjects factor and modality and condition between-subjects factors). The ANOVA revealed a significant main effect of Trial, $F(1, 36)$ 9.332, $p < .01$. There were no interaction effects across any of the variables and the test of between subjects effects revealed no main effect of either modality or condition, nor any interaction effects. Descriptive statistics for both modalities are summarised in Table 2.

**Table 2. Descriptive statistics for orienting response reinstatement**

| | TEXT BASED WARNINGS | | | | PICTURE BASED WARNINGS | | | |
| | Control | | Experimental | | Control | | Experimental | |
| Trial | Mean | SD | Mean | SD | Mean | SD | Mean | SD |
|---|---|---|---|---|---|---|---|---|
| 9 | .475 | .955 | .163 | .488 | .523 | 1.07 | .441 | 1.08 |
| 10 | .451 | .646 | 2.68 | 2.57 | 1.32 | 2.61 | 1.38 | 1.50 |

## Discussion

The hypothesis stating that repeated presentation of the standard stimulus would result in habituation was supported, as the results indicated a significant decline in skin conductance response to the standard stimulus between Trial 1 and Trial 9 across all groups. It has been specified that for habituation to be identified it must exhibit certain characteristics. For example, if a stimulus elicits a response, repeated application of the stimulus results in decreased response strength (Thomson and Spencer 1966, as cited in Petrinovich, 1973) or, habituation of the orienting response to the standard stimulus is evident during the first presentations of the standard stimulus (Zimny and Schwabe, 1965). The data collected here clearly displays those features, therefore it has been concluded that habituation is a consequence of repeated exposure to a visual warning.

The hypothesis that the introduction of a novel stimulus would reinstate the orienting response was only somewhat supported. Although the orienting response reinstatement was demonstrated in the overall effects, this was clearer for the text-based warnings than for the picture-based warnings. The findings here would support Sokolov's (1960, as cited in Lipp, 1998) scheme for orienting response elicitation, where a mismatch in the comparator system results in an increased orienting response. The results also conform to two of the hypotheses that underlie Sokolov's theory as identified by Zimny and Schwabe (1965). Specifically, that presentation of a test stimulus produces a return of the orienting response and the orienting response to the test stimulus is greater than that to the standard stimulus immediately following the test stimulus. Therefore, it could be concluded that following habituation, changing components of that warning reinstates the orienting response.

### Acknowledgements

This research was supported by a grant from the Economic and Social Research Council (ESRC).

### References

Ben-Shakhar, G., Gati, I., Ben-Basset, N. & Sniper, G. (2000). Orienting response reinstatement and dishabituation: Effects of substituting, adding and deleting components of nonsignificant stimuli. *Psychophysiology, 37*, 102-110

Lipp, O.V. (1998, August 2000). A description of Sokolov's comparator theory of habituation and the orienting response. Available http://www2.psy.uq.edu.au/~landcp/PY269/habituation/habituation.html

Petrinovich, L. (1973). A species-meaningful analysis of habituation. In H.V.S.Peeke & M.J. Herz (Eds), *Habituation I. Behavioural Studies.* (Academic Press, London)

Thorley, P., Hellier, E. & Edworthy, J. (2001). Habituation effects in visual warnings. In M. Hanson (Ed), *Contemporary Ergonomics.* (Taylor and Francis, London)

Rogers, W.A., Lamson, N.L. & Rousseau, G.K. (2000). Warning research: An integrative perspective. *Human Factors, 42*, 102-139

Zimny, G.H. & Scwabe, L.W. (1965). Stimulus change and habituation of the orienting response. *Psychophysiology, 2*, 103-115

# 2003

| Heriot-Watt University Edinburgh | 15th–17th April |
|---|---|
| **Conference Manager** | Sue Hull |
| **Chair of Meetings** | Sandy Robertson |
| **Programme Secretary** | Paul McCabe |

| **Secretariat** | Phil Bust | Charlotte Brace | Guy Walker |
|---|---|---|---|
| | Tim Dubé | Martin Robb | Ed Chandler |
| | Susanne Sondergaard | Suzanne Fowler | Nadine Geddes |
| | Judith Ashley | | |

## Social Entertainment

Tina Worthy and Ann Brooks returned to provide a few surprise rounds in the 'Quiz Night' to keep the delegates entertained in Edinburgh.

| Chapter | Title | Author |
|---|---|---|
| **Musculoskeletal Disorders** | Work-related stress as a risks factor for WMSDs: Implications for ergonomics interventions | J. Devereux |
| **Fatigue** | The impact of work patterns on stress and fatigue among offshore worker populations | A. Burke, N. Ellis and P. Allen |
| **Driving** | How does a speech user interface affect the driving task? | E. Israelsson and N. Karlsson |
| **Rail** | Driver recognition of railway signs at different speeds – A preliminary study | G. Li, W.I. Hamilton and T. Clarke |

# WORK-RELATED STRESS AS A RISK FACTOR FOR WMSDs: IMPLICATIONS FOR ERGONOMICS INTERVENTIONS

## Jason Devereux

*Robens Centre for Health Ergonomics*
*EIHMS, University of Surrey*
*Guildford, Surrey, GU2 7TE*

There is the potential for work-related mental stress (psychological stressors and strains) to reduce the effectiveness of an ergonomics intervention for preventing WMSDs. There is epidemiological and psychophysiological evidence implicating work-related mental stress in the development of WMSDs. Ergonomic interventions in the workplace are needed to reduce the risks of physical and psychosocial work risk factors for musculoskeletal disorders via organisation design changes. In addition, individual susceptibility should become an increasing concern for Ergonomists. Methods for identification and strategies for solutions need to be considered in the future.

## Introduction

This paper considers the latest scientific evidence concerning the potential relationship between work-related mental stress and work-related musculoskeletal disorders (WMSDs). The implications of such a relationship for Ergonomists making workplace interventions is also considered.

There is a major initiative across the European Union to provide good practice for preventing the leading occupational health problem in Europe – WMSDs. There is European consensus that musculoskeletal disorders can be work related and that ergonomics interventions in the workplace can reduce risks and the incidence of WMSDs (Buckle & Devereux, 1999; Op De Beeck & Hermans, 2000). Good practice is centred on reducing physical and psychosocial risk factors in the workplace (Buckle & Devereux, 2002; Cox et al., 2002; European Agency for Safety and Health at Work, 2002).

## The latest on physical factors in the workplace that increase WMSD risk

Systematic critical literature reviews regarding workplace risk factors for WMSDs have been consistent in their findings (Ariëns et al., 2000; Hoogendoorn et al., 1999; Hoozemans et al., 1998; NIOSH, 1997). For musculoskeletal disorders affecting the neck region, high postural load has been shown consistently to be a risk factor (duration of sitting, twisting and bending of the trunk).    For the upper limbs, there is strong evidence that the biomechanical load from a combination of repetition, force and posture increases the risk multiplicatively for musculoskeletal disorders affecting the elbow. The combination effects have also been shown to increase the risk of specific hand disorders, i.e. carpal tunnel syndrome and tendinitis .

In a systematic critical review of Display Screen Equipment (DSE) users, there were consistent study findings regarding increasing duration of DSE use and increasing risk of neck/shoulder and hand/wrist musculoskeletal disorders (Punnett & Bergqvist, 1997). The relationship was mainly dependent on the degree of repetitive finger motion and sustained muscle loading across the forearm and wrist.  At least 4 hours of keyboard work per day appears to increase risk about two-fold compared to little or no keyboard work.

Some Hand-Arm Vibration Syndromes HAVS (for example, vibration-induced white finger) have clearer cause-effect relationships compared to other WMSDs.  It is widely accepted that vibration is the main causal agent, however, the relationship between vibration and HAVS may also be modified by various environmental and individual variables (Bovenzi, 1998).

For the lower back, there has been consistency among critical reviews that manual handling and whole body vibration are risk factors.

## Psychosocial factors in the workplace can increase WMSD risk

Psychosocial factors in the workplace are now widely accepted as contributing to the development of clinical signs and symptoms of WMSDs.  In addition, they have also been included in HSG60(rev), the revised guidance on upper limb disorders in the workplace by the HSE, and are classified as psychosocial factors in the risk assessment (Health and Safety Executive, 2002).

Plausible models to explain the relationships between psychosocial work factors and WMSDs, as well as recent laboratory experimentation supporting the models, have provided support for an interactive relationship between physical and psychosocial risk factors in the workplace (Lundberg, 2002).

To date, only one epidemiological study has been conducted within a UK organisation to explore potential interaction effects between physical and psychosocial work factors. The study showed that high exposure to a combination of recognised psychosocial risk factors, high mental demands, low job control and poor social support, not only has an independent risk effect on WMSDs but also has an interactive effect on risk (Devereux et al., 1999; Devereux et al., 2002).

Of all the workers experiencing recurrent back problems in this study, about 15% of all cases were due to the relatively high psychosocial load, 50% of cases were due to the relatively high physical load and 15% of cases were because of the interaction effects between physical and psychosocial work risk factors.

**The potential impact of individual psychological reactions on ergonomics interventions to reduce WMSDs**

The extent of the interaction between physical and psychosocial work risk factors and WMSDs may potentially be modified by individual psychological reactions. There is evidence to support a relationship between individual psychological reactions (eg anxiety, depression and psychosomatic symptoms) and WMSDs. A review of the epidemiological literature indicates that studies have mainly been cross-sectional and subject to confounding, so it is unclear whether these so-called stress reactions are more likely to lead to the development of WMSDs or vice versa (Devereux & Buckle, 2000). There are plausible reasons why both relationships could be observed. The measures of stress reactions used in the available literature have not been similar with respect to duration, frequency or constructs used. Many of the measures used have also not been validated.

The existence of such a relationship could have a serious negative impact for ergonomics interventions that focus on physical and/or psychosocial workplace risk factors.

Interventions to reduce lifting or hand repetition rates, for example, may result in a 2 to 3-fold reduction in risk. However, despite the reduction in risk, workers may still continue to experience WMSDs because of other pathological pathways causing musculoskeletal damage, such as individual psychological reactions. These reactions may be due to prolonged exposure to psychosocial work risk factors.

Likewise, interventions that focus on reducing psychosocial work risk factors or both physical and psychosocial work risk factors may also result in about a 2-4 fold reduction in risk, but sustained anxiety and depression may mask the true impact of the intervention.

In addition, individual psychological attributes such as beliefs about the causes of job stress may also have an impact on the perceived health and health-related behaviour of workers, and may also have an impact on work organisations because of sickness absence, staff turnover etc. A recent qualitative/quantitative study identified that people possess elaborate beliefs about the causes and consequences of psychosocial work stressors, which subsequently predict psychological well-being and performance (Daniels et al., 2002).

**The latest epidemiological evidence**

An ongoing prospective epidemiological study (STRESSMSD Study led by Dr. J. Devereux), involving 8000 workers in 20 companies across 11 industrial sectors in the UK is designed to investigate the impact of mental work-related stress and lay beliefs of work stress on the development of WMSDs.

Preliminary results from the data support the following relationship. Individual lay beliefs concerning causes of work stress affect the risk of perceived job stress for workers associated with exposure to psychosocial work risk factors.

For example, workers with high exposure to perceived job demands (sometimes or often having to work very fast or intensively, with constant time pressure or pressured to work overtime) showed a 2 to 3-fold increase in risk of high perceived job stress. However, there was a 4 to 5-fold increase in risk for those workers who held a strong belief that job stress results from having to work too fast and in limited amounts of time, compared to workers who did not possess this belief.

A similar modification effect in the exposure-response relationship was observed for workers with strong beliefs concerning the following possible causes of stress:
- low managerial support
- low job control
- low job satisfaction

The data support the view that lay beliefs concerning the causes of perceived job stress can form an individual susceptibility to psychosocial work risk factors, such that there is an increased risk of perceiving job stress if exposed to the risk factors for which a strong causal belief is held.

The study also showed another exposure-response relationship. The higher the level of job stress, the greater was the risk of experiencing recurrent musculoskeletal problems in the previous year affecting the lower back, neck and hand/wrists (odds ratios 1.30-1.61 for moderate levels of job stress, 1.85-2.37 for very high or extreme levels of job stress, 95% confidence intervals greater than one). These results are from the cross-sectional base-line, however results from the prospective study will available next year.

The STRESSMSD study is now the only prospective multi-company study in the UK. The study is due to be completed in October 2003 and is funded by the Health and Safety Executive. A cohort such as this is needed to evaluate the effect of interventions that address physical and psychosocial work risk factors and individual psychological reactions for reducing WMSDs.

## Implications for Ergonomic Interventions

Research which attempts to reduce risk to both work-related mental stress and WMSDs is still in its infancy (Pransky et al., 2002). The available literature has not addressed changes at organisational level to reduce risk to both work stress and WMSDs, but has primarily focused on specific job design changes, for example training and rest breaks. Other studies have focused on individual interventions, for example stress reduction and cognitive-behavioural techniques.

A case study is summarised in order to exemplify the meaning about changes at organisational level. A participatory ergonomics study in a delivery driver work system showed that interventions at organisational level were needed to reduce exposure to both physical and psychosocial work risk factors where the exposure levels were high (Devereux & Buckle, 1999). An intervention was designed around a model in a previous paper by the same authors (Devereux & Buckle, 1998). Modifications were made to work system goals by changing the "Delivery to the point of use" customer service package. The organisation made it clear to customers that a delivery driver would only provide this service if safe to do so. The decision was made by the delivery drivers. The intention was to reduce psychosocial work risk factors by giving delivery drivers greater control over manual handling behaviour for each delivery, greater decision-making authority concerning exposure to risk and greater managerial social support to eliminate hazards.

In addition, poor quality communication between call centre staff and delivery drivers could increase the daily load lifted, the exposure time for whole body vibration and also put the delivery driver behind schedule. The latter increased perceived time pressure and also increased the biomechanical load via increased lifting velocity and asymmetrical working postures. Therefore, recommendations were made to retrain call centre staff and for delivery drivers to meet call centre staff and discuss issues in communication. These interventions contributed to manual handling injuries being significantly reduced over the following two years.

The results provided in this paper indicate that there are two levels of primary intervention. Firstly, organisational level interventions are needed to influence work system design with respect to the work environment, specific job design, the tools and technology. Secondly, individual susceptibility is important and there is a need to identify individuals who are more likely to develop ergonomic injuries such as work stress and WMSDs. Improving self-recognition of exposure and effects and methods of health surveillance, particularly for those with high risk beliefs about work stress, may be useful in a work culture that is constructive and geared towards humanistic and encouraging norms.

The mismatch between the demands of the work system and individual capacities is more likely to be minimised if both sets of factors can be assessed and included in the work system design.

## References

Ariëns, G. A. M., Van Mechelen, W., Bongers, P. M., Bouter, L. M., & van der Wal, G. (2000), Physical risk factors for neck pain, *Scandinavian Journal of Work Environment and Health*, **26**, 7-19

Bovenzi, M. (1998), Exposure-response relationship in the hand-arm vibration syndrome: an overview of current epidemiology research, *International Archives of Occupational and Environmental Health*, **71**, 509-519

Buckle, P. & Devereux, J. (1999), *Work-related neck and upper limb musculoskeletal disorders*. (European Agency for Safety and Health at Work, Bilbao, Spain)

Buckle, P. W. & Devereux, J. J. (2002), Work-related neck and upper limb musculoskeletal disorders: reaching a consensus view across the European Union, *Applied Ergonomics*, **33**, 207-217

Cox T., Randall, R., and Griffiths, A. (2002), *Interventions to control stress at work in hospital staff*. HSE Books, Sudbury

Daniels K., Harris, C., and Briner, R. B. (2002), *Understanding the risks of stress:A cognitive approach*. HSE Books, Sudbury

Devereux, J. and Buckle, P. A participative strategy to reduce the risks of musculoskeletal disorders. Hanson, M. A., Lovesey, E. J., and Robertson, S. A. *Contemporary Ergonomics* , 286-290. 1999. London, Taylor & Francis.

Devereux, J. J. & Buckle, P. W. (1998). The impact of work organisation design and management practices upon work related musculoskeletal disorder symptomology. In P. Vink, E.A.P. Koningsveld, & S. Dhondt (Eds.), *Human Factors in*

*Organizational Design and Management – VI* (pp. 275-279), Amsterdam: North-Holland.

Devereux, J. J., Buckle, P. W., & Vlachonikolis, I. G. (1999), Interactions between physical and psychosocial work risk factors increase the risk of back disorders: An epidemiological study, *Occupational and Environmental Medicine*, **56**, 343-353

Devereux, J.J. and Buckle, P.W. (2000), Adverse work stress reactions-a review of the potential influence on work related musculoskeletal disorders. *Proceedings of the IEA 2000/HFES 2000 Congress*, July 29-Aug 4, San Diego, U.S.A. 457-460

Devereux, J. J., Vlachonikolis, I. G., & Buckle, P. W. (2002), Epidemiological study to investigate potential interaction between physical and psychosocial factors at work that may increase the risk of symptoms of musculoskeletal disorder of the neck and upper limb, *Occupational and Environmental Medicine*, **59**, 269-277

European Agency for Safety and Health at Work (2002), *How to tackle psychosocial issues and reduce work-related stress* (Office for Official Publications of the European Communities, Luxembourg)

Health and Safety Executive (2002), Upper limb disorders in the workplace. HSE Books, Sudbury

Hoogendoorn, W. E., van Poppel, M. , Bongers, P. M., Koes, B. W., & Bouter, L. M. (1999), Physical workload during work and leisure time as risk factors for back pain: a systematic review, *Scandinavian Journal of Work Environment and Health*, **25**, 387-403

Hoozemans, M. J. M., Van der Beek, A. J., Frings-Dresen, M. H. W., Van Dijk, F. J. H., & van der Woude, L. (1998), Pushing and pulling in relation to musculoskeletal disorders: a review, *Ergonomics*, **41**, 757-781

Lundberg, U. (2002), Psychophysiology of work:stress, gender endocrine response and work-related upper extremity disorders, *American Journal of Industrial Medicine*, **41**, 383-392

NIOSH (1997), *Musculoskeletal disorders and workplace factors: a critical review of epidemiologic evidence for work-related musculoskeletal disorders of the neck, upper extremity, and low back* (DHHS (NIOSH) Publication No. 97-141, Cincinnati)

Op De Beeck, R. & Hermans, V. (2000), *Research on work-related low back disorders* (European Agency for Safety and Health at Work, Bilbao, Spain)

Pransky, G., Robertson, M. M., & Moon, S. D. (2002), Stress and work-related upper extremity disorders: Implications for prevention and management, *American Journal of Industrial Medicine*, **41**, 443-455

Punnett, L. & Bergqvist, U. (1997), *Visual Display Unit Work and Upper extremity musculoskeletal disorders. a review of epidemiological findings* (Swedish Institute of Working Life, Stockholm)

# THE IMPACT OF WORK PATTERNS ON STRESS AND FATIGUE AMONG OFFSHORE WORKER POPULATIONS

**Ailbhe Burke, Neil Ellis and Paul Allen**

*Centre for Occupational and Health Psychology,*
*Cardiff University,*
*63 Park Place,*
*Cardiff,CF10 3AS*

This study examined the effects of tour length on stress and fatigue in seafarers in the coastal and short sea shipping industry, in terms of both self report and objective measures. Firstly, a brief outline of the sample and measures used will be given. Then, some background on the issue of tour length is provided. This will be followed by analysis of length of tour for this study in terms of its impact on various measures used in testing. These included self-reports of sleep quality, fatigue, stress levels and mood and performance on reaction time and attention tasks and objectively measured sleep quality. These findings are then outlined and discussed, and the role of tour length in seafarers stress and fatigue is evaluated.

## Assessment of Fatigue Onboard Ship

The unique combination of stressors present in the offshore environment - e.g. extreme weather conditions, noise, motion and demanding work schedules - mean that research findings from other transport industries and onshore populations cannot automatically be applied to seafarers. As well as collecting survey data, it was felt important to actually go onboard ships to gather more detailed information. In this part of the research, a variety of objective indicators and subjective reports were used in assessing seafarers' fatigue.

## Sample

177 participants were recruited in total by researchers who visited seven ships, operating in the UK sector. These consisted of 3 small oil tankers, 2 passenger ferries, a freight ferry, and a fast ferry. This sample was compared with the survey sample and the two were found to be generally similar, although the onboard sample were younger on average, which may be attributable to the higher proportion of officers in the survey sample, or to the comparatively young crew of the fast ferry. This is compared with the

phase one onboard sample of 144 workers from the offshore oil industry in order to assess generalisability of findings between phases.

*Age*

Participants were generally older in the phase one sample. This may again be partially accounted for by the relative youth of the fast ferry crew, and also the relative youth of those working on ferries compared to those working on the offshore oil support ships studied in phase one (see Table 1)

**Table 1: Mean ages of subjects by vessel type**

| Group | N | Mean | SD |
|---|---|---|---|
| Phase 1 | 144 | 41.31 | 9.82 |
| Pipe Layer | 18 | 40.78 | 10.14 |
| Dive support vessel | 81 | 42.04 | 8.63 |
| Shuttle tanker | 19 | 38.84 | 12.85 |
| Supply Vessel | 12 | 44.00 | 8.16 |
| Standby/supply Vessel | 14 | 38.86 | 12.51 |
| Phase 2 | 177 | 36.07 | 11.40 |
| Freight | 27 | 39.11 | 10.45 |
| Tankers | 24 | 41.83 | 12.67 |
| Passenger Ferries | 71 | 37.28 | 9.90 |
| Fast ferries | 55 | 30.49 | 11.07 |

There were more mixed nationality crews in phase 2, with only 63.8% (n=113) of crews being from the British Isles, in comparison to 91.2% (n=134) in phase 1. Other nationalities in phase 2 included Spanish (20.3%, n=36), Polish (13.0%, n=23), and Canadian (2.8%, n=5).

*Length of tour*

The typical tour length was shown to differ between the two phases, with the majority of participants in phase 1 (68.3%, n=99) working 4 weeks on/4 weeks off tours, in comparison to phase 2 in which the majority worked 1 week tours (34.4%, n=61) (see Table 2). However, again this was skewed by tour length on the fast ferry, which never exceeded seven days.

**Table 2:  Tour length**

| Tour length | Phase 1 | Phase 2 |
|---|---|---|
| 1 week | ---- | 34.4% (n=61) |
| 2 weeks | 2.8% (n=4) | 15.3 (n=27) |
| 3 weeks | 4.1% (n=6) | 6.2% (n=11) |
| 4 weeks | 68.3% (n=99) | 3.4% (n=6) |
| 5 weeks | 6.9% (n=10) | 0.6% (n=1) |
| 6 weeks | 2.1% (n=3) | 1.1% (n=2) |
| 7 weeks | 9.0 (n=13) | ---- |
| 8 weeks | 6.9 (n=10) | 10.2% (n=18) |
| 8+ weeks | ---- | 29.0% (n=51) |

Phase 1 and phase 2 participants were tested at a similar stage into the tour, with the highest proportion of subjects being tested in week 1 (43.7% in phase 1, and 48.0% in phase 2) (see Table 3).

### Table 3 . Weeks into tour at testing

| Weeks in tour | Phase 1 | Phase 2 |
|---|---|---|
| week 1 | 43.7% (n=62) | 48.0% (n=85) |
| week 2 | 26.1% (n=37) | 13.0% (n=23) |
| week 3 | 16.2% (n=23) | 6.8% (n=12) |
| week 4 | 4.2% (n=6) | 5.6% (n=10) |
| week 5 | 2.1% (n=3) | 4.0% (n=7) |
| week 6 | 1.4% (n=2) | 4.5% (n=8) |
| week 7 | 5.6% (n=8) | 3.4% (n=6) |
| week 8 | 0.7% (n=1) | 5.1% (n=9) |
| week 8+ | ---- | 9.6% (n=17) |

## Procedure

Volunteers participated in four sessions overall, which were scheduled before and after work on the first and final day of their testing, typically an interval of 5-7 days. During these sessions, before and after work questionnaires (henceforth 'logbooks') recording food intake, medication, breaks, caffeine consumption, smoking, sleep, symptoms of fatigue and perception of work related issues were completed. Performance tasks measuring reaction times, errors and lapses of attention as well as subjective reports of alertness, hedonic tone (happiness, sociability) and anxiety were also administered during these sessions.

Further objective measures involved sleep recording to assess sleep quality, noise measurements from different areas of the ship, and measurements of the pitch, roll, and heave dimensions of motion. The survey was completed by participants onboard ship in their own time during this testing interval.

## Tour Length - Background

Seafaring may be regarded as a very important occupational area for study, having high accident rates and more deaths per capita than any other industry in Britain. It is therefore necessary to examine the issues that make this so. However, studies of seafarers have, until now, paid little attention to potentially important factors such as days-into-tour. Intuitively the expectation is that longer tours of continuous duty would be more detrimental in terms of cumulative effects leading to more fatigue and poorer health. It is indeed the case that some research on installation workers (Collinson, 1998; NUMAST, 1992) has indicated that tours exceeding 2 weeks show increased injury rates, and that adverse physiological changes may be related to tour lengths exceeding one week.

However Forbes' (1997) study found that accident frequency among installation workers was greatest at the beginning of a tour, specifically during the first tour week, and then declined steadily over the course of a tour. Thus clearly, these mixed findings indicate that this intuitive sense that longer tours are more detrimental in health and fatigue terms requires re-examination.

Generally indications from phase one (Smith *et al*, 2001) also affirmed that shorter tours are not necessarily better in terms of seafarers' well-being. Although the accident data is limited, most accidents were found to occur in the first week of tour regardless of actual tour length. The logbook data confirmed that the impact of their work and work environment on seafarers may change over the course of the tour. For example, sleep duration was reduced for the first night offshore but improved with days into tour. Also, seafarers showed improvements in alertness across the week, and analyses of effects of days into tour on those working night shifts showed that those more than 5 days into tour (average 18 days) made fewer errors on performance tasks than those less than 5 days into tour (average 3 days).

It was found that many phase one respondents reported feeling 'below par' on returning to their vessel/installation after a period of leave. Approximately half of all respondents felt that adjusting to life offshore took at least 2-3 days and felt their performance to be affected during this period of adjustment. This may partially account for these findings, since the period of adjustment takes up a smaller proportion of longer tours than of shorter tours.

**Tour Length – Phase Two**

This analysis was conducted to yield information about changes in seafarers' stress, fatigue and performance levels over a discrete period of time, in order to assess the impact of offshore work on seafarers as a function of time into tour. This was done using a mixture of data from both the logbooks and the objective measures. Survey data were not relevant to this analysis since the measures within the survey were only completed once, not at the specific onboard testing intervals.

The period of analysis matched the period of the performance testing onboard, i.e. typically 5-7 days, and the analysis itself had two layers. Firstly, the aim was to reveal any fatigue and performance related differences there may be as a result of a specific period – that is to uncover what, if any, effect working over approximately a week long period has on self-reports of fatigue, work-related variables, and objectively measured performance and sleep. Secondly, these effects were analysed as a function of time into tour, to assess whether longer tours mitigated or exacerbated these effects. This analysis was restricted to the logbooks only.

Approximately forty participants (the fast ferry sample) were excluded from this analysis since their entire tour was only 6-7 days long, and it was felt that including them would bias the data set, by confounding ship type with tour length.

*Key findings:*

- Levels of job stress, job effort, alertness and outcomes on some of the performance measures were more negative further into tour
- Habituation to noise levels aboard ship seems to occur fairly consistently as a function of time into tour.
- Sleep appears to improve further into tour, which may help account for the relatively low levels of fatigue in this sample.

Both the logbook and the objective measures provide evidence that the cumulative effect of work, both across the day and across the working week, may influence both levels of fatigue and performance. Across the working week, job stress was found to increase and this was mirrored by slower task reaction times and lower levels of alertness. This may indicate that over longer periods, seafaring work has a detrimental effect on individual fatigue and performance.

There is also some evidence from the logbooks that seafarers' sleep improves as a function of time into tour. Although an actual improvement in sleep was not recorded by the objective measures, no impairment in sleep was identified either so at the very least these do not contradict the logbook findings. Also, generally habituation to noise levels onboard was observed as a function of days into tour.

*Key differences on the first fortnight vs. after two weeks analysis:*
First fortnight of tour:
- Physical effort significantly lower after seven days
- General health significantly worse after seven days

After first two weeks of tour:
- Almost no change in sleep across the testing interval
- Physical effort significantly higher after seven days
- Weather becomes more of an issue across the testing interval, despite no change in actual weather conditions
- Support from fellow workers, self-regulation of own work, and work satisfaction are all affirmed more highly after seven days for this sub-sample

This further analysis of the logbook data indicates that any cumulative effects over the testing interval vary as a function of weeks into tour. There is some evidence of habituation, and some evidence of cumulative negative effects of time at sea, e.g. fewer effects of noise are observed further into tour, whereas the subjective impact of motion increases. This first fortnight/after second week of tour split is supported by the manifest differences between the two sub-samples. For example, from day one to day seven of the first fortnight of tour, there is a significant increase in self reported work stress and lack of sleep, and even though physical effort decreases over the seven day period, general health is reported as worse. After the first fortnight of the tour there are fewer negative day one day seven differences. Again, general health was found to be worse on day 7. Stress is mostly the same, sleep appears stable over a week long period after the first fortnight of the tour. There are some indications that longer tours may be better in that there is higher affirmation of receiving support from fellow workers, self-regulation of work and work satisfaction further into tour.

## Conclusion

Thus tour length seems to be an important factor in stress and fatigue at sea, and also to affect sleep and other exposure variables. Furthermore, the effects of work and work related issues on seafarers over the work period vary as a function of time into tour. These time into tour differences may even indicate that in some ways, longer tours are actually less detrimental in terms of fatigue and work related exposure variables than shorter ones. This finding is supported by analyses of the survey data, indicating that the longer the tour, the lower the fatigue as measured by the PRFS scale and the fatigue at/after work factors.

Thus it seems that tour length may in fact prove quite an important factor in addressing issues of fatigue and health, and indeed accident and injury in seafaring. Further analysis is currently being undertaken to extend our understanding of the importance of this issue.

## Acknowledgements

The research described in this article is supported by the Maritime and Coastguard Agency, the Health and Safety Executive, NUMAST and the Seafarers' International Research Centre. We would also like to acknowledge the contribution made by the ship owners and seafarers who have participated in the research.

## References

Collinson, D.L. 1998, "Shift-ing lives": Work-home pressures in the North Sea oil industry. *Canadian Review of Sociology and Anthropology.* **35** (3), 301-324

Forbes, M.J. 1997, *A study of accident patterns in offshore drillers in the North Sea.* Dissertation prepared for the Diploma of Membership of the Faculty of Occupational Medicine of the Royal College of Physicians.

NUMAST 1992, *Conditions for Change.* London, NUMAST.

Smith, A.P., Lane, A.D. and Bloor, M. 2001, *Fatigue Offshore: A Comparison of Offshore Oil Support Shipping and the Offshore Oil Industry.* Seafarers International Research Centre. ISBN 1-900174-14-6.

# HOW DOES A SPEECH USER INTERFACE AFFECT THE DRIVING TASK?

**Elisabet Israelsson[1] & Nina Karlsson[2]**

[1]*Linköping University, Linköping, Sweden*
[2]*Department of Human-System Integration, Volvo Technology Corporation, Gothenburg, Sweden*

The main objective of this study was to investigate the effect that speech user interfaces have on the drivers' mental workload in a real driving situation, and to compare this to manual interfaces.

## Introduction

In recent years, speech user interfaces have become increasingly popular in many different areas. In the automotive area, under the slogan "eyes on the road – hands on the steering wheel", the notion of speech has often been viewed as a savior to compensate for poorly designed user interfaces with deep menu structures. However, few studies have been done that investigate the impact that speech user interfaces have on the driving task, and to confirm the high expectations, especially in real driving environments (a few studies have been made in driving simulators). Driving is typically divided into primary and secondary tasks. The primary task of driving pertains to actions related to navigating and controlling the vehicle. The secondary tasks are those not directly related to the primary task, but are nonetheless performed while driving.

Due to the state of the technology, most speech user interfaces used in vehicles today are command-based and not particularly conversational, hence adding a load on the user's memory. Are these type of speech user interfaces still safer in the sense that they demand less mental workload from the driver compared to manual interfaces?

Results from a study in a driving simulator suggest that mental workload decreases when dialing a phone number on a cellular phone using voice rather than pressing buttons (Graham *et al*, 1998). Another study by Carter *et al* (2000) compares the impact different input devices have on the driving task. The respondents performed a phone dialing task using voice control, steering wheel buttons and the conventional buttons located on the centre stack. Voice control proved to have the least impact on the driving task and was most preferred by the respondents. However, using the conventional buttons on the centre stack was the fastest mode of interaction.

The main objective of this study was to investigate the effect that speech user interfaces have on the drivers' mental workload in a real driving situation, and to compare this to manual interfaces.

**Method**

The general method of the study is described below.

*Respondents*
Nineteen respondents participated in the study, thirteen men and six women.

All respondents were Volvo employees, none of them having worked with anything related to navigation systems or speech user interfaces. Age range was from 23 to 44 years old, with a median of 30 years. All respondents were experienced drivers, i.e. they drove more than 5000km/year. None of them had any prior experience with speech user interfaces; eight had prior experience of Volvo's navigation system.

*Equipment*
A vehicle equipped with a voice controlled navigation system was used in the study. The navigation system's display is situated on top of the dashboard, and is controlled by using a set of buttons placed on the backside of the steering wheel. The buttons can be reached without letting go of the steering wheel when holding a regular "ten-to-two" grip. Pressing a push-to-talk (PTT) button on the front side of the steering wheel activates the voice control system.

A peripheral detection task (PDT) was used with the visual stimuli placed on top of the dashboard in a manner that the 16 diodes were reflected on the windshield. A video camera was used to record the test sessions.

*Experimental set-up*
The respondents were seated in the driver's seat. They were informed of the procedure of the test, and were given a chance to familiarize themselves with the navigation system for approximately 10 minutes. Additionally, they were given a map of the route. Each respondent drove the route three times. The first lap was considered a baseline-lap. During the first lap, no tasks were performed except the PDT. During the second and third laps, the respondents either performed speech tasks (using the speech user interface) or manual tasks (using the conventional buttons on the steering wheel) in the navigation system. The order of the second and third lap was balanced. After each lap the respondents filled out the NASA-TLX questionnaire. After the last lap they also filled out an attitude questionnaire.

*Measures*
Four dependent variables were measured: i) hit rate on the PDT, (ii) reaction time on the PDT, (iii) NASA-TLX score, and (iv) time to solve a task.

PDT is a secondary task measurement used to measure workload while driving (van Winsum *et al*, 1999). The instrument consists of a plate with size of 16*5 centimetres, onto which 6 red randomly blinking diods are placed. Each light remained lit for one second. Each time the respondents detected a red light they were instructed to press a button mounted on their left index finger. The number of correct responses, hit rate, and the reaction time is measured.

NASA-TLX is a subjective measurement of workload. It consists of a multidimensional scale with 6 dimensions of factors related to mental workload (Hart, 1999). Since the term mental workload can be interpreted somewhat differently among the respondents, their personal opinion on what mental workload means for them is taken into the final calculation of the NASA-TLX score. This is done by subjective weighting of each scale.

## Results

### *PDT – hit rate*
A 3*2 SPANOVA (Split-Plot Analysis of Variance) was performed to find out whether there were any effects of task type (manual or speech) or interaction. There was a significant difference between manual and speech task [$F(2,34)=20,753$, $p<0,0001$].

A least significant difference test revealed the hit rate was lower for manual tasks than baseline [$p<0,0001$] and for speech tasks [$p<0,0001$]. However, there was no significant difference between the speech tasks and baseline regarding hit rate.

Additionally, there was no effect of experience and no interaction effects.

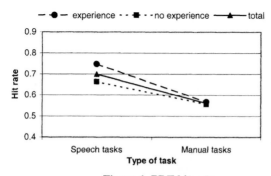

**Figure 1. PDT hit rate**

### *PDT – reaction time*
No effect of type of task, that is modality, was discovered. Furthermore, there was no effect of experience or any interaction effects. All respondents basically had similar reaction times.

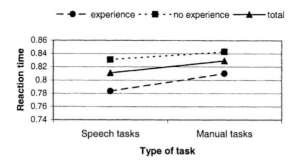

**Figure 2. PDT reaction time**

*NASA-TLX*

There was a significant difference between the manual tasks and speech tasks [$F_{(2,34)}$=53,438; p<0,0001]. Pairwise comparisons revealed that the respondents experienced the manual tasks to have a significantly higher demand on workload compared to the speech tasks [p<0,0001] and baseline [p<0,000]. However, there was no significant difference between the speech tasks and baseline.

Finally, there was no effect of experience or any interaction effects.

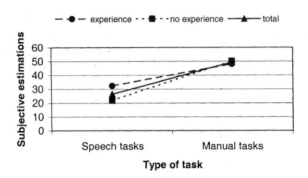

**Figure 3. NASA-TLX score**

*Task completion time*

There was a significant difference between the manual tasks and speech tasks [$F_{(1,17)}$=176,855; p<0,0001]. No effect of experience or any interaction effects was detected.

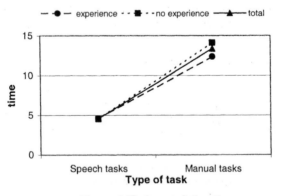

**Figure 4. Task completion time**

*Errors*

Even though all respondents successfully completed all tasks, they performed a number of errors along the way. The most frequent types of error when performing the speech tasks were not saying the correct utterance and command, forgetting to press the PTT-button, waiting too long before uttering the command (resulting in that they were timed out), or speaking too unclear.

When performing the manual tasks, the most frequently occurring errors were mixing up the button, that is moving backwards instead of forward, and getting lost in the menus.

*Attitudes*

Nearly all respondents experienced the speech user interface to be more fun and easier to use than the conventional manual interface, with only one respondent preferring a manual interface. They could also see themselves buying a vehicle with voice control in the future. In addition, they believed that using a speech user interface would actually increase safety.

## Discussion and Conclusions

The results from the peripheral detection task (PDT) and the NASA-TLX ratings show that performing tasks manually in the navigation system requires a higher workload than performing the same tasks using voice control. These results were somewhat expected, since prior studies conducted in driving simulators reveal the same tendencies.

Surprisingly however, there was no significant difference between baseline driving (no secondary tasks) and driving while operating the navigation system using voice control. This was unexpected since using voice control requires the user to remember a specific command (e.g. press the PTT-button and utter the command) and to acknowledge feedback from the system in order to make sure he/she was correctly understood. This finding could be a result of the selected tasks being relatively simple. Only one-command tasks were used in the study, due to limitations in the prototype navigation system. If longer dialogs had been used, requiring the user to interact with the system in two or more iterations before one task is completed (such as entering a new street address), the difference might have been greater.

Additionally, the order in which the respondents drove the baseline route was not varied due to safety reasons. The baseline was considered a chance for the respondents to familiarize with the vehicle, route, and PDT-task. Giving the respondents a chance to do this before driving the three laps would have required too much time and might have been considered too tedious for the respondents.

The reaction time on the PDT-task was similar during all conditions for all respondents. This might have been a result of the tasks being difficult. The respondents either saw the red light blink and reacted to it or did not see it at all.

Prior experience of using the navigation system did not significantly affect the respondents' performance on the secondary tasks. This might imply that the experienced users were not really that experienced (cannot be compared to experienced users of a radio for example) or that the interface differ to such an extent that experience of one does not affect the other. However, it is natural to believe that an experienced user is familiar with terms and functionality in a way that he/she will more easily remember for example a command.

A closer look into the general mean scores on the NASA-TLX score shows that the respondents experienced the manual interface as twice as demanding as the speech user interface. This fact is probably related to the positive remarks given in the attitude questionnaire, where almost all of the respondents preferred the speech user interface over the conventional interface.

The time to complete a task was significantly longer for the manual tasks compared to the speech tasks. This is not surprising since using the manual interface requires one to wander around in menus while the speech user interface only requires one to utter a command. However, some of the manual tasks were actually faster than some of the speech tasks. These were the tasks that required only one or two button presses. This could imply that speech user interfaces are best suited for current manual tasks requiring more than a single push of a button.

Further studies are needed in order to fully understand the impact of voice control on the driving task. The next step is to compare simple commands with longer dialogs and to measure vehicle data and eye movement. Is the driver really keeping his eyes on the road?

## References

Graham, R., Carter, C., Mellor, B. 1998, *The use of automatic speech recognition to reduce the interference between concurrent tasks of driving and phoning,* in 5th Conference on Spoken language Processing 1998, Australia

Carter, C., Graham, R. 2000, *Experimental comparison of manual and voice controls for the operation of in-vehicle systems,* in Proceedings of the IEA 2000 / HFES 2000 Congress, HUSAT Research Institute, Loughborough University, UK.

Hart, S. G. 1988, *Development of NASA-TLX (Task load index): Results from Empirical and Theoretical Research,* Aerospace Human factors Research Division, NASA-Ames Research center, Moffett Field, California.

van Winsum, W., Martens, M., Herland, L. 1999, *The effects of speech versus tactile driver support messages on workload, driver behaviour and user acceptance,* TNO-report, TM-99-C.

# DRIVER RECOGNITION OF RAILWAY SIGNS
# AT DIFFERENT SPEEDS - A PRELIMINARY STUDY

**Guangyan Li[1], W. Ian Hamilton[1] and Theresa Clarke[2]**

*[1]Human Engineering Ltd.*
*Shore House, 68 Westbury Hill*
*Westbury-on-Trym, Bristol BS9 3AA*

*[2]Ergonomics Group, Railtrack (part of the Network Rail Group)*
*Railtrack House, Euston Square, London, NW1 2EE*

Experimental trials were conducted using computer simulation to investigate the influence of train speed on driver recognition of railway signs. Twenty-four professional train drivers participated in the study and thirty-six types of lineside signs were tested under the simulated approach speeds of 100, 200 and 300km/h respectively.

Based on driver performance data, critical train speed was estimated for each type of sign. Beyond these speeds, the signs could no longer be correctly identified within a minimum approach time. The study also indicated that with increasing speed a larger visual angle is required in order to correctly identify a sign. In addition, more reading errors are related to signs containing two or more pieces of information, with each information item containing two or more letters or numbers.

## Introduction

With advancing technology and increasing demand on railway transport, trains are travelling faster t han t hey w ere d ecades a go. H owever, a m ajority o f e xisting r ailway signs are still based on traditional designs that have been used for many years. Faster approach speed reduces the time that a sign stays in the driver's view, which increases the chance for the signs to be misread. There is thus a need to better understand whether these signs can still be correctly identified by drivers while travelling at different speeds.

The scientific basis for the present study has been reported elsewhere (HEL, 2002), which reveals that, despite much research that has been carried out to date regarding the design and safety issues for traffic signs, most published studies are in connection with road transport. Little is known about the effect of train speed on driver responses to different types of lineside signage, and no satisfactory answers could be found to some of the important questions. For example, how much information can be correctly identified while travelling at different speeds? Are all the current lineside signs, especially those designated for high-speed use, readable for high-speed train operation? If not, what are the speed 'cut-off' points beyond which the current signs are no longer effective? (thus an in-cab information system may be required to present such information); What is the maximum acceptable complexity level of a lineside sign in relation to a particular train speed?

Experimental trials were conducted to test the effects of train speed on driver recognition of lineside signs, using a computer simulation programme which was developed by HEL and successfully used in previous studies (e.g. Li et al, 2002). The tests covered the current operational (or potential) train speeds in the UK (from 100km/h or 62mph up to 300km/h or 186 mph) and included most standard lineside signs.

## Experimental studies

### Representation of Railway Signs

The lineside signs were shown at full apparent size (based on their current standard design and calculated visual angles). They were presented at an equivalent starting distance of 800 meters and 'moved' (expanded) towards the subject at an apparent speed of 1 00, 2 00 or 3 00 k m/h (one at a time, in random order). The starting distance was chosen to ensure that the drivers should not be able to see the signs at the start of the trials.

Thirty-six railway signs were tested, which covered most of the standard lineside signs that are currently used in the UK for this speed range. Figure 1 shows some examples.

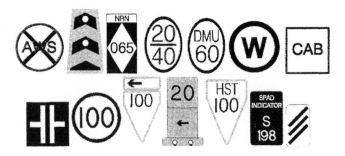

**Figure 1. Examples to show some of the railway signs tested**
(Drawings are for illustration purpose only, simulations were based on standard designs GK/RT0033, 1996)

### Testing Theoretical Signs

In addition to testing the current standard realistic signs, a theoretical test session was also conducted to investigate how much information (and presented in what pattern) the drivers can effectively identify while approaching a sign at different speeds. The theoretical s igns w ere d esigned s uch t hat e ach s ignboard contained from one to three items of information (combined letters and numbers), and each information item also had a complexity level from one to three (Figure 2). The size of the signboard was kept constant at 750x1050mm which is a standard size for most combined numeric/text signs.

It was hoped that the results would help to develop guidelines for the design of new signs which may have to convey multiple messages.

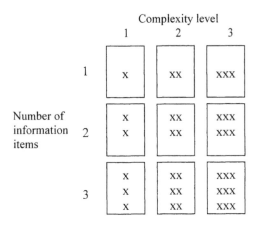

**Figure 2. Patters of theoretical signs**

*Participants*
Twenty-four train drivers from nine Train Operating Companies (23 male and 1 female) participated in the trials. Their average age was 40.6 years (Range=27-64, SD=9.3), and their average duration of driving experience was 12.1 years (Range=0-42.0, SD=10.6).

*Apparatus*
Two Intel Pentium®4 desktop computers with 512 MB RAM and 3D graphic card (Geforce2) were used for the trials. This enabled two parallel trials to be run with two participants at the same time (in two separate rooms). The monitor size was 17 inch; screen area (resolution) was set at 1280 by 1024 pixels and the colours were set for 'true colour' (32 bit). The screen background was set to medium grey (128). The computers were running MS Windows 2000 system.

*Procedures*
The participants were seated at the simulator and were allowed to adjust the chair to their preference. Their eye-to-screen distance was maintained constant at approximately 50cm. The simulation settings ensured that this distance would achieve a correct visual angle to the eyes such that the simulated signs viewed on the screen were equivalent in size to how they would appear in the real world.

The participants were required to press a key (e.g. the 'space bar') as soon as they could positively identify the sign, and at this point the computer recorded the equivalent sighting distance/time to sign. At the end of each run the participants gave their answers on what they had seen and the results were recorded by the experimenter.

It took approximately 60 minutes for each participant to complete the trials. The room lighting was dimmed to approximately 10-15 lux throughout the trials.

## Results

*Findings of the Railway Sign Tests*
The influence of train speed on driver response to lineside signs was tested by calculating an equivalent visual angle at the eye at the moment when a particular type of sign was correctly identified while approaching the sign at one of the speeds tested. The results show that in order to correctly identify a certain type of sign at a certain approach speed, the sign has to achieve a minimum visual angle at the eye. The pattern of visual angle varied depending on the types of signs. In general, ANOVA and multiple comparisons (Tukey's HSD test) showed a trend which was that the faster the approach speed, the greater the required visual angle (or the larger the sign should be) at the point when its content could be correctly identified ($p<0.00001$ for all three speed levels tested), as shown in Figure 3. This is probably due the fact that subjects need a minimum amount of time to process the sign's information content.

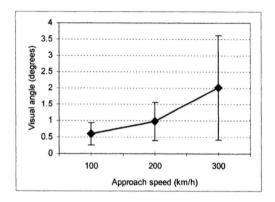

**Figure 3. Visual angles required to correctly identify the lineside signs tested**

The relationship was also estimated using linear regression between approach speed and time remaining to reach the sign at the moment when it is correctly identified. This gives an equation to estimate a speed limit beyond which the sign can no longer be correctly identified within a set time remaining before it is passed: $V = a - bT$

Where: V is approach speed (km/h); 'a' is the maximum speed limit when the signs can be read correctly at any time reaching/passing the sign (T=0s); 'b' is a constant showing the amount by which a change in time remaining (T, measured in seconds) will cause a reduction in speed limit in order to identify the sign correctly. Table 1 gives the estimated maximum speed corresponding to some of the signs tested, and beyond these speeds the signs are no longer readable.

**Table 1. Estimated speed limit based on approaching time remaining**

| Type of signs | Equation for speed estimation | Max speed km/h (When T=0s) |
|---|---|---|
| Whistle board | 301.1-65.4 T | 301 (187mph) |
| Cab signalling and AWS gap warning | 257.7-97.3 T | 257 (159mph) |
| Countdown marker 100 yards | 261.7-175.3 T | 261 (162mph) |
| Countdown marker 300 yards | 232.8-184.1 T | 232 (144mph) |
| Level crossing, temp AWS cancelling/cab signalling | 273.7-34.2 T | 273 (169mph) |
| Electrification neutral section warning | 277.7-77.1 T | 277 (172mph) |
| Emergency indicator | 270.3-43.1 T | 270 (167mph) |
| Permissible speed indicator (triangle) | 293.0-94.8 T | 293 (141mph) |
| Permissible speed indicator (circle) | 320.1-104.9 T | 320 (198mph) |
| Temp differential speeds | 275.3-121.6 T | 275 (170mph) |
| Standard differential speeds | 297.5-122.3 T | 297 (184mph) |
| Permissible speed–standard differential | 253.0-159.2 T | 253 (156mph) |
| Permissible speed–arrow diverging route (red circle) | 347.4-130.2 T | 347 (215mph) |
| Permissible speed–arrow diverging route (yellow triangle) | 278.1-190.9 T | 278 (172mph) |
| Temp speed restriction–diverging route (warning lights) | 247.9-93.1 T | 247 (153mph) |
| Permissible speed – HST or DMU trains (red oval) | 270.4-256.8 T | 270 (167mph) |
| Permissible speed – HST or DMU trains (yellow triangle) | 196.0-115.2 T | 196 (121mph) |
| Radio channel change markers and SPAD indicators | 270.0-150.5 T | 270 (167mph) |

*Findings of the Theoretical Sign Test*

Driver response errors were significantly affected by sign complexity at $p<0.000000001$ level, but the effect of speed on reading error was insignificant. However, there was a significant interaction between complexity and speed ($p<0.001$) with respect to their influence on sign reading errors. Multiple comparisons indicated that the major differences lay only between sign designs 3x2 and 3x3 (i.e., 3 pieces of information each having 2 or 3 digits) and the remaining sign patterns. The results suggest that lineside signs should not contain information which is at or above 3x2 complexity level, especially if the sign is to be used for an approaching speed of over 100km/h.

## Discussion

This study found a significant influence of approach speed on driver recognition of railway s igns i n t erms o f t he v isual a ngle r equired a t t he e ye a nd t he t ime remaining when the sign could be correctly identified. Using the trial data, a critical train speed can be estimated beyond which the contents of the corresponding sign may not be correctly identified within a given time. It must be understood, however, that these results were based on only 24 drivers' performance data in an impoverished simulated environment, it is not yet known to what extent driver responses/behaviour in this environment represents that in the real world. In addition, the image display ability of a PC screen is limited by the smallest pixel size, therefore, it can be assumed that, as far as the sighting distance is concerned, if a sign can be viewed at a certain distance on the simulator, it should be seen farther in the real world, which also allows a faster approach speed.

However, it should also be understood that the simulated environment in the impoverished simulator is much simpler and visually cleaner than what the drivers see in the real world, in which for example, the information can be cluttered, the signs can be covered with dirt, the windscreen of the train may be smudged or there may be glare from the sun. In these situations, the real speed limits could be even lower if the sign is to be correctly identified.

The present study tested the driver sign reading performance under three speed levels only. Information is thus limited for the development of a more realistic relationship b etween s ign r eading p erformance and approach speed based on the data collected. Further data analysis showed the accuracy of the linear models of around $R^2=0.56$ on average (Range: 0.22-0.77, SD: 0.12), suggesting that the relationship between speed and driver recognition of lineside signs may be to some extent non-linear.

Drivers commented in the present study that they often do not need to recognise the signs before reaching the signpost. Therefore, it may be reasonable to assume that the remaining time (T) can be less than 4 seconds (the minimum sighting time required by standards) for the recognition of signs, even up to the point of reaching it (i.e., at T=0).

Speed did not significantly affect driver response errors for the identification of most of the railway signs tested. One of the reasons is possibly due to the fact that during the trials, the drivers were given a free choice not to respond until they were sure that the content of the sign had been positively identified, resulting in a low reading error rate, but a shorter sighting distance (or equivalently larger visual angle at the eye).

## Conclusion

This study indicated a dynamic relationship between a visual angle required to correctly identify a sign and the approach speed, suggesting that larger signs would be required for higher train speed in order to ensure a minimum reading time. The maximum speed limit relating to each type of lineside signs (current standard designs) can be estimated using the linear models developed on the basis of the experimental data; and beyond the maximum speed the corresponding sign may not be correctly identified. However, due to the limitations of this study, as discussed earlier, the results at this stage should be regarded as being only indicative rather than conclusive.

Beyond speeds of 100km/h more reading errors occur with signs containing two or more pieces of information, with each information item containing two or more letters/numbers. Therefore, when designing new lineside signs, consideration should be given to how to avoid presenting the drivers with more than two pieces of information with each information item containing three letters or numbers. Similarly, for signs containing up to three pieces of information, each information item should not contain more than two digits.

## References

Human Engineering Ltd., 2002, *Driver recognition of lineside signals and signs during the operation of high-speed trains – Scientific basis for the study.* Technical report to Railtrack (part of the Network Rail Group), prepared by Guangyan Li and authorised by W. Ian Hamilton, HEL/RT/01651/RT1.

Li, G., Rankin, S. and Lovelock, C., 2002, *The influence of backplate design on railway signal conspicuity.* In: Contemporary Ergonomics 2002, (ed. P.T. McCabe), London: Taylor & Francis, 191-195.

Railway Group Standard GK/RT0033, Issue Three, July 1996, *Lineside Signs.*

# 2004

| **Swansea University** | 14th–16th April |
|---|---|
| **Conference Manager** | Sue Hull |
| **Chair of Meetings** | Sandy Robertson |
| **Programme Secretary** | Paul McCabe |

| **Secretariat** | Phil Bust | Charlotte Brace | Gerry Newell |
|---|---|---|---|
| | Lorraine Rogers | Martin Robb | Ed Chandler |
| | Susanne Sondergaard | Dominic Furness | Lisa Doddington |
| | Nam Loc | Karin Gibberd | Sandy Thomson |

## Social Entertainment

Stormy weather prevented people from taking advantage of the beach at Swansea which put pressure on the conference staff to keep everyone happy. This year the 'Quiz Night' became themed and all the questions were based on 'Sound and Vision'.

| Chapter | Title | Author |
|---|---|---|
| **Slips, Trips and Falls** | Fall causation among older people in the home: The interacting factors | C.L. Brace and R.A. Haslam |
| **Inclusive Design** | Designing for people with low vision: Learnability, usability and pleasurability | C.M. Harrison |
| **Occupational Health and Safety** | Process ownership and the long-term assurance of occupational safety: Creating the foundations for a safety culture | C.E. Siemieniuch and M.A. Sinclair |
| **General Ergonomics** | Development of a Crowd Stress Index (CSI) for use in risk assessment | K.C. Parsons and N.D. Mohd Mahudin |

# FALL CAUSATION AMONG OLDER PEOPLE IN THE HOME: THE INTERACTING FACTORS

## C L Brace and R A Haslam

Health and Safety Ergonomics Unit,
*Department of Human Sciences,*
*Loughborough University,*
*Leicestershire,*
*LE11 3TU*
*UK*

Falls in the home are a major problem for older people. Although personal and environmental risk factors for falling among this group are well understood, less is known about how these risks are influenced by behaviour. Focus groups and interviews were carried out with 207 older people. The findings of this investigation suggest that there are a variety of interacting factors which affect risk of falling, including intrinsic, extrinsic and peripheral influences and that behaviour is an overarching control over all of these. It can be concluded that older people who are at risk of falling need to be better educated on the individual risk factors and on the help that is available to support them in healthy aging.

## Introduction

It has been well documented over the years that a third of individuals over 65, and nearly half of those over 80, fall each year. Approximately half of all recorded fall episodes that occur among independent community dwelling older people happen in their homes and immediate home environments (Lord *et al*, 1993). Fall related incidents are influencing factors in nearly half of the events leading to long-term institutional care in older people (Kennedy and Coppard, 1987). Clearly, if the incidence of falls can be reduced, people can live longer, more healthily and more independently in their own homes, with a better quality of life. Falls pose a threat to older persons due to the combination of high incidence with high susceptibility to injury. The tendency for injury because of a high prevalence of clinical diseases (e.g. osteoporosis) and age-related physiological changes (e.g. slowed protective reflexes) makes even a relatively mild fall dangerous.

Over 400 potential risk factors for falling have been identified, which are commonly split into categories of intrinsic and extrinsic risk. Intrinsic factors are age and disease related changes within the individual that increase the propensity for falls, e.g. decreased balance ability, disturbed gait, etc. Extrinsic factors are environmental hazards that present an opportunity for a fall to occur, including floor surfaces (textures and levels), loose rugs, objects on the floor (e.g. toys, pets), poor lighting etc. However, individual fall incidents are generally multifactorial.

Personal and environmental risk factors for falling among this group are well documented, although it is only recently that the influence of behaviour has been investigated in relation to these risks (Hill *et al*, 2000; Brace *et al.*, 2003). Important behavioural factors which affect the risk of older people falling in the home have been established, e.g. rushing, carrying objects. It has also been found that behaviour patterns change after a fall episode; general psychological state and experience can have an effect on the individual, affecting confidence and fear of falling, and general behaviour.

Of further interest is the extent to which the design of domestic products and areas of the home might be factors in falls. Although environment-related risk factors are reported to be causal in around one third of falls, it has only been lately that detailed work has been done to look at the design of some areas of the home environment in relation to older people, falls and independent living (Brace *et al.*, 2002).

## Method

Preliminary focus groups (5) were conducted with older people (30 participants in total) to gain insight into the problem. The discussions were used to collect preparatory information on patterns of behaviour likely to affect risk of falling, informing the design of materials for the subsequent interview survey.

The main part of the study involved semi-structured interviews with 177 older people (150 households), in their own home. Quota sampling was used, based on age and gender using estimated population figures from the UK, and according to type of accommodation. Properties were selected both by age and type of housing, using national estimates of housing stock. Issues explored by the interviews included respondents' perception of factors affecting risk of falling in the home, understanding of immediate and longer term consequences of having a fall and the value and acceptability of preventative measures. The interviews involved detailed discussion of different areas of the home, and the interviewee's fall history. In addition, standard anthropometric dimensions of interviewees were recorded, along with other measurements including grip strength, ability to get off a stool without using hands, spectacle wear and measures of visual acuity and depth perception.

Interviewees were briefed both verbally and in writing about the study prior to participation. They were informed that the discussions would consider falls in the home (including the garden), examples of falls, and risk factors and safety issues that might be involved. However, they were not given any further information prior to the discussion, to avoid leading responses in any particular direction. Each interview lasted approximately two hours, with all interviews conducted by the same researcher.

## Results

*Intrinsic influences*

Mean age of participants was 76 years (range 65-99), of whom the majority (73%) were female. Half the sample (47%) lived alone and 93% had at least one health problem related to falling, including problems with vision (35%). One or more medications were taken daily by 79% of the sample and 4 or more taken daily by one quarter (23%) of interviewees. Half the individuals (48%) had fallen at least once in the last 2 years, and

21% had experienced 2 or more falls in this period. Participants were of varying health status and inhabited a range of differing accommodation.

Although there were no significant relationships between falls and physical measures, qualitative evidence attributed intrinsic factors as the primary cause in just under one fifth (18%) of falls. Age-related factors thought to lead to increased risk of falling included the negative effects of decreased mobility, reduced balance and strength, and weakened vision. Individual capability affected by such behaviour was discussed as amplifying risk of falling. The inability to cope with chosen footwear and spectacles, the use (or non-use) of lighting and prescribed medication, and a lack of regular exercise, were examples discussed by interviewees.

*Extrinsic influences*
Nearly half (44%) of reported falls were attributed primarily to extrinsic factors. The design of buildings and gardens were reported to introduce risks. This was apparent when examining the areas of the home where falls were reported to occur, e.g. garden (40% of falls), stairs (23%) bathroom (8%), and kitchen (6%), due to the nature of the tasks performed in these places and subsequent behaviour (bending, reaching, etc.) and the environmental hazards present Slippery floors alone were reported to have caused nearly one fifth of falls (17%).

Choice of footwear was perceived to be a factor in fall safety, particularly the quality, thickness, grip and durability of the sole. Choice and use of footwear was reported as contributory in 10% of falls.

The design of some domestic products were reported to have directly contributed to falls (6% of cases), including oven and dishwasher doors that open downwards forming a trip hazard, or cleaning equipment that is heavy and difficult to hold.

One quarter of users of walking aids reported problems with their design and use that affected risk of falling; such devices were reported to be directly causal in 4% of incidents, and were often stated to be unsuitable for use in the home environment, due to the changes in floor surface and texture, and limited room for manoeuvre.

Combined with these extrinsic factors, behaviours involving direct use of the home environment were reported to affect fall risk. These included aspects of house maintenance, e.g. changing light bulbs, using stepladders, and 'clutter' (25%). Lack of storage space was a problem highlighted, resulting in objects being left on the floor (which was causal in 13% of falls). Often when storage was available, it was difficult to access, such as kitchen cupboards that are too high to reach without using steps etc.

*Peripheral influences*
Comments were made about the importance of support from family, friends and health professionals. However, this is dependent on socio-economic issues, and the proximity of family, friends and falls services. On the other hand, it was emphasized repeatedly by interviewees that older people do not want to be, or to be seen as, a 'burden on society' and wish to remain independent in their own homes for as long as possible.

It was clear that the majority of the cohort had little knowledge about the help and support that was available to them in their local area. This is something that urgently needs to be addressed, as without the advertising and subsequent awareness of fall related health and community services, many older people are missing out on useful opportunities.

## Discussion

From the findings of the research, a model of the interacting influences has been proposed, Figure 1.    Although an individual has little choice over their general physical state and subsequent abilities, they do have some facility to maintain their health at its current level, e.g. by exercising, and cutting down (with help from their GP) on polypharmacy effects. However, an individual may choose to move about their home in a way that increases fall risk, due to their specific capabilities, e.g. rushing, carrying etc. Additionally, a person may choose to design and keep their home in a certain way, or impose an option on themselves, that affects their personal limitations.    The model demonstrates how falls arise from an interaction between  an older person's physical capabilities (intrinsic influences), and the design, condition, suitability and use of their home environment and of aids and equipment (extrinsic influences).    The home environment and equipment are in turn influenced by the interaction with health professionals, family etc., and the older person's socio-economics (e.g. the ability to be able to afford to make changes to the home or equipment) and knowledge and understanding of fall risk.  These are the peripheral influences.  The latter also impact on intrinsic influences, e.g. in terms of medical support from health professionals.

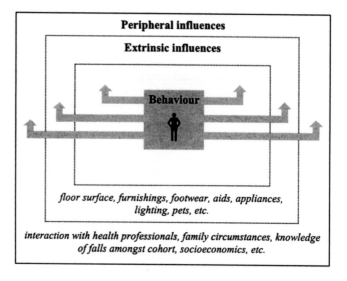

Figure 1. Influences affecting older people's risk of falling in the home

Any of these choices may be due to a lack of awareness of their personal limitations and failure to adjust their behaviour accordingly.    However, these behaviours are all dependent on physical and psychological health, fall history, socio-economic status, pressure, support from family and health professionals, and product and equipment interaction. These findings have been drawn together in the proposed model, detailing the influences in falls among older people in the home. The environment in which an older person resides and the equipment and products that are used should be designed appropriately for the individuals' capabilities in order to keep the demands of the environment and equipment as usable as possible. An individual's behaviour affects intrinsic, extrinsic and peripheral influences, each of which in turn have an impact on behaviour and each other.

## Conclusions

Behaviour has been highlighted as an overarching control in fall risk and it is evident that further information is needed to direct the precise course of action for health promotion. This could involve further analysis of the specific health beliefs that older people exhibit with respect to fall risk. However, in order to combat negative health behaviours, efforts should be made to reduce the demands and challenges of products and equipment for older people to use, and the homes in which older people live in.

Most importantly, it appears that there is a need to raise awareness of the falls epidemic, amongst all stakeholders and to provide practical fall prevention advice. This approach must encourage individuals to realise that falling is not an inevitable and uncontrollable part of ageing.

## Acknowledgements

Katherine Brooke-Wavell and Peter Howarth collaborated in the initial ideas for the study. The authors wish to acknowledge the support of the Department of Trade and Industry (DTI) who sponsored part of this research. The views expressed, however, are those of the authors and do not necessarily represent those of the DTI.

## References

Brace, C.L., Haslam, R.A., Brooke-Wavell, K., Howarth, P. 2003, *The Contribution of Behaviour to Falls Among Older People In and Around the Home.* (Department of Trade and Industry: London)

Brace, C.L., Haslam, R.A., Brooke-Wavell, K., Howarth, P. 2002, Reducing Falls in the Home Among Older People - Behavioural and Design Factors. In: McCabe, P.T. (ed.) *Contemporary Ergonomics 2002,* (Taylor and Francis, London), pp 471-476.

Hill, L.D., Haslam, R.A., Howarth, P.A., Brooke-Wavell, K., and Sloane, J.E., 2000, *Safety of Older People on Stairs: Behavioural Factors.* (Department of Trade and Industry: London). DTI ref: 00/788.

Kennedy, T.E. and Coppard, L.C. 1987, The prevention of falls in later life. *Dan Med Bull* 1987; **34**: 1-24.

Lord, S.R., Ward, J.A., Williams, P., and Anstey, K.J., 1993, Physiological factors associated with falls in older community-dwelling women. *Australian Journal of Public Health*; **17** (3): 240-5.

# DESIGNING FOR PEOPLE WITH LOW VISION: LEARNABILITY, USABILITY AND PLEASURABILITY

**Chandra M Harrison**

*Human Interface Technology Laboratory*
*University of Canterbury*
*Christchurch*
*New Zealand*

A new low vision reading aid is under development that uses advanced technology and aims to improve usability and pleasurability while maintaining the learnability of existing reading aids. Existing video magnifier users, other low vision people and normally sighted matched-age controls completed a reading test using existing machines and a mock-up control panel of the redesigned unit to determine if the aims had been reached. Improved usability was achieved in the redesigned product and once users mastered the more complex interface most found the redesigned product more pleasurable. However, learning time was greater than with the existing technology. Results also indicate that while people with low vision experience a similar level of technology-related anxiety to the control group, they are more likely to engage with technology that enhances their quality of life.

## Introduction

Low vision reading aids such as video magnifiers have remained essentially unchanged for 20 years despite improved computer technology. The traditional and predominant product is a Closed Circuit Television (CCTV) system that magnifies printed text placed under the camera and displays the magnified text using a computer monitor or television screen. Users navigate the text by means of an x-y table under the camera. These reading aids provide independence for people with low vision who would otherwise have to rely on other people to read them any printed material.

CCTVs have a simple and learnable interface; they also have several limitations. Discomfort can result from using the product for extended periods and viewing can be physically, visually and mentally exhausting. Physical requirements for dexterity and coordination to use the x-y table can be further complicated among the elderly, the predominant user group, who may also have other health issues such as arthritis. Due to the magnification required reducing resolution and viewing on a screen, the visual load can be high (Harpster *et al*, 1989). Users also need to remember the beginning of the previous line for long periods, which can cause a heavy mental workload.

Technological innovations currently available could eliminate these limitations. Optical Character Recognition (OCR) software has now progressed to a stage where it could enhance text to reduce the visual fatigue. Processed text could be presented automatically, in either a single line or as a single column of text, eliminating the need for the x-y table and assisting with the flow of text. Synthesised speech could also be utilised and would prolong the life of the machine by accommodating worsening eye conditions. However, increasing functionality should not be at the cost of usability, pleasurability or learnability. The goal then is to develop a new system that takes advantage of new technology, improves usability in terms of reading speed and comprehension, reduces physical demands by eliminating the x-y table, and maintains the learnability of the existing technology.

Recent product development literature has focused on enhancing the pleasure of products by assessing users' needs (Jordan, 2002). To have pleasure, however, products must not only appeal to the user but there must also be an absence of negative emotion. It is therefore important to not only determine what aspects make a product pleasurable for a specific user group, but also to determine what aspects of a product elicit negative emotion. Technology-related anxiety or frustration with consumer electronic products is common for many people, especially the elderly (Rosen & Weil, 1995). Identifying what causes the frustration or anxiety can help reduce it, making a more pleasurable product. In addition, identifying why some people avoid technology may assist with designs that encourage use.

The main user group for low vision products is elderly (75% of CCTV users are elderly, 15% are in employment and 10% in education). There is an increase in the elder population and the elderly are more likely to experience technology-related anxiety. Therefore, it is important to make the experience of using accessibility products such as low vision reading aids as pleasurable as possible. One example of how to increase pleasurability is to determine if participants could use the redesigned machine for longer than the CCTV. People with low vision do not usually read text for prolonged periods. Instead, they use talking books and use the reading aids intermittently for smaller tasks. However, surveys show that they would like to read books but cannot because of the limitations of the CCTVs or the severity of their eye conditions. In addition, identifying specific actions that induce frustration or anxiety and altering the design to reduce negative emotions will enhance the pleasure.

Learnability, usability and pleasurability are reliant on each other. Pleasurability is not possible if frustration results from using a product. If a product takes an excessive amount of time to learn users may not preserver to determine if the product is usable. Therefore, the amount of time taken to achieve competency, the level of task achievement, frustrating actions and the degree of positive emotion that accompanies the tasks needs to be assessed to determine if a product is worthwhile. While other aspects of the control panel were assessed (e.g. optimum control layout, biomechanical issues) during a study for the company developing this product, research discussed in this paper looks specifically at whether learnability was maintained and whether usability and pleasurability were increased in the redesigned low vision reading aid.

## Method

Participants included 16 expert users of the existing CCTV technology (experts), 14 people with low vision who were not current users or who had only recently begun using the machines (novices) and a matched aged control group of 15 people whose sight was sufficient to qualify to drive a car in New Zealand (control). Participants ranged in age from 9 to 90 years old with the average age being 71 years, with 16 males and 29 females. Three participants were unable to participate in the testing due to the severity of their eye conditions (1 expert and 2 novices), three others were unable to complete both tests (1 novice, 1 expert and 1 control). Their partial results were excluded from the reading speed analysis as it was possible to obtain only one reading rate. Their responses to the questionnaires are included.

Once recruited, participants completed a telephone questionnaire to gather demographic information, current usage of CCTV machines and other low vision aids, exposure to technology and current levels of negative emotion towards technology. The questionnaire was developed during a pilot study and combined aspects from Rosen and Weil (1995) along with questions to gauge technology exposure such as "how many hours do you spend using a computer per week." It also included questions to assess negative emotion such as "have you ever felt frustrated when using a new electronic product."

Each participant was visited twice in their own home where a reading test was administered, once using a CCTV machine, either their own if they were experts or with a *SmartView 8000*, and once using a mock-up of the control panel of the redesigned product now known as *myReader*, running off simulation software on a Compaq Evo laptop. *MyReader* incorporates processed text, automatic scrolling and uses a Liquid Crystal Display. The x-y table is eliminated and navigation is via trackball or scroll wheel. The order of product exposure was counterbalanced.

Learnability was assessed by timing how long participants took to achieve a predetermined level of competency. Novice and control participants were instructed in the function of each control of both machines, how to navigate text and asked to read two paragraphs of text. They were then tested on their knowledge of the controls and if 100% accuracy was not obtained they were asked to read a further piece of text and tested until they were competent. Expert CCTV users were asked to read two paragraphs of text and name the various controls on their CCTV machines and provided with the same instruction as others on the *myReader* machine.

Initial assessment of usability was on a performance only basis,[1] using comparisons of reading speed and comprehension with an existing CCTV product and the *myReader* mock-up using a similar paradigm to Harland *et al* (1998). Once competency was attained, participants were asked to read one of two chapters of a children's book, *Swallows and Amazons*, by Arthur Ransom (counterbalanced exposure). They were asked to read at their normal speed and told that they would be given several multi-choice comprehension questions at the end of each 10-minute period. Reading speed was determined as the average number of standard-length words per minute (wpm). Standard-length words were determined by the total number of characters including spaces and punctuation, divided by six (Carver, 1990). Comprehension was the percentage of correct answers on multi-choice questions.

Pleasurability was based on the subjective assessment of participants, problems encountered with actions and their willingness to continue reading (endurance). In a post-test interview, participants were asked to discuss the experience of using both machines and at the completion of both tests were asked for their preference and reasons for it.

---

[1] Further assessment of biomechanical elements will be conducted at a later stage, using behavioural analysis of video footage of the interactions between participants and the two machines. Data logging from *myReader* will also be used to assess biomechanical differences and confirm optimum button placement.

Repeated errors and negative comments regarding frustrating functions were analysed to determine aspects of the interface that could be altered to enhance pleasurability. Endurance was determined by the difference in time participants were willing to read. Given that average use of existing CCTVs does not exceed 10 minutes at one time, it was envisaged that people may have been able to read for longer with the redesigned machine and this measure of endurance would also gauge pleasurability. Their level of technology-related anxiety gathered from the questions was also used to analyse their preference

## Findings

### Learnability

The time taken to achieve competency was greater for the CCTV machines (approximately 5 minutes) and the *myReader* (approximately 10 minutes). A difference was to be expected given the increased functionality and complexity of the interface and that experts would have very low learning times for the CCTV. Due to problems experienced with the simulation software and mock-up, further analysis of the video footage is required to eliminate time not directly related to achieving competency. While this may reduce the difference between CCTV and *myReader* times, it is expected *myReader* will still have a greater learning time. The design has also been modified to eliminate the number of steps required to initiate reading which could also reduce the learning time.

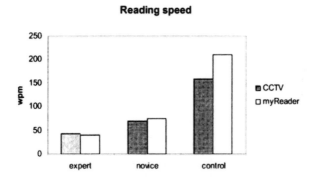

**Reading speed**

**Figure 1. Reading speed as words per minute**

### Usability

Three participants were excluded from the reading speed analysis as it was only possible to obtain only one reading rate. There was an overall increase in reading speed of 18% for the *myReader* unit (see figure 1). The speeds achieved are comparative to those found in previous research (Harland *et al,* 1998). Experts' speed decreased 3%, novices increased 8% and controls increased speed with *myReader* by 25%. Those unfamiliar with CCTVs show an improvement in reading speed after a relatively short exposure. It is likely that with further exposure experts may also show an increase in reading speed.

Comprehension for CCTV was 62% and 63% with *myReader* (see figure 2). While an increase in comprehension was not experienced, there was also no degradation with the automatic scrolling, which in itself supports the increased functionality. The overall level of comprehension achieved was not as good as reported in previous studies (Harland *et al*, 1998). As there were no significant differences between low vision and control participants, we could assume this may be due to the questions.

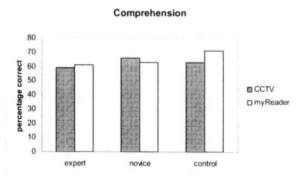

**Figure 2. Percentage of correct comprehension questions**

*Pleasurability*
In the post-test interviews 31 of the 42 (74%) participants tested preferred the redesigned product (figure 3). The group with the greatest preference for the simpler interface of the CCTV was the control group, who, it could be argued, have the least to gain from the more complex technology. Two of those stating they preferred the CCTV technology had reported discomfort during its use, highlighting the limitations of subjective assessment.

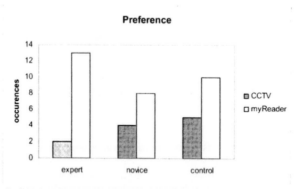

**Figure 3. Subjective preference for CCTV or myReader reading aid**

Analysis of the questionnaires revealed that people with low vision experience a similar level of technology-related anxiety as the control group. Many reported that they preferred to do without technology if they did not require it and had limited exposure. However, 15 participants had a cell phone, mostly for emergencies. Participants with low vision were more likely to engage with technology that would enhance their quality of life. Those with a support person or not interested in reading were less likely to engage.

Only two of the 42 people tested were willing to continue after 10 minutes with either product indicating endurance has not increased. However, many reported being less fatigued by the *myReader*. It is suggested that being observed may have reduced participants' willingness to continue and longer-term usage of the machine may result in increased endurance. Beta testing of a working *myReader* prototype is planned. This testing will involve obtaining a base line performance, leaving the equipment with participants for four weeks and then retesting to determine if gaining greater proficiency may also increase the length of time users are comfortable reading at one sitting.

User comments revealed that frustration was caused by a lack of understanding of terminology used in labelling of controls and simulation software notices. While the computer processes the text the message "loading" appeared. This was changed to "please wait" which was better received. A control labelled "next" was misunderstood causing some frustration. Increased memory load caused by an extra step to access processed text also caused frustration. The company has since eliminated this step. Other issues that effected the pleasurability of the control panel included a high profile control which was often bumped. The company has since lowered the profile of the dial.

This research forms the basis of a larger study to determine optimal design characteristics of the *myReader*. Because *myReader* was under initial development at the time of the study only a part of the functionality was assessed. The reading task was chosen due to the ease of simulation and the desire expressed by low vision people to be able to read faster and for longer.

## Conclusion

The redesigned low vision reading aid offers greater functionality than the existing CCTV machines. While it does take longer to master the control panel of the machine, users achieve faster reading, their comprehension is unaffected and they report greater pleasure with using the *myReader*. This sample did appear to avoid technology for a variety of reasons. However, technology such as low vision reading aids can assist these people in maintaining independence and those in need were willing to accept the new technology. It is important to further investigate why this demographic (the elderly) in particular avoids the technology to assess if accessibility products can be designed to avoid these issues. The findings will assist the company in the further development of the product.

## References

Carver, R.P. (1990). *Reading rate: A review of research and theory*. (Academic Press, San Diego).

Harland, S., Legge, G.E. and Luebeker, A. 1998, Psychophysics of reading: XVII. Low vision performance with four types of electronically magnified text. *Optometry and Vision Science*, **75** (3), 183-190.

Harpster, J., Frievalds, A., Shuman, G. and Leibowitz, H. 1989, Visual performance on CRT screens and hard-copy displays. *Human Factors*, **31**, 247-257.

Jordan, P. 2002, *Designing pleasurable products*. (Taylor & Francis, London).

Rosen, L.D. & Weil, M.M. (1995). Adult and teenage use of consumer, business, and entertainment technology: Potholes on the information superhighway? *Journal of Consumer Affairs*, **29**(1): 55-84.

# PROCESS OWNERSHIP AND THE LONG-TERM ASSURANCE OF OCCUPATIONAL SAFETY: CREATING THE FOUNDATIONS FOR A SAFETY CULTURE

## C.E. Siemieniuch[1], M.A. Sinclair[2]

*[1]Dept of Systems Engineering, [2]Dept of Human Sciences,
Loughborough University
LE11-3TU*

This paper addresses the longer-term (i.e. 10 years or more) assurance of safety, from an organizational perspective. The problem addressed is that of the general, imperceptible trend towards unsafe conduct of company operations, as first enunciated by Rasmussen (2000), and discussed by Amalberti (2001). An example is given to illustrate the difficulties of detecting this before disaster strikes. The paper goes on to discuss one way in which this problem could be addressed; essentially by good corporate governance, knowledge management and ownership of processes. Links are made to the literature on these topics, and a blueprint to help organizations to gain the benefits is outlined. The paper ends by outlining how an organization may be readied for knowledge management, without which the rest of the measures suggested are vitiated.

## Introduction

This paper is specifically NOT concerned with day-to-day management of safety, but with the assurance of safety over the longer term – a decade, or longer. The problem of interest has been identified by (Rasmussen 2000); his elegant diagram is shown in Fig 1 below

In this diagram, the central region indicates the area of safe operation for the company's systems. As the organisation drifts towards the boundaries, so business death becomes more likely. But management, in response to changing circumstances in their market, will always be seeking greater efficiency allied to reduced costs – the 'down-sizing' issue. Concomitantly, operators and managers together, although not operating in concert, will be exercising the Principle of Conservation of Energy, whereby short cuts, whether procedural or cognitive, will be discovered and will gradually become standard practice.

The net effect of these two aspects is to create a gradient from right to left, down which the organisation drifts, until it crosses the 'disaster' boundary. The issue is to f8ind countervailing pressures to flatten the gradient, and it is this which we discus in this paper.

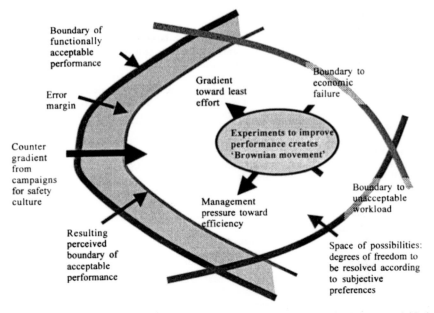

Fig 1: Rasmussen's diagram, redrawn, illustrating how organisations can drift into disaster.

In addition to this, Amalberti (2001) has discussed the paradox that essentially safe systems can be made less safe by the efforts of safety managers, over the longer term (a decade or more). Amalberti's argument hinges on two observable aspects of organisational behaviour; firstly, managers move on from one job to another in periods of less than a decade; and safety programmes to make safe systems even more safe tend to use as their starting point the elimination of such errors that occur, by procedural changes, allied to hardware changes. Safety managers make their reputations by demonstrating improvements in safety performance; a 5% improvement will take about 306 years to be statistically significant, whereas a 15% improvement will show in about 32 years. The plausibility of procedural changes, leaning largely on face validity, will serve as proof of a manager's efforts.

But what is to ensure that these safety 'improvements' will be maintained? It should be noted that in 40 years, all of a company's current employees will have left, and new generations will be in place. Theirs will be different interests, probably serving different goals. Clearly, we are discussing a knowledge management issue here, under the title of organisational learning.

Does this happen in practice? Well, yes; Piper Alpha, Bhopal, and Columbia are egregious examples of this, and there are many more.

## Corporate governance and disasters

If there is one consistency among disaster investigations ons, it is that either the necessary systems and procedures (for training, auditing, control, etc.) were not in place, or that they were in place but their operation was of insufficient quality. Both of these issues are issues of corporate governance, and we discuss this briefly.

We use the OECD definition:

> "Corporate governance is the system by which business corporations are directed and controlled. The corporate governance structure specifies the distribution of rights and responsibilities among different participants in the corporation, such as, the board, managers, shareholders and other stakeholders, and spells out the rules and procedures for making decisions on corporate affairs. By doing this, it also provides the structure through which the company objectives are set, and the means of attaining those objectives and monitoring performance" (OECD, 1999)

This definition includes both the internal and external affairs of the organisation; it implies the need for risk management; it embraces the whole organisation; and, although it is well-hidden within this definition, it does imply that the adequacy and quality of processes are significant aspects of good corporate governance .

These aspects are summarised in the following set of questions:

- Are the strategic goals appropriate for the company, given its history and competitive context?
- Has the company assessed adequately the risks associated with these?
- Has the company created appropriate processes, alliances, partnerships, etc. to deliver the goals, bearing in mind the risks?
- Have the processes been engineered as 'best in class'?
- Are the processes operated as 'best in class'?
- Does the company have processes for self-renewal (e.g. process auditing; capability acquisition; change management)
- Does the company audit itself regularly and transparently?

It follows fairly swiftly, from consideration of these points, that the organisation would have to pay co-ordinated attention to all of the following, as components of corporate governance:

- Structure (e.g. allocation of responsibility and authority; autonomy)
- Infrastructure (e.g. IT&T networks; security; access)
- Resources (e.g. time, money, people, knowledge & skills, equipment, and the distribution of these)
- Leadership (e.g. commitment to goals, support, clarity of communications)
- Culture (e.g. trust, willingness to learn, tolerance & retrieval of errors)
- Policies (e.g. resource management, change management, evaluation, suppliers, customers)
- People (e.g. selection, training, appraisal, knowledge, commitment)
- Processes (e.g. maturity, simplicity, metrication, controllability)
- Technology (e.g. maturity, deployment, utilisation, replacement)
- Knowledge (e.g. formal, tacit; organisational configuration; lifecycle)

Considering in particular the 'drift to disaster', it would appear that the 'safety case' approach being adopted in the developed world will have to consider all of these aspects. However, it is not apparent that attention to all of these would prevent a series of decisions by different people, all acting in the best interests of the stakeholders, from precipitating a disaster, as Amalberti (2001) has indicated. What is missing from this list is continuity, and wisdom. Continuity refers to the accumulation of knowledge and experience of a process, so that the decisions in the example above are not made in isolation, but are made in the context of prior decisions, and provide a path into the future. Wisdom is the ultimate goal of knowledge management, and is a blend of experience and knowledge. Given these, it is probable that most disasters could be obviated.

Hence, we now need to consider how continuity and wisdom can be provided within the organisation, and how these can be expressed effectively in control.  One way in which this could be accomplished is by the notion of Process Ownership, and we turn to this next.

## Process ownership

First, we define a 'Process', for the purposes of this paper:
- A Process has customers
- The Process is made up of partially-ordered sequences of activities
- The activities create value for the customer
- The activities are carried out by combinations of technology (machines, software) and people
- A Process can involve several organisational units, either within a company or across several companies
- A Process is instantiated by the allocation of goals, resources, responsibilities, and authority, and by the acceptance of appropriate metrics for measuring the performance of the process
- Resources typically comprise space, money, machinery, software, communications, people and knowledge

Process Ownership was introduced in the 1990s (e.g. Hammer 1996), with the notion of value chains. However, this early conception saw the role of the Process Owner more as an operational role than as a governance role. If, however, the role is re-defined to place emphasis on the latter, we may have better control over the drift to disaster.  The intention here is that the process is 'owned' by a given individual, a 'Process Owner', responsib le for the safety and integrity of the process, and who 'leases' the process to a Process Manager who is responsible for the day-to-day operation of the process and making a profit, etc..  A given process might be owned by a nominated Manufacturing Systems Engineer as a generic process, and instantiations of the process (including one located in Las Palmas, as well as others in China and France) are operated by local managers and operators.

The Process Owner's responsibilities may be defined as:
- documenting the process as 'best current practice'.
- maintaining the integrity of the capability within the process (tools, procedures, skills, and the health and safety of its stakeholders).
- authorising improvements to the process, to ensure it continues to be 'best current practice'.
- ensuring that process changes do not have bad effects on related processes (and *vice versa*).
- Supporting the change process for making process improvements.
- Authorising physical instantiations of the process in a given geographic location, and ensuring that any changes necessary for the process to fit the local context do not harm the integrity of the process
- Ensuring that the process metrics are properly used, and the results are made accessible.

The Process Owner thus becomes distinct from the Process Manager, who is responsible for process performance goals. The Process Owner now maintains corporate governance over the process, and is the repository of process knowledge and process history, both of which are fundamental to continued process safety as discussed above.

The advantages of this approach are:
- it focusses management attention on the prime assets of the organisation – its

knowledge, and the efficient deployment and utilisation of that knowledge.
* it engenders a focus on strategic considerations
* it provides a basis for a thorough understanding of process capabilities with their related safety issues within the enterprise
* it presents a coherent structure for good governance for safety, and for the maintenance of any 'safety cases', as demanded by regulatory authorities;
* it provides a built-in bias against the 'slow drift to danger' identified by Rasmussen and others (though it does not eliminate it).

The disadvantages are:
* Process owners move on, too.
* the role will become ineffective unless the Process Owner is supported and resourced from the highest levels of the organisation, especially with regard to sufficient time to execute the role properly, and to have sufficient authority to stop the process should the Process Owner have cause for alarm about the state of the process in a given instantiation.
* Process Owners will be unable to perform their roles effectively unless the organisation has reached a high level of process maturity
* it divorces direct responsibility for process integrity from process performance.
* there will be differences between the goals for Process Owners and Process Managers/Operators, with the potential for considerable conflict.
* Where a process consists of many sub-processes, each with a process owner, it is possible that conflict will occur between the sub-process owners, leading to delays in innovation and capability acquisition.
* the inevitable creation of an hierarchy of roles for process ownership may be seen to be an exercise in over-staffing, to be resisted fiercely in the interests of profit and efficiency
* the pool of people competent to undertake the role in a given organisation appears to be small (see, for example, the list of components of corporate governance outlined above); the pool of those capable of giving training and acting as experts in process ownership appears to be much, much smaller.

Nevertheless, this approach seems to offer a means of addressing the drift to disaster problem, which is not particularly evident in current organisational scenarios. It also appears that companies are gradually moving towards this approach (unfortunately, this is not the stuff of experiments), and it is hoped that case evidence may follow in the next decade.

Finally, we hope that this paper will help thinking in this domain.

### References

Amalberti, R. (2001). "The paradoxes of almost totally safe transportation systems." Safety Science 37, 2-3, 109-126.

Hammer, M. (1996). Beyond re-engineering, Harper Collins.

OECD (1998). Policy implications of ageing societies. Paris, Organisation for Economic Co-operation & Development: OECD Working Papers Vol VI, no 21.

Rasmussen, J. (1997). "Risk management in a dynamic society: a modelling problem." Safety Science 27: 183-213.

Rasmussen, J. (2000). "Human factors in a dynamic information society: where are we heading?" Ergonomics 43(7): 869-879.

Turner, B. and T. Kynaston-Reeves (1968). The concept of Temporal Disjunctive Information. Private Communication.

Turner, B. A. (1978). Man-made disasters, Wykeham Publications.

# DEVELOPMENT OF A CROWD STRESS INDEX (CSI) FOR USE IN RISK ASSESSMENT

Ken Parsons and Nor Diana Mohd Mahudin

*Department of Human Sciences,*
*Loughborough University,*
*Loughborough, Leicestershire, LE11 3TU, UK*

A Crowd Stress Index for use in risk assessment was developed in four parts: a literature review to identify relevant factors and understand the terms 'crowd' and 'crowding'; a study on an underground train; personal interviews; and a climatic chamber experiment in which physiological and subjective measures were taken from people in a crowd. Eight components were identified: physical, social, personal, crowd characteristics, crowd dynamics, crowd behaviour, location and psychological. Ninety-two questions in a risk assessment questionnaire provide the data for scores on 42 themes, leading to scores on the eight components in terms of contribution to crowd stress and finally a single index value ranging from 'no crowd stress' to 'extremely high crowd stress'. The index appears sufficiently well developed for validation in practical applications.

## Introduction

A person in a crowd can be exposed to stress due to the proximity of other people. This stress can cause both physiological and psychological strain that can range from discomfort and dissatisfaction to concern, panic and death. There may be positive reactions to being in a crowd. A feeling of purpose, belonging, excitement or simply warmth. This paper presents the development of a Crowd Stress Index. It identifies factors that can contribute to crowd stress and proposes a method that integrates them into a single number that relates to the magnitude of that stress, and hence likely strain, on people in the crowd. A detailed description is provided in Mahudin (2003).

## Method

An appropriate structure for the Crowd Stress Index and factors that can contribute to it were not obvious at the outset and hence information was collected in four parts:

1. *A literature review* to determine the nature of a stress index of this type, identify how terms such as 'crowd' and 'crowding' have been used and to review studies of peoples' reaction to stress caused by crowds.

2. *A study on an underground train* to determine the physiological and subjective response of one person in context and in a range of 'crowd' levels.
3. *Personal interviews* with a range of people to identify their perceptions of crowds and their responses to being in a crowd.
4. *A climatic chamber experiment* to identify physiological and subjective responses of people while being exposed to a 'crowd' under controlled conditions.

## Part 1 : Literature Review

Le Bon (1895) prophesied that we were entering the era of crowds. They represent a primary datum of social existence. A crowd is more than a mass of people, each crowd is unique and has a life of its own. Dickie (1995) stated that ebullient crowds with inadequate management have within them their own seeds of disaster. Berlonghi (1995) emphasised that those involved in crowd management must foresee the nature of the crowd that will be in attendance, be able to observe the behaviour of a crowd whilst an event is taking place and make timely decisions for effective action.

Mahudin (2003) identified four areas of research: animal research; experimental research on people; conceptual research; and human behaviour research. She considered definitions of 'crowd' and 'crowding' from different perspectives (demographic, phenomenological, social) and crowd stress as 'the feeling of having insufficient space due to the proximity of other people'.

A Crowd Stress Index will therefore predict the extent to which a person in a crowd will feel that they have insufficient space due to the proximity of other people. A high Crowd Stress Index (CSI) value will therefore predict high levels of strain with consequences such as panic, injury and death. A low CSI value will predict low levels of strain and no unacceptable consequences. A particular aspect of crowd stress is strain caused by heat. Griffitt and Veitch (1971) found that people were less friendly to each other on hot days and that even moderate temperature combined with crowding led to aggressive behaviour. Parsons (2003) suggested that examples of aggressive behaviour (e.g. security 'forces' providing a confrontational social context) may encourage aggressive behaviour. Braun and Parsons (1991) conducted a laboratory study into crowding and found that the inability to disperse metabolic heat, including restrictive evaporative loss due to sweating, can cause significant strain even in moderate temperatures and low density crowds.

## Part 2 : Crowd stress on an underground train

To obtain information about strain caused by crowding in a practical context, skin temperatures, heart rate and subjective responses were measured on a standing passenger while travelling on an underground train. In addition, information on crowd characteristics was gathered by an observer who was seated nearby. The male passenger was a 24 year old undergraduate of Asian origin. The experiment was conducted from 16.15 hrs to 18.30 hrs in summer with outside (above ground) temperature of 28°C and relative humidity of 47%. Conditions inside the carriage increased over the recording period from 26.6°C and 44%rh minimum to 30.1°C and 52%rh maximum. Mean skin temperature (Ramanathan, 1964) rose from 31°C to 33°C over the session and heart rate increased slightly from 61bpm to around 80 bpm. Heart rate measurement was however unreliable as the recording instrument appeared to receive interference from train operation. Results showed that in general the subject was under little strain. He generally reported his experience in the crowds as pleasant and not tense, not irritable, tolerant, not sticky and had little difficulty in moving around. Crowd density was however generally light. In more dense crowds higher skin temperatures were recorded and the subject perceived the situation as 'very crowded', squashed, warm and slightly uncomfortable.

*Part 3 : Personal interviews*

The results of the literature review and preliminary study on an underground train indicated that more detailed personal interviews were required to identify a fuller breadth of factors that would influence crowd stress. Individual interviews were therefore carried out with a selected range of ten male and ten female subjects (aged 20 to 45 years). They included students, university staff and the general public. One male subject was a person in a wheelchair. The interview was semi-structured and in six sections. A. Definitions of crowd and crowded; B. General experience in crowds; C. Discussion of scenarios (trapped in a lift, going in and out of a stadium and standing close in a cash dispenser queue); D. Describe feeling if in a crowd presented in photographs (swimming pool, football match, concert); E. Discussion of similarities and differences between pictures of crowds; F. Final comments. Interviews were private, recorded and lasted 30-40 minutes. The results provided seven components and a total of 48 factors within the components (see Figure 1).

*Part 4 : Climatic chamber experiment*

To complement the studies presented in Parts 1, 2 and 3, a laboratory experiment under controlled conditions was conducted to determine detailed responses of people while in a crowd. One male subject (23 years, 1.70m, 63kg) had subjective and physiological (skin temperature and heart rate) responses recorded while surrounded by nine male subjects in a climatic chamber set to 30°C and 50%rh and 0.15ms$^{-1}$ air velocity. All ten participants and a thermal manikin stood in a space 1.0m x 1.5m representing a tightly packed crowd. Subjects were of Asian origin and did not appear to be disturbed by the crowd. Over the one hour session mean skin temperature rose from 33.5°C to 34.7°C and heart rate from 71 bpm to 93 bpm. The subject initially felt very crowded, uncomfortable, slightly irritable and slightly intolerant. As he knew the others in the crowd well he did not feel that he had to avoid interaction. The 'crowd' around the subject generally felt very crowded, squashed and uncomfortable. All agreed that they needed more space. Behavioural measures indicated that there was a tendency for individuals to reduce crowd density by subtle movements away from those around them. Interviews with crowd members suggested that factors such as density level, space satisfaction, heat, air, lighting, adaptation, room volume, duration of exposure, focus of attention, relationships and social atmosphere all affect crowd stress levels. In particular, the expectation about the crowd, especially knowledge of when they could leave the crowd, influenced strain experienced.

## A Crowd Stress index

Figure 1 shows the structure of the Crowd Stress Index that was developed from Parts 1, 2, 3 and 4. Forty-two factors contributed to eight components that lead to a final Crowd Stress Index for use in risk assessment.

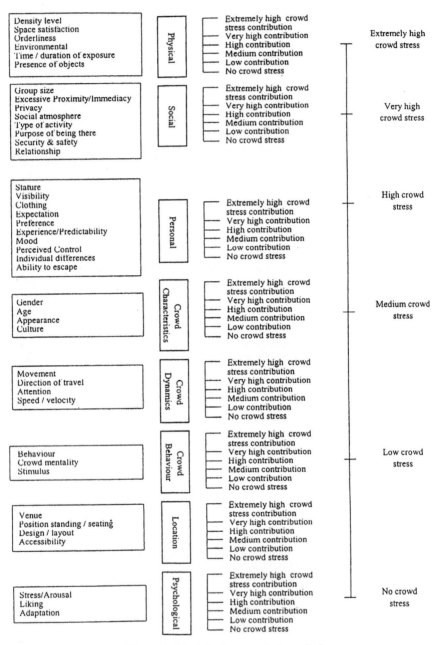

**Figure 1. Model of a crowd stress index**

## Conclusions

A four-part method investigating crowd stress from different perspectives has allowed the development of a Crowd Stress Index. The index is a simple tool that could be used for risk assessment as well as in the management of crowd control. It is sufficiently well developed to allow validation over a range of practical applications.

## References

Berlonghi, A.E. 1995, Understanding and planning for different spectator crowds, *Safety Science,* **18,** 239-247

Braun, T.L. and Parsons, K.C. 1991, Human thermal responses in crowds. *Contemporary Ergonomics. Proceedings of the Ergonomics Society's 1991 Annual Conference, 16-19 April* (Taylor and Francis, London) 190-195

Dickie, J.F. 1995, Major crowd catastrophe, *Safety Science,* **18,** 309-320

Griffitt, W. and Veitch, R. 1971, Hot and crowded: Influences of population density and temperature on interpersonal affective behavior. *Journal of Personality and Social Psychology,* **17**(1) 92-98

Le Bon, G. 1895, *La Psychologie des foules* Translated : The crowd. (Unwin, London, 1903)

Mahudin, N.D. Mohd. 2003, The development of a Crowd Stress Index, *MSc project report,* Loughborough University

Parsons, K.C. 2003, *Human Thermal Environments,* Second edition, (Taylor and Francis, London) ISBN0-415-23792-0

Ramanathan, N.L. 1964, A new weighting system for mean surface temperature of the human body, *Journal of Applied Physiology,* **19,** 531-533

## Conclusions

## References

# 2005

| | |
|---|---|
| **University of Hertfordshire** | 5th–7th April |
| **Conference Manager** | Sue Hull |
| **Chair of Meetings** | Martin Anderson |
| **Programme Secretary** | Paul McCabe |

| **Secretariat** | Phil Bust | Charlotte Brace | Gerry Newell |
|---|---|---|---|
| | Lorraine Rogers | Martin Robb | Ed Chandler |
| | Susanne Sondergaard | Yasamin Dadashi | Michael Quek |
| | Emmy Graham | Etienne Abrahams | |

## Social Entertainment

The title of the 'Quiz Night' was "I'm an Ergonomist, get me out of here". A considerable amount of 'networking' was carried out at the first formal 'Late Night P arty' sponsored by Marshall Ergonomics Ltd.

| Chapter | Title | Author |
|---|---|---|
| **Applications of Ergonomics** | Reducing the risk of musculoskeletal disorders in construction workers | M. Hanson and T. Barrington |
| **Cognitive Ergonomics** | Human factors issues in airport baggage screening | A.G. Gale |
| **Inclusive Design** | Models for inclusion evidence for choice and innovation | J. Mitchell, R. Chesters and J. Middleton |
| **Transport** | Violence in the workplace – Designing out the problems | S. Broadbent, Z. Mack and L. Swinson |

# REDUCING THE RISK OF MUSCULOSKELETAL DISORDERS IN CONSTRUCTION WORKERS

## Margaret Hanson & Taran Barrington

*Hu-Tech Ergonomics, 91 Hanover Street, Edinburgh,*
*EH2 1DJ, www.hu-tech.co.uk*

Construction is in the top three industrial sectors for prevalence of work related musculoskeletal disorders (WRMSDs) in the UK. The Health and Safety Executive (HSE) estimates suggest that up to 30% of the construction workforce are significantly affected by WRMSD injuries. HSE commissioned this research to identify practical solutions that could be applied to reduce the risk of musculoskeletal injury occurring within the trades of bricklayers and plasterers – two trades that are identified as having a particularly high prevalence of WRMSDs. A series of case studies illustrating practices that reduce the MSD risks are presented.

## Introduction

The HSE estimates that up to 30% of the construction workforce are significantly affected by WRMSD injuries. This has an obvious cost to the industry as well as to the individual. The UK construction industry has an aging workforce, with fewer young people entering the industry. Skilled trades people are in short supply, and the industry has some difficulty in recruiting, partly due to it being seen as a heavy industry to work in, with people at risk of experiencing discomfort.

Recent research has indicated that MSDs are particularly prevalent in the construction trades of bricklayers and plasterers (Reid et al. 2001). Tasks in the construction industry can involve handling of heavy components, repetitive movement, awkward postures, working under pressure and in poor weather (cold, wet); all these are known to be risk factors for MSDs. Problems are particularly common in the lower back, shoulders and knees for plasterers, and in the lower back, neck, shoulders, elbows, wrists and hands for bricklayers.

Because of the high prevalence of MSDs, HSE has identified construction as one of their target industries for reducing prevalence of musculoskeletal disorders (as part of their Priority Programme). They are also focussing on MSDs generally within this programme. With this attention on MSDs and construction, the HSE required practical guidance for the industry on what could be done to reduce the risk of these problems.

The study reported here, commissioned by HSE, sought to identify practical ways to reduce the risks of MSDs in these trades. All stages of the construction process were considered, from design and planning, to site management and the trades people's tasks. The aim was to develop a series of practical examples and case studies which could be disseminated to HSE Inspectors in the construction industry, and ultimately, to the public.

## Methods

Internet searches and literature reviews were undertaken to identify tools and equipment available that could reduce the risk of injury. Site visits, to a range of construction sites (small, medium, large), were undertaken to observe the tasks, identify good working practices, and discuss the feasibility of implementing risk reduction measures on other sites. Discussions were also held with architects and designers to understand how these risks could be avoided through appropriate design. Most risk reduction solutions identified were available and in use in the UK. A small number of solutions were identified as best practice in other countries; these were included where they would be practicable in the UK.

## Risk reduction solutions

The case studies and examples that were developed were categorised into the Pre-construction phase and the Construction phase. Separate case studies were developed for bricklayers and plasterers. Each case study contains a summary, a description of where the solution could be used (e.g. small, medium or large sites, new-build or refurbishment), a description of the problem and the solution, a testimonial from users where available, where the solution could be obtained, and an indication of the cost.

A sample of the solutions are presented below.

### Pre-construction phase: ensuring good knowledge and communication

Many of the musculoskeletal risks identified on construction sites arise due to a lack of consideration of the trades people who will actually build the building. If due thought is given to the construction process, and if different stakeholders are aware of the implications of their decisions, it is possible to reduce the risk in the design.

Duties are placed on the client, designer, principal contractors and subcontractors under the Construction (Design and Management) Regulations (1994) to reduce the risk of injury in the construction, maintenance and demolition of the building. Manual handling and musculoskeletal risks need to be considered, assessed and reduced through appropriate planning, work organisation and specification of materials. Awareness training for designers in their duties and of different risk reduction options can be helpful, as can ensuring that there is good communication between different stakeholders in the construction process. This may be achieved through regular (e.g. weekly) team meetings where the client, principal contractor and all subcontractors meet together and identify solutions to reduce any risks that have been identified in the construction process.

### Pre-construction phase: considering the materials specified

The materials that are specified will have an impact on the people who handle them. Weights can be reduced, and materials made easier to handle if this is considered when the materials are specified. Examples are given below.

Standard plasterboard is 1.2 m wide, and 2.4 m long. It typically weighs over 25 kg (depending on thickness). Its width makes it awkward to handle (arms have to be abducted if holding it vertically), and awkward to see past. Narrower plasterboard (0.9 m wide) is available. This is 25% lighter than the same length/thickness of 1.2 m wide board, and is therefore typically less than 20 kg. It is also easier to see past.

Foundations can be specified to be concrete poured rather than block laid. This eliminates the risk of musculoskeletal injury that arises from laying blocks in narrow trenches with limited access, where twisted and stooped postures are often required.

It should be ensured that materials are not over-specified in relation to the structural strength required for the building. Materials of a higher structural strength specification are usually heavier and may not be required.

Metal lintels can be specified instead of concrete lintels as these typically weigh about 3–4 times as much as metal lintels for the same length / structural strength.

Hollow or cut-away concrete doorsteps should be specified rather than solid ones, as this will reduce the weight significantly (approximately by 50%). Hollow or cut-away steps usually still provide the structural strength required.

Blocks with handhold cut-outs are available that facilitate the manual handling of blocks.

Special thin-joint mortar is available that requires a 2 mm layer of mortar between blocks rather than the standard 10 mm. This will reduce the amount of mortar that has to be handled on site, and the weight of the mortar that has to be placed on the blocks. However, it is acknowledged that more blocks may have to be handled during the rise of a wall.

Prefabricated units (e.g. of toilet facilities) or use of wall panels (as has occurred with some of the newer structures, e.g. the Swiss Re building in London) eliminates the need for bricks and blocks to be laid; materials can be mechanically handled into place and attached to the super-structure. This eliminates the risks associated with brick and block laying.

## *Pre-construction phase: planning the construction*

The design of the building and methods and equipment that will be used to construct it can have a significant impact on the risk of musculoskeletal injury.

Consideration should be given to the delivery of materials to ensure that maximum use is made of mechanical handling devices available on site. This may include laying roadways into a development prior to digging foundations, so that tele-handlers can be driven along these roadways to deliver materials mechanically. The position of loading bays should also be considered in relation to the movement of materials within a building, such that the distance to be travelled internally is reduced. The sequence of delivery of materials should also be considered, so that items can be mechanically handled into place before internal structures prevent this.

Buildings should be designed with no change in floor level for each storey so that trucks and trolleys can be used to deliver materials internally during the construction.

Mast climbers can be used to lay bricks from. They provide a working platform whose height can be adjusted precisely; brick can then be laid at waist height all the way up the rise of the wall.

Scaffolding is available with hop-ups (2 boards wide); these provide a working platform 500 mm below the main run of the scaffolding (5 boards wide), and can then be raised to the scaffolding level as the wall is built.

## *Construction phase: site management*

Much can be done in the construction phase of a project to reduce the risk of musculoskeletal injury. Many unforeseen issues often arise during construction, and the site manager has a key role to play not only in planning the construction site, but in making appropriate decisions during the course of the construction.

Materials should be delivered to an appropriate point on the site so that the requirement to manual handle items is reduced. This may include scheduling the delivery of materials so that maximum use can be made of mechanical handling devices.

Materials should be kept dry so that their weight is not increased. A wet block can weigh approximately 3 kg more than a dry block. Having an appropriate materials storage area (e.g. covered) can facilitate this.

Scaffolding can be used at ground level to raise the height of the bricks and mortar worked on, in order to reduce the stooping required when accessing them.

## Construction phase: tools and equipment

A number of tools and range of equipment are available to facilitate both bricklayers and plasterers in their tasks. Although many of these tools can be useful, they should be seen as the lowest form of control after eliminating and reducing the risk (through appropriate design and site management).

The mortarboard is usually placed on two blocks, meaning it is about 200 mm above ground level, and bricklayers have to repetitively stoop to it. A lightweight platform for the mortarboard is available which raises it to a more appropriate height (approximately 650 mm) and thus reduces the bending required to access the mortar.

Mechanical lifts are available that can be used for raising lintels into position above a doorway.

Pallet trucks with pneumatic tyres allow pallets of materials to be manually moved within buildings without the need to carry them. Note that buildings have to be designed with no internal steps for this option to be able to be used.

Sack trucks can be used rather than wheelbarrows for moving blocks. The centre of gravity of the load can be held directly over the wheels of the sack truck, whereas the forward position of the wheel on the wheelbarrow in relation to the load means more loading is taken on the body compared with a sack truck.

Brick tongs allow 6–8 bricks to be picked up at one time. They are very simple to use, and provide a handle for lifting the bricks. This prevents the extended finger position that can occur when picking them up manually. However, with the brick tongs the weight of the bricks will still be carried, which may still present risks.

A trellis platform system is available that provides a working platform within a building. This can raise the tradesperson to the right height for working without the need for scaffolding or mobile towers.

Plasterers often have to access small areas above head height; they will often stretch or stand in inappropriate surfaces (e.g. upturned tubs). Proprietary hop-ups are available that will raise them to an appropriate height and provide a secure footing.

A range of plasterboard handling trolleys and hoists are available which can eliminate the need to manually handle boards. Some of these integrate with the delivery of plasterboard on a pallet from a truck. Some convert from being a trolley into a workbench, presenting the plasterboard at about waist height, so it can be handled and trimmed to size as required at an appropriate height. A range of hoists are available that raise the plasterboard up to ceiling height. However, they can be slow to operate, and may not be practical in small developments.

Some simple plasterboard handles are available that facilitate the handling of single sheets of plasterboard. Plasterboard can be difficult to handle due to its thickness (12–15 mm) typically requiring a pinch grip to hold it, which can contribute to discomfort. These tools provide a handle that enables a power grip to be used. However, the board still has to be manually handled.

A plasterboard clip is available for assisting with fixing the plasterboard to the ceiling, a task that is particularly difficult and forces the plasterer to adopt awkward postures, with them typically supporting the load with their head and stretching to attach nails through the edges. The clip holds one end of the plasterboard to the joists, while the plasterer is able to move to the other end of the board and attach it.

A tool is available that holds a vertical sheet of plasterboard against a wall, allowing the plasterer to move freely to position the board to the wall, and thus eliminating the need for awkward postures which occur when holding the board in place with one hand and securing it with the other.

A mechanical tool is available for mixing wet plaster; mixing by hand (using a shovel) places a significant load on the body and can be tiring. The powered tool reduces this loading.

## Dissemination of solutions

The risk reduction measures have been presented as a series of case studies written specifically for (1) bricklayers and (2) plasterers. These will be placed on the HSE intranet site, available for HSE Inspectors in the Construction Industry to use. Ultimately, they will be placed on the HSE's external website to be available for public access. The findings have been presented to inspectors and industry groups and have been favourably received.

Please note that the case study examples presented represent good practice in the authors' view, and do not constitute an endorsement of a system or product by HSE.

## Acknowledgements

The authors would like to thank HSE for funding the study, and the many companies and organisations who contributed to the development of the case studies of solutions.

## References

Reid A., Pinder A. and Monnington S. 2001, *Musculoskeletal problems in bricklayers, carpenters and plasterers: Literature review and results of site visits.* ERG/01/01 Health and Safety Laboratory, Sheffield.

# HUMAN FACTORS ISSUES IN AIRPORT BAGGAGE SCREENING

## Alastair Gale

*Applied Vision Research Institute, University of Derby, Kingsway House, Kingsway, Derby DE22 3HL*

Security operators at airports screen passenger baggage by means of X-ray imaging each item and then visually searching the resultant computer display for the presence of objects which pose a potential threat. The performance of screeners on this difficult task is variable. We have theoretically modelled the underlying visual and cognitive reasons for error occurrence in airport baggage screening which demonstrates that threat items can be missed due to failures of visual search, detection or cognitive interpretation of the image information. The key human factors issues underlying variable screener performance are described and a new UK research project introduced which aims to provide key training to improve the identification of possible terrorist items.

## Introduction

In order to secure air passenger safety it is important to examine efficiently all passenger hand luggage and other items of external clothing for the presence of potential threat items; such as knives, guns or improvised explosive devices (IEDs). Almost universally, X-ray imaging of luggage is used which results in a two dimensional image that is examined on a computer monitor. Security operators have various specialised image enhancement algorithms that they can then apply to the image in order to highlight any particular aspects of concern. The image itself is usually viewed dynamically as the bag traverses the X-ray imaging system but it can also be examined statically. Where the operator (screener) is suspicious of some area within the image then the bag can be re-imaged and examined in a different orientation, to help overcome any problems due to overlapping visual image features resulting from the three dimensional object being viewed in only two planes. Any suspicious items can then be hand searched by another security operator if needed. A technique used in airports to maintain high levels of security performance is threat image projection (TIP) where a threat item can be superimposed within the X-ray image of a passenger luggage item and when the operator detects it then s/he receives appropriate feedback.

Recent developments in X-ray imaging has led to the development of 3D imaging of baggage and such systems are now reaching the market place. The advantage of such systems is that it should be easier to distinguish between a possible threat item and something that looks suspicious but is actually simply an overlaying of visual image features on the 2D image (i.e. the approach should produce a reduction in false positive detections). However, currently it is not really known whether the skills required for examining 2D X-ray images transfer readily to the examination of 3D X-ray images or whether the two tasks are identical. What in fact are the relevant skills for examining 2D baggage images?

The airport security screening situation attracts much research but of security necessity most of this is not readily available in the public domain. However, the volume of research in the area is low when compared to similar inspection situations; industrial inspection, radar operators or medical imaging. Additionally, the number of potential threat items in hand baggage is very low and so the task for the screener is similar to many other image inspection tasks where there is a low probability of target occurrence and where the potential target, or targets, appearance is unknown. Sustaining levels of vigilance in baggage screening is therefore important and this is primarily achieved by limiting the amount of time that any one individual spends actually examining images, together with other approaches, such as regular evaluation and training.

The potential for terrorist threat to air passengers is of great current international concern. Recent indicators from the UK, and particularly the USA, demonstrate that it is still all too easy, despite enhanced security measures (e.g. from January 2003 the American Transportation and Security Administration began screening every luggage item at all 429 commercial airports), for air passengers to carry potential threat items on to aircraft.

There is a need to improve airport passenger screening which has been well documented, for instance by the Reason Foundation (Poole, 2002). Additionally, the quality of luggage screening performance has often been questioned, with the US General Accounting Office recently reporting that the average detection rates had dropped from 95% in 1993 to 85% in 1999. In 2001 the Thompson amendment to the US Aviation Security Act was proposed which requires performance standards of these security personnel in an effort to improve potential threat detection levels.

## The luggage inspection task

We have argued previously (Gale et al., 2000) that the security inspection task is similar to the task of examining a medical image for possible abnormality presence. Human visual inspection is an imperfect process and is subject to errors even when the individuals are working to the best of their abilities. In both domains both false positive (identifying a possible threat item or possible medical abnormality) and false negative decisions (failing to spot a threat item or medical abnormality) are made. However, the key interest in both areas is in the nature of any false negative errors and it is important to try to identify the causes of such errors and develop inspection techniques, as well as appropriate training methods, which act to reduce their likelihood. In medical imaging a common experimental method investigating observer imaging performance is to record the visual search process of observers as they inspect a range of images. Such work has given rise to a theoretical model (Nodine and Kundel, 1987; Gale, 1997) which emphasises the role of visual search with such images. Errors are argued to arise due to failures in search, detection or cognitive interpretative processes. Although search is a necessary and important aspect in examining the images most research points to deficiencies in detection and interpretation as underlying the majority of errors (Kundel et al., 1978). The type of error typically made may relate to the type of abnormality (Savage et al., 1994).

In a previous study of ours (Gale et al., 2000) we examined airport security operators in a simulated situation as they inspected a test series of luggage images, some of which contained known IED targets. Three viewing conditions were employed; a 200 ms "flash" presentation (too brief for the observer to change eye fixation location on the image), one second and five seconds viewing. With increasing viewing time the detectability of IEDs increased, possibly not too surprising a finding. However, even in the flash viewing case all of the IEDs were correctly detected by at least one individual, so demonstrating that observers were able to process visual information some distance away from their fixation point on the image. A similar finding has been reported several times in medical imaging studies. In this task we empirically determined the size of the observers' useful field of view and found this to be 2.5° visual angle, a similar figure having previously been found by several researchers to be applicable in the medical imaging domain – pointing to some similarities between the two inspection domains.

Overall, the different lengths of examination times only affected the true positive IED detections and the false negative responses. IED-free images were able to be classed as normal by observers even in very brief viewing times, which gives some insight into the nature of their expertise in examining such images. The false negative errors were examined in the six second viewing condition, which demonstrated that visual search accounted for 14% of errors, detection for 19% and interpretation for 67%. Such a high figure for interpretation errors had not been expected. For instance in medical imaging, Kundel et al. (1978) report circa 45% interpretation errors in a pulmonary detection task. Such a difference may point to both similarities and differences between the baggage inspection situation and the medical imaging domains.

One key difference between the two domains is that in medical imaging, whilst the nature of the abnormal appearance may not be known in advance, the general format of the image type (e.g. MRI scan of the head, mammogram of the breast, ultrasound scan) is known. Consequently the observer has some implicit initial cognitive information before looking at the image, which will aid the subsequent search process. In luggage examination each overall item is very variable in size and shape, and whilst there will be similarities in terms of everyday contents there will also be many differences.

One of the aspects of our previous research has been to try to understand the nature of expertise in the security screening domain. Seamster et al. (1997) used cognitive task analysis to examine the cognitive performance of security screeners and teased out various differences between novice and experienced screeners. However, from our own work we have found that experience in security screening does not necessarily equate with expertise, which is intriguing from both the theoretical and practical standpoints.

In medical imaging, expertise in a somewhat similar screening task (e.g. breast screening) with low rates of abnormal appearances is deemed to be a result of considerable training over years, coupled with regular examination of large number of images which then builds up an appreciation of the wide range of normal appearances possible. In the UK the minimum number of images a year that a radiologist must interpret is 5,000 cases with some examining over 20,000. In the USA the comparable figure has been much lower, circa 200 cases. In security screening it is similarly possible for an individual to build up a high number of examinations of normal luggage items but our work in this domain to date has not been really been able to determine what skills specifically underpin expertise. McCarley et al. (2004) have recently investigated aspects of the development of such domain related expertise using non-security personnel.

In medical imaging, in areas such as breast screening, the radiologists can be helped by computer aided detection (CAD) systems which have had very considerable development in recent years. CAD systems pre-process the image before the radiologist sees it and can highlight potential areas of interest to the observer by means of overlaying prompts on the image. Such systems have had a tendency to produce a high number of false positive prompts of apparent abnormal appearance which can affect the observer's behaviour who may simply choose to ignore such prompts all together as opposed to spending time addressing each prompt location and determining whether it does in fact indicate abnormal appearance. However, due to continual developments in algorithms the number of false positives per case is decreasing and systems are now available in clinical usage (e.g. R2 Image Checker). In baggage screening there exist image enhancement algorithms, which are used by the various equipment manufacturers, but these tend to enhance the image as a whole, rather than prompt the observer for specific image attributes in a particular image location.

## Epaulets

The foregoing addresses the key similarities and differences between the baggage screening scenari o and medical imaging. Building on this research background the EPAULETS project aims to take recent findings from the latter domain and apply these to airport screening. The research utilizes a detailed image database of air passenger baggage items, some of which contain a range of potential threat items of varying conspicuity. These will be presented to screeners in a number of airports across the UK in a series of empirical investigations with the aim of further ascertaining the nature of false negative decisions. Resultant data will then lead to theoretical developments, which will be used to shape new training strategies in this domain. Subsequently, the training strategies themselves will be rigorously evaluated.

## Discussion

Security at airports continues to be a challenge both in practice and for the researcher in trying to suitably apply knowledge from other domains and discipline areas. It is unlikely that there is an easy solution to correctly identifying all potential threat items all of the time. However, the EPAULETS project is targeted at reducing false negative decisions and is planned to make a significant contribution to the area.

## Acknowledgements

This research is supported by the EPSRC "Technologies for Crime Prevention and Detection" programme. The research is performed in collaboration with Tobii Technology and QinetiQ.

## References

Gale A.G. (1997): Human response to visual stimuli. In W. Hendee & P. Wells (Eds.) *Perception of Visual Information – second edition*, (New York, Springer Verlag).

Gale A.G., Mugglestone M., Purdy K.J. and McClumpha A. (2000). Is airport baggage screening just another medical image? In E.A. Krupinski (Ed.) *Medical Imaging 2000: Image perception and performance*, (Bellingham, SPIE), 184–192

Kundel H.L., Nodine C.F., and Carmody D. 1978 Visual scanning, pattern recognition and decision making in pulmonary nodule detection. *Investigative Radiology,* **13**, 175–181

McCarley J.S., Kramer A.F., Wickens C.D., Vidomni E.D. and Boot W.R. 2004 Visual skills in airport-security screening. *Psychological Science*, **15**, 302–306

Nodine C.F. and Kiundel H.L. 1987 Perception and display in diagnostic imaging, *Radiographics,* **7**, 1241–1250

Poole R.W. 2002. Improving Airport Passenger Screening. Policy Study No. 298, The Reason Foundation, http://www.rppi.org/passengerscreening.html

Savage C.J., Gale A.G., Pawley E.F. and Wilson A.R.M. 1994, To err is human, to compute divine? In A.G. Gale, S.M. Astley, D.R. Dance and A.Y. Cairns (Eds.) *Digital mammography* (Elsevier, Amsterdam) 405–414

Seamster T.L., Redding R.E. and Kempf G.L. 1997, *Applied cognitive task analysis in aviation* (Avebury, Aldershot)

# MODELS FOR INCLUSION
# EVIDENCE FOR CHOICE AND INNOVATION

## John Mitchell[1], Robert Chesters[2] & Dr John Middleton[3]

[1]*Director, Ergonova,* [2]*Inclusive Design manager for Medilink West Midlands,*
[3]*Director of Public Health, Rowley, Regis and Tipton PCT*

Thirty disabled people in Sheffield were asked about their aspiration, barriers and their ideas for living more independently at home (Mitchell, 2004). The study exposed barriers against people with different types of impairment in their homes and neighborhoods, in their domestic utilities, systems and products, in their housing services and in the marketplace for homes, contents and services.

These barriers are addressed in three papers in this edition of Contemporary Ergonomics. The first, (Mitchell, 2005a) presents models for gathering evidence about the barriers that restrict disabled home users. The third (Mitchell, 2005b) presents models for using this evidence to enable consumers and providers to make informed choices and, in so doing, help to establish an informed marketplace for inclusion. This paper proposes a model for eradicating barriers during home refurbishment in the West Midlands the "Smart, Inclusive Homes" initiative under the leadership of the Rowley, Regis and Tipton Primary Care Trust and Medilink West Midlands Ltd.

The Model for Inclusive Refurbishment has the following stages

(a) Gathering systematic local user evidence
(b) Achieving an inclusive home carcass
(c) Achieving inclusive home contents
(d) Achieving inclusive smart technological backup
(e) Achieving inclusive home services

## Background and barriers

Despite recent advances in knowledge and technology and the potential benefits of inclusion, new homes and contents still contain historic barriers against users with impairments. During the 19th and 20th centuries, large numbers of "terraced", "council" and "high-rise" homes were built during times of rapid industrial and social change. Although they met their objectives with considerable success, time has exposed flaws in their designs which reflect the prevailing levels of knowledge and technology. If their builders had had access to modern technology, if they had understood barriers and how they were caused and if they had had to make good the costs of excluding people who were disabled by their homes they might well have built much more inclusively. Similarly, if the need for inclusion had been recognised when gas, water, electricity and telephones were introduced into homes, it is quite possible that they might have been easier to find, reach, adapt and use as and when required. There may now be an important opportunity for revisiting the designs of these homes and finding out if they could become fully inclusive during refurbishment.

The government's Decent Homes 2000 program is intended to encourage Local Authorities to renovate their existing housing stock and bring it up to modern constructional standards. The Home Market Renewal Pathfinder Program, on the other hand, is aimed at revitalising areas of underused and derelict housing in order to re-establish its social and economic viability. The Joseph Rowntree Foundation has developed its "Lifetime Homes" guidance for achieving a higher degree of inclusion in new build and refurbished homes. Its aim is to produce homes that have few barriers and that are easy to adapt when required. However, these guidelines are focused mainly on people with physical rather than sensory or cognitive impairments.

Disabled peoples' aspirations, barriers and ideas for greater inclusion at home have emerged from a study of thirty people with a range of cognitive, physical and sensory impairments in Sheffield (Mitchell, 2004). Participants said that barriers restricted their lives and activities in their neighbourhoods, their homes and gardens, their domestic utilities, systems and products, their housing services and in the markets in which they were seeking to choose good quality homes, contents and services.

The Rowley, Regis and Tipton Primary Care Trust is preparing to bid to develop "Smart, Inclusive Homes" in Sandwell to act as a model for inclusive home refurbishment. The team of disabled home users, builders, suppliers, and providers intend to acquire a stock of "terraced", "council" or "high-rise" homes, strip them of their barriers and fittings and refurbish them to the highest available standards of inclusion.

## The challenge

The challenge for the "Smart, Inclusive Homes" project will be to use evidence-based choice and innovation (Bennington et al. 2000) to eliminate barriers during home refurbishment.

## Model for inclusive home refurbishment

The purpose of this five part model is to develop methods for achieving inclusive home refurbishment to the benefit of consumers, providers and commerce.

### *Gathering systematic local user evidence*

Inclusion cannot be achieved without good quality, authentic evidence from disabled people about the barriers they face, about their ability to use one or another product or system and whether new approaches help them achieve inclusion. The role of user and provider evidence is outlined in Table 1 over the page.

**Table 1.    User and provider evident for inclusion at home.**

The team will:

- Establish partnerships and feedback networks with local users and providers
- Audit and expose the barriers in local homes
- Develop performance specifications that define inclusive practice and establish the penalties that these barriers impose
- Calibrate local findings with those from Sheffield
- Assess the scope for replacing existing barriers with "inclusive" alternatives
- Compare inclusivity of existing and prototype designs for homes and systems

## Achieving an inclusive home carcass

It is important to ensure that the carcasses of homes are barrier-free because the shells of buildings tend to be solid, durable and resistant to adaptations. Factors that can limit the flexibility and adaptability of homes are shown in Table 2 below.

Home adaptations are often carried out on a "one-off" basis as previously non-disabled people find they can no longer cope with the barriers in their homes. Home adaptations have tended to stigmatise their users and to be removed as quickly as possible. Once users have set out the barriers they face in finding, reaching and using their homes the team will consider how these are caused and how they might be overcome by adapting the home carcasses. The team will work with users, providers and the building industry to develop inclusive, systematic and mutually advantageous ways of enabling users to find and reach every part of their homes and gardens. Table 3 sets out possible approaches to overcoming these barriers.

## Achieving inclusive home contents

A barrier-free home carcass will eliminate many of the most intractable barriers to inclusive living at home but it cannot prevent barriers from being "imported" in the shape of non-inclusive mainstream fixtures, fittings and contents. Table 4 provides examples of some of the barriers that may be imported in this way.

Attempting to design out the barriers in these examples would be fruitless and impractical during any short term project. However, it is quite practical to "choose out"

### Table 2.    Home features that can limit inclusion.

It can be difficult to remove barriers to inclusion if homes have:

- Different surface levels inside and outside homes
- Steep, narrow stairways
- Restricted circulation spaces in rooms and passages
- Internal walls that are difficult to relocate and cannot bear the loads of subsequent adaptations

### Table 3.    Systematic approaches to inclusive home carcasses.

Systematic, modular approaches may be required for:

- Equalising surface levels inside and outside the home
- Tactile way finding
- Step free access
- The ready re-arrangement of doorways, room spaces and passageways
- The reinforcement of fragile room walls

### Table 4.    Barriers in non-inclusive home contents.

Barriers to finding, reaching, understanding, relating to and controlling any of these items will disable some potential users:

- Doors, windows and double-blazing fittings
- Electrical, gas, water and central heating outlets and controls
- Fixtures and fittings such as bathrooms and storage facilities
- Domestic products such as cookers, washing machines and refrigerators
- Leisure and communication systems such as telephones, TV and sound systems, security and alarm systems

barriers by using user feedback to find out which items contain the fewest barriers for users with different impairments. This feedback could also reveal areas in which innovation could prove most valuable in the longer term.

### Achieving inclusive smart technological backup

Evidence-based choice and innovation can be seen as ways of "Designing-out" and "Choosing-out" barriers in the carcasses and contents of homes. If these are used successfully, they will greatly reduce the numbers of barriers that restrict disabled peoples' home lives and activities. Inevitably however, some systems will continue to include barriers which reflect existing technical limits or the failure to exploit technical resources fully to achieve inclusion. "Smart technology" can be used to provide the monitoring, communicational, sequencing, co-ordination and control functions that disabled people require when they attempt to use non-inclusive products and systems. For example, disabled home users with a range of impairments (Mitchell, 2004) described the following barriers that prevented them from using their front doors as fully as they wished.

These barriers could be overcome by developing an "inclusive front door". They could also be overcome by assessing the barriers that particular users experience and by choosing and installing packages of smart technology that would make good the deficiencies in their existing front doors as suggested in Table 6 below.

Essentially, control of each of these items involves receiving information in a usable format, choosing a response or setting and activating the desired actions. Each of these functions is actually or potentially within the reach of smart technology and it is not difficult to support functions similar to those listed in Table 7 below.

**Table 5.  The "non-inclusive" front door.**

People with a range of impairments reported that their front doors are difficult to:

- Find (non-inclusive signage)
- Reach (non-inclusive access and turning space for wheelchair users, door controls located too high or too close to door framework)
- Understand (non-inclusive means of identifying and communicating with visitors)
- Control (handles and keys difficult to operate)

**Table 6.  Barriers to controlling home environments and systems.**

People with impaired sight, hearing, use of their hands, mobility, understanding and health said that they had difficulty in finding, reaching, understanding, relating to or controlling their:

- Central heating systems
- Security and monitoring systems
- Visually and/or audibly mediated communication systems
- Kitchen and laundry facilities

**Table 7.  Inclusive home control through smart technology.**

Individual control at home could be achieved through a smart, portable device that could:

- Receive emergency and status signals from all domestic sources
- Present these in whichever format individual users require
- Provide appropriate sequencing and guidance as required
- Transmit the desired commands or settings to the controlled element

## Achieving inclusive home services

All home users require a range of services, for example, from local and housing authorities, from specialist care and adaptation agencies and from those who provide the choice and acquisition of homes and their contents. User feedback about the quality and appropriateness of these services will be gathered and used in the same way as described under 3 achieving inclusive homes, above.

## References

Bennington et al., 2000, *Autonomy for Disabled People* (King's Fund)
Joseph Rowntree Foundation, *Lifetime Homes*, (York)
Mitchell et al., 2005a, *Evidence for Inclusion at home and elsewhere* (Contemporary Ergonomics), in press
Mitchell et al., 2005b, *Evidence for an Inclusive Home Marketplace* (Contemporary Ergonomics), in press
Mitchell et al., 2004, Wh*y Can't We Have a Home with nothing wrong in it?* (Inclusive Living Sheffield)

# VIOLENCE IN THE WORKPLACE – DESIGNING OUT THE PROBLEMS

## Suzy Broadbent, Zoë Mack & Lucy Swinson

*CIRAS, WS Atkins House, Birchwood Boulevard, Birchwood, Warrington WA3 7WA*

The Railway Group Safety Plan 2003/4 states that "88% of staff say they have experienced assault in the past six months" and that many of these incidents go un-reported. A considerable amount of reports received by CIRAS (The Confidential Incident Reporting and Analysis System for the UK Rail Industry) relate to assault or the abuse of staff by members of the public. These events vary in severity and also in the factors underlying them. When these events are analysed for root causes, it transpires that a significant number can be traced back to design issues, either physical design of the workplace or job design. This paper looks at how such issues can affect the likelihood of violence in the workplace and discusses methods of reducing this risk. A background of the CIRAS taxonomy and methodology will be presented, along with examples of relevant reports and suggestions for improvements in the future.

## Introduction

In 2002/03 it was estimated that workers in England and Wales experienced 849,000 incidents of violence in the workplace. The Health and Safety Executive (HSE) defines work-related violence (WRV) as, "Any incident in which a person is abused, threatened or assaulted in circumstances relating to their work". It is estimated that on average 1.2% of the working population experienced physical assault and 1.7% were threatened at work during the year 2002/03 (TSSA website). In the public transport sector, this figure rises to 2.8%.

The Railway Group Safety Plan 2003/04 states that "88% of staff say they have experienced assault in the past six months" and that many of these incidents go un-reported. The cost of such incidents is not only physical, but also psychological and financial. The threat of assault can lead to anxiety and stress that in turn can result in low staff morale and high staff turnover.

The TUC state that certain jobs have higher risks, inherent within the design of tasks they are expected to carry out. For example, jobs such as those that involve handling money, provide care, advice or information, work with violent people, deal with complaints, work alone or work unsocial hours.

Certain work places also have higher risks inherent within their design. For example, if you work on your own, away from other staff, in the community, in a workplace that is badly lit, not on your own employer's premises or in multi-occupied premises (TUC website).

## Methodology

CIRAS is the UK Rail Industry's Confidential Incident Reporting and Analysis System. It offers staff an alternative route through which to report their safety concerns.

The system was set up to capture safety-critical information that may go unreported through company channels for fear of recrimination.

Trained CIRAS analysts conduct short interviews with staff when they make reports, in order to establish the nature of the concern and the issues surrounding it. The issue is written up in a de-identified fashion and fed back to the relevant company for comment. The details of the concern and the transcript of the interview are entered onto the national database and analysed by the CIRAS Core Facility.

The CIRAS Core analysis method enables the assignment of up to three actual, near miss and/ or potential events to each safety issue that is reported by staff. The events in the taxonomy were developed in consultation with RSSB. Using the taxonomy, each safety issue is coded by trained CIRAS analysts based on the information in the report and on the company response(s) provided to that report.

An event is an undesirable occurrence related to railway operations (e.g. a derailment or a collision). These can be actual, potential or near miss events. In addition, operational failures can be assigned to indicate that an error or violation has been or could have been made by front line operational personnel. Operational failures are only assigned when stated explicitly by the reporter or commented upon in the company response. They may be actual failures (have happened) or potential failures (could have happened).

Up to three causal factors can also be coded for each report. Causal factors (precursors to the event), such as assault/ abusive behaviour, can also be assigned to describe the problem and help us understand why the event occurred. The causal factors analysed in this report are "Malicious Act – Assault" and "Welfare/OHS problems – Inadequate Security".

## Reports received by CIRAS

*Overview of data*

In the period June 2000 to June 2004, a total of 3972 reports were received by CIRAS from all sectors, including LUL. A search of the CIRAS database found 238 reports that related to violence. Figure 1, below, shows the root causes associated with the causal factors of "Malicious Act –Assault" and "Welfare/OHS problem – Inadequate Personal Safety (Threat of Assault)". Of these reports, 120 of them had "Design" as the root cause. These reports are analysed in the following section.

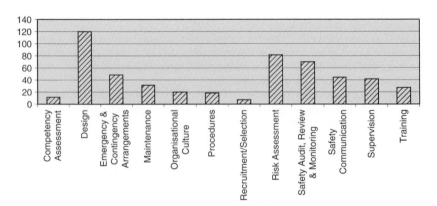

**Figure 1.    Root causes relating to assault.**

*Design issues*

Figure 2 shows the job categories of the staff that made reports regarding risk of assault and design issues. Figure 2 shows that Train Crew made the most reports to CIRAS (n = 40), followed closely by Station Staff (n = 36). This is to be expected as it is these job groups that deal with the public most frequently and therefore they are more likely to encounter violent situations.

*Safety consequences*

Figure 3 shows the safety consequences associated with violence problems. It is possible to have more than one event associated with each report, for example one reporter was concerned that certain trains were overcrowding and that this could lead to passengers being caught in the doors and also passengers becoming violent with staff. Figure 3 shows that reporters believe major (n = 57) and minor injuries (n = 92) are a possible safety consequence of violence. Considering that those reports relate to Malicious Acts and Welfare/OHS Problems it follows that the main concern of reporters would be an injury. However, as Figure 3 shows, these are not the only potential ramifications of assault. For example, there are 8 cases of "person struck/crushed by train". This could occur when an attack breaks out on a platform and someone is pushed into the path of an oncoming train.

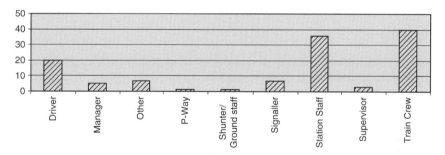

**Figure 2.   Job categories of reporters.**

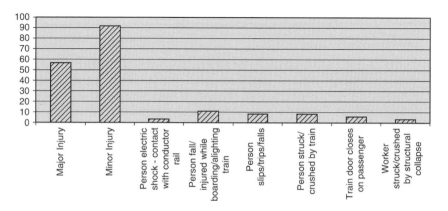

**Figure 3.   Potential safety consequences of assault.**

## Job design

Reports that relate to job design and assault often also relate to rostering. Numerous concerns have been raised to CIRAS regarding staff working alone at specific times of the day, such as early morning or late at night, as can be seen in the extract from a report synopsis below.

> "Concern has been raised regarding the security of staff whilst carrying out station inspections at unmanned stations. The reporter believes as these inspections are carried out after peak hours (often about 7pm) there is the risk of assault on staff by members of the public."

This report also highlights the further risks of working alone. Other issues raised relating to job design and violence include concerns over handling money.

> "...Well it's the threat of it, because we do come under, quite a few threats just doing our regular duties as revenue inspectors, but it's erh, it's just the threat of carrying money, whatever some, the attacker obviously doesn't know, but if they see you there for a few hours they will presume that you have money on you"

Both of these issues would hopefully be overcome by highlighting problem areas and altering the job design. For example, by altering the time that the inspections need to be carried out or by sending staff to carry out these inspections in groups, the risk of assault could be reduced.

## Physical design

The risk of assault due to the physical design of the workplace (be it a station, a train or a worksite) can be seen in the following examples.

> "...a guard reports that this service is often dangerously overcrowded and, as a result, passengers are frequently left at station platforms on route. Recently on this service a passenger became verbally abusive and threatening to staff in relation to the poor provision of service information and the overcrowded conditions. The guard making this report suggests that by putting a 4-car set on this service, train overcrowding and passenger pressure on Staff could be reduced."

> "It is stated that Drivers must use a public pathway that is often frequented by gangs of youths, and in an area where there have been serious attacks on members of public and also on a member of staff. It is said that Drivers are seriously concerned about using this pathway and would like action to be taken quickly to provide a different access route."

> "A CIRAS report has highlighted that for over three years this station has been without a passenger lift to all platforms...when disabled passengers are unable to gain access to the platforms, staff are often verbally abused, with suggestions of discrimination in relation to the poor provision disabled access facilities."

Overcrowding, lack of security and inadequate facilities are just some of the concerns raised by staff that can result in assaults against them.

## Reducing the risk

More than half of the reports made to CIRAS that relate to assault have design as a root cause, be it job design or physical design. Often when making such reports, reporters themselves suggest solutions for mitigating risks and often believe that it is the cost of implementing such changes means that nothing is done.

Alterations to job design, such as removing the need for train crew to carry cash, increasing the number of staff working during certain periods and ensuring staff are trained in conflict resolution, may reduce the risk of assault.

Similarly, simple changes to actual physical design and layout can make a huge difference in the probability of bring attacked. For example, ensuring that walkways are well lit, that CCTV is installed at places of high risk and that the physical environment on trains and in stations is at a comfortable level (HSE Website).

By involving Ergonomists and Human Factors Consultants when first considering the redesign of a station, train or worksite and indeed when designing jobs or shift patterns, it would be hoped that some of this risk could be mitigated.

## References

*Railway Group Safety Plan 2003/04.* Railway Safety.
www.hse.gov.uk/violence
www.tssa.org.uk/advice/hs/hs09.htm
www.tuc.org.uk/h_and_s/index.cfm?mins=30

# 2006

| Robinson College Cambridge | 4th–6th April |
| --- | --- |
| **Conference Manager** | Sue Hull |
| **Chair of Meetings** | Martin Anderson |
| **Programme Secretary** | Phil Bust |

| **Secretariat** | Ed Chandler | Nora Balfe | Gerry Newell |
| --- | --- | --- | --- |
| | Igor Zakhleniuk | Rosemary White | Abi Searle |
| | Susanne Sondergaard | Yasamin Dadashi | Lisa Doddington |
| | Nam Loc | Alice Madeley | |

## Social Entertainment

Cambridge seems to favour the Ergonomics Society. In 2002 the sun shone and this year the 'Quiz Night' was firing on all cylinders with it's 'Universally Challenged' themed quiz and the annual dinner although not to the taste of everyone was the best for some years.

| Chapter | Title | Author |
| --- | --- | --- |
| **Control Rooms Symposium** | CCTV in control rooms: Meeting the ergonomic challenges | J. Wood |
| **Defining Ergonomics** | Ergonomics advisors – A homogeneous group? | C. Williams and R.A. Haslam |
| **Design – Engage Project** | Safety semantics: A study on the effect of product expression on user safety behaviour | I.C.M.A. Karlsson and L. Wikström |
| **HCI Symposium – Access and Inclusivity** | A technique for the client-centred evaluation of electronic assistive technology | G. Baxter and A. Monk |

# CCTV IN CONTROL ROOMS: MEETING THE ERGONOMIC CHALLENGES

## John Wood

*CCD Design & Ergonomics Ltd,*
*95 Southwark Street, London, SE1 0HX*

The paper examines the ergonomic issues associated with the presentation of Closed Circuit Television (CCTV) images. Management expectations about users performance in using CCTV systems are compared to laboratory findings.

The development of a CCTV task taxonomy is discussed and the impact on operator performance of picture organisation is reviewed.

## Introduction

Recent events in the UK have shown how much we now rely on the capture and playback of CCTV images. Scarcely a day passes without our news services screening images of robberies, acts of terrorism, dangerous driving or a town centre fight recorded on CCTV. Expectations have been raised that CCTV will provide a key to solving many of our society's problems. The media, however, give scant attention to those who have to work with these systems, monitoring their outputs and forensically analysing recordings. In this paper some of the ergonomic issues raised by the use of CCTV systems in control rooms are discussed.

Whilst all the ergonomic features highlighted in Figure 1 have a role to play in overall performance this paper will concentrate on some of the lessons we are learning

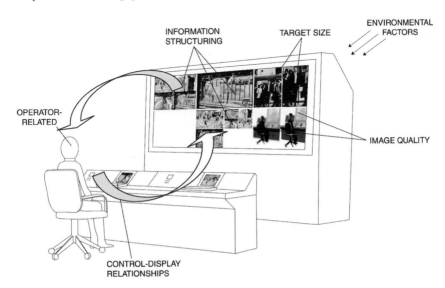

**Figure 1. Ergonomics & the Control of CCTV Systems.**

**Table 1.    Accuracy versus Monitor Numbers.**

| Monitor Numbers | 1 | 4 | 6 | 9 |
|---|---|---|---|---|
| Accuracy Scores % | 85 | 74 | 58 | 53 |

about performance at CCTV picture observation, task taxonomies and the presentation of CCTV images.


## Management aspirations & performance

Work by Tickner & Poulton, carried out in the 1970's, on the use of video images to monitor prisons highlighted operator limitations when looking at television scenes showing either a great deal of movement or little movement (Tickner et al, 1972 and Tickner & Poulton, 1973). Earlier work by the team had concluded, for short duration incidents on motorways, that the observation of more than one television picture would be likely to lead to failure to detect significant incidents (Tickner & Poulton, 1968) – note results reported in 1968!

The lessons from this early work seem to have been conveniently forgotten in the rush to install multiple camera systems whose output is typically presented on a wall of monitors. A bank of 30 monitors was recently counted during a visit to a town centre control room – to which must be added the 7 monitors on the workstation itself. The underlying potential for failure is compounded by management setting impossibly high standards where no faults are permitted – often by not actually specifying performance levels but then blaming operators when any failures occur!

Tickner & Poulton's work has been repeated by the Home Office, also under laboratory conditions, and examined the accuracy in target detection against the number of monitors being viewed. The target used was a man carrying crossing a busy road with an open umbrella at three possible locations in the picture – foreground, mid ground and in the distance. The open, large coloured golfing umbrella presented a clear and distinct target.

Subjects were seated 2 metres in front of a bank of 9 colour monitors. Subjects were asked to shout out when they observed a target, on any monitor and in any of the 3 locations in the picture.

Table 1 summarises the results from tests on 38 subjects who completed a 10 minute trial. Notably even with just a single monitor the best performance that can be achieved is 85% and this drops off rapidly to only 58% with 6 monitors (Wallace et al, 1997).

Management is deluded if it feels that operators will be able to reliably detect targets when viewing banks of 30 monitors for shifts of up to 12 hours!


## CCTV task taxonomy

By grouping the observed use of CCTV from various applications CCD is developing a framework for a task taxonomy upon which measures of image "complexity" can be overlaid (Wood, J. & Cole, H., 2006). In ergonomic terms four primary uses have been identified for CCTV systems, Figure 2.

In a further breakdown of CCTV tasks activities such as "detection", "verification" and "recognition", are being linked to these primary tasks. The aim of the programme is to create a structure which will allow for the logical classification of CCTV tasks.

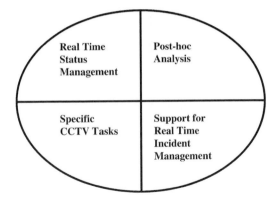

**Figure 2.    Primary Classification for CCTV Task Taxonomy.**

**Figure 3.    Alternative CCTV Image Presentation Formats.**

## Picture organisation

Evidence is emerging that the way in which images are organised has an effect on operator performance.

In a programme to examine the most efficient way of looking at groups of CCTV images CCD tested different ways of presenting motorway pictures. The operator's task was to identify whether the "hard shoulder" was clear of obstructions before diverting traffic onto it. A 98% performance level had been set for this task. The ergonomist was tasked to determine the most efficient strategy for viewing images whilst also achieving the required levels of performance. With over 200 camera images to check a regime which required an excessively long time to complete ran the danger of hindering rather than aiding this congestion easing procedure.

The solution had to be based around the use of a single, control room operated monitor. Three alternative presentation methods were considered, see Figure 3.

Under laboratory conditions subjects were asked to view each image in turn and to record when a vehicle had been detected. The software was designed to record "response times" as well as "positive", "false positive" and "false negative" detections of vehicles.

The single image solution did provide very high levels of accuracy but the viewing sequences were excessively long. With 9 simultaneously presented images on a single screen the distraction caused by the surrounding images reduced detection performance. The "quadded" solution was that which was finally selected.

The final example of picture organisation is taken from a non-control room application. Train drivers are increasingly being asked to check that no one is trapped in the

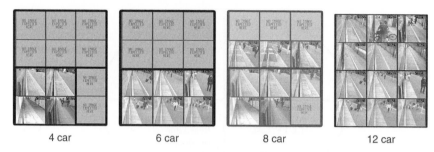

|  |  |  |  |
|---|---|---|---|
| 4 car | 6 car | 8 car | 12 car |

**Figure 4.    Alternative arrangements of "Patches" on two Screens for 4, 6, 8 and 12 Carriage Configurations.**

**Table 2.    Arrangement of CCTV Images and Detection.**

| Rolling Stock Configuration | Target detection reliability (%) |
|---|---|
| 4 car | 99 |
| 6 car | 98 |
| 8 car | 93 |
| 12 car | 98 |
| Overall | 97 |

closing doors – a job that used to be carried out by the guard or station platform staff. This is made possible by bringing CCTV images of the platform – train interface into the cab itself. The train cab is an exceptionally constrained environment into which adding any extra equipment is a challenge. It was within this context that CCD was asked to examine the potential for introducing additional pictures ("patches") onto the two available flat screens.

The experiment examined the effects on driver performance of different "patch" arrangements on a set of cab-mounted CCTV screens. With two screens up to a 12 carriage formation could be catered-for providing the existing rail standards limiting the number on each screen to 4 could be relaxed. Without being able to prove the safety of this alternative arrangement only an 8 carriage train could be run within the existing constraints.

The images for train configurations of up to 12 carriages were presented in differing arrangements on the two monitors. "Patches" which were not being used were blanked off so that the driver could immediately identify a failed camera, Figure 4.

Under laboratory conditions drivers ran through simulations involving different length trains and using the CCTV images to detect a range of targets of varying difficulty. The study found that there would be no significant deterioration in target detection, as the number of scanned views increased to 12 though the time required to scan the images increased.

There was, however, an anomaly in the results which illustrates how the organisation of the patches could have a direct impact on accuracy in a detection task. It was found that with the 8 car arrangement of images performance was worse than for the 12 carriage presentation, see table 2. In retrospect it is clear that 4, 6 and 12 car presentation offered a tighter grouping and allowed for an easier series of fixations. It is proposed to examine this in more detail in future.

Mouchel Parkman and the Railway Safety and Standards Board (RSSB) are thanked for allowing work conducted on their research programmes to be cited.

## References

Tickner, A.H., Poulton, E.C., Copeman A.K. and Simmonds D.C.V. 1972, *Monitoring 16 Television Screens Showing Little Movement*, Ergonomics, Vol. 15, No 3 279 – 291

Tickner, A.H. and Poulton, E.C. 1973, *Monitoring up to 16 Synthetic Television Pictures Showing a Great Deal of Movement*, Ergonomics, Vol. 16, No 4, 381 – 401

Tickner, A.H. and Poulton, E.C. 1968, *Remote Monitoring of Motorways Using Closed Circuit Television, Ergonomics*, Vol. 11, No 5, 455 – 466

Wallace, E., Diffley, C., Baines, E., and Aldridge, J., *Ergonomic Design Considerations for Public Area CCTV Safety & Security Applications*, Proceedings of 13th Triennial Congress of the IEA, Tampere, Finland, 1997

Wood, J., and Cole, H., *Image Complexity and CCTV operation*, Submitted to IEA Congress, 2006

# ERGONOMICS ADVISORS – A HOMOGENEOUS GROUP?

## Claire Williams & Roger Haslam

*Health & Safety Ergonomics Unit, Department of Human Sciences, Loughborough University, LE11 3TU*

A number of different professions apply "physical" ergonomics principles as part of their practice. In the drive to define and explain ergonomics as a field of practice, what professionals do in the name of ergonomics is important. This paper reports on a focus group study with professionals known to give advice on ergonomics issues as part of their work. Eight focus groups; two each of Ergonomists; Occupational Health Advisors; Health and Safety Advisors and Specialist Furniture/Equipment Suppliers were undertaken. Results show that there are differences between the professional groups in their aims, attitudes, perspectives and methods, when undertaking ergonomics activities. This paper focuses particularly on differences between the professions in their understanding of and their aims in using ergonomics. The implications of these differences are discussed.

## Introduction

It is widely accepted that ergonomics is not solely the realm of Ergonomists but that a broad spectrum of different professions will apply ergonomics as part of their practice, and might be considered "Ergonomics Advisors". In the UK, it will certainly include Ergonomists, Health and Safety Advisors, Occupational Health Advisors and Physicians, Physiotherapists, Occupational Therapists, and Specialist Furniture Suppliers.

Some will carry out activities, such as work station assessments, or manual handling risk assessments as part of their ergonomics undertakings. This begs the question "will they all be delivering the same 'product', containing the same message, of the same quality and with the same outcomes?" Anecdotally, at least, the answer to this question is "no". In the ever present requirement to define the ergonomics field of practice (Wilson 2000; Ahasan & Imbeau, 2003) what other professions do in the name of ergonomics is an important aspect. In order to begin to build an answer to this question, the more fundamental issues underpinning the differences between these practitioners is also pivotal.

Given this situation, the aims of this study were twofold. First, to gain an appreciation of the different professionals' understanding of ergonomics and what ergonomics activities they carry out. Second, to understand the emphases and aims of the different professionals, by analysing their responses to a number of broad questions.

## Methods

Eight focus groups were undertaken, involving a total of 54 participants. Sampling was stratified and purposeful, with each focus group containing participants from only one profession. The professions represented were: Occupational Health Advisors (OHAs) (n = 11); Health and Safety Advisors (HSAs) (n = 17); Specialist Furniture/Equipment Suppliers (SF/ESs)(n = 14); Ergonomists (n = 12).

The groups discussions were structured using a set of standard questions. The first focus group with each of the professions was recorded and fully transcribed. Template analysis (King, 2004) was the chosen method for analysing the groups' understanding of ergonomics. The template of themes was generated from the descriptions of ergonomics written by the IEA and the UK Ergonomics Society. The themes included the fact that ergonomics involves; understanding users, understanding jobs/tasks and the interactions between users and their jobs/tasks. It involves taking a user-centred, scientific, systems approach to design/assessment. It takes into account the cognitive, physical and organisational aspects with the aim of enhancing comfort, efficiency, productivity, safety and health, and can be applied in work and non-work environments. Any additional themes which emerged in response to the focus group questions were also noted.

The second focus group with each professional group was also recorded and analysed using the same template. This allowed for re-enforcement of themes which had already been covered by that profession, and for the addition of any new themes.

In addition, all 8 focus groups were asked about their aims when using ergonomics. Their responses were transcribed and the various aims given descriptive codes. These codes were then grouped under four headings. Each group was also asked to list the activities they undertook in which they used their ergonomics knowledge, in order to compare across the groups.

An external validator was present in each focus group to take notes of the key themes discussed and note emphasis. These notes were then used to check the trustworthiness of the findings represented here.

## Findings

### General

All of the focus groups responded to the question "What is ergonomics?" by giving a "text-book" definition of ergonomics, along the lines of *"fitting the task to the person"*. They all went on to make reference to most of the themes on the "ergonomics themes" template. Each of the professions cited a number of activities in common (Table 1) and each non-Ergonomist group had a strong sense of their limitations when it came to using ergonomics

> *"But again, when it's people with specific concerns, I have called Ergonomists in to come and have a look, because I've felt a little bit out of my depth...."* OHA

### Health and safety advisors (HSAs)

For the HSAs, the workplace/employer organisation was the focus of much of the discussion and the target for the majority of their aims in using ergonomics.

> *"So in designing the plant right and making sure the kit's in the right place and you can operate it....then you can control it by having your control room set up in the appropriate way with the right amount of information... You're on-line, and if you're on-line in general, that's the safest condition you can be."*

The language tended to include "risk assessment" and technical terms and the goal often included finding "solutions".

> *"And we're doing a lot of work... trying to come up with solutions for designing out the risks. Or removing those risks... from the plant... trying to come up with*

**Table 1.   Sample of responses to "What activities do you do which you consider involve ergonomics?"**

|  | Ergonomists | HSAs | OHAs | SF/ESs |
|---|:---:|:---:|:---:|:---:|
| DSE Compliance Assessments | • | • | • | • |
| Specialist Office Workstation assessment | • | • | • | • |
| Industrial Workstation Assessment | • | • | • | • |
| Site wide Ergonomics Assessment | • | • | • | • |
| Design | • | • | • | • |
| Product Sales | • | • | • | • |
| Tools/equipment advice | • | • | • | • |
| Management processes/Policy work | • |  | • | • |

> solutions on safe systems of work, basically, to prevent injury. "Cause this equipment is accessed on a regular basis, and... once you've accessed the equipment there's manual handling and ergonomic issues there. You're ducking under pipes and trying to turn valves that have all been put in at funny angles."

There was a strong emphasis on "systems thinking" in order fully to understand the nature of the jobs people were doing. There was less emphasis on understanding the person in the system, though the person was still important in their considerations.

> "So what you're looking at is not just managing the routine, but managing the non-routine and looking at various scenarios, what sort of information would come to that person in what sort of time frame. What sort of time-frame they've got to react to that."

## Occupational health advisors (OHAs)

The OHAs' spoke less about analysing the jobs people do, than the HSAs, though they did still cover this. However, their focus was very firmly placed on the person in the system. Much of the discussion and the majority of their aims were focussed around the individual.

> "I want to get them... so that they can go home at the end of the day, not feeling any worse than when they came in."

The understanding of the user was broad ranging, and included taking into account the physical, psychological and the social attributes;

> "I know I've been to places where... they're having some problems...from the way that .... they sit, but then when you actually start talking, there's a lot of office politics going on. They've been moved, they've done work and they actually end up in a smaller area to work in than they had before. So they're very unhappy."

## Specialist furniture/equipment suppliers (SF/ESs)

Like the OHAs, the emphasis of the discussions of the SF/ESs was very much on the individual in the system, though the focus was almost completely on the physical. The aims were based on identifying solutions which, understandably, were always products in their specialist ranges.

> "Primarily our work is reactive...to people who have musculoskeletal issues...part of what we do is go and assess the work environment as well as the

*individual ... and then provide solutions by way of product to try and improve the way in which they work, but also their level of comfort."*

### Ergonomists

The Ergonomists had the most comprehensive discussions. They covered all of the template themes in both focus groups, with the exception of "productivity" which was not mentioned in either group. Only the Ergonomists covered the "scientific" nature of the field in both groups; *"it's the application of science....of knowledge about people"*; and talked explicitly about the interaction between the individual and their tasks, though this was implicit in the other groups' discussions. They were alone in discussing ergonomics as a philosophical approach, rather than just as a set of methods; *"it's a way of thinking about things ... it's an approach as well as a process"*. Overall their understanding had somewhat greater breadth, and much greater depth than the other professions, and their focus was not just at an individual level but higher up;

> *"....looking at the whole systems, not just the individual at a work station...you know looking at sort of the larger side of things. Looking at... "habitability" if it's a vehicle... It's not looking specifically at one individual, it's sort of all encompassing."*

In terms of their aims when using ergonomics, the Ergonomists' focus was relatively evenly spread over aims for themselves; *"it's about doing something I enjoy doing"* and aims for the user; *"making things better for people"*. However, the majority of their responses covered aims at a broader, "societal" level; *"I want to promulgate the idea that getting the ergonomics right is important"*.

## Discussion

It is clear that the different professional groups represented in this study are neither homogeneous in their understanding, nor in their aims for using ergonomics. That said, it is noteworthy that all the professions demonstrated a relatively broad, if at times cursory, understanding of the key aspects of ergonomics.

The different emphases are likely to be the product of the role that each profession has "day-to-day". The SF/ESs have an overriding aim of selling products, and consequently their understanding and aims with respect to ergonomics are strongly biased around a physical, problem-solution model, in which their products are the solution for the individual's/employer's problems. However, for the most part, these professionals are brought into organisations when an HSA, OHA or Ergonomist has highlighted the need for an equipment based solution. Therefore, their physical model is likely to be perfectly adequate, and their superior product knowledge a great advantage.

The OHA's role in general, is one of facilitating the delivery of occupational health to individuals in an organisation, and will obviously be influenced by the fact that their background is in nursing, and is "patient-centred". This may explain why their ergonomics understanding focuses on diagnosing the problem in terms of the individual.

This contrasts with the HSAs, whose role often has more to do with ensuring the regulatory compliance and performance of an organisation or workplace as a whole. This, combined with the engineering/technical background of many of these professionals, may explain their emphasis on understanding the job and providing workplace engineering solutions, rather than on understanding the individual. There is the risk that this emphasis would prevent the true application of ergonomics with its user-centred imperative, however the HSAs participating in this study upheld this foundational tenet, whilst focussing on workplace changes.

The Ergonomists' greater depth and breadth of understanding, and more philosophical approach is likely to stem from the more intensive training they have received, and from the fact that ergonomics makes up all, not part of their role. However, rather than make the other professions seem inadequately equipped to carry out activities like DSE workstation assessments, the more extensive ergonomics knowledge of the Ergonomists seemed disproportionate to the requirements of these activities. Hignett (2000) discusses the fact that Ergonomists are skilled to deal with issues pertaining to working groups, organisations and general populations, rather than at the more individual level, and this was borne out in this study.

Some members of all of the professions stated that they might undertake broader ergonomics projects, such as site wide assessments. Given the understanding of ergonomics demonstrated during this study, these larger projects present the risk of under-representation of the field of ergonomics, thereby limiting the benefits accrued by its application, and showing ergonomics in a poor light. However, this concern should be mitigated by the fact that all groups had clear statements about when a problem required the input of an ergonomics specialist, and had mechanisms by which they sought this.

To be certain, the next step would be to look in the field at the assessments each of these groups made of similar situations, and see the outworking of their different understandings and approaches.

## References

Ahasan, R., & Imbeau, D., 2003, Who belongs to ergonomics? an examination of the human factors community. *Work Study,* **52(3),** 123–128.

Hignett, S., 2000, Occupational Therapy and Ergonomics. Two professions exploring their identities. *British Journal of Occupational Therapy.* **63(3),** 137–139.

King, N., 2004, Using templates in the thematic analysis of text, in C. Cassell and G. Symon (eds.) *Essential Guide to Qualitative Methods in Organizational Research,* (Sage Publications), 256–270.

Wilson, J. R., 2000, Fundamentals of ergonomics in theory and practice. *Applied Ergonomics,* **31(6),** 557–567.

# SAFETY SEMANTICS: A STUDY ON THE EFFECT OF PRODUCT EXPRESSION ON USER SAFETY BEHAVIOUR

## I.C. MariAnne Karlsson & Li Wikström

*Chalmers University of Technology, Department of Product and Production Development, Division of Design, SE-412 96 Gothenburg, Sweden*

This paper presents a study in which relation between the expression of a product (a bathing chair for small children) and users' behaviour was investigated from a safety point of view. The results show that the evaluated products were perceived as expressing "safety" to different extents and that this expression could be attributed different details in the product's design (material, shape etc). The results show furthermore that a strong expression of "safe" may result in an unsafe behaviour, contradictory to any warnings provided.

## Introduction

Each year a large number of accidents occur where people are injured. In Sweden, approximately 1/4 of these accidents happen at work or in traffic while 3/4 occur in people's homes and during leisure activities, games or sports (www.vardguiden.se). A similar distribution has been reported from, e.g., Finland (www.redcross.fi) while statistics from the Netherlands indicate that 50% of all accidents occur in or around the home (Weegels 1996). In many of the accidents is at least one consumer product involved.

Different explanations exist to the emergence of accidents. Accidents are considered to have one cause (e.g. the risk homeostasis theory), to have multiple causes (e.g. the domino theory), or to be the result of the interaction between agent – host – environment (e.g. Heinrick 1959, Haddon et al 1964, Wagenaar & Reason 1990 in Weegels 1996). Yet another approach is to consider accidents from a strictly technical viewpoint (e.g. technical failures) while behavioural approaches entail models of information processing, decision-making and cognitive control (e.g., Rasmussen 1983, 1986).

With the purpose of understanding and preventing accidents that involve consumer products, the different approaches briefly mentioned above have met critique. For instance Weegels (1996) argues that the approaches address other situations than the use of consumer products, that the diversity of the user population in such situations have not been considered, nor the amount of freedom as to how, where and when to use a consumer product. Furthermore, the interplay between user and product has not been taken into consideration other than from a more traditional "human error" point of view.

The appearance of a product may, however, be just as relevant in the occurrence of accidents in that it may provoke different behaviours and ways of use (cf. Weegels 1996). The product can look safe when it is not, or it can invoke a safe way of use (Singer 1993).

This paper summarizes an explorative study investigating the relation between product appearance and users' actions from a safety point of view, in particular the feasibility of product semantics analysis as a tool to further understand this relationship.

## Product semantics

The theoretical basis for the study is product semantics. Product semantics has been defined as "…. *the study of the symbolic qualities of man-made forms in the cognitive and social context of their use and application of knowledge gained to objects of industrial design.*" (Krippendorff & Butter 1984). Product semantics concerns, thus, the relationship between, on the one hand, the user and the product and, on the other, the importance that artefacts assume in an operational and social context.

Monö (1997) has, based on product semantics theories, chosen to describe the product as a trinity. The first dimension, the *ergonomic whole* includes everything that concerns the adjustment of the design to human physique and behaviour when using the product. The *technical whole* stands for the technical function of the product, its construction and production. The third aspect, the *communicative whole*, designates the product's ability to communicate with users and its adjustment to human perception and intellect (Monö 1997). Through the product gestalt, i.e. the totality of colour, material, surface structure, taste, sound, etc. appearing and functioning as a whole, the product communicates a message which is received and interpreted by the customer/user. This message is, according to Monö (1997), "created" by four semantic functions. The semantic functions are:

- *To identify*. The product gestalt identifies, e.g., its origin and product area. A bowl can be identified as part of a specific china set; a company can be identified by its trademark or by a specific design philosophy apparent in its products.
- *To describe*. The product gestalt can describe the product's purpose and its function. It can also describe the way the product should be used and handled. For instance, a doorknob can describe the way it should be gripped and turned.
- *To express*. The product gestalt expresses the product's properties, for instance "stability", "lightness" or "softness".
- *To exhort*. The product gestalt triggers a user to react in a specific way without contemplating or interpreting the product's message. For instance the user is triggered to be careful and to be precise in his/her operation of the product.

From a product safety point of view, at least the last three of the above functions may play an important role in the way users interact with and use a product. The study described in this paper has, however, focused in the function "to express".

## The empirical study

### Introduction

Small children are often involved in accidents. From a product semantic perspective, one reason may be that the children's parents have assumed that the products used in caring for the children, e.g. nursing tables, bathing chairs etc., are safe products. The products may express "safety" in a way that make the parents less aware of possible risks associated with the products' use, the products may communicate a false message of safety.

One product, a particular type of bathing chair, was noted to have been involved in a number of drowning accidents. Investigations showed that the parents had left the child in the chair, placed in the bathtub, for a short while to, e.g., open the front door. They had done so even though a warning label read: "Warning! To prevent drowning never leave the child unattended!" During the parents' absence, the chair had tilted and the child had fallen forwards, face down in the water.

**Table 1.   Number of subjects who indicated what were the
desired properties, undesired properties, and properties of
no consequence (n = 24).**

| Property | Desired | Undesired | No of consequence |
|---|---|---|---|
| Active | 16 | – | 8 |
| Functional | 23 | – | 1 |
| Durable | 22 | – | 2 |
| Clinical | – | 19 | 5 |
| Amusing | 15 | – | 9 |
| Soft | 24 | – | – |
| Unpleasant | – | 23 | 1 |
| Frightening | – | 24 | – |
| Safe | 23 | – | 1 |
| Restful | 12 | 1 | 10 |

## Method, procedures and results

The fundamental method used for the evaluation and comparison was Product Semantic Analysis (Karlsson & Wikstrom 1999, Wikstrom 2002).

The first step in the analysis is the identification of key concepts and the construction of a semantic scale and evaluation instrument. The assumption is that desired and undesired product expressions can be verbalized by consumers. The words used for describing the qualities of a design are, however, for each type of product. Therefore the words describing the particular product were generated in two focus group interviews, the first consisting of eight parents with children aged 0–1 year old, the second consisting of individuals who did not have any small children. According to the focus groups, bathing a small child was associated with "play" and "activity". At the same time one could never feel altogether "certain" or "safe", one had to be "careful". Almost identical lists of words were generated in the two groups, including (in translation from Swedish) "safe", "useful", "comfortable" and "soft". In order to acquire the individual user's appraisal of the product expression, two semantic instruments were constructed on basis of the results from the focus group interviews. One was used for the assessment of desirable and undesirable qualities of the product while the second was used for the assessment of the specific product's expression (see Table 1). The semantic scale used was a visual analogue scale ranging from 0–100. The one pole indicated the maximum value of the property, expressed in terms of a word describing this certain property; the midpoint designated a "neutral" value for the property, while the opposite pole indicated the opposite of the maximum value (the meaning left to the individuals themselves).

The second step concerned the evaluation of different product designs. Five different bathing chairs (products A-E) were evaluated (see Figure 1). Altogether 24 subjects participated in the study of which 16 were parents with children aged 0–1 year old and 8 had no children of their own or had grown up children. Eight were men and 16 were women, their average age was 33 years. The five chairs were evaluated, one at a time in a randomized order, in two different contexts; in a neutral environment (white background, on white cardboard) and in the intended use environment (placed in the subject's own bath tub, in 15 cm deep water). The subjects were allowed to see and touch the product but not try it out by placing a small child in it.

The subjects' indications of desired and undesired product expressions show that "safe" was a desired expression, as was, e.g., "soft" (see Table 1). No differences could be noticed between subjects with or without children.

 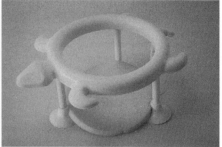

**Figure 1.    Products A and B. Product A (left) is the product that was documented to have been involved in a number of drowning accidents.**

**Table 2.    Ratings of product expressions (median values, n = 24). The scale ranged from 0 = not at all to 100 = maximum.**

| Property | Product | | | | |
|---|---|---|---|---|---|
|  | A | B | C | D | E |
| Active | **85** | 61 | 27 | 50 | 52 |
| Functional | 67 | 67 | 74 | 59 | **81** |
| Durable | **81** | **76** | **88** | 50 | **79** |
| Clinical | 18 | 40 | **86** | 56 | 44 |
| Amusing | **80** | 56 | 21 | 20 | 50 |
| Soft | 57 | 59 | 76 | 75 | **81** |
| Unpleasant | 50 | 50 | 54 | 50 | 22 |
| Frightening | 21 | 24 | 50 | 12 | 22 |
| Safe | *81* | 56 | 64 | 50 | 60 |
| Restful | 38 | 41 | **82** | 78 | **82** |

The subjects were asked to rate the expression of each product (see Table 2). The different designs were perceived differently in terms of what they expressed and how "strong" the expression was. The ratings did not appear to be influenced by context.

The subjects were also asked to describe what in the product gestalt that caused a specific expression. For instance, products A and B were considered to express the properties "functional" and "durable". The expression "functional" in product A was achieved by the fact that the child was considered to be able could sit on its own in the chair without any support from the parent. Product C was considered to have a "clinical" expression, i.e. a not desired expression. Associations were clinical experiments and medical examinations, the product was therefore considered "frightening". Product D was, on the other hand, considered to express desired properties, such as "soft" and 'soothing. These expressions originated mainly from the choice of material (terry cloth). The results show further that product A was perceived to have a strong expression of safety. Compared to the other four products the rating of "safe" was the highest and with the least deviations. The product was explained to be safe mainly because of its form; a large flat surface on which the child sits, a T-bar to keep the child in position, and large suction cups in each corner of the chair to keep it firmly attached to the bathtub. Only one out of the 24 subjects commented on the possible risk of the suction cups becoming unattached.

In an additional study participated another 12 parents with small children aged 0–1 years. Four were men and 8 women, their average age was 30 years. In this study, the subjects was first asked to rate the expression of product A. Then two different scenarios were presented and the subjects were asked to describe the way they would act given the described situation. The expression of "safe" was given a median value of 87. Ten out of the 12 subjects would consider doing "other things" in the bathroom while the child was having a bath placed in the bathing chair. Six would feel safe enough to leave the child alone for a few minutes. The reasons given corresponded to those in the previous study.

## Conclusions

The study shows that a product's appearance may influence a user's safety behaviour. Different product designs may result in different expressions, including more or less "safe". An expression of "safety" may, through the product's gestalt, carry stories of use in which no risks or reasons for caution are present. These stories may be what determine people's behaviour and use of products. Product semantics may be an important complementary tool when designing products and when analysing the causes for accidents.

## References

Heinrick H.W. 1959, Industrial accident prevention. A scientific approach. McGraw-Hill, New York.

Haddon W., Suchman E.A., & Klein D. 1964, Accident research. Methods and approaches. Harper & Row, New York.

Karlsson M. & Wikstrom L. 1999, Beyond aesthetics! Competitor advantage by a holistic approach to product design. In proceedings from the 6th international product development conference, EIASM, Cambridge, July 5–6.

Krippendorff K. & Butter R. 1984, Product Semantics. Exploring the symbolic qualities of form. The Journal of the Industrial Designers' Society of America. Spring.

Monö R. 1997, Design for product understanding. Liber, Stockholm.

Singer L.D. 1993, Product safety and form. In: Interface '93., 84–88.

Wagenaar W.A. & Reason J.T. 1990, Types and tokens in road accident causation. Ergonomics, 33 (10/11), 1365–1375.

Weegels M.F. 1996, Accidents involving consumer products. Delft Technical University, Delft.

Wikstrom L. 2002, The product's message. Methods for assessing the product's semantic functions from a user perspective. Department of Product and Production Development, Chalmers University of Technology. Gothenburg. (In Swedish)

# A TECHNIQUE FOR THE CLIENT-CENTRED EVALUATION OF ELECTRONIC ASSISTIVE TECHNOLOGY

## Gordon Baxter & Andrew Monk

*Centre for Usable Home Technology, Department of Psychology,
University of York, Heslington, York YO10 5DD*

Electronic Assistive Technology (EAT) provides assistance and assurance for an increasing number of elderly and disabled people who wish to live independently. The technique described here aims to optimise the use of EAT by ensuring that it impedes as few aspects of everyday life as possible. The Post Installation Technique (PIT) is designed to be used by people with little technical or human factors knowledge to provide a client-centred evaluation of a recently installed EAT application. It systematically probes for aspects of their daily life that have been negatively affected by the technology. These problems are prioritised and passed to the EAT service provider so that it can be better tailored to client's needs.

This paper describes the development of the PIT through application in two small field studies and an expert evaluation.

## Introduction

The potential benefit of assistive technologies to increase the quality of care of older people and reduce the associated costs is widely accepted within the UK (The Audit Commission, 2004). Electronic Assistive Technology (EAT) is increasingly used to enable older and disabled people – clients – to live independently in their own homes. Clients' are assessed on their ability to perform routine daily activities, such as eating, bathing, and moving around, to determine if EAT could help them. The provision of EAT is often technology-led, however, rather than needs-led (Sixsmith & Sixsmith, 2000). Although the prescribed EAT may make it easier or give clients more confidence to do some activities, after installation it may not properly support the task at hand – e.g., window opener switches located nowhere near the relevant window – and it can adversely affect other tasks in ways that are irritating, or problematic, and are not immediately obvious. When a client washes their kitchen floor, for example, they need to move the flood sensors beforehand – to avoid a false alarm – and remember to replace them afterwards. Such small details can affect the successful operation of the overall system.

Solving the problems described above requires a detailed knowledge of the minutiae of the client's everyday life. The obvious person to provide this knowledge is the client, although they cannot be expected to have a deep understanding of the technological constraints. In a post installation evaluation, however, this is not necessary. The client can provide details of any aspects of their daily life that have been negatively affected by the EAT which can then be passed to the EAT provider to make appropriate adjustments.

### Instantiating the risk management framework

The Post Installation Technique (PIT) is an instantiation of a framework for investigating the risks of introducing and using technology in the home (Monk et al, Accepted for publication). The framework gives social and psychological harms the same level of importance as physical harms (injury, cost and so on.). The risk is assessed by considering the likelihood of occurrence (*high, medium, low*) of a generic harm (*injury, untreated medical condition, physical deterioration, dependency, loneliness, fear or costs*) arising from everyday activities, and the generic consequences of that harm (*distress, loss of confidence, a need for medical treatment, death*). The importance of the harm (e.g., injury) is conditioned by its consequences (e.g., distress, medical treatment).

The PIT allows clients (possibly with the help of a carer) to systematically consider how EAT affects their everyday living. It embodies a client-centred approach, identifying problems from the client's point of view in a four step process.

Step 1 lists the EAT installed in the client's home. This list of equipment determines which checklist type questions are asked in Step 2, to elicit the benefits and problems of the EAT when clients perform everyday activities. The client simply ticks the appropriate box for each question.

Once all the relevant questions have been answered, Step 3 summarises the benefits, and details the problems for further analysis. For each problem the client: selects the potential generic harm that could result from the problem; assesses the chances of that harm occurring; and decides what the consequences of that harm might be. Clients are also asked where the problem arises and how important it is to fix it.

In Step 4 clients suggest how they would like to see each of the problems solved. The completed forms are then handed on, nominally to the EAT service provider, for action.

## Evaluating the post installation technique

The PIT is being iteratively developed and has been evaluated three times so far. For the first and third evaluations the PIT was used with clients in the field; for the second a semi-structured interview technique was used during an expert review of the PIT.

### Field study 1: West Lothian

The PIT was initially trialled in West Lothian with three participants, all elderly females. The purpose of the visit and the PIT questionnaire were first explained to the client, before asking the questions in the order that they appear in the questionnaire.

The first participant lived in her own flat, and used a walking stick to get around her home. She had had EAT installed for about a month, and mainly used the carephone to set the security alarm when she went out. She had had one false alarm which she had cancelled without any problem.

The second participant lived in her own sheltered accommodation flat. She could walk a little using a Zimmer frame, but generally used a wheelchair. The EAT had been installed for just over a year when the client moved into the flat.

The third participant lived in her own home, and had limited mobility. She had had EAT installed since September 2003, but had had similar equipment installed in her previous house. She had had a couple of false alarms (once when her granddaughter leant on the pendant button, and once when she burnt the toast!).

### Results and discussion

No problems with the EAT were uncovered, but the interviews revealed some problems with the PIT. The main problem concerned the original intention of working through the PIT walking round the home with the client. The mobility problems of the clients in West Lothian suggests that this may often not be a viable option.

The PIT systematically poses simple (yes/no) questions to identify problems. Without a full appreciation of the bigger picture, however, clients were concerned about the correctness and usefulness of their answers. They often answered in general terms, rather than focusing on specific activities and particular technology in a particular room.

Just having the EAT made all the clients feel more safe and secure. Even though no problems were identified for these clients, discussions with the support worker highlighted the existence of problems with EAT, such as extreme temperature sensors placed too close to the cooker, and the potential problem of clients falling if they rush to get to the carephone to clear any false alarms within the 15 second time limit.

The PIT was changed so that clients do not have to walk around the house. The structure and purposes were made clearer to the clients by revising the preamble for the PIT. The questions were also amended to focus first on client activities, and then on the room(s) where they are performed.

In addition, the last three questions in each section, which related to the aesthetics and general defectiveness of the system, were placed into a single general section at the start of the PIT. Getting these general issues out of the way first should help clients to focus their attention better on the specific issues raised in subsequent questions.

## Expert review: Belfast

Copies of the PIT were passed to a service manager from Belfast. This was followed up with a semi-structured interview focusing on the PIT and the assessment process that they used.

After installation a follow-up visit is conducted about two weeks later, mainly to check that clients understand how to use the EAT. The other purpose is to identify any initial problems such as sensors that need to be relocated, or to add extra devices to the system. It is rare for any devices to be removed at this stage: assessors try to persuade clients to at least attempt to work with the EAT. Further reassessments are carried out at six month intervals. As part of the process, a record for each installed system is kept, covering: the client's call history; which sensor initiated any call; what type of response was generated; and details of any system maintenance or upgrades.

The service manager suggested that clients would respond better if questions were asked face to face, rather than having them complete the forms themselves. The multi-part questions were regarded as a little too long, and could possibly be reduced to two or three parts. It was also pointed out that questions about washing and toileting will have been asked during the original assessment; it may be a bit too personal to ask clients about them again.

It was felt that the PIT could be useful during the six monthly reassessments. The listed categories of harm, chances of harm and possible consequences could also be used with their existing assessment process. More generally, it was suggested that some of the material from the PIT could be used in training staff.

The preamble to the PIT was revised to make it clear that only the relevant parts of questions should be answered. The answer boxes for the questions were also flagged as relating to either Benefits or Problems, to make it clearer which form they should be copied to for step 3 of the PIT.

## Field study 2: Durham

The first client had a basic EAT package installed in her home, and had recently been given a fall detector, and a bed sensor with a lamp attached. She had stopped wearing the fall detector – a common problem, especially among women – and had instead become more reliant on her pendant, wearing it all the time. She had had a couple of problems with the carephone. One was attributed to a fault on the line, although none were found subsequently. The other was its loudness when the phone dials through to the call centre.

Any other problems appeared to be isolated incidents, often light-heartedly dismissed by the client, apart from the bed sensor which she said could be removed as far as she was concerned. The second part of the interview therefore focused solely on this issue.

The first problem was that the lamp never came on when she got out of bed in the night; she had resorted to using the bedroom light instead. The second problem was that the client did not think that the sensor was working. She reported that the device had been programmed for her being in bed by midnight. One night, however, she said she was sat in her living room at 12:15 and no alarm was ever raised.

The second client had a trial lifestyle monitoring system installed (in August 2004). Such systems collect data from strategically placed sensors and upload them for analysis so that inferences can be made about the client's state of health. She identified four problems with the system. The first two are really installation problems (making sure that sensors are securely fitted so that they cannot be dislodged accidentally or otherwise by the client or their pet(s), and making sure that door sensors are fitted to doors that are likely to be regularly opened by clients). Such problems can easily be avoided by talking more to the clients prior to installation.

The third problem was that the electrical plugs for the equipment are large and heavy. The fourth problem was that the flood detector was not properly sensing when a flood occurred, because it was not positioned on a level floor. These two problems are more indicative of possible design flaws that would have to be addressed by the equipment manufacturer, although the client viewed them as irritants, rather than major problems.

The client had also had problems with the bed sensor failing to detect that she did not get up during a period of illness. This was attributed to the device's timing parameters having been incorrectly set.

## Results and discussion

This was the first time the PIT had been used in analysing the identified problems, and it uncovered some shortcomings in how the client does this. The main concern is how the PIT deals with harms and the likelihood of harm. In industrial risk assessment methods, such as HAZOP (Kletz, 1999), the harms and the associated likelihood of those harms occurring are assessed by a panel of qualified experts. The client-centred nature of the PIT means that judgements about possible harms and the likelihood of the occurrence of those harms are made by the client (and carer).

The first client was very explicit about the difficulty of determining the likelihood of the harm occurring, saying, "Your guess is as good as mine, dear." The second client mostly regarded the problems as largely unquantifiable irritants, although she had deeper concerns about the problem with the electrical plugs. A better method is therefore needed to enable the client (and carer) to appropriately assess the likelihood of harms occurring.

In thinking about the problems, the clients seemed to focus on a specific incident. One-off incidents may be perceived as temporary glitches, whereas persistent incidents are more likely to be considered *real* problems. Clients may still not be able to express the likelihood of the identified harm occurring, however. The PIT was therefore revised to ask clients to focus on their personal experiences to identify any real harms that have been "caused" by the equipment. Clients are now also asked to consider whether anything worse could happen given the same problem, e.g., "What do you think is the worst thing that could happen to you if this problem happened again?"

The problem of identifying the likelihood of harm was rephrased to reflect the client's personal experiences. So clients are now asked how often the problem has arisen or does arise (Does the problem arise daily/weekly/monthly? and so on.). This can subsequently be translated into a qualitative equivalent (high, medium and low).

**Summary and future work**

Our experiences of developing the PIT to date suggest that it is a useful and worthwhile exercise. The PIT was designed to be used either as a standalone instrument, or as part of the client reassessment process. Most EAT service providers routinely reassess clients approximately six months after installation. One service provider has already expressed interest in using the PIT as part of this reassessment.

Evaluating the PIT has been a lengthy process. The main reason for this is the need for access to suitable clients with appropriate EAT. This often requires delicate negotiations with care providers or social services. Although clients should ideally be randomly selected, opinionated and loquacious clients tend to provide more extensive feedback on the EAT and the PIT, which helps to improve both.

Whilst the PIT has proved to be useful for identifying and analysing benefits and problems of EAT, further evaluation is required to test out the latest revisions which should improve the analysis of problems in particular. Once the PIT reaches a steady state, the intention is to release it for use by clients. The latest revision of the PIT can be downloaded from http://www-users.york.ac.uk/~am1/ftpable.html.

**References**

Kletz, T. A., 1999, *HAZOP and HAZAN*, 4th edition, (Institute of Chemical Engineers, Rugby, UK)

Monk, A., Hone, K., Lines, L., Dowdall, A., Baxter, G., Blythe, M. and Wright, P. Accepted for publication, Towards a practical framework for managing the risks of selecting technology to support independent living, *Applied Ergonomics*

Sixsmith, A. and Sixsmith, J. 2000, Smart care technologies: meeting whose needs?, *Journal of Telemedicine and Telecare*, **6**, (Supplement 1), 190–192

The Audit Commission 2004, *Assistive technology: independence and well-being 4.* (The Audit Commission: London)

# 2007

| Nottingham University | 17th–19th April |
|---|---|
| **Conference Manager** | Sue Hull |
| **Chair of Meetings** | Martin Anderson |
| **Programme Secretary** | Phil Bust |

| **Secretariat** | Ed Chandler | Nora Balfe | Gerry Newell |
|---|---|---|---|
| | Damien Livingston | Rosemary White | Tom Griffin |
| | Stephven Lemalu Kolose | Yasamin Dadashi | Polly Shelton |
| | Gemma Huddy | | |

## Social Entertainment

"The Degeneration Game" was the theme of the 'Quiz Night' which included the 'Egg Drop' challenge. The drinks reception, sponsored by the Osmond Group Limited, was rescued at the last minute by the secretariat when the drinks were delivered to the wrong building. The annual dinner was held at the home of the Duke and Duchess of Rutland, Belvoir Castle.

| Chapter | Title | Author |
|---|---|---|
| **Ergonomics and Security** | How visual skills and recognition ability develop with practice in airport luggage inspection | X. Liu and A.G. Gale |
| **Ergonomics in Education** | One brief: Four concepts adjustable furniture for schools | P. Magee and A. Woodcock |
| **Patient Safety and Medical Ergonomics** | Why do student nurses continue to use the draglift? | L. Allen, D. Stubbs and S. Hignett |
| **Sitting at Work** | Seating problems – The missing link? | E.N. Corlett |

# HOW VISUAL SKILLS AND RECOGNITION ABILITY DEVELOP WITH PRACTICE IN AIRPORT LUGGAGE INSPECTION

## Xi Liu & Alastair Gale

*Applied Vision Research Centre, Loughborough University, Loughborough, LE11 3UZ, UK.*

Twelve naïve observers took part in three practice sessions and two test sessions of a task requiring them to search for terrorist threat items in airport passengers X-ray luggage images. Their eye movements were recorded remotely and they rated their confidence in whether or not a potential threat item was present. Results showed that sensitivity was improved and reaction time was decreased with practice. Observers could fixate on a target area earlier and the speed of information processing increased as a result of practice. However, this kind of stimulus-specific learning did not transfer to unfamiliar targets so that performance degraded and the decision time increased in the test sessions. Perceptual learning and visual recognition skills are discussed with regard to airport security screener training.

## Introduction

What is learnt from practice and under what conditions does transfer of learning occur? Learning in many perceptual tasks has shown that performance improved with practice but was stimulus-specific (e.g. Karni and Sagi, 1991). Ahissar and Hochstein (1996) summarized that learning is specific for orientation, size and position. Sowden *et al.* (2000) found that naïve observers' sensitivity to low-luminance contrast dots did not transfer in the same task with the opposite direction of contrast stimuli. However, learning transfers in some specific conditions. They found learning on searching micro-calcification clusters in positive contrast (i.e. targets lighter than background) mammograms can positively transfer to the same task with negative contrast (i.e. targets darker than background). Sowden and his colleagues indicated that sensory learning is specific to the direction of contrast and fails to transfer. Only these independent directions of contrast such as shape, size and location of abnormalities transferred to the same task with opposite contrast. In the task of searching for knives in airport passengers' X-ray luggage images, learning partly transfers to unfamiliar knife targets (McCarley *et al.*, 2004). They suggested that practice had little effect on the effectiveness of visual scanning but did enhance the observers' ability to recognize targets.

The aims of the present study are to investigate how search and recognition skills developed with practice and whether these kinds of skills can transfer to novel stimuli in the task of searching for terrorist threat items in air passengers' X-ray luggage images. Knives, guns and improvised explosive devices (IED) served as target objects to counteract stimulus-invariant benefits. In practice sessions, participants were required to search for threat items in X-ray luggage images using two sets of targets. In test sessions, participants performed the same search task for threat items where each participant was more familiar with half of the threat target items than the other half. The evolution of visual search with the course of practice was analysed to understand how perceptual expertise in inspecting security X-ray luggage images developed. The results contribute to understanding the training screeners require to attain expertise.

## Method

### Participants

Twelve people (6 male, 6 female) took part in this study. All of them had normal or corrected-to-normal vision. None of the participants had any experience of searching for threat items in X-ray luggage images.

### Stimuli

Target stimuli were sixteen guns (eight guns from two different viewpoints), eight knives and eight improvised explosive devices (IED) which were selected from a large set of threat items. In a pilot study, these targets were classified into four sets of four guns (two guns from two viewpoints), two knives and two IEDs each, on the basis of visual similarity which was defined as meaning 'similar objects share visual characteristics such as colour, texture, size, orientation and shape'. Visual similarity was compared between objects of the same kind (e.g. a knife was only compared with a knife, not with a gun or IED).

The same kind of objects in each set had high visual similarity. For the same kind of objects, high visual similarity was scored between Set 1 and Set 3, and between Set 2 and Set 4, while low similarity was scored between Set 1 and Set 2, and between Set 3 and Set 4. Set 1 and Set 2 targets images were electronically inserted into 60 normal bag images and served as target objects for training. Likewise Set 3 and Set 4 targets were inserted into 16 normal bag images and served as target objects for testing. There was only one threat item in each luggage image.

### Design and procedure

Participants attended experimental sessions on two consecutive days. On Day 1, participants completed three practice sessions (session 1, 2 and 3) and one test session (session 4). On Day 2, the test session was repeated again (session 5). There were three minutes break between each training session and ten minutes break before testing on Day 1.

There were 40 X-ray luggage images of 16 target-present images and 24 target-absent images in each session. In practice sessions, half of the participants were assigned to the group searching for targets from Set 1, and half were assigned to the group searching for targets from Set 2. All of the participants were tested in the same session in which half of the targets were from Set 3 and the other half of targets were from Set 4. Therefore, half of the target objects were familiar and half of the targets were novel for each participant in the test sessions.

X-ray luggage images were displayed on a 21-inch (53 cm) 2560 × 2048 monitor and viewed from 70 cm distance. Eye position was calibrated before each session and eye movements were recorded by a Tobii X50 eye tracker. For each image participants searched for the threat item and pressed the keyboard space bar to indicate that they had finished searching. They then rated their decision confidence with a five-point rating scale. Other than decisions of 1 or 2, participants were asked to indicate the location of potential threat items. In the practice sessions feedback was provided immediately. Participants were asked to complete each image thoroughly and quickly. Before the test sessions, participants were told the targets in the test sessions were different from the targets used in the practice sessions and no feedback would be given – otherwise the task was the same as the practice sessions.

## Results

### Performance analysis

For each image the location of the threat item was considered so that a 'false location with positive response' was considered to be a false negative decision (miss). Only 'correct location with positive response' was scored as a true positive decision (hit). For target-absent images, only false positive decisions (false alarm) and true negative decisions (correct rejection) were possible. Hit rates, false alarm rates, area under the ROC curve ($A_z$) and the decision time of the correct responses were calculated to measure individuals' performance.

Performance was expressed as an $A_z$ value (the area under the ROC curve). Figure 1 shows the mean overall performance in pooled ROC curves with the accuracy measure ($A_z$) ranging from 0.80 for session 1 to 0.96 for session 3, which intuitively shows the performance increased with practice. An analysis of variance (ANOVA) revealed that it was significant, $F(2, 20) = 23.829$, $p < 0.001$, reflecting an increase of the overall hit rate from 0.71 for session 1 to 0.89 for session 3 and a decrease of the false alarm rate from 0.33 for session 1 to 0.08 for session 3. ROC curves of session 1, session 4 and session 5 were very close and differences among them were not significant. It indicated that the performance on test sessions was similar to the first practice session and significantly worse than the final practice session.

In the practice sessions, immediate feedback was provided and each threat item was displayed two times in different viewpoints and backgrounds so that the detection performance would be enhanced after participants learned them after the first presentation. In order to get the detection baseline of participants and the degree of transfer, the hit rate of threat items for the first presentation ($H_{first}$) in session 1, similar targets ($H_{similar}$) and unfamiliar targets ($H_{unfamiliar}$) in test sessions were calculated separately. The $H_{similar}$ of session 4 and 5, 0.68 and 0.70, were both better than $H_{first}$, 0.65, but not significant. Hit rates decreased significantly as novel targets were introduced, 0.28 of $H_{unfamiliar}$ in session 4 and 0.38 of $H_{unfamiliar}$ in session 5,

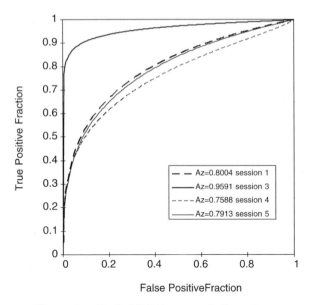

**Figure 1.    Pooled ROC curve of all sessions.**

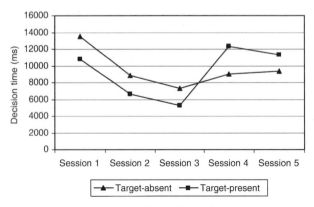

**Figure 2.    Mean decision time of each session.**

which were worse than H$_{first}$, F (1, 10) = 68.820, p < .001, and F (1, 10) = 30.179, p < .001 respectively. However, False alarm rates significantly decreased from .33 for session 1 to .13 for session 4, F (1, 10) = 19.929, p = .001; and .14 for session 5, F (1, 10) = 20.180, p = .001.

*Decision time*

Figure 2 showed an apparent decrease of decision time over the course of training, F (2, 20) = 35.742, p < 0.001 for the target-absent images and F (2, 20) = 19.928, p < 0.001 for the target-present images. Decision times for target-absent images were always longer than that of target-present images during session 1 to session 3, $p < 0.05$.

Decision times for target-absent images of session 4 and session 5 were longer than that of session 3 but shorter than that of session 1. Furthermore, analysis showed that these were significantly shorter than session 1, $F$ (1, 10) = 36.586, $p < 0.001$ for session 4 and $F$ (1, 10) = 29.813, $p < 0.001$ for session 5 respectively. For target-present images, Figure 2 showed that the decision times of session 4 and session 5 were longer than all practice sessions and an ANOVA showed these were longer than the decision time of session 3, $F$ (1, 10) = 26.026, $p < 0.001$ and $F$ (1, 10) = 13.432, $p = 0.004$ respectively.

*Eye movement data analysis*

Eye movement analysis further explained the performance improvement. The area of interest (AOI) was defined for each target on each image to measure the visual development (Liu *et al.*, 2006). With practice, participants were inclined to focus on the threat item area (AOI) quickly, $F$ (2, 20) = 4.432, $p < 0.05$, reflecting a decrease in the time to first enter the AOI from 2252 ms of session 1 to 1088 ms of session 3 for false-negative (FN) decisions and from 1131 ms of session 1 to 670 ms of session 3 for the true-positive (TP) decisions. Participants spent less time in the AOI with practice, $F$ (2, 20) = 4.045, $p < 0.05$. Also the number of fixations on the AOI decreased, $F$ (2, 20) = 3.298, $p = 0.058$. This all meant that participants could fixate on targets and recognize them quickly after training. Eye movement data for the target-present images of the first presentation in session 1 were compared with images with similar and unfamiliar targets in the test sessions in order to assess the training benefits. The time to first enter the AOIs for sessions 4 and 5 were shorter than that of the first presentation targets in session 1; analysis showed the differences were significant for both similar and unfamiliar targets in session 4 and 5, $p < .05$. The time to first enter the AOIs of sessions 4 and 5 were longer than that of session 3 but were not significant. For the eye dwell times on AOIs of session 4 and 5, these were longer than that of the first presentation targets in session 1 [not significant]; but they were significantly longer than that of session 3, $p < 0.05$.

## Discussion

The main aim of this study was to investigate the development of recognition and search skills with training and explore whether these kinds of skills are stimulus specific. Our results are consistent with previous work (McCarley *et al.*, 2004), which indicates that performance increased significantly with practice and participants were inclined to fixate on AOIs more quickly and were more likely to recognize targets. This study extends these findings to a larger stimulus set. Moreover, our results agree with other studies demonstrating that learning transfers more to relatively similar objects than to unfamiliar objects with novel shapes (Tarr and Gauthier, 1998; Furmanski and Engel, 2000). Although the hit rate did not improve significantly for familiar targets and hit rate decreased badly when novel target objects were introduced, eye movement data revealed more detail about specific-stimulus practice benefits. In comparison with the first practice session, not only was visual search for similar targets in the test sessions more effective (longer dwell time on target areas but less search time on images), but also sensitivity to potential threat items was improved such that participants were faster to fixate on both similar and unfamiliar targets. More practice might be the way to overcome the problem of lacking target experience.

The hit rate declined severely following the introduction of unfamiliar targets, and was even worse than the hit rate of the first presentation of targets in session 1. One possible interpretation was that participants obtained some target object knowledge from the immediate feedback of the previous image and then applied this to the next one, due to the high visual similarity between the same kind of objects in the practice sessions, so that the hit rate on the first presentation of targets in session 1 was then better than that of the unfamiliar targets of the test sessions. The false alarm rates of two test sessions were both low and indicated that learning features about normal X-ray luggage images transferred to new normal images. If perceptual expertise was accumulated with repeated exposure to targets, then generic knowledge of luggage X-ray features obtained from practice was very helpful in enabling rejection of non-target items.

In conclusion, the current study showed that the skills of recognition and search for threat items in X-ray luggage images were enhanced with training of frequency exposure stimuli by using immediate feedback in a real visual search task. However, learning in this visual search task for threat items was stimuli specific. Participants cannot successfully recognize novel target objects although they located upon the appropriate target areas quickly and fixated on these with long eye fixation durations. Inadequate training and lack of generic knowledge about potential X-ray threat items could be possible explanatory reasons for these findings. At least, the results indicated that familiarity with stimuli is the source of performance improvement. Therefore, very large numbers of X-ray threat objects should be employed for airport security screener training so as to enlarge their object knowledge and enhance their recognition ability.

## Acknowledgement

This research is supported by the EPSRC.

## References

Ahissar, M. and Hochstein, S. 1996, Learning pop-out detection: specificities to stimulus characteristics, *Vision research*, **36**(21), 3487–3500

Furmanski, C.S. and Engel, S.A. 2000, Perceptual learning in object recognition: object specificity and size invariance, *Vision Research*, **40**, 473–484

Karni, A. and Sagi, D. 1991, Where practice makes perfect in texture discrimination: Evidence for primary visual cortex plasticity, *Proceedings of the National Academy of Sciences of the United States of America*, **88**, 4966–4970

Liu X., Gale A.G., Purdy K. and Song, T. 2006, Is that a gun? The influence of features of bags and threat items on detection performance. In Bust P.D. (ed.) *Contemporary Ergonomics*, London, Taylor & Francis, 17–22

McCarley, J.S., Kramer, A.F., Wickens, C. D., Vidoni, E.D. and Boot, W.R. 2004, Visual skills in airport-security screening, *Psychological Science*, **15**, 302–306

Sowden, P.T., Davies, I.R.L. and Roling, P. 2000, Perceptual learning of the detection of features in X-ray images: A functional role for improvements in adults' visual sensitivity? *Journal of Experimental Psychology: Human Perception and Performance*, **26**(1), 379–390

Tarr, M.J. and Gauthier, I. 1998, Do viewpoint-dependent mechanisms generalize across members of a class? *Cognition*, **67**, 73–110.

# ONE BRIEF: FOUR CONCEPTS ADJUSTABLE FURNITURE FOR SCHOOLS

## Paul Magee[1] & Andree Woodcock[2]

[1]*Design4Advantage*
[2]*The Design and Ergonomics Applied Research Group, Coventry School of Art and Design, Coventry University, Coventry, UK*

Poor fitting school furniture may contribute to the discomfort of children and impaired musculoskeletal development. One possible solution is adjustable furniture which enables a wider range of children to be accommodated. However, in focussing primarily on anthropometry, the context in which the furniture is used is sometimes overlooked. Bearing this is mind; undergraduate industrial design students were given a brief to design adjustable furniture for schools which took account of the schoolroom context as well as considering the anthropometric characteristics of the end user. Over 20 different design concepts were developed. This paper examines the design rationale behind four of these, to exemplify the need to address the wider issues of educational ergonomics in the design of school equipment.

## Introduction

Educational ergonomics addresses the design of educational systems, processes, environments and equipment that are fit for purpose, and that enable all users to achieve their maximum potential (Woodcock, 2006). Schools are the work environment for children in the UK from around 5 to 16 years of age. During that time it has been estimated that they spend approximately 15,000 hours sitting down. It follows, therefore, that the design of the furniture plays a crucial part of a child's immediate environment. It can also contribute significantly to their levels of comfort and general health and well being (for example, Marschall *et al.*, 1995). This is being recognised globally (in Germany, New Zealand, US and UK).

Previous research has addressed the need to produce furniture which can accommodate, or have sufficient adjustability to accommodate, children of varying sizes (e.g. Legg *et al.*, 2003). More recently systems have been developed for teachers and children to help them select the best fitting chair (e.g. Kane *et al.*, 2006). Teachers and school managers increasingly emphasise the comfort and fit of the child to the seat during purchase, and this may now be a priority over durability, stackability and cost. However, some of the resultant designs, although anthropometrically appropriate, seem to follow an office/computer workstation model, rather than providing a creative and exciting learning environment. 'The office has the office dynamic – the school needs the school dynamic' (Breithecker, 2006). The aim of this study was to develop a series of achievable, anthropometrically appropriate concept designs (desk and chair) for schools of the future, which would have sufficient adjustability and scalability to accommodate 4 to 14 year old, would meet the needs of changing classroom environments and move away from the office aesthetic.

## The design process

To accomplish this, the authors worked with 4 undergraduate interns, who undertook the brief as part of their work placement. It was hoped that this would provide them with opportunities to work individually and in a group, apply both their experience and imagination, and design for populations anthropometrically dissimilar to themselves.

The design brief was to develop concept designs for adjustable school furniture giving special consideration to the need to accommodate a range of children from nursery to secondary school), manufacturability and cost, user experience, context of school work – in terms of teacher/student interaction, private and group work, and the classroom context – student requirements, storage and accessibility. The brief therefore concerned the design of multi purpose teaching spaces, rather than just furniture. The programme was based on four stages. Examples of the type of designs produced at these stages are shown in Figures 1–4:

- Stage 1: Concept sketch work. A series of black ink sketches were rapidly produced, discussed and passed between members of the group in an intense iterative cycle, which both reduced the number of possible design candidates and enhanced the most plausible. Although the concept stage was about thinking outside the box, a design party was organized in which local children were invited in to the studio to comment on the designs and provide their own ideas, and a film showing 'a day in the life of a school' (produced by Woodcock and Slater as part of Woodcock, 2006) to ground the designs in the reality of school life. Towards the latter part of this stage, designers were allowed access to internet and other resources.
- Stage 2: Resolution of four selected routes. The 12 designs which survived the internal review processes were reduced to four on the basis of feasibility, cost, manufacturing and potential to support all requirements. These were subsequently developed by all interns.
- Stage 3: 1/4 scale models of furniture for photographic reference using appropriately scaled manikins ranging from 4 year old 5th-le females to 14 year old 95th-le males.
- Stage 4: Full size mock up of the tumble and easel desks, with detailed drawings.

## The designs

As all designs could accommodate the range of adjustability required to accommodate the population, a discussion of this has been omitted, in favour of a discussion on the wider issues embodied in the designs. The designs are shown in Figure 1, a –d.

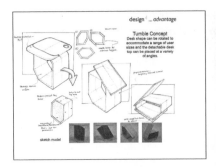 

**Figure 1.   Tumble table (a) Concept sketches (b) Configured for drawing.**

## Cockpit design

Different lessons require children to work alone, or in groups. This is enabled by grouping desks together in the required configurations. However, the result can be visually boring and cluttered.

The aim of this design was to provide individual workplaces, which could also be grouped together to provide interesting spaces having a unified identity, whilst providing the large work surfaces, storage and places to display work (side of desk) required. The triangular design of the table (Figure 2) allows them to stand alone or to be reconfigured as pairs, triads or in groups of four depending on the requirements of the lesson. The resultant forms are more visually interesting than groupings achieved by standard furniture. The long sides also mean that teachers have adequate space to inspect the work.

Children in the UK are normally obliged to carry coats and bags around the school from lesson to lesson. Whilst children are encouraged to deposit such items in specified areas at the start of the lesson, they still carry materials with them. The sides of the cockpit desk hide this from view. The sides also provide opportunities for students to display work (thereby giving a group working together in double periods a group identity; to provide colour coding to reinforce the identity of particular areas of the school (for example, tables in the science block could be blue); or to display the school logo, again reinforcing identity and belonging.

Although children may be required to work together on some tasks, they can be very sensitive about other children seeing work in progress, or that they consider to be of a non public nature. Children are still seen learning over or covering their work, or hiding it with their arm. The need to preserve privacy, whilst at the same time allowing work to be shared across tables, has been achieved by providing a small rim at the end of the desk. This provides a degree of privacy and a visual reminder to co-workers that the desk and material on it are in the private domain.

## Tumble table

'In order to make a working environment a creative one, it ought to provoke thought, argument, and disagreement if necessary. The result is edgy, it is contentious but most of all it isn't, ever, dull...' Magee. The 'tumble table' emerged from the requirements that the school environment should be fun, creative and

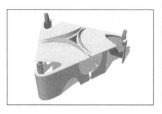  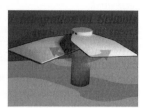

**Figure 2. Cockpit desk.    Figure 3. Easel.        Figure 4. Leaf desk.**

accommodating to children of varying sizes, engaged in a number of activities, in different locations, with preferred or different working styles.

The design consists of a six-sided lateral shape, with internal cavity that can provide a framework to integrate potential power and communication along with applying structural integrity. The table has a distinctive profile that can be rotated along its central axis to give selectable desk height and working surface. This provides four modes (shown in Figure 1a), the use of which will depend on lesson type and size of child; sit down desk (for small and larger children), standing easel position, remote working using the worksurfaces as a detachable drawing tablet (Figure 1b).

Learning, especially in primary schools should be fun. In the tables this element is provided by the interesting, irregular shape, the way the tables can be combined together, and also the opportunity to use the internal cavity as a play/work environment, or corridor between different parts of the classroom.

The design also anticipates the need for greater IT integration in classrooms and the need to extend the learning environment out if the classroom. To enable this, the top of the table is formed from a graphics tablet with handwriting recognition (Figure 1b). The tablet can be carried out of doors, or used as a drawing tablet with contents unloadable on to a central whiteboard. This means that technology and computer supported learning are more seamlessly introduced in the primary school.

*Easel desk*

This desk (Figure 3) was inspired by the traditional, flip top school desk. The tilt can be adjusted depending on the current activity. It would be accompanied by a kneeling bar which could be adjusted vertically and horizontally to accommodate differently sized children. The design also considers requirements for under-desk storage and the private/public viewing.

Although not shown in Figure 1c, the lips of the desk would be used to house individual lighting systems which would flood the work surface with different coloured lights. Jordan (2006) has demonstrated the educational advantages certain children (such as those wit autism and dyslexia) may derive from different coloured lights. The problem lies in providing the optimum lighting for each child. This could be achieved through different coloured lenses, lamps or, it is proposed by flooding individual desk tops with light. This is considered to be an important innovation, as it will enable pupils and teachers to tailor lighting configurations to individual needs.

*Leaf desk*

Space is a valuable commodity in schools. Technological developments such as wireless networks, laptops and video conferencing will lead to the evolution of different forms of classrooms. The leaf desks (Figure 4) are height adjustable work tops designed to accommodate a notebook, textbook and appropriate stationary. They are located around columns which can either be pulled down from the ceiling or up from the floor as required. The columns would provide the technological infrastructure. Each surface could be positioned at any height, providing a range of surface heights, and fold away when not required. Different types of spaces could be created by pulling down different columns, for example for auditorium-like lessons, group work (with three 'leaves' attached to different pillars), or an open space in the middle of the classroom, with columns round the sides of the classroom.

This concept also has wider benefits brought about by the technological infrastructure being part of the building. The columns, when retracted, could be locked into position thereby increasing security. The open floor space provided could also allow better access for cleaning.

## Educational ergonomics and the production of furniture design concepts

The concept designs explored in this paper have evolved from a top down, rather than bottom up approach to educational ergonomics. Looking at wider usage scenarios enables different issues to be brought into the design solution space and to be considered in parallel with other aspects of the problem.

In these designs, the students looked at issues idea of adjustability at the work environment task/level rather than workstation level. This brought to the forefront the need to design furniture which would appeal to young children's sense of fun, the need to incorporate technology, to reconfigure classroom environments depending on lesson type, to support group and private work and other forms of tailorability (such as the immediate sensory environment).

At present full scale models are being developed of the easel and tumble table. These and other concept designs will be presented to teachers and children. As an exercise in design and ergonomics the students have benefited from designing for populations unlike themselves. The project has also shown that we need to consider not just the anthropometry of the children, but what it is like to be a child, and to design school environments which capture children's imagination rather than reproduce an office aesthetic.

## Acknowledgements

Intern designers: Amy Chubb, Pip Davis, Ashley May and Bryan Moss. Research supported by Coventry School of Art and Design.

# References

Breithecker, D. 2006, Ergonomics a subject for the school as a place of work' statement issued by the 'Federal Working Group of Posture and Exercise', downloaded from http://www.bag-haltungundbewegung.de/fileadmin/bag/binary/Statement_Ergonomics.pdf on November 20th 2006.

Jordan, I . 2006, Visual dyslexia at http://www.orthoscopics.com/ accessed December 1st, 2006.

Kane, P.J., Pilche, M. and Legg, S.J. 2006, Development of furniture system to match student needs in New Zealand schools, presented at IEA 2006, Maastricht.

Legg, S J., Pajo, K., Marfell-Jones, M. and Sullman, M. 2003, Mismatch between classroom furniture dimensions and student anthropometric characteristics in three New Zealand secondary schools, IEA 2003, Seoul.

Marschall, M., Harrington, A.C. and Steele, J.R. 1995, Effect of workstation design on sitting posture in young children, Ergonomics, 38, 1932–1940.

Woodcock, A. 2006, Re-evaluating Kao's model of educational ergonomics in the light of current issues, Contemporary Ergonomics, 2006.

# WHY DO STUDENT NURSES CONTINUE TO USE THE DRAGLIFT?

## Linda Allen[1], David Stubbs[2] & Sue Hignett[3]

[1] *Manual Handling Coordinator, School of Health & Social Care,
University of Teesside, Middlesbrough, Tees Valley, TS1 3BA*
[2] *Professorial Research Fellow in Ergonomics, Robens Centre for Health
Ergonomics, European Institute for Health & Medical Science,
University of Surrey, Guildford, Surrey, GU2 7TE*
[3] *Director: Healthcare Ergonomics & Patient Safety Research Unit,
Department of Human Sciences, Loughborough University,
Loughborough, Leicestershire, LE11 3TU*

A patient handling technique known as the 'draglift' has been widely condemned as dangerous to both patients and staff for well over two decades. Assuming this message to be well know, and that practitioners do not knowingly set out to injure themselves or others, the question must therefore be asked 'Why is the draglift still being used?' This paper reports on the use of Grounded Theory to identify the influences on student nurses that shape their continued use of the draglift. Focus groups were used for data gathering with 30 student nurses taking part. The following 'categories' for continued use of the draglift were identified; (i) vicarious learning, (ii) groupthink, and (iii) the theory-practice gap. The socio-psychological mechanism of 'belonging' is proposed as the substantive theory that integrates the emergent categories, their properties and dimensions.

## Introduction

A particular patient handling technique known as the draglift has been widely condemned as dangerous to both patients and staff for well over two decades. The principle text used to publicise this message is *The Guide to the Handling of Patients* in its various editions (Troup *et al.*, 1981; Smith, 2005). In addition, all qualified nursing and midwifery practitioners received copies of a practice standard in which the poor application of patient handling technique was cited as 'physical abuse' (Nursing & Midwifery Council (NMC), 2002). Despite this publicity however, Crumpton *et al.* (2002) in a survey to measure compliance with the RCN *Safer Patient Handling Campaign* report 77% of respondents as having observed the use of condemned patient handling techniques.

The traditional approach to the management of musculoskeletal injury amongst healthcare personnel and by implication the adoption of safer patient handling methods has been to provide technique training. This approach originates from the belief that injury occurs as a result of selecting the incorrect technique and that teaching the correct way to lift will therefore reduce injury. Technique training also utilises a behavioural approach to learning based on the assumption that learners react passively to the transfer of knowledge and skills. Success of this approach is therefore dependent upon (i) whether or not the draglift is unintentional, and (ii) whether or not learners are passive in response to the transfer of knowledge and skills.

It is the authors' contention that this is not the case as not all decisions follow a 'rational' decision-making model (Furnham, 1997). In fact the 'irrational' decision-making model more closely fits with continued use of the draglift, as there is (i) a lack of ownership of safer technique; (ii) desirability of a quick solution; (iii) acceptance of a low quality solution; (iv) familiarity with the draglift; (v) workload and patient pressure; (vi) no prior knowledge or experience in the generation of alternatives (especially in the case of student nurses); and (vii) patient handling is complex.

The evaluation of risk is also important in selecting a 'rational' decision making approach. In summary, there are two categories of risk, 'objective' and 'subjective'. Traditionally governments, industry and experts, consider objective risk as the most valid whilst for individuals subjective risk is more influential. The cost or importance given to the decision also has influence, as 'experts' describe risk in terms of expected annual mortality and individuals recognise 'dread risk' as the most significant (Adams, 2001).

Given these influences the origin of the draglift lies within the definition of violation behaviour as proposed by the Human Factors Reliability Group (HFRG), (1995). If this is accepted, then intentions to perform behaviours of different kinds can be predicted by measuring attitudes towards the behaviour, subjective norms in relation to the behaviour, and the individuals perceived behavioural control over the desired behaviour (Ajzen, 1991). Interventions that take account of these processes are therefore proposed as more likely to be effective in reducing occurrence of the draglift relative to the behavioural strategies currently advocated.

## Method

Grounded theory is derived from the theoretical framework of 'symbolic interactionism' which is the study of the 'experiential' aspects of human behaviour (Glazer and Strauss, 1967). The underlying belief is that individuals initially shape their behaviour on that of 'significant others' eventually being able to play a number of roles simultaneously i.e. the 'generalized other'. The methodology involves the generation of theory from emergent data, in that it starts from the research question, moves through a joint process of data sampling and analysis and ends with a theory that is 'grounded' in the emergent data. The principle strength of this approach is that it enables the 'expert' to give up their own preconceptions in favour of the 'user' perspective. The method of data gathering was through focus groups with 'participant validation' (Strauss and Corbin, 1998) being achieved by returning the analysis from each focus group to the original participants and asking for confirmation of the analysis and continued consent. The study aims were:

1. To describe the phenomenon of 'continued use of the draglift' from the perspective of the student nurse, and
2. To add to the understanding of practitioners involved in the preparation and support of student nurses in relation to safer patient handling.

## Results

### The sample population

A total of 30 participants attended 6 focus group sessions, 3 males and 27 females with ages from 20 to 48 years. All of the participants had at least one practice experience so had either used a draglift or seen one in use, two-thirds having previous experience of patient handling before commencing their student nurse training. Although it was intended to recruit students from all four nursing pathways i.e. adult, child, mental health and learning disabilities, the adult pathway were over represented in the sample with the other three pathways being under represented. Due to the inexperience of the author in terms of the method no new cases entered the sample once the focus group sessions commenced i.e. snowball samples, confirming and disconfirming cases, which has implications for 'trustworthiness' (Bryman, 2001) of the emergent data. The following questions were 'theoretically sampled':

1. Who do students learn their patient handling skills from?
2. What is the value of patient handling in the practice area?
3. Does the quality of the nurse patient relationship influence patient handling?
4. What are the characteristics of a good safety culture?
5. What is the students' perception of a 'caring' relationship?
6. What is a draglift?
7. What is the status of student nurses in practice? And finally
8. How significant is the need to belong?

### The emergent categories

In respect of student nurses continued use of the draglift, pressures to conform can be broadly attributed to 'vicarious learning' (Bandura, 1977); 'groupthink' (Janis, 1982); and the 'theory-practice gap' (Ashworth and Morrison, 1989). These are not however independent variables but interact with each other around the socio-psychological mechanism of 'belonging' (Maslow, 1954).

*Vicarious Learning* – The emergent data provided evidence of use of a draglift by all grades of staff from Health Care Assistants (HCAs) to Ward Managers. It was also identified that if those in a more senior or specialist role i.e. ward based manual handling assessors, chose to do a draglift it would have a greater influence on the student. The reverse was also true as some students reported positive role models, especially amongst HCAs, as influencing the adoption of safer methods. This is of particular relevance to the future preparation of student nurses, as the majority of patient handling tasks are completed by HCAs and student nurses. The intentional selection of an alternative but equally dangerous technique i.e. the 'sheet lift', in order to avoid conflict with practice staff was also reported. On exploration student nurses view this method as the lesser of two evils as they could not always offer safer alternatives with confidence. The absence of reinforcement of university learning in practice and its subsequent impact on the transfer of learning from the short to the long-term memory was influential in terms of 'forgetting', this being more prominent in 1st year student nurses than 3rd years, as they received less learning in terms of 'maintenance rehearsal'.

*Groupthink* – Janis (1982) explains that whenever a member deviates from the group norm, the group members initially increase their communication with the deviant in an attempt to talk them around. If this fails then they remove their communication in order to exclude the deviant and restore group unity thereby minimizing any doubts about their own actions and any merits of the counter argument. This process of groupthink was highly prevalent in the emergent data, the need for 'speed', and perceptions of the draglift as more 'caring' relative to the use of equipment being frequent responses. For example, one student explained that 'everything was done by the clock' yet she had 'never drank so much tea'. Another gave the example of the perceived need by practice staff to complete all the work in the ward diary in the morning, yet the handling practices did not improve in the afternoon even though more time was available. The use of equipment to move patients was also perceived by practice staff as more time consuming which led to student exclusion from patient handling activities if they suggested alternative methods. When asked for examples of when 'speed' was felt to be justified 'emergency situations' were the immediate response, as were patient falls. On further exploration this included uninjured patient falls as it was perceived to be an indicator of poor nursing care for a patient to be seen to be on the floor. Speed was identified as more valued amongst lower grades of staff as was constant activity relative to process and communication skills. For example, students were frowned upon for attempting to help patients through explanation of normal movement to stand up from a chair. The same was also the case if they attempted to plan the patient handling activity i.e. move the patient using some form of equipment. Practice staffs expectations of a 'caring role' also had a strong bearing on the method of patient handling and the subsequent student expectation. If a patient had unusual needs i.e. a 'broken hip', it was more likely that a written handling plan would be completed and adopted. If the patient was however of a light weight with no unusual care needs, and the staff felt the patient was within their capabilities to move manually, then the draglift was the norm. The fact that the draglift has a 'name' was also highlighted and illustrates its 'habituation'. Interestingly in areas where the term 'resident' was used to describe the client group as opposed to 'patient', then more time was given to planning the handling activity and more involvement was expected from the resident.

*Theory-practice gap* – Melia (1984) makes a distinction between 'caring about' i.e. the theory, and 'caring for' i.e. the practice, and adds that these different types of 'caring' are attributed to different groups of staff, the 'theory' being a qualified nurse's role, and the 'practice' belonging to the unqualified. The student nurse however sits somewhere in between and must work with both groups in order to acquire and demonstrate for assessment purposes the necessary cognitive, practical and attitudinal skills. In terms of its presentation in practice, the phrase 'welcome to the real world' is often thrown at the student nurse in an attempt to undermine their argument for the adoption of safer patient handling methods. The temporary position the student holds in practice, illustrated by the fact that they have no name being referred to as 'the student', also has influence in that it allows practice staff to exclude the student from the nursing team with no long term repercussions on their behaviour. In terms of reducing the theory-practice gap, practice staff who were able to support student nurses in their learning came from both qualified and unqualified groups and were usually those who were undertaking some form of 'Continuous Professional Development (CPD)' leading to a recognized qualification. HCAs undertaking National Vocational Qualifications were particularly influential with 1st year student nurses.

## Conclusion

Maslow's (1954) 'hierarchy of needs' proposes that 'motives' at the lower level in the hierarchy must be partially satisfied before those at a higher level become determinants of action. As acceptance and belonging are at a lower level relative to cognitive needs, the need to belong supersedes the need to apply safer patient handling techniques. Social influence produces compliance even when the target individual publicly conforms to the group norms but does not change their underlying belief. When a group obtains compliance through setting an example it is referred to as 'compliance' but when compliance is obtained by wielding authority it is referred to as 'obedience'. In both cases the individual complies because the group has the power to administer rewards and punishments usually of a social nature. 'Belonging' was therefore chosen as the 'socio-psychological mechanism' that integrates all the categories, their properties and dimensions, as it is the mechanism that mediates the students need to conform to use of the draglift.

The following recommendations for the future preparation and support of student nurses are proposed:

1. The adoption of a 'constructivist' approach to learning over the full period of student nurse preparation.
2. The development of an active role for Health Care Assistants in terms of skills development especially for 1st year student nurses.
3. The undertaking of shared learning opportunities between Universities and their practice area so as to demonstrate the integration of theory and practice.
4. The development of decision making and communication tools that reduce the complexity of patient handling assessment, and
5. The enforcement of patient focused care in the practice setting.

## References

Adams, J. 2001, *Risk*, (Routledge, London)

Ajzen, I. 1991, The theory of planned behaviour, *Organizational Behaviour and Human Decision Processes*, **50**, 179–211

Ashworth, P. and Morrison, P. 1989, Some ambiguities of the student's role in undergraduate nurse training, *Journal of Advanced Nursing*, **14**, 1009–1015

Bandura, A. 1977, *Social Learning Theory*, (Prentice Hall, Englewood Cliffs, New Jersey)

Crumpton, E., Bannister, C. and Maw, J. 2002, Survey to Investigate NHS Trust Compliance with the RCN Safer Patient Handling Policy 1996

Furnham, A. 1997, *The psychology of behaviour at work: the individual in the organization*, (Psychology Press, London)

Glaser, B.G. and Strauss, A.L. 1967, *The Discovery of Grounded Theory: Strategies for Qualitative Research*, (Aldine de Gruyter, New York)

Human Factors Reliability Group. 1995, *Improving Compliance with Safety Procedures: Reducing industrial violations*, (HSE Books, Sudbury)

Janis, I.L. 1982, *Groupthink: Psychological Studies of Policy Decisions and Fiascos*, (Houghton Mifflin, Boston)

Maslow, A.H. 1954, *Motivation and personality*, (Harper and Row, New York)

Melia, K.M. 1984, Student nurses' construction of occupational socialization, *Sociology of Health and Illness*, **6**, 132–150

Nursing & Midwifery Council (NMC). 2002, *Patient-client relationships and the prevention of abuse*, (Nursing & Midwifery Council, London)

Troup, D., Lloyd, P., Osborne, C. and Tarling, C. 1981, *The Handling of Patients: A guide for nurse managers*, (Back Pain Association in collaboration with the Royal College of Nursing, Teddington)

Smith, J. 2005, *The Guide to the Handling of People*. 5th Edition, (Backcare in collaboration with the Royal College of Nursing and the National Back Exchange, Teddington)

Strauss, A. and Corbin, J. 1998, *Basics of Qualitative Research: Techniques and Procedures for Developing Grounded Theory*, (Sage Publications, London)

# SEATING PROBLEMS – THE MISSING LINK?

## E.N. Corlett

*The Institute for Occupational Ergonomics,*
*University of Nottingham*

Most workplaces have fixed height work surfaces, anthropometric differences are dealt with by providing adjustable height chairs. These introduce difficulties with shorter people due to knee interference under desks and the need for footrests.

A major problem is the flattening of the lumbar curve due to the horizontal seat surface, introducing a constantly maintained pressure on the lumbar discs. A different design of seat is needed to reduce under thigh pressure, maintain the lumbar curve and let users of whatever stature sit at the appropriate height and keep their feet on the floor. Evidence of the benefits of such a seat will be described.

## Introduction

Increasing numbers of workpeople are sitting at work, and there is no evidence to suggest any lessening in this increase. Large numbers of these seated workers suffer musculoskeletal problems. The major ones are back pain and upper limb disorders.

Both these problem areas have been under intensive mexamination for several years. The causes of each phenomenon are complex and investigations have proposed many procedures for mitigating the injuries incurred. Yet the prevalence of back pain, in particular, remains and it would appear to be increasing in the population at risk.

Working height and distance, work intensity, psychosocial factors, lumbar support, seat height and depth are just some of the factors which have been analysed. Recommendations have been set out to improve the situation. Improvements have, indeed, been evident, for example Upper Limb Disorders (ULDs), appear to be less now than formerly. Yet back pain is still a major, and increasing, problem. Are we missing something?

## The Missing Link?

One factor, evident for at least fifty years, is rarely taken into account in the design of current work chairs. In 1953 JJ Keegan illustrated that the lumbar curve flattened as the thighs were raised to be at right angles to the trunk. The first seventy or so degrees of rotation occurred at the hip joint, but the last twenty degrees or so were achieved by a backward rotation of the pelvis, hence the flattening of the lumbar curve.

This phenomenon has been confirmed since then, see the review by Bridger and Bendix (2004). Different body shapes show differences in their responses, but all are affected. The consequences of this flattening are twofold. Firstly, the faces of the adjacent lumbar vertebrae tilt towards each other, stretching the erector spinae muscles as well as exerting pressure on the forward edges of the discs. Secondly, because the flattened curve has placed the vertical gravity load from the trunk etc. forward of the lumbar spine, the muscles at the back of the spine are active to stop the trunk from folding over, thus creating increased pressure on these same discs. The gravity load still exists, so the total spinal load is considerably increased.

Over short periods, and given opportunities to recover, this loading of the discs is beneficial in the sense that variations in disc pressure aid the nutrition of the discs. But, when sitting at work, these loads are substantially constant and are maintained for hours every day over the years of seated work activities. Some 300 years ago Ramazzini (1700/1983), documented that constant loads on the body, from a lifetime of work, will lead to both distortion and injury. We cannot assume that these constant loads imposed on the lumbar spine will have no effect.

But, in chair design, that is what we do. Mandal's (1974) work, after over thirty years of proselytising, is only now being recognised, (Gardner and Kelly 2005, Corlett, 1994, 2006). Desks and chairs for schools using his research are appearing, as detailed in the following paper by Gardner (2007). School children will benefit.

## The adult workplace

In the adult workplace the scene is not so good. A major problem is that the many manufacturers of work seats have their investment in horizontal seats. True, some are tweaked to tilt a few degrees forward, they rock, twist and have many extras to add to their 'ergonomic' quality. Most of this is peripheral to dealing with the problem. Yet, if they change their designs to ones using Mandal's ideas, they are implicitly saying that their previous seats were not so satisfactory. So there is little incentive to change.

Bearing in mind that desks and other work surfaces are not usually adjustable in height, what would a basic seat, which would protect the spine from the above adverse loadings, look like? It would have to be shaped so that the sitters' thighs were able to slope at approximately 20 degrees or more downward. It would also have to adjust in height to cope with variations in stature. Sitters would have to be able to keep their feet on the floor, whilst support for the lumbar region should be available. Importantly, the sitters should not slide forward off the seat, but be stable and well supported.

To fulfil such a specification a seat, curved from front to back and with the peak of the curve about one third of the way from the rear of the seat, is the basic requirement. It would require adjustable forward tilt over a range of about 15 degrees or more. Thus an open angle at the hip of some 110 to 115 degrees would be possible for all sitters, whilst sitting on the top of the curve would mean that they are stable, with no tendency to slide off. They would have their feet on the floor, since as the height increased the tilt could also be increased to increase the slope of the thighs, so there would be no need for footrests.

Most workplaces have fixed height work surfaces. With the tilting and curved seat, if users need to sit higher, they can get their knees under the work surface without pressure on the thighs from the under side of the desk, as can happen when a horizontal seat is raised. Shorter users can raise their height to sit above their work without the need for footrests. Tall males would still be at a disadvantage from a conventional height work surface, but they require a higher work surface in any case if their cervical and thoracic spines are not to face damage. There is no doubt that adjustable height work surfaces are necessary to create satisfactory workplaces which would suit all sedentary workers. As yet their supply in this country is rare.

If the workplace has an adjustable height desk, the seat design can be simpler. The seat can have a horizontal portion at the rear third, with a sloping front portion steep enough to allow the 110–115 open angle at the hip. No tilt facility is needed on the seat as adjustment of the height of the seat to accommodate the leg length of the sitter can then be followed by adjusting the desk height to suit the sitter's stature. The school furniture discussed by Gardner, (2007) demonstrates this situation.

## Comment

We spend large amounts of money and time to try to prevent, as well as cure, the back pain suffered by sedentary workers. Yet we impose a sitting posture which puts a serious initial load on the spine before any work is done. Those who use these seats have to sit for their whole working day with these imposed and static loads. The loads from the work, which are often less than these imposed loads, are additional. We will not see the reduction in back pain which we all desire until we stop using seats which actively create bad backs.

## References

Bridger, R.S. and Bendix, T. (2004) Section 7.3 in Chapter 7, The Pelvis. In: *Working Postures and Movements*. Eds. Dellman, N., Haslegrave, C. M. and Chaffin, D. B. CRC Press, Boca Raton, USA.

Corlett, E.N. (2006) Background to Sitting. Ergonomics **49**, 14, 1538–1546.

Corlett, E.N. and Gregg, H. (1994) Seating and access to work. Chapter 25 in: *Hard Facts about Soft Machines*. Eds. R. Lueder and K. Noro. Taylor & Francis, London.

Gardner, A. (2007) Back Pain in Schoolchildren, will they be fit to work? Ergonomics Society Annual Conference, 2007.

Gardner, A. and Kelly, E. (2005) Back Pain in Children and Young People. BackCare, 16 Elmtree Road, Teddigton, TW11 8ST. UK.

Keegan, J.J. (1953) Alterations of the lumbar curve related to posture and seating. Jnl. of Bone and Joint Surgery, **35A**, 589–603.

Mandal, A.C. (1974, 3rd edn. 1985). *The Seated Man, Homo Sedens.* Dafnia Publications, Klampenborg, Denmark.

Ramazzini, B. (1700) *De Morbis Artificum.* Text of 1713 revised with translation and notes by Wilmer Cave Wright. New York. The Classics of Medicine Library, Division of Gryphon Editions. 1983.

# 2008

| Nottingham University | 1st–3rd April |
|---|---|
| **Conference Manager** | Sue Hull |
| **Chair of Meetings** | Martin Anderson |
| **Programme Secretary** | Phil Bust |

| **Secretariat** | Gerry Newell | Nora Balfe | Lauren Morgan |
|---|---|---|---|
| | David Hayes | Nastaran Dadashi | Tom Griffin |
| | Richard Parker | Yasamin Dadashi | Polly Shelton |
| | Jemma Taylor | Amy Crawford | |

## Social Entertainment

The use of mobile devices during last year's 'Quiz Night' might have had a bearing on this year's theme – "Mission Impossible". This included a challenge to make a tiny self propelled vehicle to travel the furthest over the bar floor. The annual dinner was held at Trent Bridge Cricket Ground.

| Chapter | Title | Author |
|---|---|---|
| **Ageing Population** | Understanding workplace design for older workers: A case study | P. Buckle, V. Woods, O. Oztug and D. Stubbs |
| **Health and Well Being of Construction Workers** | Maintenance workers and asbestos: Understanding influences on worker behaviour | C. Tyers and S. O'Regan |
| **Methods and Tools** | Laptops in the lecture theatre: An ergonomic focus on the critical issues | R. Benedyk and M. Hadjisimou |
| **Transport** | Ergonomics issues in a jet car crash | M. Gray |

# UNDERSTANDING WORKPLACE DESIGN FOR OLDER WORKERS: A CASE STUDY

**Peter Buckle, Valerie Woods, Ozhan Oztug & David Stubbs**

*Robens Centre for Public Health, Faculty of Health and Medical Sciences University of Surrey, Guildford GU2 7TE*

The ageing workforce presents a significant challenge to ergonomists. Whilst our knowledge of changes in physical and cognitive capacities with age has been researched, there has been little attention paid to the perceptions of the older workforce regarding their experience of ageing in the modern workplace. This case study has used qualitative methods to gain new insights into the motivations, health issues, coping capacities and self perceptions of ageing workers in a food processing industry.

## Introduction

Few workplaces or work organizations have been designed with the needs of the 65 year old (and older) worker in mind Huppert 2003, Woods and Buckle 2002). Despite the "Inclusive Design" research agenda there is little knowledge of what this workforce sector requires. There is also little known on what help organizations need to accommodate older workers.

This case study focused on identifying organizational issues that older workers believe need to be addressed for a healthy working environment that meets their abilities, capacities and expectations. The study is one of four workplace case studies undertaken as part of the research councils' Strategic Promotion of Ageing Research Capacity (SPARC) initiative.

## Methods

The study took place at the manufacturing centre of a food processing organization. Focus group interviews were undertaken with workers from age groups of 40–49, 50–59, 60+ years and a sample of recently retired workers. In addition, two occupational health professionals and a representative of the human resource management team were interviewed. In total 14 workers participated; these were mainly from the shop floor but a number of office workers also took part in the study. "Older workers" refers to the participants who were aged 50 years and above and "younger workers" to those aged 40–49 years. Although both male and female workers took part in the focus groups all participants are referred to as male in the following summary to preserve anonymity. All the interviews were transcribed verbatim and coded line by line using qualitative data analysis software QSR NUD*IST (Richards and Richards, 1991). Codes were arranged into coherent themes using the approach advanced by Taylor-Powell and Renner (2003).

## Results

The issues raised during the interviews were categorized under the following main themes: "perceptions of organizational culture", "motivations to leave or continue working", "health issues", "perceptions of own capacity", and "coping with job demands". A brief summary of the issues raised and examples of what participants said are presented below.

*Perceptions of organizational culture*: This category comprised the issues that the participants perceived to be important with respect to their age and the organizations' culture. As they aged, some participants reported that their career opportunities were limited and that their accumulated knowledge was superseded "I'm thinking more of a permanent fixture thing and I personally don't think that industry now rewards the older person with the knowledge, it's almost like we get to the burn-out stage and then we are put out to grass."

There was a consensus that limited career opportunities affected their motivation for personal development; however it appeared that some were happy with their existing knowledge, that they had accumulated over the long years of service, and did not wish for further development. One of the participants expressed his resistance for progress as follows: "If I'd wanted to, I think I should be able to be given the opportunity of doing something, but I feel that it's not there. They may contradict me and say 'Yes, it is.' But it's not that apparent. It's definitely not apparent to me, but as I say my choice is that I wouldn't want to at this stage."

Similar concerns were raised by the younger workers. They believed that as workers aged they were likely to be perceived as less adaptable in dynamic work settings in comparison with younger workers. One of the workers expressed this as follows: "I think part of people's perception is as you're getting older that they can't think of you in the same way as the younger person coming into the business and that you're not as flexible."

*Motivations to leave or continue working*: The participants expressed a range of factors as motivators for continuing to work. "Finance" was the most frequently reported motivator for older workers to stay and continue working. Overall both younger and older workers reported that they were quite satisfied with their earnings. One financial reason mentioned by an older worker for continuing to work was better investment in pension plans: "As I said, 18 years ago, I took out an investment plan. If you remember, 18 years ago insurance companies weren't allowed to give you advice like they do nowadays. If hindsight had prevailed, I could have been retired now, but my investment plan takes me until I'm 65."

Similar views were expressed by a younger worker: "I said 60 [previously thought age for retirement] but you just don't know, do you? You're on a pension, you see the changes being made now, we could be here until we're 65 because monetary wise you could be worse off if you don't. You just don't know."

Some of the older participants reported social reasons for continuing to work. One of them thought that coming to work provided a structure to his life. Another reported that he was enjoying coming to work as he had good social networks there: "It's the discipline isn't it really? Getting up every day, knowing that you've got to do certain things . . ."

The participants also expressed some factors as motivators to leave the job. One of the older participants, a shift worker, said having regular sleep was a motivator to leave the job: "But what I have also noticed as well is that on some occasions, some of the blokes that you see now, that you haven't seen for two or three years, they look younger now than when they worked here because . . . They're getting sleep. They're getting sleep; they're getting regular sleep . . ."

Another older worker thought that the attitudes of the management had changed and that it was more difficult to communicate problems to the senior management. He expressed this as a motivator for leaving the job earlier than the normal retirement age of 65: "I must admit, if you'd have asked me 10 years ago whether I wanted to retire, I'd have said no. I'd have been quite happy working as long as I could, but it's the atmosphere, I think, in the factory that's changed. I wouldn't want to stay on."

Some of the younger workers expressed health status as a reason for possible early retirement. One of them thought that working for many years might lead to ill-health that might result in premature death: "My dad took early retirement when he was 59 and he always said to me he'd lost a lot of money with his ABCs and different things, but he wanted to be at home with my mum. Good job because he died when he was 67. If he'd have stopped when he was 65 he'd have only had two years . . ."

There was a consensus among workers that the shift patterns were difficult to follow: "Some of the things that are killing me off with age . . . pressures, I don't take them as I used to, and the hours. That is what I find difficult as I've got older, the pressure of the work and the physical hours of the shift work."

One of the younger workers suggested that shift work was a source of stress as it made working conditions difficult and was one of the reasons people were choosing to retire early and changing to less stressful jobs: "But you look at the people that have actually retired from . . . . . . . . . . that took early retirement from . . . . . . . . . . and got other jobs and are enjoying themselves because they've got different jobs without any stress related to them . . . . People have gone. They want to get away from this after they've been in it for such a long time."

*Health issues*: The older workers perceived regular health checks as a positive aspect of working as an older worker. One of them perceived this as being cared for. There was a consensus among all the workers that the Occupational Health Department was looking after older workers well: "Certainly the Health & Safety departments look after the older worker. They monitor our health, I think more frequently than others . . ." and "As I said, the only positive I see [about getting older in this organisation] is the occupational health, as far as the fitness thing goes, and the physio."

One of the participants on the other hand raised a concern regarding the environmental conditions and believed that older workers might become more sensitive to environmental exposures such as heat, dust and noise: "The working environment at the factory as a whole is quite good. I mean, it [dust, noise, heat] can get terrible out on the Process." (Note: "Process" here refers to the food processing area of the organization.) When asked if these factors got worse as they got older "Oh yes . . . Oh yes. I certainly find now that things like that would aggravate me far more now than they would have done. It can make you feel unwell. I mean, you can probably hear that I've got a problem with my throat anyway, which the dust aggravates. It does aggravate it. I should wear a dust mask, but unfortunately with that, these hair nets, you feel as though you're in a helmet and it's hot out there. It's dirty, it's wet."

*Perceptions of own capacity*: Two issues emerged under perceptions of older workers' own capacity. These were cognitive ability and patience. Some workers thought that as they aged their cognitive ability to learn and recall information was declining. One of them expressed this as follows: "The other thing I find wrong is that as you get older, although you might get there eventually, it takes a bit longer to take things in. You can't be as quick. Twenty years ago I was quite sharp in the brain and now it takes a bit longer." Another older worker suggested that as he was getting older he perceived himself as becoming less patient: "I just find that at my age now, whoever said 'With age you get patience' is a liar. You don't get patient. You get less patient." In contrast, one of the younger workers thought that as individuals grow older they become more mature in their views: "I don't know. I think you get a more rounded view on things, don't you? I think when you're young you tend to just hone in on specifics and treat them as the centre of the earth at that particular time."

*Coping with job demands*: The results demonstrate that there might be a trade-off among some of the older workers and their younger colleagues. A number of the older workers said they found the physical aspect of their jobs difficult to carry out and sometimes they asked for help from their younger colleagues: "It's very difficult on the process because the lines that we run – I'm on the process – you've got to do every job on that line to be able to . . . be an active part . . . It just doesn't work any other way. I'm lucky that the two people I work with will take a fair bit off my shoulders on the heavy stuff."

A number of the younger workers confirmed that they sometimes helped their older colleagues in carrying out physically demanding tasks. One of them expressed this as "carrying older people": "You definitely feel that you're carrying the older people? Yes, not with the technical bit. The fellow I work with, he's as technically advanced as anyone and he's 62, but the physical nature of it, there comes to a point where people find it more difficult to do the job that we're doing." "Like we say, we carry the older people and it happens a lot on the night work."

## Discussion

The ageing workforce is an inevitable consequence of demographic trends and current pension policies. The challenge this presents to ergonomists can only be met if our knowledge of the difficulties faced by the older workforce are understood and acknowledged. This study has gained insights into the perception of work by older members of the workforce and, importantly, has provided insights into teamwork issues, physical demands and cognitive capacities that will need to be considered in future work system design.

## References

Huppert, F. 2003 Designing for older users. In: J. Clarkson, R. Coleman, S. Keates and C. Lebbon (Eds) Inclusive Design: Design for the Whole Population. London: Springer Verlag.

Platman, K. and Taylor, P. 2004 Themed section on age, employment and policy, Social Policy and Society, 3, 2, 143–200.

Richards, T. and Richards, L. 1991 The NUDIST System, Qualitative Sociology, 14, 289–306.

Taylor, P. 2002 New Policies for Older Workers. Joseph Rowntree Foundation: The Policy Press. ISBN 1 86134 463 5.

Taylor-Powell, E. and Renner, M. 2003 *Analyzing qualitative data.* University of Wisconsin, Cooperative Extension. Available: http://learningstore.uwex.edu/pdf/G3658-12.PDF

Woods, V. and Buckle, P. 2002 Work, Inequality and Musculoskeletal Health. Norwich: HSE Books, CRR 421/2002. ISBN 0 7176 2312 2.

# MAINTENANCE WORKERS AND ASBESTOS: UNDERSTANDING INFLUENCES ON WORKER BEHAVIOUR

## Claire Tyers & Siobhán O'Regan

*The Institute for Employment Studies*

This paper reports on a selection of the findings from a qualitative study on the attitudes, knowledge and behaviour of maintenance workers to working with asbestos. Although supply, import and use of asbestos has been bannned for some years, maintenance workers remain at risk of new exposure and therefore asbestos related diseases (ADRs). The study was commissioned by the Health and Safety Executive, UK and completed in 2006.

## Introduction and background

Asbestos is the single largest cause of work related deaths from ill-health. The long latency period between exposure and ill-health means the death rate has not yet peaked but the current annual toll is estimated at 4,000 deaths. It was used extensively as a building material between 1950 and 1980, and was used widely until as late as 1999. Asbestos is dangerous when the fibres are released into the air, which could happen during application, removal or maintenance. The occupational groups now at risk of new exposure, and therefore asbestos related diseases (ADRs), are maintenance workers. Research has indicated that this group of workers underestimate their exposure to asbestos (Burdett and Bard, 2003). The HSE has therefore targeted maintenance work in a bid to reduce the number of new cases. The duty to manage asbestos contained in current legislation[1], requires dutyholders (eg the building owner) to identify and record the location and condition of asbestos, assess the risks, and take steps to reduce exposure. The regulations mainly pertain to non-domestic premises; the HSE also ran a campaign in the autumn of 2006 to capture work not covered by the dutyholder regulations. IES was commissioned by the HSE to conduct a research project to examine maintenance workers' awareness, attitudes and behaviour towards asbestos risks including an examination of the barriers which discourage such workers from taking appropriate action.

---

[1] Control of Asbestos Regulations 2002, and more recently the Control of Asbestos Regulations 2006, which amalgamated all the existing asbestos legislation into one set of regulations.

## Method and participants

The research involved 60 in-depth interviews with maintenance workers. Interviewees were recruited using a variety of methods, including through the contacts of trade unions and colleges, press advertisements, and liaising with large facilities management companies. The most successful method was the use of an opt-in survey of sole traders using a commercial database, and more than half of the achieved sample came through this route. Sole traders are a particularly interesting group as they are able to offer a dual perspective from their experiences of working on jobs where they self-manage and also where they have worked as virtual employees (ie as sub-contractors for other firms). All participants were given a £20 gift voucher and asbestos safety information at the end of the interview.

The obtained sample cannot claim to be representative of the maintenance worker population, however, the research did include an approximately even distribution of electricians, plumbers/heating engineers, carpenters/joiners, painters/decorators and workers in other areas of maintenance. Around half of the interviewees were aged over 50, and the majority were speakers of English as a first language.

The majority of interviews were conducted face-to-face and took place outside the work environment such as at the worker's home, or a local café or public house. The interviews were semi-structured and included a brief review of their work and training history and an exploration of their attitudes towards working with asbestos. Interviewees were then asked to rate their own knowledge about the various aspects of working safely with asbestos. Finally, a behavioural event interview technique was employed to explore their experience of working with asbestos. Transcriptions of interviews were analysed with the aid of Computer Assisted Qualitative Data Analysis Software.

## Research messages

### Understanding of asbestos risks

A number of the interviewees had taken part in formal asbestos training, although this was often part of a broader health and safety course rather than a specific training episode. Training messages for many, however, had not been updated since their apprenticeship or early in their career. Therefore, much of the training discussed pre-dated 1992 when the asbestos regulations came into force. In addition, interviewees, who at the time of the interview were working as sole traders, were often relying on training which had taken place in a very different context, such as when working for a large employer. Most respondents supplemented their knowledge of asbestos obtained through training in an informal manner, through their work experience, and/or the knowledge and experiences of colleagues and family. Almost all those interviewed had picked up on the general message that asbestos was a dangerous material, with clear messages not to touch it, and to not work with it. However, their knowledge was much more limited when it came to identification or safe working procedures. In the face of high anxiety and a lack of specific knowledge about how to deal with asbestos, individuals are likely to listen to, or give credence to, sources of knowledge that reduce their anxiety, even where they know these sources to be less reliable.

The generic awareness about the dangers of asbestos exposure were not always supported by more specific, action-oriented knowledge of how to determine when exposure may occur. Many respondents lacked confidence about their ability to recognise and identify asbestos during their work, or lacked awareness of the range of materials that potentially contain asbestos. Workers tended to focus on the different types of asbestos, relating more to colour descriptions (ie brown, blue, white), and relative dangers posed by different materials, rather than any detailed understanding of the full range of asbestos containing materials (although there were a number of exceptions).

Additionally, few felt confident about the specific procedures to follow once asbestos had potentially been identified (eg what to do , who to notify, how to test etc). Older workers were generally more confident about their ability to identify asbestos, but their methods of identification could involve disrupting asbestos and therefore potentially causing harm to themselves or colleagues. Examples of professional testing of potential asbestos containing materials were fairly rare. These options were generally only available to workers when working on larger sites, commercial or public buildings. So some workers were aware, theoretically, that they needed to report their suspicions and get some testing done (because that is what they had learned from training on a large site) but were lost in practical terms when they found themselves working on their own in the private domestic sector.

Older workers are often strong role models for younger colleagues in a more general sense, shaping social norms on work sites. This can act as a positive influence where older workers reinforce messages about the importance of safe behaviour, and older workers can be powerful champions for asbestos awareness and safety in their workplaces. However, in some cases, older workers felt that it was too late to protect themselves from the dangers of asbestos exposure and therefore engaged in behaviours which could put themselves and others at risk. There were also examples given where older colleagues had discredited safety messages about asbestos, playing down the risk. In addition, given that training on asbestos tends to be a fairly infrequent event, knowledge of what constitutes safe practice and the information to support this has changed over time, and over the course of some older workers' careers. Their ability to provide appropriate information and support for less experienced colleagues can therefore be limited, even given good intentions.

### Barriers to safe working

To even begin to take appropriate action workers need a basic awareness that asbestos exposure is a risk to them and that they may come across asbestos containing materials (ACMs) in the course of their work. There does appear to be some tolerance for what is perceived as 'low level' exposure, and some confusion about how exposure levels are linked to the onset of asbestos related diseases (ARDs). There was a widespread belief that the potential levels of exposure in the maintenance trades is now very low. This was driven by the knowledge that asbestos is not placed in new builds, but also, in some cases, by a misperception that it was no longer present in the majority of older buildings due to widespread removal. The asbestos message was also diluted by concerns about other risks. Safety risks, such as falls or other accidents, for example, were felt to be of more immediate concern, whilst 'new' materials (eg MDF or fibreglass) were often felt to pose more of a health risk to today's worker than asbestos.

As a result, many workers don't feel they need to know more, particularly about some aspects of working with asbestos that they don't perceive as relevant to them (eg decontamination, disposal or risk reduction). This appears driven by the general perception that in their working lives, the tasks they complete and the job they do, asbestos isn't really a risk. Therefore, whilst asbestos is acknowledged as a dangerous material in a general sense, individuals tend to be reluctant to accept, or be dismissive of, the actual risks to their own health. This can manifest itself in inaccurate estimations of the risks facing those conducting maintenance work. It is therefore entirely rational for workers who do underestimate the risks in this way to dismiss the need to learn anything other than the very basics about working with asbestos.

There are a range of other factors which contribute to how they actively view or manage these risks. In order to take the decision to behave safely around asbestos, workers not only need to know about the risks, but also take these risks seriously enough to cause them to change their behaviour. Reasons given for **not** adopting safe practices included financial pressures and constraints, concerns about speed and timescales, work cultures and peer pressures. Individuals were able to rationalise unsafe behaviour through their calculation of the risks of the job versus the financial benefits. Often, people were prepared to 'take a chance' if the job was felt to warrant it through a sufficient financial pay off. This was further exacerbated by the fact that a fairly common view was that there is somehow a random element to whether ARDs are contracted after exposure, often described as 'a lottery'. Therefore, individuals are able to focus much more on concrete, commercial concerns rather than less visible and less clear issues around their own health. Prospect theory details the asymmetry of human choices, that how we assess risks of losses differently from how we assess risks of gains (Kahneman and Tversky, 1979).

The general safety culture of a site (eg rigorous risk assessments, use of personal protective equipment) plays a role in how seriously workers take all risks, and certainly how they assess the risks posed by asbestos. The attitude of employers towards the safety of their staff has a very powerful effect. Where they are seen to take the risks seriously, and express their willingness to stop the job if workers' health is at risk, this is extremely effective in preventing risk taking.

There is also likely to be a complex interplay between these factors. For example, on sites where corners are cut, there is likely to be a combination of negative peer pressure, a strong drive for cost reductions and tight deadlines. In such an environment the extent to which individuals feel able or sufficiently 'in control' to take appropriate action in relation to any risk could be limited. However, given the lack of understanding about the risks posed by asbestos and how to deal with them, it is likely that individuals feel even more powerless in relation to asbestos risks than many others. This could be a particular issue for those working on a temporary or casual basis, or the self-employed working as sub-contractors.

*Experiences and work behaviour*

Detailed knowledge of how to actually reduce risks when working with asbestos were uncommon, largely because people didn't identify their job as involving this type of contact. A few interviewees did have detailed knowledge on this issue, however, and were able to identify a range of actions that could be taken. Decontamination was seen as an emergency procedure only, rather than something that affected their day to day working lives, and was also seen as difficult to undertake properly in practice. Disposal often involved using the council, and in a small number of cases individuals discussed examples of illegal disposal that they were aware of. Options to report or test asbestos were not common except by those working for larger employers or as subcontractors on a big site. Walking away from the job appeared to be related to an individual's perception of the strength of their market position and choice of other work.

Individuals were able to discuss a range of experience of working with asbestos. Some of these related to situations where asbestos exposure was likely, either through unwitting or deliberate dismissal of the risks. More common were instances where, although some procedures had been followed, these were in some way incomplete or incorrect, and therefore may have led to exposure. Despite this, the vast majority of more recent experiences were viewed by individuals as relatively safe, and their own behaviour appropriate. The most commonly discussed safeguards were: to pause or stop work; avoid unnecessary fibre release; use personal protective equipment or damp down materials. Overall, workers' experiences were characterised by a lack of detailed procedural knowledge of their own, and many were reliant on others to inform them about correct procedures. There were a number of examples where individuals themselves had tried to follow correct procedures, but where this required them to break with the overriding culture on a site, or stand up to colleagues. It is likely that individuals wanted to share examples where they felt they had, themselves, behaved appropriately, so it is unclear how widespread unsafe behaviours are more widely.

*Influences on knowledge and awareness*

Workers were positive about the effect that large scale, national, marketing campaigns could have upon their knowledge and awareness of the issue. When showed materials produced by the HSE, individuals reacted well to them, but most could not recall having seen them before. For information to effectively reach all those working within this sector, it will be necessary for a variety of formats to be adopted and information developed to fit the varied needs of a heterogeneous community.

It could be useful to move away from the basic message that 'asbestos kills', which has successfully reached this population of workers, to a more detailed message about how many and how. This clarity is what causes people to take the risks seriously. Previous marketing and awareness raising campaigns have been immensely successful in communicating basic messages about how risky it can be to work with asbestos in an unsafe way. The priority for the future should be to continue to focus on providing workers with more detailed knowledge on how to protect themselves.

*Conclusions*

There are a range of issues which affect how likely an individual is to behave safely around asbestos. This research suggests that they can be broken down into four main categories:

- Technical issues, relating to the complexity of messages about asbestos, its effects and how to deal with it effectively.
- Psychological issues, concerning an individual's attitudes towards risk, health and the specific risks posed by asbestos.
- Cultural factors such as pressures from their employers, clients, co-workers etc., which are largely driven by economic as well as social pressures.
- Control factors, namely the extent to which individuals feel that they are able to control their work environment. These are linked to the nature of the employment contract an individual has, and their labour market capital.

Attitudes towards behaving safely are likely to be affected by an individual's internal perception of whether the potential negative impact of exposure is outweighed by other factors. In essence, whether the economic or social costs are outweighed by the health benefits. However, even intending to behave safely is not enough if an individual lacks sufficient knowledge to support this intention.

# References

Burdett, G. and Bard, D. 2003, Pilot study on the exposure of maintenance workers (industrial plumbers) to asbestos, HSL MF/2003/15

Hope, C. 1999, The impact of procurement and contracting practices on Health and Safety: a literature review, HSL report RAS/99/02, in White, J. 2003, *Occupational Health in the Supply Chain: A literature review*, HSL/2003/06

Kahneman, D. and Tversky, A. 1979, Prospect Theory: An analysis of decision making under risk, *Econometrica,* **47**, 2

This paper is based on research reported in full as:

O'Regan, S., Tyers, C., Hill, D., Gordon-Dseagu, V. and Rick, J. 2007, Taking Risks with Asbestos: What influences the behaviour of maintenance workers, HSE Research Report 558

# LAPTOPS IN THE LECTURE THEATRE: AN ERGONOMIC FOCUS ON THE CRITICAL ISSUES

## Rachel Benedyk & Michalis Hadjisimou

*UCL Interaction Centre, University College London, Remax House, London WC1E 7DP, U.K.*

A previous survey showed evidence that the casual use of laptops by students in university lecture theatres causes a significant level of discontent for both students and staff. Typical complaints related to sightlines, poor seating, and environmental intrusions. Lecture theatres are characterised by fixed furniture arranged in tiered rows, often in an arc around a low dais for the lecturer, with some form of information projection behind, for the audience. Whereas most teaching in such theatres is didactic in style, there remains a critical need for direct interaction between audience and speaker; does the student laptop compromise this interaction in a poorly integrated theatre environment? This study set out to use two complementary ergonomic methods (Soft Systems and the Hexagon Ergonomic Model) to identify in a structured manner the range of issues and conflicts that influence the integration of the student laptop into the lecture theatre. The implications for future lecture theatre use and design are discussed.

## Introduction

A previous paper (Benedyk et al, 2007) discussed the recent trend for university students to bring personal laptops onto campus to support their study and for private use. Despite the fact that universities provide dedicated computer suites for student use both on campus and in halls of residence, and despite the fact that ergonomics would advise against such intensive use of laptops, students find the portability and immediacy of laptops to be advantageous. Universities need to set themselves up to support their use.

The survey reported by Benedyk et al (op cit) identified 19 different settings in which students use their laptops, including libraries, classrooms, bedrooms, cafes, bars and parks. Only 6% of the respondents reported using a laptop in lecture theatres; yet lectures are the mainstay of university study activity. Why is it that students seem wedded to their laptop almost everywhere except in the lecture theatre? This study set out to evaluate the student-laptop-lecture theatre dynamic from an ergonomic perspective.

While university teachers are encouraged increasingly to use laptops themselves for demonstration, data projection and using on-line resources during teaching sessions, they do not often show the same encouragement to their students. The survey by Benedyk et al uncovered a mainly negative attitude by teachers to student use of laptops. The two main objections were environmental intrusion (flashing screens and noise from key-tapping, start-up tunes and fans) and disruption to engagement between teacher and student: either apparent (eg lack of eye contact) or real (15% of those students who do take a laptop to lectures admitted they use it for recreational activities not connected with the lecture!). Three teachers even admitted they had banned the use of laptops in lectures. 10% of students reported that other students had also voiced objections to laptop use in lectures, mostly because of noise. These negative attitudes may well be part of the reason for the lack of laptop use in lecture theatres, but not the whole story.

29% of students surveyed in fact reported that use of a laptop would be helpful in lectures – what are the various factors that are hindering them?

## The lecture theatre

The main use for a university lecture theatre is to project the spoken word from a teacher, together with illustrative material such as slides, to a large audience of students; typically, they are designed more for information flow from the teacher to the student than from student to teacher, or student to student, than might be the case in other teaching spaces such as classrooms. Architecturally, the acoustic arrangement thus facilitates this direction of sound projection, and the seating arrangement (which is typically fixed) facilitates the sightline of students towards teacher and projection screens. Current trends (JISC, 2006) would suggest design of teaching spaces is moving towards much more flexible furniture and facilities, to reflect the increase in active and collaborative learning etc, but lecture theatres are still the near future on many campuses.

Ergonomically, then, we often see fixed multiple seated workstations, arranged in tiers, sometimes arced, around a centre-frontal teaching station. Lighting often excludes daylight, and prioritises dimming facilities to enhance projection slides, and sometimes downlights for student note-taking and spotlights for the speaker. Power sources tend to be concentrated at the dais, for use with projection equipment. The teaching station frequently is organised to accommodate a laptop computer for the speaker, which can be connected in to the theatre data-projector.

The student workstation, however, facilitates only the tasks of listening, watching, and usually note-taking – for which a narrow fixed table or a chair arm-plate is provided at each seat. The seat itself is often of the tip-up type, to allow access and egress of the multiple occupants of each row. These workstations are not inherently designed to accommodate student laptops; as recently as 2006, university minimum standards for the technical equipping of lecture theatres make no mention of facilitating student laptops (Victoria University, 2006).

## Ergonomic evaluation of laptop use in lecture theatres

This study used two complementary ergonomic evaluation techniques to identify in a structured manner the factors in lecture theatres likely to come into effect when

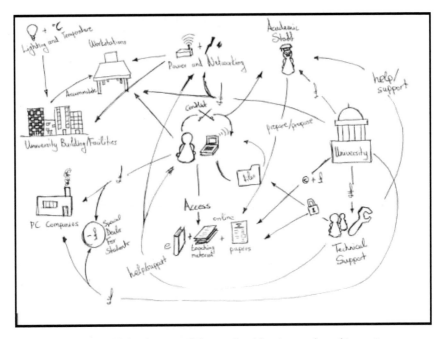

**Figure 1.    Rich picture of the student-laptop-university system.**

students try to use laptops there. First, the known information on university students using laptops was used with the Soft Systems Methodology (SSM) (Checkland and Holwell, 1998) to help identify comprehensively the key actors involved in this complex scenario, and the possible conflicts between them. Then the Hexagon model of educational ergonomics (Benedyk, Woodcock and Harder, 2006 and in press) was applied to focus in on the lecture theatre and to extract the ergonomic issues arising from student/laptop interaction in a lecture theatre setting.

*SSM summary*

The survey data from Benedyk et al (2007) was used to create a rough rich picture, in the middle of which the student is represented working on a laptop, which itself can cause ergonomic conflict (Heasman, 2000). The student interacts with the workstations in university settings (both when using and not using the laptop) and with the PC companies (buying a laptop). The student also interacts and shares files with other students and with the academic staff, with the possibility of further conflicts.

The university facilities are characterised by their configuration, space, lighting, temperature, ventilation etc and they accommodate the workstations. The university has a budget allocated for buying these workstations and for making power and networking available for the students using them. The university also employs the lecturers and the technical support people. Finally, the university has to pay to get copyright for the e-Books and papers accessed by its students.

| | Organisational Sector | | Contextual Sector | | Personal Sector | |
|---|---|---|---|---|---|---|
| | **Management Factors** | **Infrastructure Factors** | **Task Factors** | **Tool factors** | **Individual Factors** | **Social Factors** |
| **External Level** | Government funding to universities; DfES policy | Developments in integration of IT facilities into buildings | Best practice in university teaching and use of computing | Trends in use of computers in university teaching and learning | Student income level (for buying a laptop); peer pressure | Cultural and family expectations of university facilities |
| **System Setting Level (University)** | Allocation of budget for teaching space; costs and benefits of having lecture rooms suitable for laptop users | Provision of wireless connectivity; facilities spec for lecture theatres; technical support | Syllabus and content of lectures | Age and antiquity of university facilities, flexibility of use of space | Attitudes of students at this university to needing to own and use a laptop in lectures | Attitudes of lecturers to student use of laptops; trust that student will not use laptop for leisure when in lecture theatre |
| **System Workplace Level (Lecture Theatre)** | Procurement and planning choices for design and facilities in this lecture theatre; room size and layout | Connectivity and power supply in overall space; accessibility, temperature, acoustics, lighting | Style of delivery of teaching; type of learning | Suitability of workplace configuration; environmental issues eg key-tapping noise; adjustability | Crowding factor; relation of this workstation to neighbours | Attitudes of neighbouring students and of this lecturer to the use of this student's laptop; rake of tiered seating |
| **System Workstation Level (Laptop use)** | Procurement choice of individual workstation eg adjustability and size-suitability for supporting a laptop | Connectivity and power supply in individual workstation; task lighting | Task factors influencing frequency of, and reason for, use of a laptop in the lecture | Laptop design; fit to this work-station; potential for docking station; intrusion eg loudness of fan, volume of sound | Posture and flexibility of posture factors imposed by current configuration | Sightlines and engagement between lecturer and student |
| **Interaction Level** | | | | | | |
| **System User Level (Student)** | Personal characteristics, abilities, preferences, expectations and knowledge of particular student population in relation to laptop use in lecture theatres | | | | | |

**Figure 2. Hexagon model analysis table.**

The academic staff prepare or propose online teaching material, as well as delivering lectures. The technical support people are responsible for securing the shared files and the online material and are responsible for maintaining the networks, but not the laptops.

After drawing this rich picture of the system it became obvious that the key actors are the students and lecturers, with technical support services playing a secondary role. The conflict representation indicates that student–student and lecturer–student interaction might be problematic when the student uses their own laptop in a lecture theatre not designed to accommodate or support laptops. This picture was then examined in more detail for the lecture theatre setting, using the Hexagon model.

## Hexagon model analysis

This analysis (Figure 2 shows part of this) pointed up the likely factors, in all sectors and at all levels, potentially intrusive to successful student/laptop interaction in lecture theatres. From the organisational sector, we see the design and facilities of the teaching space potentially limiting the interaction through poor connectivity, unsuitable workstation design, or lack of power sockets. From the contextual sector, the style of teaching and learning will encourage or discourage the use of the laptop, and the laptop itself will contribute potential ergonomic limitations. The personal sector highlights attitudes and expectations from the lecturer and other students, and we see clearly how apparent disengagement between lecturer and student might occur during laptop use.

If lecture theatres are to accommodate students using laptops successfully, there are a host of ergonomic issues that evidently need to be addressed. For example, what is the right lighting for a room where students either use a laptop or take notes? How can all laptop users have network and power supply? How can a laptop user maintain real engagement with the lecturer, and is any such disruption perhaps related to the rake of the seating? Will a large number of laptops in a lecture room cause too much heat and noise? Is the workstation configuration suitable for both taking notes and using a laptop? Can lecture theatre workstations integrate laptop docking stations? Can the crowding of students into fixed workstations be relieved to allow sufficient flexibility of posture?

Purpose-designed flexible teaching spaces can happily address all these ergonomic points and more, and there is ample evidence that students can interact with laptops in a supportive way given the right facilities and positive attitudes (JISC, 2006). However, many current lecture theatres fall sadly short of this on multiple fronts; it is unsurprising therefore that many students do not use their laptops in lectures, despite wishing to do so.

## References

Benedyk, R., Harder, A., Hussain, R., Ioannidis, A. and Perera, M., 2007, The ergonomic impact of student use of laptops in university settings. In P. Bust (ed.) *Contemporary Ergonomics 2007*, (Taylor and Francis, London).

Benedyk, R., Woodcock, A. and Harder A., 2006, Towards a new model of Educational Ergonomics. *Pro. 2nd Conf on Access and Integration in Schools.* (Ergonomics Soc).

Benedyk, R., Woodcock, A. and Harder A., in press. *The Hexagon model of educational ergonomics.*

Checkland, P. and Holwell, S. 1998 *Information, Systems, and Information Systems: Making Sense of the Field.* (Chichester: John Wiley).

Heasman T., Brooks A. and Stewart T., 2000 *Health and safety of portable display screen equipment*, (HSE Contract Research Report 304/200).

JISC, 2006 *Designing spaces for effective learning*, JISC e-Learning programme, (HEFCE).

Victoria University Australia, *Minimum acceptable standards for teaching spaces*, July 2006 htp://tls.vu.edu.au/SLED/ETSU/TSAG/classroom%20standards-%20Rev%20Doc%20Draft%201-July%2006. PDF accessed November 2007.

# ERGONOMICS ISSUES IN A JET CAR CRASH

**Mike Gray**

*Health and Safety Executive, Edgar Allen House,*
*Sheffield, S10 2GW, UK*

A crash occurred in September 2006 during filming for a television programme of a jet car being driven at speed. As part of the ensuing investigation by the Health and Safety Executive, the role of some of the human factors issues was assessed.

## Introduction

During filming of the driving of a jet powered car in September 2006 for the BBC TV programme Top Gear the car left the track while travelling at high speed. The driver, presenter Richard Hammond (RH), was fortunate to survive the crash. The Health and Safety Executive (HSE) carried out an investigation into the circumstances surrounding the incident and as part of this the human factors issues were studied.

The incident occurred at Elvington airfield where the jet car "Vampire" was driven down the runway at speeds up to 300 MPH. The car was a purpose built dragster powered by a Rolls Royce Orpheus jet engine; it held the current Outright British Land Speed record of 300.3 mph.

In preparation for the filming RH had received a written briefing describing the arrangement of the car's controls. On the morning of the day of the incident he received training in the form of a series of briefings and in-car instruction. In the afternoon a series of runs with increasing jet power were then undertaken. It was on the seventh run at about 5:25 pm that the crash occurred, initiated by a front tyre failure.

## Ergonomics issues

The investigation was lead by two HSE inspectors with assistance from North Yorkshire Police. They collected information from the various parties involved in planning and carrying out the filming of the jet car runs. They produced a report (HSE 2007) analysing the events that lead up to the incident and recommending actions to help minimise risks in the future. Human factors input to the investigation was provided by the author. This concentrated on the physical arrangements of the driver's cab, the operation of the controls, the time available for action and the training requirements.

The arrangements for driving this jet-powered car are somewhat different from those found in conventionally powered vehicles. During any run the left foot has to be kept pressed on the hold to run pedal or the engine will cut out. The right foot operates a brake pedal but it is primarily used for holding the car on the start line, while the engine is started by the crew using a separate starter. A hand throttle is set by the driver during starting and adjusted to a predetermined engine speed setting once the jet engine is running. Higher engine speeds were used as the day progressed. While the engine is coming up to speed the driver has to keep pressing on the foot brake to ensure the car stays at the start. For afterburner use it is necessary to depress a switch attached to the steering wheel for a second or so. The run then commences when the right foot is released from the brake. The driver then has to look forward out of the cockpit and use the steering wheel to maintain a straight course down the runway. This in practice means keeping an angle of about 30° on the steering wheel with the right hand high, apparently this counters the built in tendency for the car to veer to the right.

Once the car reached the area of the runway designated with a green cone on either side the driver should pull the parachute release lever back, this deploys the parachute while simultaneously cutting the engine. The parachute provides a significant deceleration and the driver keeps steering while speed is lost.

The overall operation of the car is thus to hold the brake on the line, keep the hold to run pedal depressed, adjust the throttle for start and then increase to the required power. Hold the steering wheel, activate the after burner switch. The car will then rapidly accelerate during which time the driver has to steer in a straight line. The driver needs to view the runway to maintain a straight line and activate the parachute when the green cones are reached. The time from the start to the release of the parachute is around 20–30 seconds.

## Cockpit layout

The layout of the area where "Vampire's" driver sat was assessed. This was based on a visual examination and measurement of the car in the garage to which it was moved after the crash had occurred. The roll cage had been cut from the car at the scene of the accident and the steering column disconnected but the positioning of pedals, levers, seat etc. was reasonably clear. The compact layout of the cockpit of Vampire means that the various hand-operated controls should be within reach of most people. Adequate reach is important because the driver is restrained by a full 5-point harness; this removes the possibility of leaning forward to gain extended reach. There was no provision for adjustment of either the seat or the controls. The width of the cockpit is around 485 mm that will allow most people to fit in but it will be a tight fit for larger individuals.

The parachute deployment lever is positioned about 780 mm from the back of the seat, at the side of the cockpit just inboard of the chassis tubing. To operate the lever the driver has to remove his hand from the steering wheel and move it sideways and slightly forward. The shaped end of this control makes it readily distinguishable by touch. The hand should readily be able to grip the end of the lever, a 38 mm diameter ring, the driver needs to pull the lever back at least 200 mm to move the rods which release the sprung cap over the parachute and cut off the fuel to the engine. About 70% of men would be able to fully grip the lever when restrained in the car, but the majority of women would find this reach difficult.

The positioning of the foot controls are such that virtually all people should be able to fully reach the pedals. Taller people will find that there is insufficient space for their legs.

Generally, the lack of adjustment for the position of the seat, pedals and steering wheel means that tall or heavy people would have difficulty fitting in the car. Fortunately RH at about 1.7 m tall is of similar stature to the regular driver of the car and consequently should be able to reach and operate the important controls while strapped into the seat, though the parachute deployment lever could be toward the end of his reach.

## Driver expectations

When driving a car people have expectations as to how the controls used will operate. Because of this, standardized arrangements of pedals and hand controls have been developed which facilitate someone moving from one vehicle to another. Thus, the accelerator is provided to the right of the brake pedal. People can get used to other arrangements but will be more likely to select the wrong action if they have to deal with a novel arrangement, especially in an emergency. Systems, which take account of people's expectations and are tolerant of errors are more likely to be used successfully. (see (HSG48) – HSE (1999)).

There have been examples of accidents involving fork lift trucks where the layout of the pedals was different to the normal car arrangement. Some designs of fork lift truck use separate forward and reverse accelerator pedals which require a driver to use a pedal on the right to go backwards while the pedal to its left accelerates forward. A brake pedal is available for use by the left foot. If the driver is reversing and wishes to stop, then moving the foot to the left and pushing that pedal would cause the truck to slow but then accelerate forward. The failure of the truck design to conform to common expectations has resulted in some fatal accidents. (Gray 2000).

The jet car has many similarities with normal car operation, especially single seater cars. However from a control point of view they are quite different. There is no foot control for increasing speed, the left foot is continually required to push down, the steering requires quite a lot of lock to keep in a straight line and the stopping mechanism is hand operated as a single event rather than the continuing action required by a foot brake. In addition, there is no real control of speed; the driver will not feel as if they are in the control loop in the same way as with most other cars. The noise, vibration, acceleration and the knowledge that record breaking speeds were being reached provide additional factors which would be expected to influence human performance of a task.

A set sequence of actions have to be performed during each run. The expert who normally drove the car talked RH through each of the pre-run checks and ensured they were done in order. On early runs RH's visor steams up and he complains that he cannot see. Subsequently he keeps the visor open until the last few seconds to try and keep it clear. A hand signal is given to remind him to put the visor into place once the engine is running at full throttle. Ensuring that actions are completed and in the right order requires concentration without distraction. Typically people use checklists to ensure they do not lose their place in a sequence. It appears from the video footage that verbal direction as well as the briefing sessions are used to make sure RH does things correctly. Once the foot brake is released however he has to deal with things by himself.

The main action he takes once the run has commenced is to try and steer in a straight line while maintaining pressure on the left pedal. It is evident from the audio recording that he finds it difficult to see well out of the cockpit, mainly because of the misting of the visor, though this is said to be better on the later runs. The seating arrangements appear to give RH a clear view forwards though the g force when being accelerated forwards will push the head backwards which may make it more difficult to gain clear sight of the runway and cones. There also appears to be some uncertainty about where he is meant to align the car on the runway. On an early run he says that he is trying to get midway between the cones, this however is not the centre line of the runway, which is cambered and so not the best line – one cone is moved after the first two runs to help with alignment. The telemetry for the penultimate run indicates a steering correction at high speed; this apparently is because the car was deviating from the centre line. A driver's ability to maintain a straight line at high speed will increase with experience of the steering system, briefing sessions and training while stationary cannot provide this type of experience.

The action of pulling the parachute lever to stop the car is one that is peculiar to this type of car. RH is given a series of sequential instructions during the briefing session which requires him to pull the lever back when the green cones are reached. Thus for normal driving he has to keep left foot depressed, keep the car on a straight line using the steering and watch for the green cones. He has to contend with noise, vibration (recorded comment that it is very bumpy), and the g force. The instruction to pull the lever at the end of the run is clear and the sequence of events leading up to it is straightforward.

It is however when the driver has to get involved with another task that the progression of actions can become disturbed. Once the car starts to move away from the straight line and not respond to the steering as it has previously then the driver will focus on dealing with that situation. An experienced driver will have amassed a number of strategies for dealing with this type of situation, and in what is perceived as an emergency a competent driver will bring these to bear. Less experienced people might well freeze and not be able to take action. RH, as someone who has driven many fast cars on tracks, is likely to have developed a feel for how a car that is going out of control can be brought back under control. This will involve the use of the steering and foot brake.

Bailey (1996) in discussing human error suggests that "People are frequently the victims of unwanted releases of automatic performance", i.e. a well practiced sequences of actions can be undertaken unintentionally if an event which seems like the start of a sequence occurs. Where a habit is strong enough then cues that even only partially match the situation can trigger its performance. Thus while RH is driving the jet car and it is behaving as expected he is dealing with a novel situation but one for which he has been given a set of rules. As soon as the situation changes to dealing with a loss of control his habits of trying to feel the situation with the steering and bring the speed under control by use of a foot brake can be expected to take over. If the car does not seem to be responding to these controls then a good driver will re-evaluate and try to find other means for dealing with the situation.

## Time aspects

The maximum speed on the final run is 288 MPH, a 100 metres would be covered in only 0.7 seconds at this speed.

The time interval between becoming aware of a dangerous situation and taking defensive action against it is the reaction time. Many studies (e.g. Green 2000, Triggs and Harris 1982) on reaction times have been carried out which have explored the factors which affect reaction time, and the components which make it up. Reaction time can be broken down into a number of elements:

- sensation time (detecting something),
- perception time (recognising the meaning of the sensation),
- response selection time (deciding what to do),
- movement time (time for muscles to move the body) e.g. moving foot onto the pedal and pushing,
- device response time (e.g. for the brakes to start to slow the vehicle).

The action of removing the right hand from the steering wheel and feeling for a lever to pull back is quite different from the response for keeping control of a skidding car. Overcoming a practiced response will take further time. Deciding to take such action will probably require the driver to process information from the sound and feel of the car, as well as any visual cues. In any case deploying the parachute is not a fast action in the same way as using a foot brake.

According to telemetry the car was going well until 14.25 s from the start when a change in vertical velocity is detected. 0.35 seconds later the vehicle starts to turn right at 2.1g, this continues to build for a further 0.66 seconds to 5.1g. The elapsed time from the indication of a problem until a large impact is 1.25 s, a further 0.67 s elapses before telemetry is lost.

A skilled driver might be able to apply the foot brake in about half a second, even faster if anticipating the need to brake. The time taken to operate the parachute would be appreciably longer, partly because of the additional mental processing and partly because of the additional physical movement required to relocate the hand and pull back the lever in order to activate the parachute and fuel shutdown controls. An experienced jet car driver who is practised at using the parachute to stop the car may be able to operate the control in about 1 second after an alerting event. A less experienced driver, working in attentive processing mode, could take an additional 0.5 to 1 seconds to carry out this action.

Video of earlier runs suggests that the parachute takes about 1 second to inflate from the time the cover over the drogue chute is released. However even if the driver is able to activate the parachute control within 1 second of an alerting event substantial braking will not occur until 2 seconds after the event. The fact that the parachute was deployed during Richard Hammond's crash but did not have time to fully inflate suggests that he must have moved to activate the parachute control very soon into the sequence of events.

## Conclusions

Although the car setup was not adjustable it was probably suitable for someone of Richard Hammond's size.

The means for controlling the jet car are different to those of a conventional track car. The differences are such that the skills built up by an experienced driver might result in inappropriate actions in the event of an emergency.

The time available for a driver to take appropriate action is short and an inexperienced jet car driver might travel around 100 metres further than an experienced driver while deploying the parachute.

## References

Bailey R.W., 1996, Human Performance Engineering, 3rd Edition, (Prentice Hall, New Jersey).

Gray M.I., 2000, Controls, Displays and Ergonomics Standards – Ensuring Health and Safety in Machinery Design, Ergonomics for the New Millennium. Proceedings of the XIVth Triennial Congress of the IEA, San Diego, California, USA, July 29–August 4, 2000.

Green M., 2000, "How long does it take to stop?" Methodological analysis of driver perception-brake times, Transportation Human Factors, 2(3), 195–216.

HSE, 2007, Investigation into the accident of Richard Hammond, http://www.hse. gov. uk/foi/releases/richardhammond.pdf

HS(G)48, 1999, Reducing error and influencing behaviour, Health and Safety Executive.

Triggs T.J. & Harris W.G., 1982, Reaction time of drivers to road stimuli, Human Factors report No. HFR-12, Monash University Accident Research Centre, Victoria.

# Author Index